Urban Operations

Plus

Physical Security

Department of Defense

Pentagon
Publishing

Field Manual
No. 3-06

Headquarters
Department of the Army
Washington, DC, 26 October 2006

Urban Operations

Contents

***This publication supersedes FM 3-06, 1 June 2003.**

Figures

Historical Vignettes

Preface

Doctrine provides a military organization with a common philosophy, a language, a purpose, and unity of effort. Rather than establishing a set of hard and fast rules, the objective of doctrine is to foster initiative and creative thinking. To this end, FM 3-06 discusses major Army operations in an urban environment. This environment, consisting of complex terrain, a concentrated population, and an infrastructure of systems, is an operational environment in which Army forces will operate. In the future, it may be the predominant operational environment. Each urban operation is unique and will differ because of the multitude of combinations presented by the threat, the urban area itself, the major operation of which it may be part (or the focus), and the fluidity of societal and geopolitical considerations. Therefore, there will always exist an innate tension between Army doctrine, the actual context of the urban operation, and future realities. Commanders must strike the proper balance between maintaining the capability to respond to current threats and preparing for future challenges.

PURPOSE

This manual provides the analytical tools for evaluating an urban operation to determine if the operation is necessary for overall mission success. It also provides the means to understanding and determining the impacts of the urban environment on military operations and provides information on managing, taking advantage of, and mitigating the effects of those impacts as appropriate. As such, this manual demonstrates how to apply the doctrinal principles in FM 3-0 to this unique environment.

SCOPE

Chapter 1 introduces theoretical and historical perspectives of urban operations that serve as the underlying basis for the rest of the manual. While this manual has incorporated lessons learned from recent and ongoing operations, it has deliberately taken a broad and varied historical perspective in order to remain relevant to future threats and circumstance differing from those that the Army currently faces. Chapter 2 discusses the characteristics of urban centers and populations as well as their impact on operations. It is unlikely that Army forces will ever operate in a benign urban environment; therefore, Chapter 3 discusses the varied nature of potential urban threats. Chapter 4 describes the effects of the urban environment on warfighting functions and tactics. An understanding of the complexities of the urban environment, the nature of the enemy as an adaptive, learning opponent, and the effects of the environment on warfighting capabilities and skills is essential to sound decision making. Chapters 5 and 6 discuss the potential costs of urban operations and risk reduction measures that the commander and his staff must consider early in their planning. These chapters also outline an urban operational framework and specific urban considerations that create the foundations necessary for successfully applying operational doctrine to an urban environment.

The second part of the manual (Chapters 7 – 10) discusses how urban operations are conducted and resourced. Urban operations include major offensive and defensive operations in urban environments as well as stability or civil support operations ranging from peace operations and counterterrorism to disaster relief and humanitarian assistance. For the different types of operations—offense, defense, and stability or civil support—the purpose, characteristics, organization, and considerations are discussed. However, commanders consider that most urban operations will normally involve the simultaneous execution of offense and defense with stability or civil support (although proportional emphasis will shift over time) and plan accordingly.

APPLICABILITY

This manual is intended for Army commanders and their staffs at the brigade level and above. It addresses full spectrum operations that Army units will execute in urban settings. However, users should also consult JP 3-06 for specific joint information. Additionally, users should be familiar with FM 3-06.1, FM 3-06.11, TC 90-1, and urban operations chapters, appendices, or sections found in other infantry, armor, combined arms, and proponent field manuals for the tactics, techniques, and procedures (TTP) and appropriate proponent information necessary to conduct tactical urban operations at the brigade level and below. This publication applies to the Active Army, the Army National Guard (ARNG)/Army National Guard of the United States (ARNGUS), and United States Army Reserve unless otherwise stated.

ADMINISTRATIVE INSTRUCTIONS

When this FM 3-06 was approved, doctrine was incomplete for incorporating stability operations and civil support operations in place of stability operations and support operations. This manual incorporates stability and reconstruction and civil support operations doctrine found in FM 1 and FMI 5-0.1; it will be revised to reflect the future updates of FM 3-0 and FM 3-07 and, when published, FM 3-28. (Note: Stability and reconstruction operations has been redesignated stability operations to comply with Department of Defense Directive 3000.05.)

Chapter 2 defines "city" according to a population size. However, in historical vignettes and accounts, the term "city" may be applied in its common usage without specific regard to size to maintain conformity with most other historical reports.

In this manual, the term "threat" is applied broadly to include an enemy force (conventional or unconventional), an armed belligerent in a peace operation, antagonistic or unfriendly elements of the civilian population, or some other hazardous condition in the urban environment that negatively influences mission accomplishment.

The term military operations on urban terrain (MOUT) is replaced by urban operations (UO). MOUT is an acronym from FM 90-10, which is superseded by this manual.

Otherwise, the glossary lists most terms used in FM 3-06 that have joint or Army definitions. Where Army and joint definitions are different, (Army) follows the term. Definitions for which FM 3-06 is the proponent manual (the authority) are marked with an asterisk (*). The proponent or amplifying manual for other terms is listed in parentheses after the definition.

The manual attempts to incorporate a broad range of historical vignettes into each chapter where the account supports the doctrinal line of reasoning. Two historical vignettes, however, were included as appendices (A and C) because of their longer lengths.

Unless this publication states otherwise, masculine nouns or pronouns do not refer exclusively to men.

This publication contains copyrighted material.

The proponent for this publication is the United States Army Training and Doctrine Command. Send written comments and recommended changes on DA Form 2028 (Recommended Changes to Publications and Blank Forms) directly to Commander, U.S. Army Combined Arms Center and Fort Leavenworth, Combined Arms Doctrine Directorate, ATTN: ATZL-CD, (FM 3-06), 201 Reynolds Avenue, Fort Leavenworth, KS 66027-2337. Send comments and recommendations by e-mail to webb-cadd@leavenworth.army.mil. Follow the DA Form 2028 format or submit an electronic DA Form 2028.

This page intentionally left blank.

This page intentionally left blank.

Chapter 1
The Urban Outlook

Today's security environment demands more from Army leaders than ever before. Army leaders must not only be able to lead Soldiers but also influence other people. They must be able to work with members of other Services and governmental agencies. They must win the willing cooperation of multinational partners, both military and civilian. But ultimately, the Army demands self-aware and adaptive leaders who can compel enemies to surrender in war and master the circumstances facing them in peace. Victory and success depend on the effectiveness of these leaders' organizations. Developing effective organizations requires hard, realistic, and relevant training.

FM 1

Given the prevalence of large cities throughout the world, Army forces will likely be required to conduct operations in, around, and over large urban areas. These operations will normally be in support of a joint force commander (JFC) conducting military operations pursuant to U.S. national security policy. This manual is designed to facilitate the planning and conduct of the full range and spectrum of land operations in a complex urban environment. Each urban environment and urban operation is unique; prescribing a specific doctrinal "solution" for each situation is impossible. Instead, this manual provides a framework to commanders and their staffs for understanding the urban environment, for analyzing and deciding whether or not to initiate urban operations (UO), and for applying operational doctrine to this complex environment. It also provides a broad base of historical vignettes and examples to help develop a refined analytical perspective and stimulate thought. The manual also includes some planning points and tactics and techniques to assist in preparing for and conducting UO. This information provides a foundation for approaching major UO. Combined with other joint and Army doctrine, this information also will help commanders and their staffs learn to adapt and succeed in this challenging environment.

THE PROSPECT OF URBAN OPERATIONS

1-1. The world is undergoing massive urbanization. Although exceptions exist, an overall trend of migration from rural to urban areas is occurring throughout the globe. (Australia, one of the world's most urbanized countries, is actually becoming less urbanized.) This trend is especially evident in developing nations. Combined with the exponential growth of the global population in the last quarter century, this migration has created massive urban areas that hold the centers of population, government, and economics in their respective regions. In many cases, rapid urbanization has overburdened already weak infrastructures, scarce resources, and a fragile economic base. As urbanization has changed the demographic landscape, potential enemies recognize the inherent danger and complexity of this environment to the attacker, and may view it as their best chance to negate the technological and firepower advantages of modernized opponents. Given the global population trends and the likely strategies and tactics of future threats, Army forces will likely conduct operations in, around, and over urban areas—not as a matter of fate, but as a deliberate choice linked to national security objectives and strategy, and at a time, place, and method of the commander's choosing.

> **Army Urban Operations**
>
> Army forces conduct UO either as one component of a larger operation or as a single operation focused totally on a specific urban environment. Major Army UO are often part of a joint and multinational effort requiring rigorous interagency and civil-military coordination that typically includes the full spectrum of military operations. Commanders of Army major operations must determine if UO are essential to mission accomplishment. If so, commanders must carefully integrate the operations into campaign planning to support the operational objectives of the JFC.
>
> Army leaders conducting UO must—
>
> - Understand the urban environment to determine decisive points.
> - Shape the operation to set the conditions for success.
> - Precisely mass the effects of combat power to thoroughly engage the decisive points that lead to centers of gravity.
> - Continually consolidate gains essential to the retention of the initiative.
> - Transition the urban area to the control of another force or agency or, ultimately, back to legitimate and functioning civilian control.

AN URBAN PERSPECTIVE

1-2. As a subset of all Army operations, UO are operations focused on an urban environment. UO include full spectrum operations—offensive, defensive, and stability or civil support—that may be executed, either sequentially or (more likely) simultaneously, during the conduct of a single urban operation (see figure 1-1). UO may be the sole mission of the commander or one of several tasks nested in a larger operation. Regardless of the types of operations conducted or whether the urban area is the single focus of the operation or only one component of a larger operation or campaign, the complex urban environment significantly affects the overall conduct of the mission.

1-3. When conceptualizing urban operations, commanders must understand two important terms: urban area and urban environment. The first is a subset of the second. An ***urban area*** **is a topographical complex where man-made construction or high population density is the dominant feature**. Focusing on urban areas means concentrating on the physical aspects of the area and their effects on weapons, equipment, line-of-sight, and tactics, techniques, and procedures. The ***urban environment*** **includes the physical aspects of the urban area as well as the complex and dynamic interaction and relationships between its key components—the terrain (natural and man-made), the society, and the supporting infrastructure—as an overlapping and interdependent system of systems.**

1-4. Importantly, commanders must also understand and consider that critical elements of the infrastructure may lie far beyond the area's physical confines. For example, the generating source providing power to the urban energy system is part of that system but may be located well outside of the urban area. Similarly, effects of the interaction between components of the infrastructure, located both inside and outside the urban area, extend well into smaller, neighboring urban areas and surrounding rural areas and often form their political, economic, and cultural focus. Understanding the total urban environment is essential to planning and conducting full spectrum urban operations.

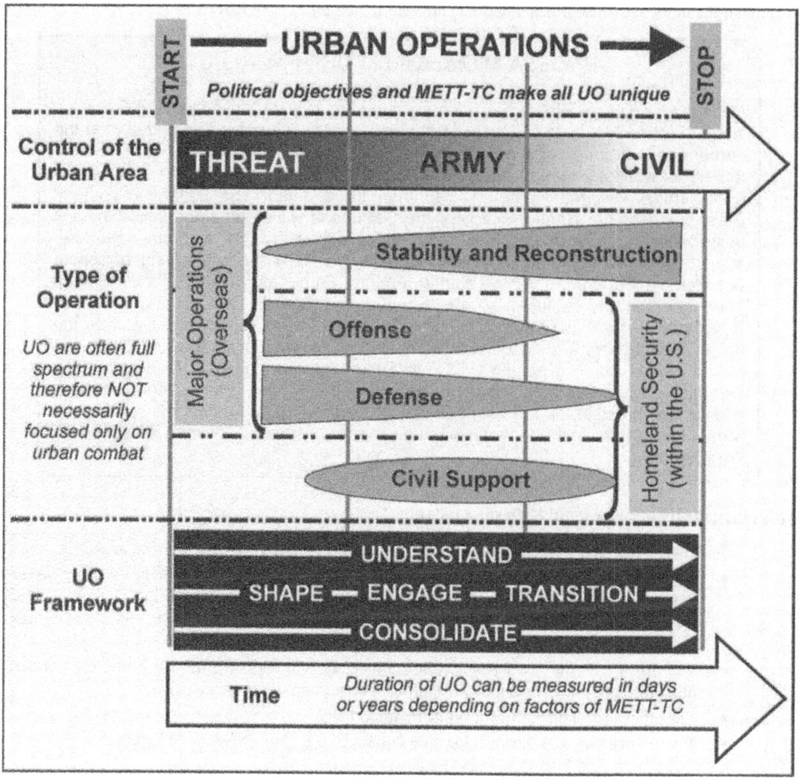

Figure 1-1. Full spectrum urban operations

HISTORICAL SIGNIFICANCE OF URBAN AREAS IN WARFARE

1-5. Urban areas always have been central to, or have significantly influenced, military operations. One of the first urban-centered battles was the siege of Troy at the beginning of Greek history. Moreover, much of the history of early Greece revolved around wars between its city-states or with Persia and centered on the conquest, siege, or blockade of cities. Five hundred years later, the Roman Empire replaced Greece as the dominant world power; although, urban areas remained central to the Roman method of warfare. Even Rome's history can be viewed as a microcosm of urban warfare over the past two thousand years. Though military operations within the physical confines of many of these historic urban areas were not the norm, the focus of these operations was their conquest or control.

1-6. Although Rome last saw combat in 1944, urban areas have been no less prominent in warfare since that time. Seoul in Korea, Beirut in Lebanon, Panama City in Panama, Grozny in Chechnya, Sarajevo in Bosnia-Herzegovina, Kabul in Afghanistan, and Baghdad in Iraq have been centers of conflict in the last 50 years. Urban areas, now more pervasive than ever before, will continue to be essential to successful operational and strategic warfighting. Today, armies cannot expect to execute major military operations without the influence of the urban environments within their area of operations.

> **Rome: A Microcosm of Urban Warfare**
>
> During two millennia, Rome has been the center of at least 12 battles. The Gauls lay siege to Rome first in 387 BC. That first siege lasted six months and ended after the barbarians burnt much of the city. The surviving patrician families paid a ransom for the withdrawal of Brennus' army. From 408 to 410 AD, the Goth leader, Alaric, successfully besieged Rome no less than three times. The Byzantine General Belisarius captured Rome twice from the Goths and withstood siege inside the city once between 536 and 549. Five hundred years later in 1084, Norman adventurer Robert Guiscard captured medieval Rome and sacked the city during a dispute between the Pope and the Holy Roman Empire. Forces of the Holy Roman Empire again stormed and captured the city to punish the Pope in 1527. During the Italian Revolution in 1849, a French army supporting the Pope captured the city from the Italian revolutionary army under Garibaldi. In 1944, the last military action took place in and around Rome when the U.S. Fifth Army captured the city from the retreating German army. Rome's turbulent history—fought over ethnic and religious differences, prestige, and military necessity—demonstrates the importance of urban areas in warfare and the various causes and combatants within this complex environment.

STRATEGIC IMPORTANCE OF URBAN AREAS

1-7. Several reasons have attracted (and continue to attract) armies to combat in urban areas:

- A military force chooses to position itself in an urban area to capitalize on the perceived defensive advantages offered by the environment. In contrast, an opposing force, by analyzing the factors of the situation, determines that it must enter the urban area to attack and destroy its enemy (or devote essential combat power to their isolation).
- The urban environment's people (their allegiance and support), infrastructure, capabilities, or other resources have or can be of significant operational or strategic value.
- The urban area has significant symbolic importance.
- The urban area's geographical location dominates a region or avenue of approach.

1-8. Russia's 1994 experience in Chechnya illustrates an increasingly important motivation for conducting urban operations. The Chechen rebels, after failing to engage Russian forces outside the city, chose to turn Grozny into the main battlefield. Leaders of the defeated Chechen conventional forces recognized that fighting in the urban area provided them their best chance for success. The complexities of urban combat and the perceived advantages of defending an urban area mitigated their numerical and technological inferiority. The urban area provided the Chechens protection from fires, resources, interior lines, and covered and concealed positions and movement. Given such advantages offered by the environment, smaller or less-sophisticated military forces have similarly chosen to fight in urban areas either as a deliberate strategy or to escape certain destruction in open terrain.

1-9. Such advantages of operating in an urban environment also prompt forces to conduct an urban operation to facilitate a larger campaign plan and decisive battle in another location. The urban operation can focus the enemy on the urban area and allow other forces to conduct operations elsewhere. From a defensive perspective, an urban defense may gain time and space to reorganize forces in new defensive positions, to divert enemy forces from other critical tasks, or to prepare to conduct offensive operations. To some extent, these reasons motivated Soviet forces defending Leningrad and Stalingrad from the Germans in World War II. The stubborn defense permitted the Soviets to reorganize for later offensive operations. From an offensive perspective, an attack on an urban area may be a shaping operation used to divert enemy resources from the decisive operation that will follow.

1-10. Armies also fight in an urban area to obtain some critical feature or resource in the area, such as a port facility. The desire to control an important seaport and access to the Persian Gulf largely motivated the Iranian and Iraqi struggle for Basra in the 1980s. Earlier, in 1944, British forces fought German units in Arnhem for control of the Rhine River Bridge. Other key infrastructure of the urban environment may have operational or strategic significance and can compel military forces to attack or defend the area. As urban areas account for an increasing share of a country's national income, often generating over 50 percent of gross national product, the strategic implications for their control or influence become even greater.

1-11. Urban areas are often located on terrain that dominates a region or an avenue of approach. In these cases, offensive armies capture these areas to proceed with security to another objective. Conversely, defensive forces commonly defend the area to deny the area of operations. To illustrate, Cassino, Italy during World War II stood astride the critical highway approach up the Liri valley to Rome. The allies had to attack and capture the monastery to facilitate the allied offensive north. Cassino's location made bypassing virtually impossible. Likewise, in the early 1980's, Israeli army urban operations in Beirut were a result of its strategic location near the Israeli security zone; various Arab insurgent and terrorist groups used Beirut as a base for attacks against Israel. Beirut evolved as the major base of the Palestine Liberation Organization, a major opponent of Israel. Beirut's location made it a security threat to Israel and thus compelled several major Israeli operations in the urban area (see Appendix A).

1-12. Another reason for engaging in urban operations is the symbolic—historical, cultural, political, and even economic—importance of many urban areas. Often, capital cities—such as Rome, Paris, Seoul, Berlin, and Baghdad—are identified as the strategic centers of gravity of their respective nations. Possessing or threatening these urban areas may impact directly on the outcome of a conflict. The objective of Germany's wars with France in 1870 and 1914 was ultimately Paris. Napoleon's 1812 campaign had as its objective Moscow, as did Hitler's 1941 offensive into Russia. The objective of the Soviet 1945 offensive was Berlin, and the North Vietnamese 1975 offensive had as its objective the South's capital of Saigon. Still, history also reminds us that commanders must assess the sustainability and decisiveness of operations directed toward these "prestige" objectives. For example, in 1812, Napoleon captured Moscow but had to evacuate it within 30 days. He lacked supplies and shelter, failed to destroy the Russian Army, and failed to defeat the political will of the Czar and the people. Similarly, the North Korean occupation of Seoul during the Korean War was equally indecisive.

U.S. ARMY'S EXPERIENCE IN URBAN OPERATIONS

1-13. The U.S. Army has a varied history of conducting urban operations. The American Revolution saw the Army conduct several urban operations. These operations included the unsuccessful defense of New York, the successful attack on Trenton, and the decisive siege and attack on British forces at Yorktown. The Mexican War also had a successful assault on the fortified city of Monterey and the decisive siege of Mexico City. During the American Civil War, the armies, in the tradition of Napoleonic maneuver warfare, avoided urban areas and fought in the open. However, the opposing armies frequently made urban areas their objective because of their importance as railheads. Success in the siege of several key urban areas—Vicksburg, Atlanta, and Petersburg—contributed to the Northern victory.

1-14. Following the Civil War, the U.S. Army faced no large-scale urban combat for several generations. The Indian Wars, the Spanish-American War, the Philippine Insurrection, and even World War I did not require the Army to fight in large urban areas. Between the Civil War and World War II, the U.S. Army fought in several urban areas worldwide supporting U.S. commitments. These limited urban combat operations were small but essential parts of what are currently called urban stability operations. From 1900 to 1901, the Army provided public security for a sector of Peking, China of around 50,000 inhabitants. The Army conducted UO and, in the course of the operation, the 9th U.S. Infantry suffered 20-percent casualties while fighting in Tientsin. Punitive expeditions to places such as Siberia, Cuba, Philippines, Central America, and Mexico put the Army in various urban situations that required using military power, notably, the occupation and security of Vera Cruz, Mexico in 1914. In the context of these smaller-scale contingencies, UO became a staple of U.S. Army employment.

1-15. World War II forced the Army to grapple with the issues of large-scale urban combat almost immediately. In his 1941 defense of the Philippines, General MacArthur examined how to defend Manila.

Manila represented a large, modern, friendly urban area, which was the capital city of a close U.S. ally. Defending the urban area posed numerous challenges. Ultimately General MacArthur determined that he could best conduct its defense outside the city by defeating the enemy forces in combat on the invasion beaches or shortly after they landed. When Japanese forces defeated MacArthur's Philippine Army in a series of engagements, MacArthur had to decide how best to protect the friendly populace of Manila. He had two choices: abandoning the city or waging a costly defense that would likely result in the city's destruction, thousands of noncombatant casualties, and no operational advantage. He had little choice but to declare Manila an open city and move his forces to Bataan to wage an operational defense in the vain hope that a counteroffensive could relieve his isolated force. On 2 January 1942, Japanese forces entered Manila unopposed.

1-16. Had General MacArthur decided to defend Manila, his forces would have found scant doctrine in the Army regarding how to fight in an urban area. Doctrine for urban operations did not appear until early 1944, when faced with the possibility of fighting through the larger urban areas of Western Europe. At this time the U.S. Army published FM 31-50. This manual had the first formal discussion of how the Army viewed urban combat. It was based on the Army's limited experiences in the Mediterranean theater and the study of German and Soviet experiences on the Eastern front.

1-17. FM 31-50 emphasized a deliberate pace, individual and small unit initiative, the liberal use of direct and indirect firepower, and decentralized command and execution. It focused on the urban area (as opposed to the environment); however, it did include policies towards the noncombatants. The manual was also focused at the regimental combat team level. Complementing the doctrine of FM 31-50 was the 1944 operations manual, FM 100-5. This latter manual rightly emphasized the importance of combined arms actions and the need for extensive reconnaissance of prepared and defended cities. The Army successfully implemented this doctrine in several major instances of urban combat, most notably the capture of the first German city, Aachen, and hundreds of small-scale urban assaults on cities, towns, and villages across France, the Benelux, and Germany. Army forces also successfully employed this urban combat doctrine during the liberation of Manila in 1945.

1-18. The legacy of this era of Army operations was an effective tactical solution to urban offensive combat: isolate the urban area, seize a foothold, and expand the foothold block by block until occupying the entire urban area and destroying the enemy. The doctrine's emphasis on firepower kept friendly casualties to a minimum. Unfortunately, when enemy forces stoutly defended the urban area, the emphasis on firepower resulted in the area's virtual destruction and high casualties among noncombatants.

1-19. The doctrinal approach honed in World War II remained the accepted Army approach to urban combat to the century's end. The last successful implementation occurred during the Korean War and the final liberation of Seoul. The Vietnam conflict did not offer the Army opportunities or the requirement to practice urban combat or test and refine doctrine on a large scale. The largest urban battle, Hue, was a chaotic tactical battle that validated most of the historical lessons of urban combat without generating any new doctrinal insights for large-scale urban warfare.

1-20. From the mid-1950s through the 1990s, the Army conducted UO in the U.S. in support of civil authorities during civil unrest and anti-Vietnam protests. Some operations involved numerous active and Reserve Component forces engaged in restoring public order. The Detroit riots of 1967 and the Los Angeles riots of 1992 required the commitments of active and National Guard units. In 1968, the Army deployed over 35,000 troops to Washington D.C., Chicago, and Baltimore following the death of Dr. Martin Luther King, Jr.

1-21. In the 1970s and 1980s, Army doctrine predominantly focused on urban areas and successfully fighting a conventional ground war against Soviet and Warsaw Pact forces in Central Europe. FM 90-10 (1979) described how to conduct urban operations against Soviet forces in Germany. Its concepts were never tested other than in simulation, and its approach to urban combat was not substantially different from that practiced by the Army since World War II. Despite previous doctrine's admonition to avoid cities, the Army has had to fight in them in diverse circumstances.

1-22. In the 1990s and early 21st century, commanders and planners began to more carefully consider the impact of urban environments on the overall conduct of campaigns and major operations. Maneuver warfare characterized Operation Desert Storm in 1991; however, urban environments became prominent during subsequent operations in Somalia, the Balkans, and again in Iraq. (During Operation Iraqi Freedom in 2003, for example, Baghdad was considered as a strategic center of gravity for planning.) This was a result of adversaries recognizing the asymmetric advantages to be gained by fighting superior U.S. forces in an urban environment as well as a general worldwide trend of increasing urbanization. The evolving nature of Army operations in the urban environment was recognized in the Army doctrine of full spectrum operations. No longer could combat operations be considered exclusively as a separate phase of the major operation. Instead, full spectrum operations—combat and stability or civil support—would be conducted simultaneously. It also became clear to Army commanders that while one type of operation can often dominate for a period, simultaneous, full-spectrum urban operations were now the norm and an in-depth understanding of complex urban societies essential to overall success.

MODERN ARMY URBAN OPERATIONS

1-23. Modern urban operations will be full spectrum operations executed jointly and will often contain a multinational and interagency component. They will span the entire range of possible applications of military power. At higher echelons, these separations are often viewed as levels or scales of intensity. For the tactical units conducting urban operations, these divisions appear indistinct, as the intensity is often high despite where the operation falls within the range of military operations.

JOINT, INTERAGENCY, AND MULTINATIONAL

1-24. The urban environment is too multifaceted for a single-service, single-agency, or single-dimensional solution. Generating desired effects and avoiding unintended negative consequences in this complex environment requires careful integration of joint (and often multinational) forces and interagency capabilities throughout all phases of the operation. Effective interagency collaboration will help plan effects, supporting actions, and measures of effectiveness to ensure that military actions complement diplomatic, economic, and informational activities.

1-25. Joint urban operations (JUOs) in which Army forces are a major component will be primarily land operations. These operations may take place within the context of a joint campaign conducted by a joint force land component commander or a joint task force (JTF) commander. Or they may be an Army operation under an ARFOR commander who himself operates for a JFC, depending on the organization of the theater's joint command structure. In the later case, the JFC will manage joint issues in the urban area.

1-26. The JFC conducting JUOs will focus on effectively organizing his forces for UO and tasking them in accordance with their service capabilities. His guide for the conduct of the JUO will be the joint operational tasks described in JP 3-0. JP 3 06 will provide the JTF commander specific guidance regarding the conduct of joint operational tasks in the urban environment. Army commanders will execute tasks assigned by the JFC and advise him on using Army forces and capabilities. Army commanders will also ensure that Army UO are nested within the JFC's concept of operations. Also, the ARFOR commander will request support through the JFC from other service and functional commanders who have urban capabilities critical to the success of Army UO. See Appendix D for more information on joint capabilities in an urban environment.

1-27. The military conflict will not be an end to itself. It is inevitably a means of transition from a perceived unsatisfactory state of affairs to an improved end. Therefore, urban operations will require the careful orchestration of military and civilian capabilities to achieve success. An early identification of potential requirements will allow for the proper allocation of responsibilities between military and civilian agencies and a clear understanding of what is possible for each to achieve. Such understanding will preclude the establishment of false perceptions from all concerned, but particularly from the urban population and the public at large.

1-28. Interagency cooperation will include the Army as part of the Department of Defense, elements of the Department of State, and other various governmental and nongovernmental organizations. Urban operations demand the expertise and abilities of various organizations and a synergistic unity of effort if strategic goals aimed at bringing about a more satisfactory political, social, diplomatic, economic, and military situation are to be achieved. However, there will often be a gap in time before bringing many governmental capabilities to bear that the Army and other military forces—to include coalition partners—will have to fill. Routine interagency training, planning, and coordination will help to decrease that gap in time to its minimum. As it is likely that civilian organization leaders will be in charge during some phases of the overall operation, they should lead planning at those times. A careful survey of urban requirements and rapid communication to participating civilian organizations will also help speed civilian capabilities to where they are needed most.

Winning the Peace in Iraq:
The Requirement for Full Spectrum Operations

We found that if we concentrated solely on establishing a large security force and targeted counterinsurgent combat operations—and only after that was accomplished, worked toward establishing a sustainable infrastructure supported by a strong government developing a free-market system—we would have waited too long. The outcome of a sequential plan allowed insurgent leaders to gain a competitive advantage through solidifying the psychological and structural support of the populace.

Further, those who viewed the attainment of security solely as a function of military action alone were mistaken. A gun on every street corner, although visually appealing, provides only a short-term solution and does not equate to long-term security grounded in democratic process. Our observation was born not from idealism, but because it creates the essence of true security, protecting not only our soldiers, but Iraq, the region, and, consequently, our homeland.

Major General Peter Chiarelli
Commander, 1st Cavalry Division

FULL SPECTRUM OPERATIONS

1-29. Army forces will conduct full spectrum operations within urban areas. Army commanders conduct full spectrum operations abroad by executing offensive, defensive, and stability urban operations as part of an the joint, interagency, and multinational effort described above. Army forces within the United States and its territories conduct full spectrum operations by combining offensive, defensive, and civil support operations to support homeland security. The situation will mandate that one type of operation—offense, defense, stability, or civil support—dominates the urban operation. Depending on whether the operation is overseas or within the United States, however, commanders will often find themselves executing offensive and defensive operations and stability or civil support operations simultaneously. In fact, waiting until all combat operations are concluded before beginning stability or civil support operations often results in lost, sometimes irretrievable, opportunities. Unsurprisingly, the mission determines the dominant type of operation, with the other types of Army operations conducted to shape the AO for mission success. The dominant type of operation will vary between different urban areas even in the same campaign.

Offense

1-30. Against a large conventional enemy in a major urban area with a large civil population present, offensive operations will likely require a greater commitment of Army resources than in other environments. Urban offensive operations will also incur the greatest risks to Army forces and noncombatants. Within defensive and stability or civil support operations, forces may conduct tactical offensive UO, such as counterattacks to maintain the initiative or raids to eliminate elements disrupting the stability or civil support operation.

Defense

1-31. Defensive UO are generally conducted as a shaping operation within a larger major operation. These temporary operations often set conditions for successful offensive operations, stability operations, or civil support operations. Commanders often conduct defensive UO within other types of operations to protect essential facilities in the urban area, protect flanks against counterattack, prevent the breakout of isolated enemies, or protect valuable supply bases or vulnerable convoy routes. Army forces conducting defensive UO must creatively use the environment to enhance their combat power.

1-32. In UO, essential facilities will likely include urban ports and airfields required by sister services to support ground operations. Therefore, Army commanders will need to carefully plan integrated air and base defense operations with air and naval component commanders. Protection requirements increase dramatically as air and naval assets are most vulnerable when aircraft are on the ground and ships are docked at port. Without a carefully synchronized and integrated base defense plan, Army commanders risk sporadic or total loss of support by air and naval forces—forces essential to successful UO.

Stability

1-33. Stability operations in an urban environment require offensive and defensive operations, combined with other tasks unique to each stability operation. Army forces conduct urban stability operations for various reasons, including noncombatant evacuation operations, peace operations, or support to insurgencies (see Chapter 9). Urban stability operations will require an offensive capability to destroy any military capability that overtly threatens its objectives before that military threat can adversely affect the operation. Army forces employ defensive capabilities to safeguard themselves as well as secure critical places, populations, or infrastructure in the urban area. Commanders may also employ defensive capabilities to separate and protect one faction from another. Various stability operations will also require the distribution of food or aid and the protection or assistance of agencies conducting economic or humanitarian activities.

Civil Support

1-34. Army civil support operations in an urban environment aid other U.S. agencies and organizations in mitigating the consequences of natural and man-made disasters. In response to the disaster, civil support operations require the equipment, personnel, or organizational abilities of Army forces rather than the Army's combat capabilities. In a civil support mission, these capabilities often involve Army transportation, medical, quartermaster, or engineer forces. Unless conducted in conjunction with homeland defense, urban civil support operations will seldom require combat; however, commanders must determine if force protection threats exist that could hamper Army civil support operations. During homeland security operations, defensive and offensive capabilities will be required to defeat hostile armed forces.

PREPARING FOR FUTURE URBAN OPERATIONS

1-35. To operate successfully in a complex urban environment requires a thorough understanding of the urban environment and rigorous, realistic UO training. Training should be conducted by the complete combined arms team and cover full spectrum operations to include appropriate tactics, techniques, and procedures (see FM 3-06.11 and TC 90-1). It should also replicate—

- The psychological impact of intense, close combat against a well-trained, relentless, and adaptive enemy.
- The effects of noncombatants—including governmental and nongovernmental organizations and agencies—in close proximity to Army forces. This necessitates—
  - An in-depth understanding of culture and its effects on perceptions.
  - An understanding of civil administration and governance.
  - The ability to mediate and negotiate with civilians including the ability to effectively communicate through an interpreter.
  - The development and use of flexible, effective, and understandable rules of engagement.
- A complex intelligence environment requiring lower-echelon units to collect and forward essential information to higher echelons for rapid synthesis into timely and useable intelligence for all levels of command. Understanding the multifaceted urban environment necessitates a bottom-fed approach to developing intelligence (instead of a top-fed approach more common and efficient for open terrain and conventional threats). It also emphasizes the need for intelligence reach and a truly collaborative approach to the development and sharing of intelligence.
- The communications challenges imposed by the environment as well as the need to transmit large volumes of information and data.
- The medical and logistic problems associated with operations in an urban area including constant threat interdiction against lines of communications and sustainment bases.

Developing Joint, Interagency, and Multinational Operability

1-36. UO training extends from the individual Soldier to the joint level including the integration of conventional and special operations forces. Preparedness also includes enhancing interoperability in regards to urban multinational and interagency operations. Training, as well as campaign and contingency plan creation, should include significant and sustained participation by civilian agencies. Fruitful multinational and interagency relationships must be cultivated before the onset of operations, that is, before Soldiers and their coalition counterparts are making decisions—many with strategic implications and often when they are under fire. Joint, interagency, and multinational collaboration will help design effects, supporting actions, and measures of effectiveness necessary to ensure that military actions in urban environments complement the diplomatic, informational, and economic activities necessary to achieve strategic objectives.

Conducting Live, Virtual, and Constructive Training

1-37. Force preparedness mandates integrating the actual use of urban terrain, exercises at urban training sites, simulations, or any combination into tactical- and operational-level intra- and interservice training. This type of multi-faceted training will help commanders develop a better understanding of the complexity of the urban environment and enable them to execute missions across full spectrum operations. Careful use of these facilities will also allow Army forces to fully integrate urban operations within self-development, institutional, and operational training.

Ensuring Every Soldier is an Urban Warrior

1-38. In a complex urban environment, every Soldier—regardless of branch or military occupational specialty—must be committed and prepared to close with and kill or capture threat forces in an urban environment. Every Soldier must also be prepared to effectively interact with the urban area's noncombatant population and assist in his unit's intelligence collection efforts. Each urban operation will be unique and commanders—given an opportunity to surge and conduct mission-specific post-mobilization and in-theater training—will need to conduct an analysis of the tasks requiring emphasis or modification to fit the conditions of the operational environment. In UO, every soldier will likely be required to—

- Perform advanced rifle marksmanship to include advanced firing positions, short-range marksmanship, and night firing techniques (unassisted and with the use of optics).
- Operate their unit's crew-served weapons.
- Conduct urban reconnaissance and combat patrolling.
- Enter and clear buildings and rooms as part of an urban attack or cordon and search operation.
- Defend an urban area.
- Act as a member of a combat convoy (including specific drivers training).
- Control civil disturbances.
- Navigate in an urban area.

1-39. While not all-inclusive and necessarily urban-specific, other critical individual and collective UO tasks (often modified for the urban environment) might include—

- Conduct troop-leading procedures.
- React to contact, ambush, snipers, indirect fire, and improvised explosive devices.
- Establish an observation point, personnel or vehicle checkpoint, or roadblock.
- Secure a disabled vehicle or downed aircraft.
- Call for indirect fire and close air support.
- Create and employ explosive charges.
- Handle detainees and enemy prisoners of war.
- Treat and evacuate casualties.
- Accurately report information.
- Understand the society and culture specific to the area of operations.
- Use basic commands and phrases in the region's dominant language.
- Conduct tactical questioning.
- Interact with the media.
- Conduct thorough after-action reviews.

1-40. Commanders must understand that all Soldiers will require urban-specific equipment to conduct many urban-specific tasks. Critically, commanders cannot expect supporting forces to fight alongside and support ground maneuver units in an urban environment without the appropriate equipment and training.

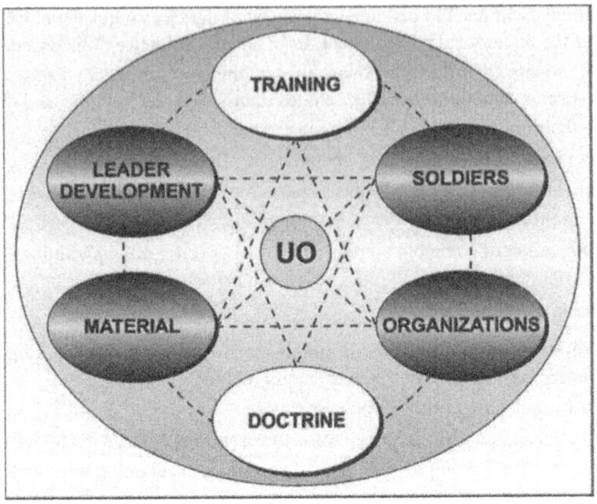

Figure 1-2. UO and DOTMLPF

Continual Adaptation and Innovation

1-41. Realistic UO training (as well as the conduct of real world operations) has the added benefit of identifying operational requirements and resultant changes necessary in our doctrine, organizations, training, materiel, leadership and education, personnel, and facilities (DOTMLPF) (see figure 1-2 and FM 100-11). While technology (materiel) and organizational changes are important, confident Soldiers (personnel) remain the decisive means for success. Technology and organizational changes will be a critical enabler to attain better understanding of the urban environment, enhance command and control on the noncontiguous battlefield, and achieve the agile, simultaneous, and precise effects required in UO. In the future, technology may lead to a radically new operational concept and approach to UO. Still, competent leaders and well-trained, disciplined, and adaptive Soldiers will remain the decisive means for the Army to succeed in this complex and multidimensional urban environment.

Chapter 2

Understanding the Urban Environment

From a planning perspective, commanders view cities not just as a topographic feature but as dynamic entities that include hostile forces, local population, and infrastructure. Planning for urban operations requires careful IPB, with particular emphasis on the three-dimensional nature of the topography and the intricate social structure of the population.

FM 3-0

Of all the environments in which to conduct operations, the urban environment confronts Army commanders with a combination of difficulties rarely found elsewhere. Its distinct characteristics result from an intricate topography and high population density. The topography's complexity stems from the man-made features and supporting infrastructure superimposed on the natural terrain. Hundreds, thousands, or millions of civilians may be near or intermingled with soldiers—friendly and enemy. This second factor, and the human dimension it represents, is potentially the most important and perplexing for commanders and their staffs to understand and evaluate. To this end, this chapter provides information essential to understanding the urban environment and conducting an effective intelligence preparation of the battlefield (see FM 2-01.3, FMI 2-91.4, and Appendix B).

Although urban areas possess general similarities, each environment is distinct and will react to and affect the presence and operations of Army forces differently. A tactical technique effective in one area may not be effective in another area due to physical differences, such as street patterns or the type of building construction. An Army policy popular with one urban group may cause resentment and hostility in another due to diverse cultural differences. All difficulties potentially exist, and they increase the complexity for Army forces operating in urban areas. These difficulties range from conventional military forces to disease and starvation (see Chapter 3) to a pervasive media—often acutely present in intricate combinations. Thus, commanders at all levels must make extraordinary efforts to assess and understand their particular urban environment to plan, prepare for, and execute effective urban operations (UO).

A COMPLEX ENVIRONMENT

2-1. Urban areas vary depending on their history, the cultures of their inhabitants, their economic development, the local climate, available building materials, and many other factors. This variety exists not only among urban areas but also within any particular area. The ever-changing mix of natural and man-made features in urban areas present commanders with some of the most difficult terrain in which to conduct military operations.

2-2. Although urban areas possess similar characteristics, no two are identical. The sprawl of Los Angeles, for example, bears little physical resemblance to New Delhi. Societal characteristics most significantly affect each area's uniqueness and complexity. While complex, information about the terrain, its potential effects on operations, and how it changes over time may be determined with some degree of certainty. However, the human dimension is much more difficult to understand and assess, particularly its effects on military operations. Like any environment, the side that can best understand and exploit the positive and mitigate the negative effects of the urban environment has the best chance of success.

2-3. Whether a large metropolis or a small village, each urban environment has identifiable components that constantly change and interact. This "system of systems" consists of the terrain, the society, and the infrastructure that links the two (see figure 2-1).

2-4. These systems are not separate and distinct categories, but rather overlapping and interdependent systems, acting dynamically with each other. Thoroughly analyzing these elements, along with the other factors of mission, enemy, weather, troops and support available, and time—

- Contributes to the accuracy of the commanders' situational understanding.
- Potentially lessens the number and cost of close combat engagements.
- Allows commanders to develop courses of action that apply appropriate resources against decisive points.

2-5. In stability operations and civil support operations, this understanding allows commanders to engage and dominate the decisive points critical to maintaining peace or restoring normalcy to the urban environment. Although each system is categorized into subordinate components or subsystems, commanders must often "step back" and visualize each system, the complex urban environment, and their area of operations (AO) as a unified whole. This "systems thinking" aids commanders in uncovering key relationships and intersections that can help reveal centers of gravity (COGs) and decisive points.

2-6. To comprehend the urban environment and its components to the fullest extent possible, commanders must carefully integrate and employ tactical reconnaissance forces, special operations forces (SOF)—to include psychological operations (PSYOP) and civil affairs units—and a myriad of other human intelligence (HUMINT) assets and regional, language, and cultural experts. The societal aspects and integrating infrastructure will challenge commanders' assessment and understanding. These aspects will also require greater dependence on nonmilitary and nongovernmental organizations (NGOs) and host-nation agencies for their information, knowledge, and expertise. This last consideration requires commanders to develop effective techniques and procedures for coordinating, interacting, and, to the greatest extent possible, synchronizing activities with these agencies.

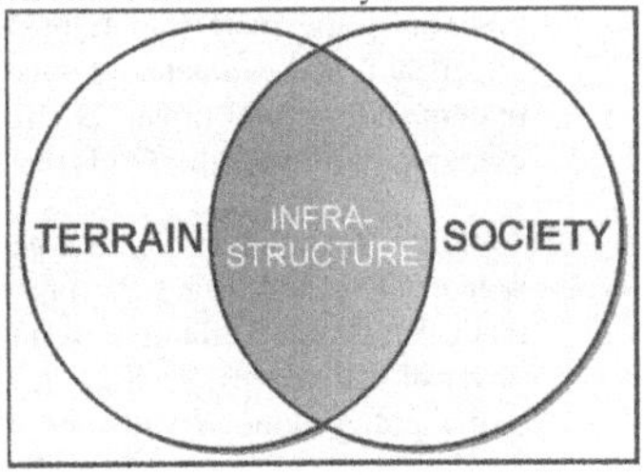

Figure 2-1. Keys to understanding the urban environment

URBAN TERRAIN

2-7. Although complex and difficult to penetrate with many intelligence, surveillance, and reconnaissance (ISR) assets, the terrain is the most recognizable aspect of an urban area. Truly understanding it, however, requires comprehending its multidimensional nature. The terrain consists of natural and man-made features, with man-made features dominating; an analysis considers both. Buildings, streets, and other infrastructure have varied patterns, forms, and sizes. The infinite ways in which these factors can intertwine make it difficult to describe a "typical" urban area. However, these various factors provide a framework for understanding the complex terrain in an urban area. Furthermore, man-made features significantly affect military systems and Soldiers, and thus tactics and operations. General effects on urban operations are discussed in this chapter. Specific effects on warfighting functions (see Chapters 4 and 10) and the spectrum of operations (see Chapters 7, 8, and 9) are interwoven throughout the manual.

MULTIDIMENSIONAL BATTLEFIELD

2-8. Urban areas present an extraordinary blend of horizontal, vertical, interior, exterior, and subterranean forms superimposed on the natural relief, drainage, and vegetation. An urban area may appear dwarfed on a map by the surrounding countryside. In fact, the size and extent of the urban area of operations is many times that of a similarly sized portion of undeveloped natural terrain. A multi-storied building may take up the same surface area as a small field, but each story or floor contains approximately an equal area as the ground upon which it sits. In effect, a ten-story building can have eleven times more defensible area than "bare" ground—ten floors and the roof. It is the sheer volume and density created by this urban geometry that makes UO resource intensive in time, manpower, and materiel.

2-9. Like natural disasters, UO can radically alter the physical characteristics of the urban terrain in ways not experienced in other environments. They may cause (either intentionally or not) uncontrollable fires or the loss of electricity. A power outage can cause flooding (especially in subsurface areas) by shutting down pumping stations. Entire buildings may be destroyed, eliminating reference points, leaving large piles of rubble, altering fields of fire, and making movement and transportation extremely difficult. Additionally, buildings and other urban structures, damaged but not destroyed, can become (or remain) effective obstacles and possible booby traps. Even without enemy exploitation, their weakened construction and unstable structure increase the risk of injury to Soldiers and civilians moving within them. (Engineer expertise will often be needed to determine whether the buildings can support occupation by Army forces or civilians.) Yet, even the total collapse of a building may not eliminate its defenders. Of additional concern, the likely presence of toxic industrial material (TIM) can create additional obstacles and health hazards.

2-10. Commanders in other environments normally address the depth, breadth, and height of their AO in terms of two areas: airspace and surface. In an urban environment, they broaden their scope to include supersurface and subsurface areas (see figure 2-2) that voluminously extend the commanders area of operations. Although spatially separated, each area may be used as an avenue of approach or mobility corridor, line of communications (LOC), or engagement area.

2-11. Supersurface and subsurface areas magnify the complexity of the urban physical environment. Commanders must consider activities that occur outside buildings and subterranean areas (the external space) as well as the activities happening unseen in buildings and subterranean systems (the internal space). This internal volume further challenges command, control, and intelligence collection activities and increases the combat power required to conduct UO. Commanders must develop methods and techniques to help themselves, their staffs, and their subordinate commanders and staffs to represent, visualize, and reference these multiple dimensions. Increasing the difficulty, such dimensions can change rapidly simply due to continued urban growth or, as described earlier, the effects of nature and UO themselves.

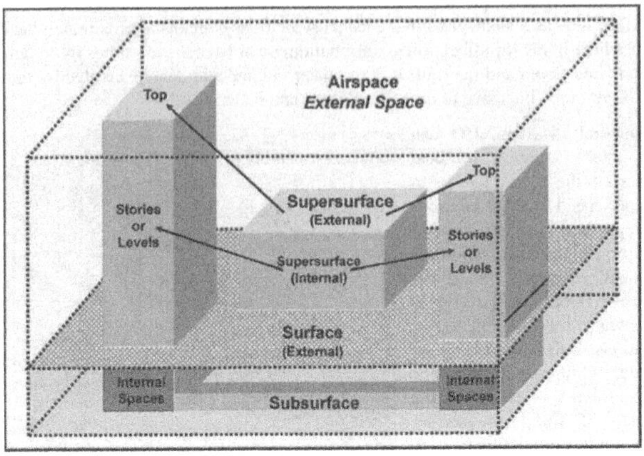

Figure 2-2. The multidimensional urban battlefield

Airspace

2-12. As in all other environments, aircraft and aerial munitions use the airspace as rapid avenues of approach in urbanized areas. Forces can use aviation assets for observation and reconnaissance, aerial attack, or high-speed insertion and extraction of Soldiers, supplies, and equipment. Some surface obstacles in an urban area, such as rubble, do not affect flight (though they may prevent the take-off and landing of aircraft). Buildings of varying height and the increased density of towers, signs, power lines, and other urban constructions, however, create obstacles to flight and the trajectory of many munitions (masking). Similarly, these obstacles can restrict a pilot's line of sight as well as physically limit low-altitude maneuverability in the urban airspace. Excellent cover and concealment afforded enemy gunners in an urban area increases aviation vulnerability to small arms and man-portable air defense systems (MANPADS), particularly when supporting ground forces. The potential for a high volume of air traffic (military and civilian) over and within urban airspace (including fixed-wing, rotary-wing, and unmanned aircraft systems) may become another significant hazard and necessitate increased airspace command and control measures.

Surface

2-13. Surface areas apply to exterior ground-level areas, such as parking lots, airfields, highways, streets, sidewalks, fields, and parks. They often provide primary avenues of approach and the means for rapid advance. However, buildings and other structures often canalize forces moving along them. As such, obstacles on urban surface areas usually have more effect than those in open terrain since bypass often requires entering and transiting buildings or radical changes to selected routes. Where urban areas border the ocean or sea, large lakes, and major rivers, the surface of these bodies of water may provide key friendly and threat avenues of approach or essential LOCs—a significant consideration for Army commanders. As such, amphibious, river-crossing, and river operations may be integral parts of the overall urban operation (see FM 3-05.212 and FM 3-90.12).

2-14. Larger open areas—such as stadiums, sports fields, school playgrounds, and parking lots—are often critical areas during urban operations. They can provide locations for displaced civilians, interrogation centers, and prisoner of war holding facilities. These areas also can afford suitable aircraft landing and pickup zones and artillery firing locations. They can provide logistic support areas and aerial resupply possibilities because they are often centrally located. Finally, large open areas (and immense or unusually shaped structures) within urban areas are often easier to see—especially from the air—and can serve as excellent target reference points from which to shift or control fires.

Supersurface

2-15. These areas include the internal floors or levels (intrasurface areas) and external roofs or tops of buildings, stadiums, towers, or other vertical structures. They can provide cover and concealment; limit or enhance observation and fields of fire; and restrict, canalize, or block movement. However, forces can move within and between supersurface areas creating additional, though normally secondary, avenues of approach. Rooftops may offer ideal locations for landing helicopters for small-scale air assaults and aerial resupply. First, however, engineers must analyze buildings for their structural integrity and obstacles. Such obstacles include electrical wires, antennas, and enemy-emplaced mines (although personnel may be inserted by jumping, rappelling, or fast roping from a hovering helicopter and extracted by hoist mechanisms). Some rooftops are designed as helipads. Roofs and other supersurface areas may also provide excellent locations for snipers; lightweight, handheld antitank weapons; MANPADS; and communications retransmission sites. They enable top-down attacks against the weakest points of armored vehicles and unsuspecting aircraft. Overall, elevated firing positions reduce the value of any cover in surrounding open areas and permit engagement at close range with less risk of immediate close assault. This area (and the subsurface area) requires commanders to think, plan, and execute ground operations vertically as well as horizontally. In this latter regard, UO share strong similarities with mountain operations (see FM 3-97.6).

Subsurface

2-16. Subsurface areas are below the surface level. They may serve as secondary and, in fewer instances, primary avenues of approach at lower tactical levels. When thoroughly reconnoitered and controlled, they offer excellent covered and concealed LOCs for moving supplies and evacuating casualties. They may also provide sites for caching and stockpiling supplies. Subsurface areas include subterranean areas such as subways, mines, tunnels, sewers, drainage systems, cellars, civil defense shelters, and other various underground utility systems. In older cities, they may include ancient hand-dug tunnels and catacombs. Both attacker and defender can use subsurface areas to gain surprise and maneuver against the rear and flanks of a threat and to conduct ambushes. However, these areas are often the most restrictive and easiest to defend or block. Their effectiveness depends on superior knowledge of their existence and overall design. Army commanders may also need to consider potential avenues of approach afforded by the subsurface areas of rivers and major bodies of water that border urban areas. This particularly applies when operating as part of a joint task force (JTF) task organized with SOF or when opposing a threat with similar capabilities.

Major Urban Patterns

2-17. Four major urban patterns (satellite, network, linear, and segment) can influence UO (see figure 2-3). Central to two of the patterns (satellite and network) is the hub or dominant urban area or pattern around which outlying urban areas or patterns radiate. (A segmented urban area, because it tends to be a larger urban area, can often be a hub.) In offensive and defensive operations, the hub serves as a pivot or strong point; as such, it can become a major obstacle to an attacker. If the attacker chooses to bypass the urban area (hub) located along his axis of advance without first isolating the area, he may expose his flank or LOC to attack from the hub as well as dependent urban areas or subordinate satellite patterns. Because the focus of stability and civil support operations is normally on people, commanders should understand the value and influence of the hub to the economic, political, or cultural well being of the surrounding area. Generally the larger the hub, the greater influence it has on satellite urban areas and surrounding rural areas. Commanders must remember that urban areas are not islands; all are connected to the surrounding rural (and other urban) areas through fluid and permeable boundaries and LOCs.

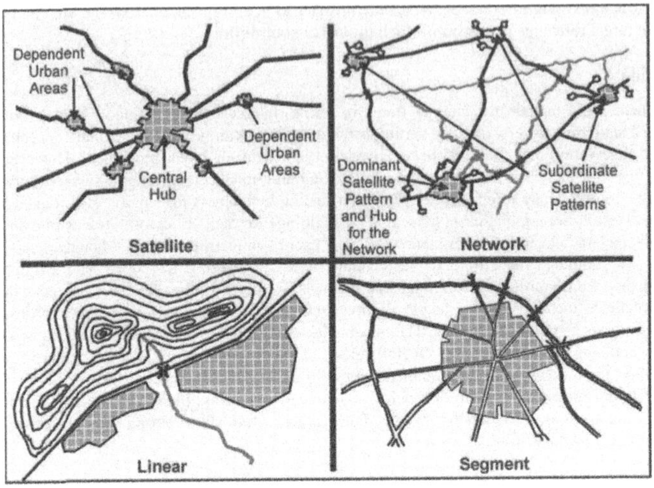

Figure 2-3. Major urban patterns

Satellite Pattern

2-18. This common pattern consists of a central hub surrounded by smaller, dependent urban areas. LOCs tend to converge on the hub. Outlying areas often support the principal urban area at the hub with means of reinforcement, resupply, and evacuation. In some instances, they may serve as mutually supporting battle positions. Commanders should consider the effects of the outlying urban areas on operations within the hub, and, conversely, the effects of operations within the hub on outlying urban areas. Information operations (IO), for example, targeted primarily at key leaders and other civilians located within the hub of a satellite pattern may subsequently influence civilians in outlying urban areas and achieve necessary effects without having to commit specific resources to these outlying areas.

Network Pattern

2-19. The network pattern represents the interlocking of the primary hubs of subordinate satellite patterns. Its elements are more self-sufficient and less supportive of each other, although a dominant hub may exist. Major LOCs in a network extend more than in a satellite pattern and take more of a rectangular rather than a convergent form. Its natural terrain may vary more than in a single satellite array. Operations in one area may or may not easily influence, or be influenced by, other urban areas in the pattern.

Linear Pattern

2-20. Potentially a subelement of the previous two patterns, the linear pattern may form one ray of the satellite pattern or be found along connecting links between the hubs of a network. Most frequently, this pattern results from the stringing of minor urban areas along a confined natural terrain corridor, such as an elongated valley, a body of water, or a man-made communications route. In offensive and defensive operations, this latter form of the linear pattern facilitates developing a series of strong defensive positions in depth, effectively blocking or delaying an attacking force moving along the canalized terrain.

Segment Pattern

2-21. When dominant natural terrain, such as a river or man-made features (canals, major highways, or railways), divides an urban area, it creates a segmented pattern. This pattern often makes it easier for commanders to assign areas of operations to subordinate commanders. However, this pattern may fragment operations and increase risk to an operation requiring mutual support between subordinate units. Still, the segmented urban areas may allow commanders to isolate threats more easily in these areas and focus operations within segments that contain their decisive points. Although an integral part of the whole (the urban area), each segment may develop distinct social, economic, cultural, and political characteristics. This social segmenting may benefit commanders faced with limited assets to influence or control the urban populace. After thoroughly analyzing the society, they may be able to focus IO and populace and resources control measures against only specific segments that affect decisive operations. Commanders may need only to isolate other segments or may need to just monitor for any significant changes in the attitudes, beliefs, or actions of the civilians located there.

LESSER STREET PATTERNS

2-22. Lesser patterns in the urban area result from the layout of the streets, roads, highways, and other thoroughfares. They evolve from influences of natural terrain, the original designer's personal prejudices, and the changing needs of the inhabitants. Street patterns (and widths) influence all warfighting functions; however, they greatly affect movement and maneuver, command and control, and sustainment. (In some portions of older Middle Eastern urban areas, the labyrinths of streets were designed only to allow two loaded donkeys to pass each other; tanks are too wide.) Urban areas can display any of three basic patterns and their combinations: radial, grid, and irregular (see figure 2-4).

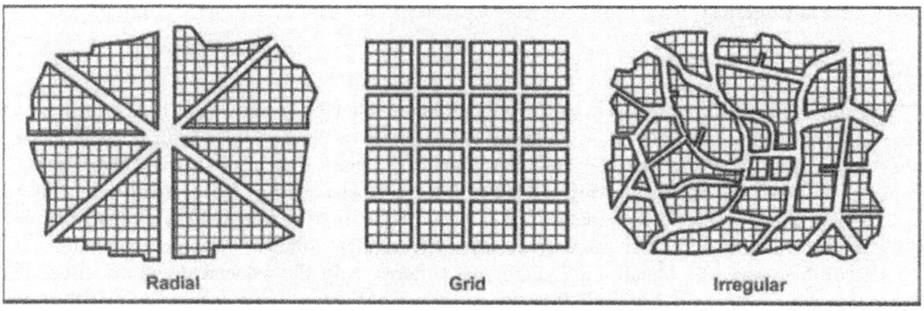

Figure 2-4. Basic internal street patterns

Radial

2-23. Societies of highly concentrated religious or secular power often construct urban areas with a radial design: all primary thoroughfares radiating out from the center of power. Urban areas with this design may signal an important historical aspect in the overall analysis of the urban society. Terrain permitting, these streets may extend outward in a complete circle or may form a semicircle or arc when a focal point abuts a natural barrier, such as a coastline or mountain. To increase mobility and traffic flow, societies often add concentric loops or rings to larger radial patterns. Unless commanders carefully plan boundaries, routes, and axes of advance, their subordinate units' movement or maneuver may be inadvertently funneled toward the center of urban areas with this pattern resulting in congestion, loss of momentum, and an increased potential for ambush or fratricide.

Grid

2-24. The most adaptable and universal form for urban areas is the grid or rectangular pattern: lines of streets at right angles to one another forming blocks similar to the pattern of a chessboard. A grid pattern can fill in and eventually take over an original radial pattern. Grid patterns often appear to ease the assignment of boundaries for subordinates units. However, commanders also consider how the natural terrain influences operations and the establishment of graphic control measures. They also consider the influence of the buildings and other structures lining these streets, such as their height and construction, before assigning boundaries and developing other control measures. Commanders should also consider the following when developing urban graphic control measures:

- Describing boundaries, phase lines, checkpoints, and other graphic control measures by easily recognizable features is as important in urban areas as elsewhere. While easily identifiable urban structures such as unusually tall or oddly-shaped buildings, cemeteries, stadiums, or prominent rail or highway interchanges can be useful references, available natural features are a better descriptor than man-made features that may be altered or unrecognizable. As an aid to air-to-ground coordination, commanders should select features that can be identified by both ground and air forces. Those that help in controlling ground forces may not be easily visible from the air and vice versa.
- Commanders should also consider whether a boundary along an easily-recognizable terrain feature, such as a river, will also be easy to identify by the threat who may seek to "find the seam" and exploit the likely control and coordination difficulties associated with boundaries, especially between higher-level units. This often requires commanders to carefully position a control measure away from the key feature to provide a designated subordinate force with the terrain and space necessary to control the feature. On the other hand, commanders working closely with local authorities during stability and civil support operations may not need to thoroughly understand the physical effect of street patterns on the assignment of boundaries as they might for combat UO. Instead, commanders may choose to assign boundaries overlaid on existing geopolitical boundaries used by local agencies to increase interoperability and aid in unity of effort.

IRREGULAR

2-25. In most urban areas, regardless of the original intent, plan, or vision, existing street patterns emerge from successive plans overlaid one on another. Some are well planned to fit with previous plans while others a haphazard response to explosive urban growth. The result may mix patterns. Urban engineers and planners may specifically design irregular patterns for aesthetic reasons (as in many suburban housing developments) or to conform to marked terrain relief. Irregular street patterns may alert commanders and analysts that the underlying natural terrain may exert greater influence over operations than in other portions of the urban area. Finally, irregular street patterns make the movement and maneuver of forces less predictable. However, a labyrinth of irregular or "twisting" street patterns may increase the possibility of fratricide particularly for units that are trained or accustomed only to grid patterns.

AN URBAN MODEL

2-26. Throughout the world, urban areas have similar form and function. In form, urban areas contain like characteristics, readily divisible into distinct sections or areas. Functionally, they tend to be the centers of population, finance, politics, transportation, industry, and culture. While urban areas may be modeled by several different means, figure 2-5 illustrates the general forms and internal functions. Some forms and functions may overlap. For example, high-rise buildings are located in core areas as well as in outlying areas and may be used for residential purposes. With the rapid urbanization associated with developing nations, the areas displayed in this urban model often manifest themselves less clearly there than in developed nations.

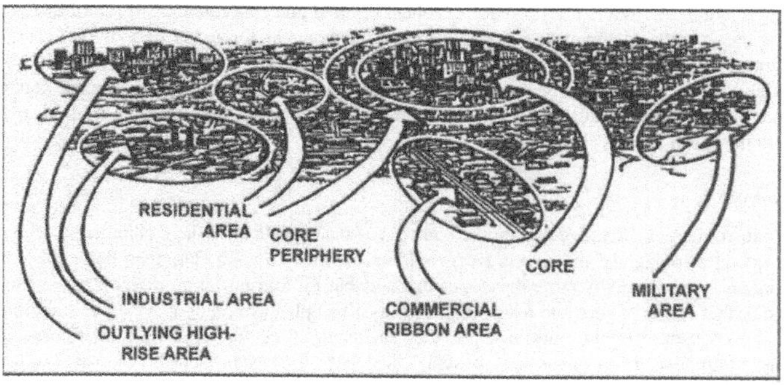

Figure 2-5. Urban functional zones

2-27. This analysis helps to determine, in general terms, potential advantages and disadvantages each portion of the urban area may have toward accomplishing the urban operation. However, construction materials and methods can vary drastically. Commanders and their staff will often need to identify specific building types and construction and understand weapons effects on them. If a commander desires precise effects, the chosen munitions or weapons system must be sufficiently accurate, capable of penetrating the target structure (without exiting the other side), and achieve effects within. Often noncombatants, critical infrastructure, or protected targets are in the vicinity. Commanders may need to determine if the surrounding walls or structures will sufficiently absorb or negate the blast or thermal effects of the weapon. Regardless, understanding the structure and composition of buildings and other structures in the urban AO may be necessary to allow commanders to determine the best means to accomplish the mission.

Core

2-28. The core is the heart of the urban area, the downtown or central business district. Relatively small and compact, it contains a large percentage of the urban area's shops, offices, and public institutions. Often, it houses the headquarters for commercial and financial activities and contains important cultural, historical, and governmental buildings. These activities prefer the core because of its accessibility. As the focal point of the transportation network, residents find the core the easiest part of the urban area to reach. It normally has the densest concentration of multistory buildings and subterranean features (underground parking garages, underground shopping centers, and basements).

2-29. High-rise buildings, varying greatly in height (possibly 50 stories above ground and four stories below ground), make up the cores of many of today's urban areas. Buildings routinely abut one another, with little or no setback from the sidewalks. Building height and density (except in outlying high-rise areas) often decreases from the core to the edge of the residential areas, while the amount of open areas frequently increases. Modern urban planning allows for more open spaces between buildings than found in the cores of older urban areas. Most core areas have undergone constant redevelopment resulting in various types of construction. Commonly, brick buildings abound in the oldest part of the core; framed, heavy-clad structures in the next oldest part; and a concentration of framed, light-clad buildings in the newest part. The outer edge of the core, the core periphery, has ordinarily undergone less change than the core resulting in buildings of uniform height (commonly two to three stories in towns and five to ten stories in larger urban areas).

2-30. Generally, offensive operations focused in core areas (even when effectively isolated) will require greater resources—particularly manpower, time, and information—than in many other parts of the urban area. Mounted maneuver often proves more difficult in core areas because of fewer open areas, buildings closer to the streets, and an increased density of civilian vehicles. Razed buildings in central core areas (especially high-rise buildings) become greater obstacles to mobility as they can collapse on and easily block thoroughfares. Rubble piles provide excellent covered and concealed positions for dismounted threat forces. Consequently, commanders often use more dismounted forces as part of their combined arms operations. Conversely, the core may be critical to urban defensive operations, particularly older areas of heavier construction that afford greater protection. Despite potential difficulties, the core area may be key to accomplishing many stability or civil support missions since it houses much of the human activity that occurs in the urban area.

Industrial Area

2-31. Industrial areas often develop on the outskirts of the urban areas where commercial transportation is easiest (along airfields and major sea, river, rail, and highway routes). The road networks in and around industrial areas are generally more developed and suitable for transportation assets. These areas will likely displace farther from the core and residential areas as urban planners recognize the potential threat of TIM. The dispersed pattern of the buildings provides sufficient space for large cargoes, trucks, and materiel handling equipment. These areas may provide ideal sites for sustainment bases and maintenance sites. While older heavier-clad structures may be found, new construction generally consists of low, large, flat-roofed factory and warehouse buildings with large parking areas and work yards. These structures generally have steel frame and lightweight exterior walls. Multistory structures usually have reinforced concrete floors and ceilings.

Toxic Industrial Chemical	Industrial/Commercial Uses
Ammonia	Commercial Refrigerant, Fertilizer and Food Production, Petroleum, Explosives, Other Chemicals
Arsine	Semiconductor Industry
Boron Trichloride	Organic Catalyst, Soldering Magnesium
Boron Trifluoride	Chemical Catalyst, Aluminum Refining
Carbon Disulfide	Industrial Solvent, Dry Cleaning, Agriculture, Petroleum, Electroplating
Chlorine	Potable Water, Disinfectants, Metal Treatment, Plastics & Rubber
Diborane	Plastics and Rubber
Ethylene Oxide	Industrial Alcohols, Fumigant, Industrial Sterilant
Fluorine	Uranium Processing, Rocket Fuel
Formaldehyde	Plastics, Fertilizers, Preservative/Corrosion Inhibitor, Fungicide and Germicide, Pesticide, Pharmaceuticals
Fuming Nitric Acid	Fertilizers, Explosives, Metal Processing, Pesticides, Rocket Fuel
Hydrogen Bromide	Chemical Industry, Pharmaceuticals
Hydrogen Chloride	Fabrics, Semiconductors
Hydrogen Cyanide	Pesticides, Other Chemicals, Pharmaceuticals, Electroplating
Hydrogen Fluoride	Glass Production, Chemical Catalyst
Hydrogen Sulfide	Metallurgy, Agricultural Disinfectant
Phosgene	Dyes, Pharmaceuticals, Herbicides & Insecticides
Phosphorus Trichloride	Metallurgy, Pesticides and Germicides, Gasoline Additive
Sulfur Dioxide	Paper, Food Processing, Ice Production, Disinfectant, Leather Processing
Sulfuric Acid	Fertilizers, Petroleum, Iron and Steel Production, Battery Electrolyte
Tungsten	Electronics, Other Chemicals

Figure 2-6. Some toxic industrial chemicals and their industrial or commercial uses

2-32. Toxic industrial chemicals and other TIM may be transported through an urban area (by rail, barge, truck, or pipeline) or found stored throughout. However, larger concentrations will exist in industrial areas, and their presence should concern Army forces operating near them. Some TIM are heavier than air and tend to settle in low-lying and subsurface areas.

2-33. Each year, over 70,000 different chemicals are produced, processed, or consumed globally. An estimated 25,000 commercial facilities around the world produce, process, or store chemicals that have a legitimate industrial use yet are also classified as chemical warfare agents. Many other chemicals (not classified as weapons) may still be sufficiently hazardous to pose a considerable threat to Army forces and civilians in urban areas as choking agents or asphyxiates, flammables or incendiaries, water contaminants, low-grade blister or nerve agents, or debilitating irritants. These chemicals can be released either accidentally or deliberately. For example, on 2 December 1984, nearly 40 tons of methylisocyanate used to produce pesticides leaked from a storage tank at Bhopal, India, killing thousands and injuring hundreds of thousands. Figure 2-6 contains a small sampling of other toxic industrial chemicals along with their industrial or commercial usage that commanders may encounter in an urban area. The most common chemicals that may pose an immediate risk to Army forces are highly toxic irritant gases such as ammonia, chlorine, hydrogen chloride, and sulfur dioxide.

2-34. Standard chemical defense equipment may not protect against (and chemical detection devices may fail to detect) many toxic industrial chemicals. Therefore, the risk to Soldiers operating near the chemicals may increase. Commanders must vigilantly identify these potential hazards, carefully consider them as part of their overall vulnerability analysis, factor the analysis into their risk assessment, and execute necessary contamination avoidance measures. (Local urban fire fighters may be a critical source of information for determining the likely locations of TIM.) Any assessment includes the chance that toxic industrial chemicals may be deliberately released by a threat to gain advantage or accidentally released by friendly actions (see FM 3-11.9, FM 3-11.14, FM 3-11.21, and the U.S. Department of Transportation's current version of the Emergency Response Guidebook).

Outlying High-Rise Area

2-35. High-rise areas consist of multistoried apartments, commercial offices, and businesses separated by large open areas, such as parking lots, parks, and individual one-story buildings. High-rise buildings are generally of framed, light-clad construction with thin walls of brick, lightweight concrete, or glass. The automobile, mass transit systems, and improved road networks encourage these areas to grow and function further from the urban core.

2-36. Similar to the urban core, units given the mission to clear these areas, or even portions therein, will need more resources—most notably personnel and time—to accomplish their mission. Commanders should consider courses of action that isolate these entire areas, multiple sections within these areas, or even individual buildings before assigning tasks. Without careful consideration and analysis, some tasks in these areas could unintentionally—but rapidly—drain a unit's resources or unhinge other portions of the major operation. When defending, commanders who can integrate these areas in the defense will present the attacker with similar resource problems and may be appropriate in a defense to delay. However, defending commanders must ensure that the defense is arranged so that this portion cannot be easily isolated and bypassed. Defensive positions in structures may require extensive reinforcement due to light-clad construction.

Residential Area

2-37. Residential areas can be found dispersed throughout the urban area; however, large suburban areas (or sprawl) normally form on the outskirts. Residential areas often consist of row houses or single-family dwellings set in a grid or ringed pattern in a planned development project. Yards, gardens, trees, and fences may separate the buildings in a residential area. In some areas of the world, residential areas may be located in high walled compounds with houses built right up to the edge of the street. Modern residential construction is more often of light-clad, framed wood construction, or brick, however; residential homes formed by poured or pre-cast concrete can be found throughout many parts of the world. The combined population of surrounding suburban areas often far outnumbers that of the urban area proper. Specific suburbs tend toward homogeneity based on ethnicity, religion, economics, or some other social aspect. Commanders must locate and analyze these areas to determine their impact on operations—often the most critical importance is the people located there (see the subsequent discussion in this chapter on the urban society).

2-38. In offensive and defensive operations, commanders should determine whether operations pose an unacceptable physical risk to civilians. If so, they may have to relocate civilians to a safer area, perhaps another residential area. If not, commanders may implement a "stay-put" policy for that area and attempt to isolate the effects of the operation from them. During civil support and stability operations, residential locations may be the initial focal point for operations since most of the permanent population is located there.

2-39. This area also contains an urban phenomenon known as shantytowns. These areas are commonly on unoccupied, low-value land in and around many urban areas in underdeveloped countries. Shantytowns may contain over 50-percent of the total urban population. They usually lack streets and public utilities. The lean-to structures tend to be irregularly laid out, connected by walking paths, and made of any scrap material available: lumber, brick, sheet metal, cardboard, cloth, or vegetation. The random arrangement of structures, the absence of formal street naming and numbering, and often the lack of easily identifiable buildings and terrain create challenges. These challenges include navigation, coordination, and the transmission of accurate information and intelligence. Depending on the operation, the temporary nature of the structures can also mean that mobility can be either more or less restricted than other sections of the urban area. A military force may easily knock down and traverse structures without affecting mobility at all. However, their destruction may cause unacceptable civilian casualties, in which case mobility becomes more restrictive as the narrow paths often do not accommodate vehicular traffic. Similarly, the makeshift materials inhibit weapons effects less than many other parts of the urban area built more solidly. A tank round, for example, may go much farther and injure many more noncombatants than in an area where the primary building material is stone. Regardless, commanders should consider the effects of their operations in this area, to include vehicles and weapons, as the weak structures increase the risk of fratricide, civilian casualties, and large, rapidly spreading fires.

Commercial Ribbon Area

2-40. Commercial ribbon areas are rows of stores, shops, and restaurants built along both sides of major streets that run through and between urban areas. These same types of areas often develop along the roads that connect one urban area to another (strip areas). The buildings uniformly stand two to three stories tall (about one story taller than the dwellings on the streets behind them).

Military Area

2-41. Fortifications and military installations may be found in or near urban areas throughout the world. Historically, they may have been the "seed" responsible for initiating the growth of the present-day urban area. Many countries possess long coastlines and borders with potentially hostile neighbors. To meet their defensive needs, they developed coastal and border defense works that include extensive subsurface facilities, many contiguous to urban areas. North Korea, for example, has built numerous hardened artillery, missile, and command and control facilities along both its coasts and the demilitarized zone which separates them from South Korea. While some fortifications may be inactive (particularly ancient fortifications), they can be rapidly activated and modified by threats (or Army forces) to accomplish their original role. Even if not activated, they may still serve as choke points and major obstacles to movement and maneuver.

2-42. Permanent-type fortifications can be made of earth, wood, rock, brick, concrete, steel-reinforced concrete, or any combination of the above. Some contemporary constructions have been built subsurface and employ heavy armor, major caliber weapons, internal communications, service facilities, and chemical, biological, radiological, and nuclear overpressure systems. Because they have been built specifically for military purposes, commanders and planners should carefully consider the effects of these military constructions on the conduct of UO.

URBAN SOCIETY

2-43. Although intricate, understanding the urban terrain is relatively straightforward in comparison to comprehending the multifaceted nature of urban society. UO often require Army forces to operate in close proximity to a high density of civilians. Even evacuated areas can have a stay-behind population in the tens of thousands. This population's presence, attitudes, actions, communications with the media, and needs may affect the conduct of operations. Homogeneity decreases drastically as the size of the urban area increases. Commanders must take into account the characteristics of a population whose beliefs and interests vary based on factors (see figure 2-7). Analysis and understanding of these societal factors is critical to a successful IO campaign and, thus, the entire operation.

2-44. Civilian populations continually influence, to varying degrees, operations conducted in an urban area. Thoroughly understanding these societal aspects and avoiding "mirror-imaging"—overlaying one's own values and thought processes on top of the person or group one is trying to understand—will help to accurately anticipate civilian actions and response. For example, while bribery and nepotism are not accepted norms in U.S. society, they may be customary practices in the foreign urban environment in which Army forces are operating. If commanders attempt to understand incidences of bribery and nepotism using their own societal norms and values as a basis, they are likely to misinterpret the significance of the events and implement an inappropriate course of action. Critically, however, the development of an in-depth understanding of the urban society does not mean that commanders and their Soldiers should adopt the culture of the foreign urban society. While understanding and respect of the urban area's culture is crucial to successful UO, Soldiers function best in dealing with the urban society when they act in accordance with American culture and values.

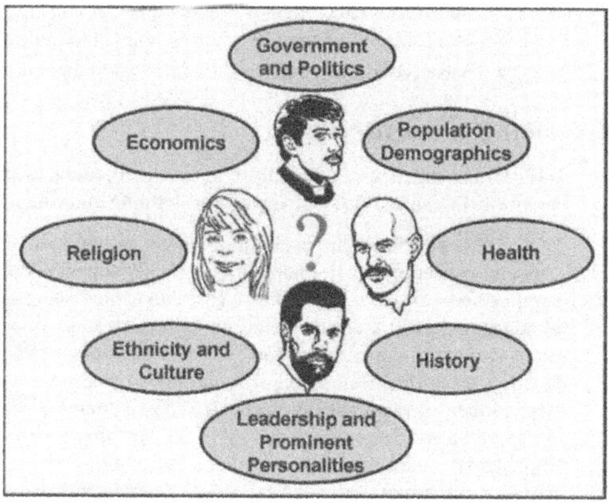

Figure 2-7. Key aspects of the urban society

POTENTIAL CENTER OF GRAVITY

2-45. A COG during an urban operation, particularly in stability and civil support operations, may be the civilian inhabitants themselves—specifically their behavior. However, supportive behavior is generally an advantage in any type of operation. Correspondingly, neutral behavior toward friendly forces is an advantage over hostile behavior. (However, a neutral population normally will not readily provide information to Army forces—often critical to understanding the urban environment and threat. In this light, neutrality can be seen as a disadvantage.) To influence or control their behavior, commanders must first understand the society's complex nature and character. Second, they must understand and accept that every military action (or inaction) may influence the relationship between the urban population and Army forces, and, by extension, mission success or failure. Lastly, they must understand that Army forces may play only a supporting (but essential) role as part of an integrated and synchronized multiagency effort focusing all aspects of national power—diplomatic, informational, military, and economic. With this awareness, commanders can take one or more actions:

- Coordinate and plan operations.
- Implement effective civil-military programs.

- Take the immediate action necessary to maintain support of a friendly populace, neutralize or gain the support of hostile or neutral factions, or do any combination of these activities to achieve precise effects and accomplish the mission.

Without this understanding, commanders increase the risk that their actions, particularly concerning the urban population, may not have the intended and necessary effects.

2-46. Although the factor of civil considerations takes on added significance in UO, it is just one that commanders evaluate. Sometimes, as in most urban counterinsurgency operations, it may be the most important factor to consider as a COG. At other times it may be the least important as to be almost negligible. Its importance is not constant; it changes over time (like all factors). At the beginning of the operation, civil considerations may not be essential to mission accomplishment, but as the operation progresses this factor's importance to success may increase. In other circumstances, the opposite may be true. Overall, commanders should consider three objectives regarding the civilians of the urban area:

- Minimize their interference with urban operations. In offensive and defensive operations this means moving them away from combat operations or establishing measures to shield them from its effects. In all operations, it often requires centralizing them in one or more locations—which may mean keeping them where they are.
- Maximize their support of Army, joint, and multinational forces and government agencies.
- Observe the necessary legal, moral, and humanitarian obligations.

General Population Size

2-47. Urban areas are commonly classified according to the general size of their population instead of landmass. Figure 2-8 lists one method of defining and categorizing urban areas by various population sizes.

2-48. These categories are useful to establish commonality and standardize terms that shape ideas, discussion, and concepts. Smaller populations usually suggest homogeneity among the inhabitants. Homogeneity can make consensus or compromise easier to achieve because fewer opposing viewpoints exist. Given this homogeneity, effects of change are more certain and often easier to determine. However, homogenous does not mean identical. If major social divisions do exist (either physical or ideological), commanders can more easily determine those divisions and their fundamental causes with smaller populations. Treating an urban population as a completely homogenous entity can lead to false assumptions, cultural misunderstandings, and poor situational understanding.

Category	Population
Village	3,000 or less.
Town	Over 3,000 to 100,000.
City	Over 100,000 to 1 million.
Metropolis	Over 1 million to 10 million.
Megalopolis	Over 10 million.

Figure 2-8. Urban areas by population size

2-49. As urban areas expand, the urban patterns begin to blur and the social complexity increases. For example, as satellite patterns continue to grow, the LOCs between a central hub and outlying urban areas may develop and begin to assume a linear urban pattern. Simultaneously, a hub and outlying urban areas may continue to expand until they merge into a single, large metropolis. On a larger scale, a network pattern can grow and unite as a single megalopolis. This growth physically unites smaller urban areas but cannot force conformity of needs and beliefs. It also increases the physical and social complexity of an urban area.

Group Size, Location, and Composition

2-50. Understanding how specific elements of the urban society affect operations (and vice versa), begins with analyzing their size, location, and composition (see figure 2-9). Because commanders must minimize civilian casualties, size and location (without regard to composition) are important initial demographic considerations. After determining the presence and numbers of civilians relative to decisive points, commanders can then decide whether civilian proximity and density represent a significant risk to their mission—refugees clogging LOCs, for example. If civilians are the primary focus of the operation, as in many stability and civil support operations, this same analysis may help to determine decisive points. In this analysis, commanders should consider that urban areas, on many levels, are in constant motion. The densities of circulating people and other traffic often vary according to cultural events or the time of day, such as religious holidays or sporting events and rush hours or market times. Therefore in planning urban operations, commanders must consider the timing, rhythms, or patterns of the population and their vehicular movements in the urban area. Identifying and understanding trends and patterns of activity (and disruptions to them) may provide critical information to commanders.

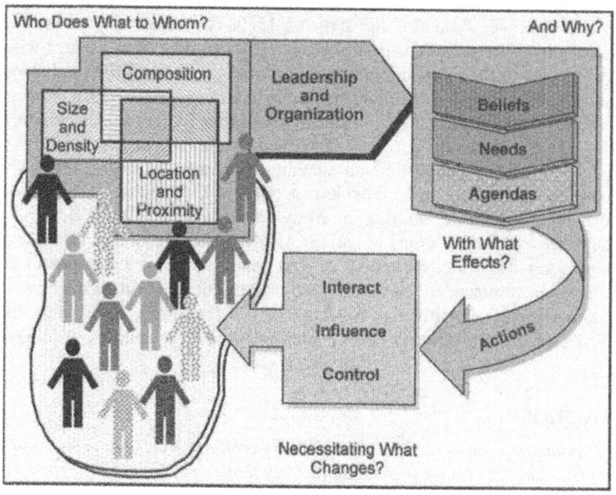

Figure 2-9. Simplified analysis of urban society

2-51. Commanders normally determine the composition of, or the identifiable groups or organizations within, the civilian urban population. Groups may be categorized by race, religion, national origin, tribe, clan, economic or social class, party affiliation, education level, union memberships, age, gender, occupation, or any other significant social demographic. Physical and ideological overlaps (and divisions) often exist between groups. Overlaps may provide early focus for analysis and suggest ways to affect more than one group simultaneously. In some cases, groups may have radically different ideologies but are (or can be) united by a single characteristic. Commanders must understand the intricacies of "who does what to whom." Such understanding furthers identifying the urban society's sources of power, influence (both formal and informal), and decisive points that hold the keys to controlling or protecting this potential COG. (See also the discussion of competing power structures in Chapter 3.) Commanders should have expert, detailed, and current knowledge and information to avoid developing simple formulas of social interaction that may actively mislead and add to a flawed course of action.

LEADERSHIP AND ORGANIZATION

2-52. Commanders must also understand how authority and responsibility is held or shared within and between each of the identified groups; they must understand leadership and the social hierarchy. For groups to exert meaningful influence, leadership provides vision, direction, and organized coherence. This leadership can be a function of personality as well as organization. Some groups depend on a charismatic leader to provide cohesion; although in some cultures, the spokesman is not the leader. Others deemphasize individual leadership and provide redundancy and replacement in decision making. Others combine elements of both these types of leadership and organization. Based solely on personality, a leader may centralize power or, while still being in ultimate control, decentralize decision making and execution to subordinates. In contrast, a single person may head a group while a ruling council actually makes and executes policy. Groups centered on one leader (which may or may not be the officially designated leader) can often produce decisions and initiate actions rapidly but are vulnerable to disruptions if key personalities are removed or co-opted. Groups with shared or redundant leadership take longer to make decisions yet are more resistant to change and outside influence.

2-53. In UO, particularly stability operations, Army commanders and leaders at all levels will devote considerable effort to identifying and cultivating relationships with civilian leaders in their AO. This civilian leadership will include political, religious, tribal or clan, ethnic, and economic leaders. Commanders should consider that their attention toward and discussion with identified leaders may increase (or, some instance, decrease) the targeted leaders' prestige and power. While this may be intentional, commanders must often ensure that the leaders that they chose to deal with are the legitimate and accepted in the eyes of the urban population. Otherwise, they may further imbalance an already weak power structure and exacerbate the situation. In unique circumstances, commanders may need to identify and interact with the leadership of criminal organizations. These civilian leaders will be the means to affect change and allow the urban populace to understand and accept the purpose behind Army operations. As importantly, they will be the conduit for understanding the urban society's sentiments, perceptions, and reactions to Army operations. As such, reliable and trustworthy linguists that can not only interpret the language but can serve as cultural advisors to Army forces will be in great demand. Keeping the local population objectively informed not only of current Army operations but also of the intent and desired end state will often be a key task for commanders. A poor communications effort and a lack of understanding of intent may alienate a commander and cause significant problems in future negotiations. It is imperative that communications be clear and effective and that all Army leaders are aware of its implications. See FM 4-05.401 and FM 3-07 for the effective use of interpreters and the conduct of productive meetings and negotiations.

Interests and Actions

> *Me and Somalia against the world, me and my clan against Somalia, me and my family against the clan, me and my brother against my family, me against my brother.*
>
> Somali Proverb

2-54. Identifying and analyzing groups also helps commanders focus on specific segments of the urban society to determine their beliefs, needs, and agendas. It also helps commanders determine how those interests motivate groups to future action (or inaction)—previous patterns of activity are critical in this regard. This analysis seeks to determine why groups (and their leaders) act as they do. Commanders should consider political, economic, cultural, and religious factors in this analysis. These factors affect all groups to some extent and often provide the basis for their beliefs, needs (actual or perceived), and subsequent behavior. Size and location considerations also apply to each group to help determine to what extent its beliefs or ideologies, needs, and actions may impact the urban operation. However, size and proximity may not accurately indicate actual or potential capabilities. Individuals, small groups, and groups located some distance from the actual conduct of the urban operation may be able to influence large portions of the population. These individuals or groups may have a capability disproportionate to their size and proximity—especially against objectives that are not terrain oriented (as in the case of many stability operations).

Interaction, Influence, or Control

2-55. Commanders must cultivate an understanding of a group's—

- Size, location (and proximity to operations), and composition (to include leadership and organization).
- Interests.
- Capabilities.
- Potential actions (intent) and their effects—if any—on operations.

2-56. Then they can develop or modify courses of action as appropriate. Certain courses of action may be needed to improve the interaction between Army forces and civilians (and between other agencies) to accomplish common goals. Others may be needed to influence favorable support, stabilize neutral groups, or neutralize hostile groups. Still others may require more forceful means to control and protect—never to punish—civilians. Courses of action may include—

- Establishing buffer zones and restricted areas.
- Setting up checkpoints and roadblocks with other travel restrictions on people and goods.
- Screening civilians.
- Conducting negotiations (directly or as a mediator).
- Providing or protecting rations, water, and other critical resources.
- Restoring or improving specific, key infrastructure.
- Enforcing curfews.
- Inspecting facilities.
- Directing amnesty programs.
- Conducting internment and resettlement operations.
- Maintaining a "stay-put" policy.

2-57. Commanders must remember that many measures will require significant resources that may initially be beyond the capabilities of the Army force to impose and enforce. (Where possible, commanders should attempt to use host-nation security forces and local urban law enforcement to control activities.) Other elements of the environment, terrain and infrastructure, may fragment efforts and make it difficult to impose controls throughout the area. A careful assessment and understanding of the urban society's interests (beliefs, needs, and agendas) is essential before implementing any populace and resources control measures. Otherwise inappropriate controls, particularly if they are improperly perceived as punishment, may only aggravate the situation. Finally, an appropriate course of action may require no specific action towards the urban society. In most cases, training and discipline, grounded in cultural understanding and sensitivity, will help mitigate many potential adverse effects resulting from military-civilian interaction. (Commanders should review FM 3-19.15 and FM 3-19.40 for additional civilian control measures and considerations.)

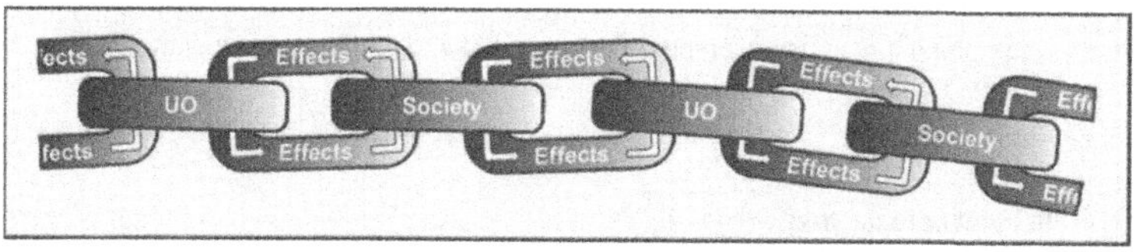

Figure 2-10. UO society chain of effects

A Chain of Effects

2-58. Since urban society is so dynamic and the relationship between various elements so complex, commanders must continually assess how their operations will affect the society and vice versa (see figure 2-10). Specifically, they assess how effectively their measures improve interaction with, influence of, and control over civilians. There is always a difference between intended and actual effects of a specific course of action. Nowhere is this more prominent than dealing with the urban society. This chain of effects frustrates understanding during UO. Therefore, commanders must continuously monitor these effects to make decisions and modifications while planning, preparing, executing, sustaining, and transitioning UO. (Commanders may conduct informal and formal polling activities to assess and understand changing civilian perceptions.) Initially certain aspects of the society, such as religion, may not affect the operation. However, if the threat successfully shapes the perceptions of the urban populace that Army forces are biased against their religion or religious beliefs (or at least critical segments are affected by propaganda), this element may become extremely important. In this instance, the urban commander may need to adjust his IO (to include PSYOP), public affairs (PA) activities, and CMO to counter this propaganda while diverting other combat power to control the populace. Overall, commanders must understand and account for second, third, and higher order effects of their actions and decisions.

Understanding the Effects of Unit and Soldier Actions in Iraq

During OPERATION IRAQI FREEDOM, an armor task force commander described his methodology as that of plotting and measuring everything. After a positive or negative event, he would have his staff evaluate all actions they had conducted before, during, and after the event. This would allow him to correlate activities with outcomes and develop tactics, techniques, and procedures for future success. While not all actions easily correlated to events, this methodological approach helped establish a base line of comparative success for his task force.

Some units assessed the Iraqi culture and developed practical guidelines or rules of interaction for their soldiers to follow as they conducted their missions. These guidelines encompassed a list of actions that their Soldiers should and should not do when dealing with the Iraqi populace. The guiding principle behind these actions was not only to "win the hearts and minds" of the Iraqi people but, more accurately, to establish legitimacy and credibility through appropriate Soldier actions that reflected a combined understanding of the mission and the Iraqi culture.

DO:

- Separate the men from the women and children during cordon and searches.
- Talk to the oldest or most senior male when civilian cooperation is needed.
- Bring something (for example, water, food, or replacement locks) to give residents to compensate for damages when the search ends up being a "dry hole."
- Learn Arabic phrases; say hello and greet people on the street.
- Use female Soldiers to search and talk to Iraqi females.
- Cordon and knock whenever possible instead of kicking in doors needlessly.
- Understand that in Iraq, women do not have the same social status as men.

DO NOT:

- Force a detainee's head to the ground or use your feet to hold someone to the ground.
- Stare at or touch Iraqi women.
- Show obvious disrespect to suspects in front of family members.
- Manhandle the Koran or other religious objects.
- Commit to anything; instead of agreeing to (or denying) an Iraqi's request, it is good practice to say "In'sha Allah" (If God wills it).

URBAN INFRASTRUCTURE

2-59. Urban infrastructures are those systems that support urban inhabitants and their economy. They form the essential link between the physical terrain and the urban society. During urban stability and civil support operations, restoration or repair of urban infrastructure will often be decisive to mission accomplishment. During urban combat operations, destroying, controlling, or protecting vital parts of the infrastructure may be a necessary shaping operation that can isolate a threat from potential sources of support. A threat force operating in an urban area may rely on the area's water, electricity, and sources of bulk fuel to support his forces. This is true particularly when his bases or facilities are physically located in or near the area. Isolating this threat from these sources may require him to generate his own electricity and transport his own water and fuel from outside the urban area. To transport supplies, the threat may rely on roads, airfields, sea or river lanes, and rail lines. Controlling these critical transportation nodes may prevent the threat from resupplying his forces. The control of key radio, television, and newspaper facilities may isolate him from the urban populace (another potential source of support).

INTERDEPENDENCE

2-60. Commanders must understand that destroying or disrupting any portion of the urban infrastructure can have a cascading effect (either intentional or unintentional) on the other elements of the infrastructure. Yet, they may be able to gain an operational advantage while minimizing unwanted and unintended effects. Commanders can control, seize, or secure an essential facility or structure by using precision munitions, electronic disruption of communications, or SOF and conventional ground forces. To gain this advantage, commanders will rely more on the expertise of Army engineer and civil affairs units; local urban engineers, planners, and public works employees; and others with infrastructure-specific expertise. After understanding the technical aspects of the area's systems and subsystems, commanders can then develop the best course of action.

A SYSTEM OF SYSTEMS

2-61. Hundreds of systems may exist. Each system has a critical role in the smooth functioning of the urban area. Simple or complex, all systems fit into six broad categories (see figure 2-11). Commanders should analyze key facilities in each category and determine their role and importance throughout all phases of the urban operation. As there is much overlap between infrastructure systems, this analysis considers each system individually and in relation to others to determine an appropriate course of action toward it.

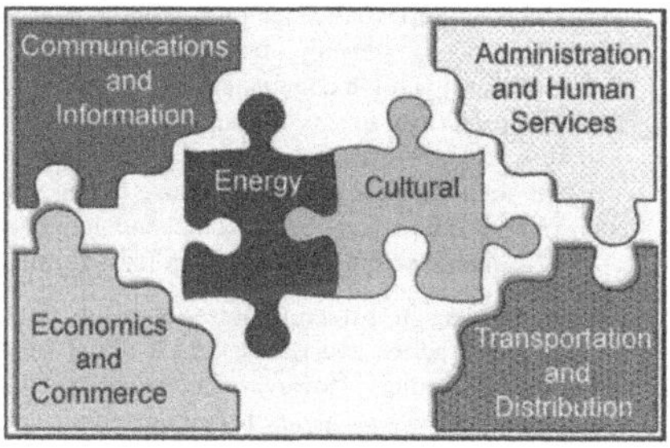

Figure 2-11. Urban infrastructure

A COMBINATION OF STRUCTURES AND PEOPLE

2-62. As depicted earlier in figure 2-1, each element of the infrastructure consists of both a physical (terrain) and human component. For example, the physical component of the electrical segment of the energy infrastructure consists of power stations, substations, a distribution network of lines and wires, and necessary vehicles and repair supplies and equipment. The human component of this same segment consists of the supervisors, engineers, linemen, electricians, and others who operate the system. Commanders must understand and recognize the physical and human components in their assessments.

POTENTIAL IMPACT ON FUTURE OPERATIONS

2-63. Destroying or incapacitating of any of these elements may impact future operations and inhabitants of the urban area. Destroying urban infrastructure during initial phases of an operation may require commanders to assume responsibility for repair, maintenance and clean up, and operation of those same facilities later. Although exceptions will exist, commanders cannot destroy or significantly damage the infrastructure of a foreign urban center during operations and expect the population to remain friendly to U.S. or allied forces. On the other hand, early repair or restoration of critical or essential infrastructure may improve civil-military relations, speed transition back to competent civilian authorities, and, overall, aid in successful mission accomplishment. Still, support from the urban society (albeit of increased importance in UO) is only one factor that commanders weigh while developing appropriate courses of action.

Resource Intensive

2-64. Requirements to protect, restore, or maintain critical infrastructure may divert substantial amounts of resources and manpower needed elsewhere and place additional constraints on subordinate commanders. Civilian infrastructure is often more difficult to secure and defend than military infrastructure. The potentially large and sprawling nature of many systems (such as water, power, transportation, communications, and government), make their protection a challenge. Yet, the infrastructure of an urban area may provide commanders with essential logistics and support. Therefore, the initial expenditure of time and other resources may be necessary to support concurrent or future operations. Legal considerations, however, may affect using the infrastructure and acquiring the urban area's goods and services. Commanders, their staffs, and subordinates (often down to the individual soldier) must know their limits concerning Army authority to commandeer civilian supplies or equipment to facilitate mission accomplishment (see the legal support discussion in Chapter 9). In stability and civil support operations, the safeguard or restoration of critical urban infrastructure for military or civilian use may be a decisive point in the overall operation.

2-65. Keys to understanding the magnitude of the resources and manpower required to restore the infrastructure are an initial infrastructure assessment and, as soon as practical afterward, a detailed infrastructure survey. An initial assessment provides the commander immediate feedback concerning the status of basic services needed to meet the urgent needs of the urban population. The systems assessed are based on mission, enemy, terrain and weather, troops and support available, time available, civil considerations, and the commander's vision of the overall end state. The infrastructure assessment, while typically performed by engineers, may be accomplished by, or in conjunction with, others with sufficient expertise to provide the type and quality of information required. These others may include civil affairs, medical, and chemical personnel. Those tasked with this assessment should routinely consult other Army and coalition forces and governmental and nongovernmental agencies currently operating in the urban area as well as the urban civilian leadership for their informed input.

2-66. While an infrastructure assessment functions to support the resolution of immediate challenges to urban reconstruction and restoration, it also provides the initial basis for determining the conditions for successful transition. However, commanders and planners must continually expand and refine their understanding. As a necessary follow-on, commanders initiate a detailed infrastructure survey. This survey is normally conducted by U.S. Army Corps of Engineers personnel assigned to forward engineer support teams. As with the assessment, the commander should incorporate other technical specialty personnel in the survey team to enhance the quality and accuracy of the product (see FM 3-34.250).

Communications and Information

2-67. This system is comprised of the facilities and the formal and informal means to transmit information and data from place to place. Understanding communication and information infrastructure of an urban area is important because it ultimately controls the flow of information to the population and the enemy. It includes—

- Telecommunications, such as telephone (to include wireless), telegraph, radio, television, and computer systems.
- Police, fire, and rescue communications systems.
- Public address, loudspeaker, and emergency alert systems.
- The postal system.
- Newspapers, magazines, billboards and posters, banners, graffiti, and other forms of print media.
- The informal human interaction that conveys information such as messengers, open-air speeches and protests, and everyday conversations.
- Other inventive informal means such as burning tires and honking horns.

2-68. Perhaps more than any other element of the infrastructure, communications and information link all the other elements in an interdependent "system of systems." It is a critical enabler that helps coordinate, organize, and manage urban activities and influence and control the urban society. Army commanders are acutely aware of the impact that a loss or degradation in communications has on their own operations. The urban environment experiences similar impacts to communication failures; however, urban governments and administrations are generally less prepared to deal with a collapsed communications and information infrastructure than are trained Army forces.

2-69. Militarily, a functioning urban communications and information system can serve as an alternate for both friendly and threat forces and can be easily secured with civilian, off-the-shelf technologies. Threats may make use of commercial systems intertwined with legitimate civilian users, making it unpalatable to prevent use of these assets. Forces can also use these systems to influence public opinion, gain intelligence information, support deception efforts, or otherwise support IO.

Increasing Impact of Computers

2-70. In many urban areas, computers link other elements of the urban infrastructure. They link functions and systems in the urban area and connect the area to other parts of the world. This latter aspect creates important implications for commanders of a major operation. Operations involving this cybernetic function may produce undesirable effects on a greater scale than initially intended. For example, commanders may be able to close or obstruct an urban area's banking system; however, this system may impact the international monetary exchange with unwanted or even unknown effects. The authority to conduct these types of IO will often be retained at the strategic level.

Pervasive Media

2-71. The media is central to the communications and information infrastructure and a critical operational concern. Compared to other operational environments (jungles, deserts, mountains, and cold weather areas), the media has more access to urban operations. This is due largely to airports, sea and river ports, and major road networks; ready access to power sources and telecommunications facilities; as well as access to existing local media structures. Hence, media presence may be pervasive and IO even more critical to success in UO than operations in many other environments.

A Complex Relationship

2-72. A complex relationship exists among information, the public, and policy formulation. Although the degree and manner in which public opinion shapes government policy are difficult to accurately determine, negative visual images of military operations presented by the media can change political objectives and, subsequently, military objectives. As important, media reporting can influence civilian activity in an urban AO to either the advantage or disadvantage of the commander.

Induce Cooperation Through Credibility

2-73. Commanders do not control the media; however, they monitor the flow of information that the news media receives and subsequently reports. Consequently, commanders should plan and execute PA operations that will induce cooperation between the media and Army forces. Successful relations between urban Army forces and the news media evolve from regular interaction based on credibility and trust. More information is usually better than less, except when the release of such information may jeopardize security and the success of the operations and threaten the safety of Soldiers. However, commanders cannot simply withhold information to protect the command from embarrassment. They consider media interests as part of the normal planning process and work to ensure that information presented to the news media is accurate, timely, and consistent with operations security. Since the media will likely arrive in the urban area before the conduct of operations, early deployment of PA assets may be critical. Commanders should synchronize PA activities with CMO and PSYOP. Such action eliminates duplicated effort and ensures a unity of purpose consistent with the IO concept of support (see Chapter 4).

2-74. Failure to provide sufficient information can hamper a commander's ability to conduct the mission. Commanders cannot refuse to deal with particular news media because they consistently report a negative image of Army forces and operations. Poor relationships with any media can result in inaccurate and even biased reporting. Such reporting can cause a public reaction that influences the ability to achieve operational objectives. During the Russian 1994-95 battle against Chechen separatists in Grozny, for example, the Russian military refused to communicate with reporters. The media reported primarily from the perspective of the Chechen rebels. This encouraged both local and international support for the rebels. It also allowed the Chechens, who lacked sophisticated information systems, to use the media to broadcast operational guidance to their forces. (During their second Chechnya campaign of 1999-2000, Russia learned this lesson well and the Russian view of the war dominated domestic public opinion.) On the other hand, successfully engaging the media can serve as a force multiplier. The Army's open and responsive interaction with the media during peacekeeping operations in Bosnian urban areas helped to explain the challenges and successes of Army forces in the Balkans to the public. This helped maintain domestic, international, and local political support for NATO operations and, with a successful command information program, helped maintain Soldiers' morale.

Transportation and Distribution

2-75. This element of the infrastructure consists of—

- Networked highways and railways to include bridges, subways and tunnels, underpasses and overpasses, ferries, and fords.
- Ports, harbors, and inland waterways.
- Airports, seaplane stations, and heliports.
- Mass transit.
- Cableways and tramways.
- Transport companies and delivery services that facilitate the movement of supplies, equipment, and people.

Similar to communications and information, this facet provides the physical link to all other elements of the infrastructure.

2-76. Army forces deploying into a theater of operations depend on ports and airfields; seizure and protection of these critical transportation nodes may impact the projection of combat power. Once in theater, transportation and distribution systems in the urban area can contribute greatly to the movement of forces, maneuver, and logistic operations throughout the entire AO. Control of decisive points in this infrastructure may be important to the military operation and to the normal functioning of the urban area (and surrounding rural areas). Supplies traveling through the transportation and distribution system may be military-specific supplies (such as ammunition and repair parts) and supplies for both the military and urban population (such as food, medicine, oil, and gas). The system may also support the movement of military forces and the urban area's population (for which it was designed). Therefore, commanders of a major operation may have to develop innovative methods that limit the transit of threat supplies and reinforcements while facilitating the movement of their own resources and those of civilians. This last consideration attempts to minimize hardship and promote normalcy in the urban area and will increase in significance as the need for legitimacy increases.

2-77. Most urban areas (particularly in developing countries) have two forms of transportation and distribution systems that exist simultaneously: a formal system and an informal or paratransit system. Large organizations, bureaucracy, imported technology, scheduled services, and fixed fares or rates characterize formal systems. Low barriers to entry; family and individual entrepreneur organizations; adapted technology; flexible routes, destinations, and times of service; and negotiated prices characterize the informal system. The informal system is more decentralized and covers a much greater portion of the urban area than the formal system. The informal transportation and distribution system often includes a waterborne element, is more likely to function through turbulence and conflict, and can extend hundreds of kilometers beyond the urban area. Accordingly, commanders should understand both systems to establish effective movement control.

ENERGY

2-78. The energy system provides the power to run the urban area. It consists of the industries and facilities that produce, store, and distribute electricity, coal, oil, wood, and natural gas. This area also encompasses alternate energy sources, such as nuclear, solar, hydroelectric, and geothermal power. Energy is needed for industrial production and is therefore vital to economics and commerce. Among many other things, this system also provides the fuels to heat, cool, and light homes and hospitals, cook and preserve food, power communications, and run the transportation necessary to move people and their supplies throughout the urban area. Loss of an important energy source such as electricity or gasoline, especially for those accustomed to having it, will become an immense area of discontent that the commander of a major urban operation will need to quickly address. Therefore some threats, particularly terrorists and insurgents, may actively target this element of the urban area's infrastructure to erode support for civilian authorities and Army forces.

2-79. Sources of energy may be tens or hundreds of miles away from the urban area itself. Therefore, commanders may exert control without applying combat power directly to the urban area itself by controlling or destroying the source (power generation or refinement plant) or the method of distribution (pipelines or power lines). With electrical energy that cannot be stored in any sizable amount, the latter may be the best means as most major urban areas receive this energy from more than one source in a network of power grids. However, control may be as simple as securing a power station or plant and turning off switches or removing a vital component that could later be restored. On the other hand, lengthy pipelines and power lines may compound security and protection of this element of the infrastructure.

2-80. The number of nations that have invested in nuclear power and nuclear research is increasing. With this increase, the potential for Army forces to operate in urban areas that include (or are near) these facilities also increases. Damage to one of these facilities and potential radiation hazards will present special challenges to commanders of a major operation. To safeguard friendly forces and civilians, commanders will need to employ a blend of peacetime and tactical nuclear contamination avoidance principles (see FM 3-11.14).

ECONOMICS AND COMMERCE

2-81. This system encompasses—

- Business and financial centers to include stores, shops, restaurants, hotels, marketplaces, banks, trading centers, and business offices.
- Recreational facilities such as amusement parks, golf courses, and stadiums.
- Outlying industrial, mineral, and agricultural features to include strip malls, farms, food processing and storage centers, manufacturing plants, mines, and mills.

2-82. An essential aspect of this area during operations may be the political sensitivity of U.S. or allied industries investing and operating in a foreign country, particularly during stability operations. An enemy or a disgruntled civilian population may attack or disrupt commercial activities as a political statement against the United States or our allies. Food production facilities also may assist commanders in Army food services and may be an essential concern during relief operations. During long-term stability operations, visible, material, and tangible economic progress consisting of the creation or restoration (and protection) of businesses, agriculture, and overall jobs will often be critical to—

- Generating or maintaining the urban population's support to Army forces and operations.
- Reducing support to threat forces and operations to include eliminating civilians as a potential manpower pool for insurgent or terrorist organizations and activities.
- Lowering other hostile civilian activities such as protests and riots.
- Transitioning the urban area back to legitimate civilian responsibility and control.

2-83. This element of the infrastructure also consists of the production and storage of toxic industrial chemicals used in agriculture (insecticides, herbicides, and fertilizers), manufacturing, cleaning, and research (to include biological agents). Fertilizer plants may be of specific concern as they contribute to providing a key material in terrorist and insurgent bomb-making activities. A thorough analysis of this element of the infrastructure may also be essential to understanding how urban insurgencies are funded and supported. This helps commanders to understand the true organization of the insurgency as well as to suggest methods to isolate insurgents from their economic or financial support. In their overall assessment of this area of the infrastructure, commanders should also consider the activities and influence of criminal organizations or elements.

ADMINISTRATION AND HUMAN SERVICES

2-84. This wide-ranging system covers urban administrative organizations and service functions concerned with an urban area's public governance, health, safety, and welfare. Together, it encompasses—

- Governmental services that include embassies and diplomatic organizations.
- Activities that manage vital records, such as birth certificates and deeds.
- The judicial system.
- Hospitals and other medical services and facilities.
- Public housing and shelter.
- Water supply systems.
- Waste and hazardous material storage and processing facilities.
- Emergency and first-responder services such as police, fire, and rescue.
- Prisons.
- Welfare and social service systems.

CULTURAL

2-85. This system encompasses many organizations and structures that provide the urban populace with its social identity and reflect its culture. (This infrastructure system overlaps with many recreational facilities included under the economics and commerce infrastructure. For example, an urban society may radically follow soccer matches and teams. Hence, soccer stadiums relate to the society's cultural infrastructure.) Some of these facilities, particularly religious structures, will be protected targets and others may require security and law enforcement protection from looting and pilferage. However, commanders will need to quickly educate, inform, and continually remind the urban populace (and the media) that cultural infrastructure may lose its protected status when used by threats for military purposes. Cultural infrastructure may include—

- Religious organizations, places of worship, and shrines.
- Schools and universities.
- Museums and archeological sites.
- Historic monuments.
- Libraries.
- Theaters.

RESTORING AND PROTECTING ESSENTIAL SERVICES

2-86. Losing the support of essential elements of the infrastructure will have an immediate, destabilizing, and life-threatening impact on the inhabitants of the urban area. In stability and civil support operations, numerous parts of the administrative and human services and energy infrastructure often rise to critical importance before all other elements. Again, however, complete restoration of these essential services is often a lengthy, resource-intensive civil-military operation. (Following the end of major combat operations of OPERATION IRAQI FREEDOM, most units developed one or more of their logical lines of operations oriented along the restoration or improvement of urban infrastructure and essential services. An acronym used by many units to focus and track critical activities within this line of operation was sewer, water, electricity, and trash (SWET). Later, other units modified this acronym to SWEAT and then SWEAT-MS to include concerns for restoring academics [or schools], revitalizing medical facilities, and establishing security [police and host-nation security forces].) Of critical importance will be a simultaneous IO campaign that includes efforts to help ensure that the urban population develops realistic expectations about Army abilities to restore their essential urban infrastructure.

Understanding the Urban Environment: Paris – 1944

The summer of 1944 confronted German General Dietrich von Choltitz with a dilemma. As military commander of greater Paris, he was to eliminate French Resistance internal to the city while defending against approaching Allied units, missions for which he had insufficient forces. (Notably, General Eisenhower, the commander of the Allied forces, wanted to bypass Paris to sustain the offensive. Seizing the French capital would task his forces with the support of tens of thousands of civilians. Eisenhower was nevertheless ordered to capture the city. A political decision resulted in civilian assistance taking precedence over combat operations.)

Choltitz's situation was further complicated by Hitler's demand that he destroy the city, an action the general saw as needlessly destructive (and infeasible given his scant resources). Choltitz's seniors directed the preparation, and later the destruction, of Paris's 45 Seine River bridges. They were the only remaining crossing points over that waterway given Allied bombing of others outside the French capital. Premature destruction would trap German forces defending to their north, a second-order effect that Choltitz used to justify his disobedience of orders demanding the bridges' demolition.

The German general also recognized that some mission-critical elements were part of Paris's social rather than physical infrastructure: the leadership of the various resistance groups and the relationships between them. Choltitz understood that he lacked resources to defeat the many separate factions; he therefore chose the unorthodox (asymmetric) approach of accepting an intermediary's offer of a truce with these groups. Such an agreement provided some measure of the stability needed while Choltitz awaited promised reinforcements. Further, he realized that the resistance factions were by no means united in their goals. Communist elements sought a much different end than those looking toward a de Gaulle-led postwar government. A truce thus set the French Communists (who sought an uprising so as to legitimize their claims to power) against others trying to buy time until Allied forces arrived, forces that included Free French units supportive of de Gaulle.

Although his defense of the capital failed, Choltitz succeeded in harboring his available resources, reducing the effectiveness of the resistance organizations fighting his soldiers, and maintaining withdrawal routes for units north of the Seine. The German commander's analysis in support of these efforts was effective in part because of his insightful (1) identification of critical points that included elements of terrain, citizenry, and infrastructure; (2) understanding of the relationships between these parts; and (3) use of an asymmetric approach to address his lack of sufficient force to otherwise handle the urban densities that challenged him.

This page intentionally left blank.

Chapter 3

Understanding the Urban Threat

... [T]he United States could be forced to intervene in unexpected crises against opponents with a wide range of capabilities. Moreover, these interventions may take place in distant regions where urban environments, other complex terrain, and varied climatic conditions present major operational challenges.

Quadrennial Defense Review Report, 30 September 2001

As the strategic environment has become less stable, more uncertain, and more dangerous, Army forces must be trained and ready to address persistent and evolving urban threats. These threats range from regional conventional military forces, paramilitary forces, guerrillas, and insurgents to terrorists, criminal groups, and angry crowds. These threats can hide in plain sight and become indistinguishable from the noncombatant urban population that may help shield, protect, and sustain them. Although uncertain about events, Army forces can be clear about trends. Increasingly, the Army will face threats that severely differ in doctrine, organization, and equipment, are skilled at developing and adapting techniques to counter Army tactics, techniques, and procedures (TTP), and can fully interact with the three other components of the urban battlefield—terrain, society, and infrastructure. In urban operations, commanders must broaden their concept of the threat to include natural disasters, hunger and starvation, and rampant disease. Further, commanders must plan to contend with many passive urban threats, such as psychological illnesses and toxic industrial materials (TIM). These threats may be found in isolation, but most likely commanders will encounter them in various combinations. Moreover, each new threat will pose a different combination and likely have new capabilities that previous opponents lacked.

ASYMMETRICAL AND ADAPTIVE

3-1. An emphasis on asymmetry to offset U.S. military capabilities has emerged as a significant trend among potential threats and become an integral part of threat objectives and tactics (discussed below). Though, this is not to imply that all future urban threats will fight and operate asymmetrically. Some may be asymmetrical; others may not. Future threats may be a symmetric conventional military; others may be paramilitary, insurgent, or even a nonmilitary threat such as a criminal organization. Asymmetry results when one opponent has dissimilar capabilities—values, organization, training, or equipment—that the other cannot immediately counter. It is not a new concept. It naturally evolves from a sound mission, enemy, terrain and weather, troops and support available, time available, civil considerations (METT TC) analysis by an intelligent, freethinking, and adaptive threat. These asymmetric approaches will include the most advanced, commercially-available technology innovatively applied and mixed with crude, simple, and unsophisticated weapons and TTP. The success of these approaches often depends upon the predictability of the forces on which they are employed. Thus, Army commanders must consider whether their operations and activities lend themselves to pattern analysis by threat forces and, if so, they may need to vary their battle rhythm and take other steps necessary to become less predictable.

WEAPONS OF MASS DESTRUCTION

3-2. A chief asymmetric means of engaging the national power of the United States is to employ weapons of mass destruction (WMD) against the United States or its allies. These weapons can be used against military forces and include high-yield explosives as well as nuclear, biological, and chemical weapons. Operations in urban areas may require concentrating forces and may create a lucrative target for a threat that possesses fewer numbers and less equipment.

3-3. A threat's WMD use will adversely affect the Army's abilities to conduct urban operations (UO) to various degrees. For example, the intervening structures and the effects of urban microclimates complicate the ability to detect and identify radiological, chemical, or biological attacks from a standoff distance. Also, the individual Soldier's ability to recognize his leaders, understand oral and visual commands, and operate increasingly sophisticated equipment is difficult when wearing protective clothing and equipment—particularly if his training proficiency is low. Despite the increased challenges and complexity, Army forces have the best (or better) training and equipment necessary to respond to such an attack compared with most armies around the world or with any civilian organization.

3-4. Although initial casualties could be high, the public can accept military casualties before those of civilians. Therefore, threats may gain an initial tactical advantage but would achieve less asymmetric benefit by directly attacking Army forces. Instead, they may attempt to achieve a strategic advantage by employing WMD against U.S. or allied civilian populations. In doing so, threats hope to use political sensitivity to high civilian casualties to reduce popular support for the United States or its allies. The chance of these attacks occurring in an urban area increases because—

- The area facilitates weapon effects and camouflages delivery means.
- The dense civilian population ensures a high casualty rate.
- The attack (or even the threat of attack) often will receive more publicity and public attention.
- The urban infrastructure is especially vulnerable to WMD, (particularly the systems of the economics and commerce infrastructure located in large urban areas) and may have far-reaching national and global effects.

THREAT OBJECTIVES

3-5. The threat will seek to achieve several key objectives when opposing Army forces operating in an urban environment (see figure 3-1). These objectives focus more on how a threat might fight in an urban area rather than specifically who the threat might be or in what region of the world the conflict might occur. These objectives are more easily achieved in an urban environment due to—

- The high costs in time, material, and manpower involved in UO.
- The limiting effects of urban areas on many technological advantages.
- The proximity of airfields and ports to urban areas.
- The potential moral dilemmas created by exposing numerous civilians to harm or injury.
- The ability of the media to observe and report the threat's version of events.

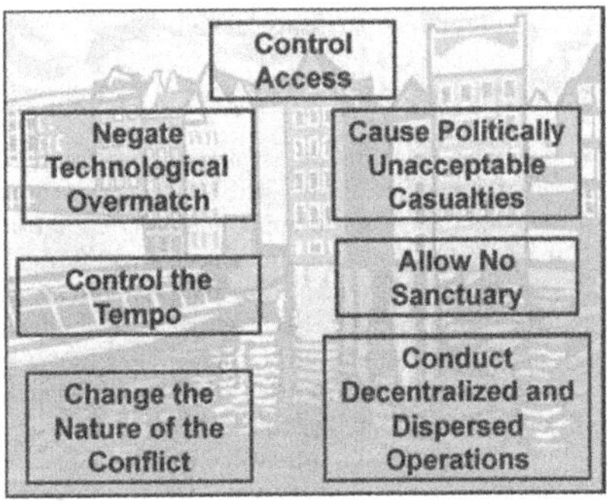

Figure 3-1. Threat objectives

3-6. These objectives complement and overlap each other; however, at their core is the threat need to defeat an enemy of superior numbers, technology, or both.

CONTROL ACCESS

> *... Iraq made no direct effort to impede the buildup in the theater. Planners should consider what might have happened if Iraq had attempted a strategy based on denying access to the region. Planners might also wonder what the outcome would have been if Iraq had attacked US forces in Kuwait before they were ready for the running start. The point is that the conditions in [US Central Command] in 2003 are unlikely to be replicated elsewhere.*
>
> *On Point: The United States Army in Operation IRAQI FREEDOM*

3-7. The Army may not be located where future conflicts are fought. Thus, the Army maintains the ability to rapidly project and sustain combat power over long distances and time spans. This capability demands that Army forces quickly gain and maintain control of seaports or aerial ports of embarkation or debarkation, particularly where the density of U.S. basing and en route infrastructure is low. Commanders gain control of these ports by unopposed (assisted or unassisted) or forcible entry operations. In either case, these phased-entry operations may present potential vulnerabilities, particularly—

- Unsuitable composition of initial or early entry forces lacking necessary combat power for immediate decisive operations.
- Initial command and control difficulties and an immature situational understanding.
- Weak initial port or base defenses without the necessary integration of naval and air component forces.
- Lack of developed logistic support.

3-8. Consequently, threats may attack during initial force projection operations to oppose, delay, disrupt, or otherwise manipulate the build-up of essential combat power into a theater of operations. These attacks may occur anywhere deploying Army forces are located, at overseas bases, at home stations, and even in military communities. Increasingly, deployment facilities such as airfields and ports exist as integral components of urban areas. Threats will invariably use the complex and concealing nature of these urban areas, coupled with the vulnerabilities, to create favorable conditions for their attacks.

NEGATE TECHNOLOGICAL OVERMATCH

3-9. Threats will always strive to force engagements at a time and place most advantageous to them. They may locate military forces and vital military capabilities in urban areas to achieve sanctuary from the effects of Army capabilities and make Army forces and systems more vulnerable to less-sophisticated weapons.

3-10. The clutter of the physical structures, electromagnetic radiation, and population diminishes Army capabilities. This clutter makes it difficult for Army forces to acquire and effectively engage targets at long ranges. In urban areas, the terrain often allows a threat to operate in closer proximity to friendly forces. Therefore, the threat may "hug" friendly forces to avoid the effects of high-firepower standoff weapon systems and degrade their ability to gain or maintain a thorough common operational picture. Additionally, this threat tactic attempts to inhibit friendly commanders from employing some weapon systems and munitions for fear of fratricide. Threats will also seek to hide and operate in culturally sensitive areas and sites, such as places of worship and schools, to establish temporary safe havens and refuge from Army forces operating within the urban environment.

CONTROL THE TEMPO

3-11. Threats will try to achieve a decisive advantage by setting and controlling the tempo. To prevent the Army's entry into theater, threats may try to create a high operational tempo to take advantage of the inherent weaknesses in power projection operations outlined earlier. As other efforts deny entry, threats may seize the initiative, achieve surprise, and exploit the tempo differential by attacking with heavy conventional forces potentially possessing greater firepower and more rapid ground mobility than the Army's initial-entry forces.

3-12. If they cannot deny entry or end the conflict quickly, threats may use any preparations made in the initial high-tempo period to prolong the event, aiming to degrade U.S. or allied commitment. (Some threats may conduct operationally planning in terms of decades rather than years.) The complex nature of the urban environment slows operations conducted in, around, and over these areas. Threats may maximize this characteristic by fighting (conventionally or unconventionally) for key urban complexes and infrastructure, forcing friendly forces to operate within these areas. If Army operations focus on one or more urban areas, the overall campaign slows. However, even when UO make up only one component of a much larger campaign, they may consume valuable resources needed for other operations and delay the entire campaign. (See also the discussion of Tempo in Chapter 7.)

Tempo

The battle for Aachen, Germany, in the fall of 1944, developed during the U.S. First Army's offensive to breach the Westwall fortifications. Aachen, the ancient capital of Charlemagne, had symbolic political and psychological significance for the Germans and Americans. Furthermore, it was the first city on German soil to face an assault by the Allies. Consequently, the symbolic importance of this first major battle in Germany ensured bitter resistance against American attackers. The Germans surrendered only after the city was destroyed. Expected to take a few days, instead, the battle took weeks. Although the Army had achieved a clear tactical victory, the German defense of Aachen cost the First Army valuable time and resources, and delayed the planned attack to the Rhine River.

Change the Nature of the Conflict

3-13. Threats may attempt to change the fundamental nature of the urban conflict to exploit ambiguous or tenuous political-military objectives. Many nations must gain and maintain domestic popular support to use their armies for political objectives. The threat may attempt to change the nature of the conflict by modifying its strategy and tactics, the environment, or any combination, ultimately hoping to reduce friendly popular support. For example, introducing an urban terrorist threat to U.S. civilians or Soldiers not directly engaged in operations changes the nature of the conflict. This type of threat may not have been an initial consideration, and this change may reverse public support for the operation. Another example, growing U.S. coalition combat power may cause the threat to switch from open maneuver warfare to urban insurgency operations to avoid decisive combat with superior forces and achieve a stalemate. Originally expecting a quick solution or victory, the political leadership may now envision a longer deployment with less chance of lasting success.

Cause Politically Unacceptable Casualties

3-14. Threat forces may gain an advantage against superior friendly forces by capitalizing on a perceived weakness of many Western nations: the inability to endure continuous losses or casualties for other than vital national interests or losses for which they are psychologically unprepared. A secondary U.S. interest may equate to national survival on the part of a threat. Therefore, the threat (particularly with fanatical leadership) may willingly sacrifice excessive amounts of money, equipment, and people (soldiers and civilians) to achieve victory. Threats may attempt to weaken U.S. resolve and national will to sustain the deployment or conflict by inflicting highly visible, embarrassing, and if possible, large losses on Army forces, even at the cost of similar losses to themselves. Many threat forces will use UO to inflict mass casualties and destroy premier Army weapon and information systems. The physical characteristics of the urban environment support these ambush techniques. Light infantry or insurgents with readily obtainable, hand-held antiarmor weapons can effectively attack armored vehicles and helicopters, no matter how sophisticated, in an urban area.

ALLOW NO SANCTUARY

3-15. Threats will attempt to deny Army forces safe haven anytime and anywhere. Terrorism and insurgency may be some of the tactics used to deny sanctuary to Army forces. They will attack Army forces anywhere, particularly while operating in urban areas where the fear from being attacked from any quarter is often greater. Threats may be or employ state-sponsored or independent terrorists, well equipped and motivated to accomplish their assigned missions.

3-16. Military buildings, facilities, and installations in urban areas are particularly vulnerable to high-yield explosive munitions as well as other clever means to create large explosions. The close-in nature of urban areas, large populations, and high volume of vehicle traffic provide a good environment for target reconnaissance, explosives positioning (conventional or high-yield), and cover for an attack. These attacks will likely be preceded by extensive, careful reconnaissance, necessitating a solid friendly counterterrorism, counterinsurgency, and counterintelligence effort.

CONDUCT DECENTRALIZED AND DISPERSED OPERATIONS

3-17. To a certain extent, decentralized and dispersed operations are an integral part of all threat objectives. However, this concept warrants separate emphasis as an objective since threat forces will likely place great significance on it on future urban battlefields. Both dispersed and decentralized approaches seek to reduce threat vulnerabilities to air power and precision-guided munitions (PGM) while increasing their agility, flexibility, and overall maneuverability in an urban environment.

3-18. Urban terrain tends to fragment and separate forces that operate in it. Threat forces recognize this characteristic, accept it, and make it work to their advantage. They conduct operations from dispersed urban locations to reduce their vulnerability to friendly decisive operations and massed firepower. Although separated, threat forces will attempt to retain the ability to assemble and mass quickly so to strike as opportunities present themselves. Once threat forces complete the operation, they will return to separate locations to avoid potential counterattack. The fluidity and seemingly disjointed appearance of these UO threats will challenge friendly efforts to conduct templating and pattern analysis. Ambushes (air and ground) will be used to deny friendly ground and air reconnaissance of their dispersed locations.

3-19. Dispersed operations normally depend on good command and control to achieve synchronization and massed effects. Threat forces also understand the debilitating effects of the urban terrain on communications and the execution of operations. When they cannot mass their forces or effects, they will depend on decentralized operations to achieve their objectives. They will operate autonomously, guided only by a higher authority's purpose and intent. These operations make them even less vulnerable to massed attacks and PGM as smaller threat forces do not present an objective or target that will allow friendly decisive operations. Again, pattern analysis and templating, although essential, will be extremely difficult. This objective may prolongs the conflict but is central to achieving the other threat objectives.

3-20. When the threat employs decentralized operations, is effectively isolated, or both, Army and coalition forces are much more likely to be able to share lessons learned and emergent TTP than a dispersed threat. Threat isolation, either imposed by Army forces or as a result of the threat's desire to remain concealed from Army intelligence, surveillance, and reconnaissance (ISR) capabilities, reduces the threat's ability to freely and easily exchange information and quickly adapt. As learning organizations, this allows Army forces to develop dominance over a threat's weapons and tactics for longer, albeit temporary, periods of time. This ability to rapidly train and learn is an advantage the Army and coalition forces must maintain while denying the same to future threats.

URBAN THREAT TACTICS

3-21. Urban areas provide a potentially casualty-producing and stress-inducing environment ideally suited for using specific urban tactics. Moreover, urban areas can provide threats with an unmatched degree of cover and concealment from friendly ISR and firepower systems. Understanding how these tactics will be used often requires a concurrent understanding of the threat's culture, customs, language, and philosophical ideals. While active urban threats may vary widely, many techniques will be common to all.

Figure 3-2 outlines a set of tactics available to potential threats opposing mission accomplishment in urban areas. Army forces may use many of these tactics, except those that violate the law, ethics, and morals, to defeat urban threats. Moreover, using asymmetric means is not the sole domain of the threat. Army commanders can also leverage capabilities, create conditions, and plan operations to develop their own asymmetric advantages to accomplish the mission.

USE THE POPULATION TO ADVANTAGE

3-22. Many urban areas may be too large to evacuate completely (if at all). Even if desirable, a military force may have no place to safeguard and secure the inhabitants. Therefore, future UO may see large segments of the populace remain. Offensive and defensive operations may be constrained not only by the terrain, but also by the presence of many civilians. Army forces involved in urban stability and civil support operations will certainly conduct missions in and amongst the residents. These residents may restrict operations and, when gathered in large numbers, may (even without initial hostile intent) present a critical force protection issue for the commander.

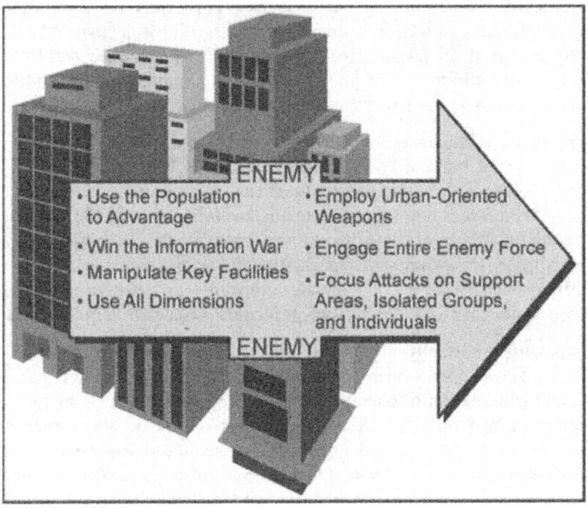

Figure 3-2. Urban threat tactics

Use as Key Terrain and Concealment

3-23. From the threat standpoint, the populace is similar to key terrain: the side that manages it best has an advantage. Threat forces may gain this advantage by using civilians as camouflage, concealment, and a means of deception. Guerrilla and terrorist elements may look no different from any other member of the community. Many foreign conventional and paramilitary troops—in addition to terrorists and insurgents—often have a "civilian" look. Western military forces originally adopted the clean-shaven and close-cut hair standards to combat disease and infection, but future opponents may not adhere to those standards. Instead, they may adopt grooming standards, civilian-looking clothing, and other "nonmilitary" characteristics to make themselves indistinguishable from the civilians. To avoid defeat or capture, threats will often throw down their weapons and attempt to blend into the urban population. Therefore, commanders must adopt effective populace and resource control measures to limit, identify, and track threat movements through the urban area of operation.

Identifying Threats from Noncombatants

During Russia's 1994-95 conflict with Chechnya, Russian forces had difficulty identifying Chechen guerrilla forces from Grozny's noncombatant population. Because their dress and ethnic appearance was identical to that of the urban populace, Chechen soldiers could freely walk around the city, suddenly disappear, and then abruptly reappear firing their weapons from basements, windows, or dark alleyways. To distinguish fighters from peaceful city dwellers, Russian forces began looking at men's shoulders to see if they were bruised (from firing weapons) and their forearms to see if there was burned hair or flesh (from the extraction of cartridges). They closely examined their clothing and smelled for gunpowder residue. To identify a Chechen artilleryman, Russian soldiers checked for glossy spots left by artillery and mortar rounds on the bends and cuffs of sleeves. They also turned pockets inside out to check for a shiny, silvery-leaden hue indicating the former presence of small arms ammunition. Russian forces also recognized a grenade launcher operator or mortar man from fibers and crumpled pieces of gun cotton on their clothing. U.S. Army commanders may need to develop similar, imaginative techniques and rapidly integrate technological means to identify and track threat forces such as—

- Military working dogs, ion spray, and electronic vapor tracers to locate explosives (or their former presence).
- Biometrics—measurable biological characteristics—such as fingerprints, voiceprints, facial scans, and retina scans.
- Electronic and chemical tagging devices.
- Monitoring hospitals for injuries that indicate participation in combat activities. For example, improper use of rocket-propelled grenade launchers may result in significant burns to the lower part of the body due to the back blast.

Gain Cover, Protection, and Increased Mobility

3-24. Threat forces may attempt to gain cover by using the urban inhabitants as human shields. With this increase in protection, they simultaneously increase their mobility. They recognize the Army's focus on developing and applying rules of engagement (ROE). They will take advantage of the restraining effects of international law and the Army ethical values to enhance their mobility in proximity to friendly positions. Knowing the Army's reluctance to cause noncombatant casualties and collateral damage, threats may operate in areas containing civilians and essential facilities to restrict the Army's use of massed or nonprecision firepower. For example, they may use school buses and ambulances to transport fighters and equipment about the urban battlefield. They may also employ "rent-a-crowds"—civilians paid or incited to demonstrate against military forces—armed only with sticks, stones, and Molotov cocktails (a potential asymmetric challenge). Threats can also hide military equipment within sensitive infrastructure—for example, caching weapons in places of worship, schools, or medical facilities. As Army and coalition forces potentially respond by targeting and firing upon these locations, threats will make use of these reactions for their possible propaganda value.

Make Moral Responsibilities a Weakness

3-25. Depending on their successes, threats may use these tactics and skillful information operations that attack national will and coalition sensitivities in an attempt to force the Army to establish more restrictive ROE. For example, a teenaged fighter coerced or enticed to fire a rifle or rocket-propelled grenade launcher can be quickly stripped of his weapon once dead and displayed at the local hospital as a child victim of coalition forces. Threat forces may also take advantage of the Army's moral responsibilities. By herding refugees into friendly controlled areas, threat forces try to make the civilians a burden on the Army's logistic and security resources. Similarly, some threats may attack to kill NGOs and force the remainder of them out of the area to achieve the same burden effect. At the same time, these threats hope to cause the population to lose confidence in the Army's (and potentially the government's) ability to protect and provide them essential services.

3-26. Threat forces, on the other hand, may not abide by international agreements, such as the Geneva conventions. Threats may not take prisoners unless they can be ransomed or made part of a local prisoner exchange. They may even execute friendly prisoners in front of the media to show their "strength" and, more importantly, to cause friendly forces to overreact and lose their legitimacy. Threat forces can then use such an overreaction to unite others with their cause. Pretending to be civilian noncombatants, they may also feign injury or pretend to need assistance (such as a disabled taxi) with the primary motive of luring Army forces into a deliberate ambush. While the threat may gain a short-term advantage by using ethical responsibilities as a weapon, the Army commander's strict adherence to moral principles will, in the long run, contribute greatly to threat defeat—particularly during long-term urban stability operations.

Acquire Intelligence and Logistic Support

3-27. Indigenous threat forces can normally use the local population for intelligence and logistic support far more effectively than can an alien army. Threat forces may manipulate local hires serving among U.S. Soldiers, such as those contracted by the Army for base operation purposes or translator duties. In addition, refugees moving through friendly controlled sectors may provide the threat with information on friendly dispositions, readiness, and intent. (Women and children may be specifically used for this purpose since U.S. Soldiers' cultural bias can create an incorrect perception that these noncombatants pose no threat to security.) Even friendly residents may become unwitting or unwilling informants, providing an enemy or a hostile with vital information on friendly activities, dispositions, and capabilities. However, a threat employing particularly cruel, abusive, or repressive measures may easily turn certain groups in the urban area against them, even when they share a common history, culture, and ethnicity with the civilians. This is more likely in those areas with higher population densities.

3-28. Threat forces may also seek to use some nongovernmental organizations (NGOs). They may try to obtain relief supplies either through the organizations' legitimate relief operations or as a target for theft. Some organizations may even be fronts for weapons, food, ammunition, money, and fighters. For example, during Russia's second conflict in Chechnya (1999-2000), documents purportedly found in Grozny by the Russians listed nations such as Sudan, Nigeria, Niger, and Ivory Coast as sending fighters to Chechnya under the guise of the International Islamic Relief Organization. (Chechen fighters also disguised themselves as Red Cross workers.) This deception increases the need for strict security and force protection measures, close coordination with NGOs operating in urban areas, and closer monitoring of suspect organizations' activities by civil affairs personnel.

Win the Information War

3-29. Threat forces will try to win the information war as much as they will directly oppose Army UO. Threat urban campaigns need not be tactical military successes. They need only to weaken legitimacy and make the opposition's campaign appear unpalatable to domestic, U.S., and world support. As a critical part of a threat's overall information operations, they will use the ever-present media to tell their story. Portable video cameras, commercial radios, and cellular telephones, available and easily concealed, will be as important to many threat actors as weapons and ammunition. Internet access, already firmly established in many urban areas, provides the means to easily disseminate threat propaganda, misinformation, and disinformation through web sites and electronic mail. The internet also allows threats to conduct their own distance learning and education. Threats will make good use of the instantaneity and global reach of television and radio broadcasts, particularly local and international media, to obtain information of immediate tactical value and broadcast their own operational and tactical guidance. Hackers, covered and concealed in the interior spaces of the urban area, may gain access to U.S. sites to manipulate and obtain information to the threat's advantage.

Information and the Media

The media coverage of the urban battle for Hue, South Vietnam, although only one of hundreds of different attacks of the Tet Offensive, affected the will of both the American people and their political leadership. On January 31, 1968, two North Vietnamese Army (NVA)/Vietcong (VC) regiments and two sapper battalions, moving rapidly and with the element of surprise, attacked and seized part of the walled city (Citadel) of Hue. It was the third largest city in South Vietnam, the former capital of a united Vietnam, the capital of Thua Thien province, and a spiritual and cultural center. Initially intending to hold the city for seven days, the NVA/VC retained portions of the city for approximately three weeks against determined U.S. and South Vietnamese attempts to retake it.

Hue marked a revolution in the coverage of war by modern media. It was the first time Americans could sit at home and watch an ongoing battle on the evening news. One of the most intense and savage battles of the Vietnam conflict, it was televised every evening for almost a month. Although the battle for Hue was a tactical victory for the United States, the North Vietnamese clearly achieved strategic success by searing the American consciousness with the high costs of urban warfare. Had U.S. leaders made winning the information war a central part of the overall campaign plan—for example, exposing the American people to the NVA's brutality by publicizing the civilian executions in Hue—civilian support for the war may have been bolstered and a different outcome achieved. See Chapter 7 for a more detailed account of the battle for Hue.

Manipulate Key Facilities

3-30. Threat forces will attempt to identify and quickly seize control of critical components of the urban area to help shape the operational environment. Urban telephone exchanges can provide threats with simple and reliable communications that can be easily secured with off-the-shelf technologies. Existing public address and loudspeaker systems can be used to incite riots or openly command and control dispersed threat forces. (Local language may provide threats with a measure of communications security.) Sewage treatment plants and flood control machinery can be used to implement WMD strategies or to make sections of the urban area uninhabitable. Media stations significantly improve the information operations abilities of the controlling force. Power generation and transmission sites provide means to control significant aspects of civilian society over a large area. Insurgents will attempt to destroy key facilities and critical infrastructure in order to shape the populace's perception that the existing government is incapable of providing for their safety and meeting their basic needs.

Use All Dimensions

3-31. Threats will think and operate throughout the depth, breadth, and height (including supersurface and subsurface areas) of the urban environment. However, threats will often choose to fight from within interior spaces rather than out in the streets and other open areas. Conventional lateral boundaries will often not apply as threat forces control some stories of the same building while friendly forces control others. The use of all three dimensions of an urban area makes identification, reporting, and targeting of enemy locations more difficult for forces accustomed and trained to acquire and engage targets in primarily two dimensions. Because of these target identification and acquisition difficulties, threats will likely use decoys in the urban area to cause erroneous assessments of their combat capabilities, strength, and disposition. They will also attempt to use these decoys to absorb expensive and limited precision-guided munitions as well as cause misallocation of other critical Army resources.

3-32. Roofs and other supersurface areas provide urban threats with excellent observation points and battle positions above the maximum elevation of many weapons. Shots from upper floors strike armored vehicles in vulnerable points. Basements and other subsurface areas also provide firing points below many weapons' minimum depressions and strike at another weakness in most armor. Sewers and subways may provide covered and concealed access throughout the area of operations.

EMPLOY URBAN-ORIENTED WEAPONS

3-33. Whether purpose-built or adapted, many weapons are more useful in an urban environment while others may have significant disadvantages. Urban threat weapons are much like the nature of urbanization and the urban environment: inventive and varied. Many threats will integrate widely available off-the-shelf technologies into their weapon systems and armed forces. However, sniper rifles and small, man-portable, fire-and-forget weapons and demolitions and other improvised explosive devices (IEDs), to include suicide and car bombs, will likely dominate the urban environment. Figure 3-3 lists examples of threat weapons favored in UO.

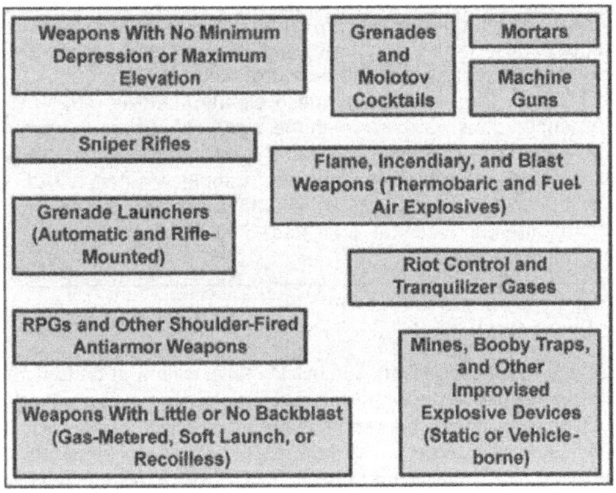

Figure 3-3. Favored threat weapons

ENGAGE ENTIRE ENEMY FORCE

3-34. Threats may attempt to keep all or significant portions of Army forces engaged in continuous operations to increase their susceptibility to stress induced illnesses. UO, by their nature, produce an inordinate number of combat-stress casualties. Continuous operations exacerbate this problem. Threat forces that employ this tactic will often maintain a large reserve or achieve respite by hiding among the civilian population to minimize the psychological impacts on their own forces.

3-35. To accomplish this key tactic, threat UO will likely involve decentralized maneuver, precision fires, and simultaneous operations involving unconventional and special purpose forces and guerilla, insurgent, and terrorist tactics. Threats may also attempt to bait Army forces by ambush or other means with a primary focus of attacking response or reaction forces. Overall, threat forces will take advantage of any exposed weakness in weapons, equipment, and TTP and they will cultivate the flexibility to engage in battles as favorable opportunities present themselves.

FOCUS ATTACKS ON SUPPORT AREAS, ISOLATED GROUPS, AND INDIVIDUALS

3-36. To supplement the previous tactic, threat forces will seek to target support areas, small groups, leaders and their headquarters, and unaware Soldiers. Their focus on resupply and convoy operations, casualty evacuation, and other sustainment activities, coupled with the compartmented terrain, navigational challenges, and multiple three-dimensional avenues of approach often makes these locations and Soldiers more susceptible to surprise raids and ambushes. Attacks on these areas and groups are conducted to erode the Army's ability to sustain UO, to inflict maximum casualties, and to induce psychological stress. These attacks can be mitigated by careful, regular evaluation of choke points and other restrictive terrain, regular awareness training for units and individuals operating in or transiting through potential incident-prone areas, well-planned and protected convoys, and thorough after-action analysis of incidents.

NEGATIVE EFFECTS OF URBANIZATION

3-37. Many urban areas are the engines for increased industrialism and economic growth as an expanding population provides the labor for manufacturing and service needs. However, rapid and inadequately planned growth can result in undesirable consequences. Uncontrolled urbanization may result in an infrastructure and economic base unable to support the growing population. A large transient, ill-housed, and idle population in a close geographic space may produce strife. Classes, cultures, ethnic groups, and races that might otherwise peacefully coexist can clash under the stress of survival. Uncontrolled urban growth has resulted in the negative effects listed in figure 3-4. In many urban stability and civil support operations, these may be the primary "threats" to mission accomplishment.

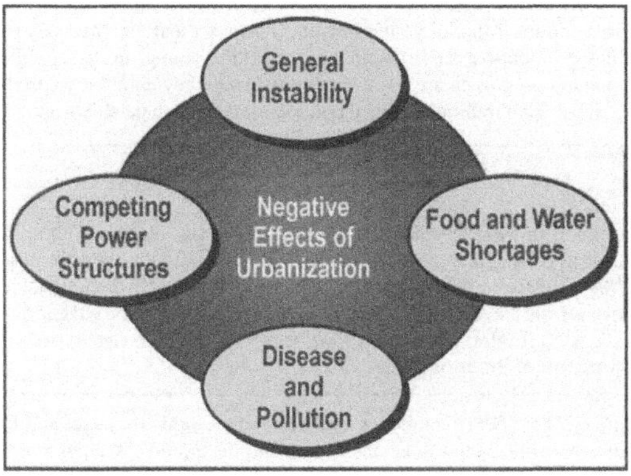

Figure 3-4. Negative effects of urbanization

3-38. Not all urban areas prevail as inherently unstable or hotbeds for unrest. Urban growth due to migration may remove sources of conflict, or it may provide the catalyst for violence. Commanders should recognize the possible effects of uncontrolled urbanization. During their intelligence preparation of the battlefield (IPB), they should determine if these conditions exist. Throughout mission analysis and the development of courses of action, commanders should consider the impact (if any) on their operations. At the same time, they must recognize that UO may create similar problems that may affect the current operation as well as the overall campaign.

General Instability

3-39. Urbanization can enhance stability by generating industrialization and economic growth resulting in more jobs, a higher overall standard of living, and an educated, relatively satisfied populace. However, the population dynamics associated with urbanization can also have an opposite, destabilizing effect. Radical population growth may create overcrowding and generate or aggravate resource and quality of life issues. Intense and destructive competition for employment, housing, and social status may develop in this climate of economic deprivation. The inability of some governments to handle these problems—

- Makes their urban areas potential sources of unrest.
- Increases the likelihood of the Army's involvement in stability operations.
- Complicates all operations conducted in such an urban environment.

Weak civil administrations—perhaps weakened by the effects of previous UO—have difficulty controlling their society, safeguarding their military armaments, and preventing their urban areas from serving as sanctuaries to terrorists, insurgents, and criminal organizations.

3-40. Urbanization in developing countries warrants more concern. Intense migration and growth, coupled with the forced closeness of people once separated by the rural countryside, may stress already struggling institutions, hasten conflict, and lead to overall instability. Commanders must understand that UO, depending on the operation, may either cause massive population movement out of or into urban areas.

3-41. Urban areas with a large youth population may also help to generate conditions for instability. Rural-to-urban migrants tend to be relatively young. In 1999, Cairo, for example, had more than 40 percent of its population younger than 15 years. Young urban populations generate enormous demands for social resources, primarily education and jobs. Even a strong urban economy may fold under the economic expectations of a tremendous influx of young migrants. Disorder and violence may result as hostiles (many nonstate actors) easily mobilize and manipulate the idle young to act politically and criminally. Urbanization and population growth are more dangerous when they combine to produce a cohort of young urban dwellers separated from traditional social controls, such as village elders and clan leaders.

> **Cultural and Religious Instability**
>
> The 1992 bombing of the Babri Masid Mosque in Ayodya, India, enflamed an already intense cultural and religious rivalry between Hindus and Muslims and led to rioting throughout many Indian urban areas. Of the 1,500 who died in conflicts and riots, almost 95 percent died in urban areas. The violence struck Ahmedabab and Bombay most seriously, with related acts of murder, gang rapes, and arson occurring months after the destruction of the mosque.

3-42. Ethnic, religious, and other social issues may become the vents for anger and frustration produced by the high tension of urban life. Major acts of violence and destruction, such as occurred in 1992 in India, can directly threaten a nation's security. Army forces may have to conduct large-scale, stability operations to promote peace and protect national interests. In these cases, all levels of command will be particularly concerned with maintaining impartiality and perceived legitimacy.

Food and Water Shortages

3-43. Rapid urbanization, primarily in developing nations, may lead to severe food shortages that could influence Army forces (or lead to their use). Such shortages may cause instability, massive migration, revolts, or increased support of armed opposition groups. Armed factions may target NGOs that supply aid as a means of furthering dissatisfaction among the populace. In effect, food may become a weapon. Deployed troops may need to provide or support humanitarian food aid networks to keep the humanitarian situation from escalating.

3-44. However, planners must be wary of the unintended consequences of well-intentioned humanitarian assistance operations. For example, providing free, safe food may alleviate starvation, but could also undercut the local agricultural system by reducing demand in the market. If the food is distributed through urban centers, urbanization could increase, further reducing the food supply and adding to the existing strains on the infrastructure. Unless they are trying to use them as a means to separate combatants from noncombatants, commanders should normally use centralized feeding centers as a last resort. Instead, Army forces should bring the food closest to the population to encourage civilians to stay in their homes. If safe areas or camps are created, they should be designed for use over as short a time as is feasible. The general rule should be to return the urban population to their homes as soon as possible. Army forces conducting domestic support or foreign humanitarian assistance operations that cannot maintain the safe food supplies may find the frustrations and hostility of the local population focused on them.

3-45. Water shortages (and quality) are becoming a serious problem in many regions. Sewage, industrial waste, other forms of pollution and deliberate contamination pose threats to existing water supplies. Commanders operating in an urban environment need to know the water supply origins and its treatment, purification, distribution, and vulnerabilities. Before beginning operations, commanders must know if they are providing water for the noncombatants as well as their own forces. Across the spectrum of operations, controlling and protecting a limited water supply is, or may become, an essential operational consideration during UO.

Food and Water Shortages

Countries as varied as Indonesia and Algeria exported their food surpluses only two generations ago but now import up to two-thirds of their basic staples. This cycle has resulted in many countries, which once exported agricultural products, facing the growing cost of imports to feed their urban populations. Estimates predict that in the 2010 timeframe, at least 65 countries (including 30 of Africa's 51 countries) may depend completely on food imports. For some countries, the outlook is even worse. The Democratic Republic of Congo (Zaire), once a net food exporter, faces severe malnutrition and utter starvation for a large portion of its population.

Other estimates predict that by 2025, 2.7 to 3.5 billion people will live in water-deficient countries. Even now in urban areas such as Jakarta, Indonesia, for example, urban authorities cannot provide the necessary water to support its growing population. Less than half of Jakarta's population is supplied by the urban area's water infrastructure; the remainder of the area's residents rely on water from wells and other groundwater supplies that lie only a few feet below the surface. These alternate sources need to be sufficiently boiled or residents risk increased disease. In the northern part of Jakarta, land is actually sinking as the urban area's groundwater supply is overused by its residents. Houses on low-lying areas have to be rebuilt every few years to keep them above sea level. Meanwhile, saltwater from the Java Sea is seeping into the land polluting the remaining groundwater and further exacerbating the situation.

Disease and Pollution

3-46. Urban areas frequently spawn epidemics; therefore, widespread disease may pose a significant threat to Army forces that operate there. In many developing nations, rapid urbanization has occurred without a corresponding upgrade, expansion, or even development of adequate sewage and water systems. Some urban areas have only one toilet for every 750 people or more. In these areas, hundreds of thousands live much as they would in poor villages, yet so confined as to ensure high transmission rates for airborne, waterborne, sexually transmitted, and contact-transmitted microbes.

3-47. In urban areas lacking adequate trash and waste management infrastructure, insect-spread diseases proliferate. Mosquitoes that breed in polluted water, open water tanks, and irrigated urban gardens carry malaria and dengue fever—the leading causes of sickness and death from infectious disease in Latin America and Africa. The problem compounds with growing numbers of bacteria resistant to various antibiotics, a shortage of trained medical personnel, inadequate or insufficient medical facilities and supplies, and unclean agricultural and food-processing practices.

3-48. Pollution also creates critical health problems in developing areas and a potential health risk for intervening Army forces. For example, urban areas in China have recorded five to ten times the levels of sulfur dioxide found in the air of urban areas in the developed world. In parts of Poland, toxic waste has so polluted the land and water that ten percent of the babies have birth defects. Pollution may cause immediate health problems but more often, the insidious effects appear months or years after exposure. As discussed earlier, UO may contribute, either intentionally or unintentionally, to an increase in pollution. Destruction of industrial complexes that use, produce, and store hazardous material may produce toxic gas and smoke pollutants and create significant health concerns to exposed Soldiers and civilians. Soldiers forced to fight within the confines of subsurface areas such as sewers and drainage systems further increases their exposure to contaminants.

3-49. Commanders should initiate force health protection planning early, including analysis of the medical threat and other critical medical information requirements during the IPB process. A medical surveillance system should monitor the daily status of Army personnel throughout the operation. In preparation, all personnel must receive a predeployment medical examination. This exam establishes an accurate baseline health status of the force and ensures that Army forces do not introduce new diseases to an urban area, possibly exacerbating the situation. Conversely, Soldiers not immune to native viruses or possessing a weakened immune system due to continuous operations and the stress associated with UO may put Army forces at a significant disadvantage. An outbreak of plague during an operation would have an effect similar to a chemical or biological attack. The closer that Army forces operate to civilians (the humanitarian assistance operations conducted in Port-au-Prince, Haiti, and Mogadishu, Somalia, for example), the more probable that these situations may occur. (See Chapter 9.)

COMPETING POWER STRUCTURES

3-50. Many groups can exist that become strong enough to rival the power of the governing officials and eventually turn the urban area into a system of divergent and competing power structures. These groups can consist of insurgent forces, a merchant class or an economic elite, criminal organizations, or some other significant source of power such as religious organizations (see Chapter 10), clans, or tribes (figure 3-5). In the absence of a legitimate authority, armed factions headed by "warlords" may vie to fill the power void. Sometimes these groups or organizations, normally at odds with each other, may form alliances to achieve specific goals. Commanders must recognize, identify, and understand these alternate urban power bases and, if necessary, develop engagement strategies to neutralize or harness them to accomplish the Army mission. Four of these groups, not previously covered or covered in a subsequent chapter, are discussed in greater detail below.

Figure 3-5. Competing power structures

Urban Insurgencies

3-51. As urban migration increases in the developing world, rural guerrillas appear to follow. This transition of insurgencies from rural to urban areas occurs because urban areas offer a rich field of targets for insurgent attacks. People immediately notice any disruption of urban infrastructure, thus having great propaganda value. A concentrated urban population whose allegiance and support are critical to the success of an insurgency is often more susceptible to propaganda, political organization, and terrorism. Insurgents can easily arrange mass demonstrations using available communications facilities, both overt and covert. Travel is effortless and large urban populations provide cover and concealment. On the whole, urban areas, particularly those suffering the effects of general instability outlined earlier, may provide a fertile environment for guerrillas to apply formerly rural insurgent strategies. However, even with a primarily rural-based insurgency, operations in urban areas will often be essential to the insurgents' strategy as they offer them distinct opportunities to disrupt, discredit, and demoralize the government (see FM 3-05.20 and FMI 3-07.22).

Merchant Class or Economic Elite

3-52. Urban areas normally possess a merchant class or economic elite as part of their social structure. In some urban areas, they may carry more power than the local or central state government. They may isolate themselves physically and socially from the sprawling poor yet wield enormous power over the country's political and economic activities. The degree of economic separation between the merchant class and the poor may be small but still socially or politically significant.

3-53. In a vastly impoverished area where the economy of the urban area is severely disrupted, the merchant class will often continue to operate and function and, as a result, achieve a measure of influence. To continue to operate under acute economic turmoil, they may form alliances in criminal organizations and secure loyalties within the government. Outside resources introduced into a crisis area (such as food, water, fuel, and pharmaceuticals) take on increased value, may replace currency as the medium for exchange, and often become the means to amass and hold wealth. One primary way to obtain wealth is to steal it.

3-54. In some turbulent situations that lead to the need for stability operations, commanders may harness the power of the merchant class as a force for peace and stability instead of one that uses crime to achieve economic goals. For example, in a relief situation, instead of competing with the merchant class by distributing food directly to the needy and possibly creating an environment of looting and black marketeering, it may be possible to monetize food. Food assistance from donor governments could be sold to merchants at an attractive price so they have a reliable source of supply. This could, in turn, create a healthy economic system and separate merchants from criminals and gangs.

Criminal Organizations

3-55. Organized criminal groups have grown common in urban areas; have also become an important part of the urban social structure (gangs for example); and can exert considerable influence on governments, people, and military forces conducting UO. These organizations can threaten the successful completion of urban operations both directly and indirectly. Criminals and criminal organizations may directly target Army forces, stealing supplies and extorting money or contracts. Criminals may also violently confront Army forces during mission execution. During many UO, particularly during or following combat, civil disturbances, or large natural disasters, looting (organized or unorganized) may become of critical concern.

For a portion of the population, crime may be the only reliable source of income and resources to support their families. Therefore, UO will often require a combined law enforcement and military response.

3-56. Some large criminal organizations relying on international connections often have better resources and equipment than their insurgent counterparts. (A prime example is drug traffickers.) Their large financial resources, long-reaching connections, and ruthlessness provide them the means to corrupt or intimidate local officials and government institutions. Their tactics parallel those of insurgents. They have developed an intuitive cultural understanding of slum neighborhoods and the ability to lure civilians into criminal activities. They have also mastered the management of mobs. They recruit teenagers and young adults in their efforts against rivals and authorities, just as insurgents muster armies from the youth of rural villages. In many developing nations, there exists an alliance between insurgents and organized criminal groups. In these alliances, the insurgents defend the criminals and the criminals fund the insurgents.

3-57. Excessive criminal activity can undermine the commander's efforts to establish a sense of security among the remainder of the urban populace. Commanders will rely on the ability of their military police (MP) units to assess and understand the potential influence that crime and criminal organizations may have on UO. MPs—in coordination with local police—will need to quickly identify law enforcement gaps to include dysfunctional police organizations and stations and inadequate prison systems. Urban police records may be an important intelligence database to assist in this assessment and may be particularly useful in counterterrorism and counterinsurgency operations. Success in ensuring that law and order is maintained or reestablishing law and order will directly affect the population's perceptions and support of Army operations. Also, MP intelligence operations will contribute immeasurably to the overall IPB effort and aid in understanding of the complexities of the urban society (see FM 3-19.50).

Crime and Criminal Organizations

Crime and poverty plague urban areas such as Rio de Janeiro, Brazil's second largest urban area, and would affect military operations conducted in their limits. Rio has some of the nation's highest negative urban indicators: the largest number of slum dwellers (1 million), the highest murder rate (1 of 700 residents per year), and the highest kidnapping rate (4 per week). In 1989, the homicide rate of the urban area was three times higher than New York City's, and the rate of urban violence has continued to rise. Therefore, law enforcement management may be a critical issue for Army forces operating in urban environments similar to that of Rio de Janeiro.

However, criminal elements or organizations may not always work against Army commanders. They can be co-opted or influenced to serve friendly objectives. For example, during World War II the U.S. Navy worked covertly with the Mafia in New York City to secure the New York harbor from German U-boats believed to be torpedoing ships there. The Mafia controlled most dock activities in New York harbor and was perfectly positioned to monitor other subversive waterfront activity. This capability provided needed information to the Navy for its counterintelligence and security tasks. New York civil authorities, therefore, agreed to permit a Navy-Mafia alliance to operate at the port for the greater good of the country. Although the Mafia was not the preferred ally of the Navy, it had the capability to protect U.S. ships and the interest (patriotism) to help in the war effort. In those circumstances, the temporary alliance worked (see also the civilian threat discussion in Appendix B).

Armed Factions

3-58. A characteristic of some stability operations has been the deterioration or complete collapse of political authority in the country or urban area in crisis. In some cases, "warlords" or the leaders of various armed factions have attempted to fill the power vacuum (see Appendix C for an example). These individuals often have no particular claim to legitimacy. Their power issues from their weapons, not necessarily from their political skills, human services provided, or popular consent (although they have some popular support to remain in their relative position of authority). Others, however, may already have a substantial base of support (for example, a religious base) and then seek to achieve their goals through violent means. In dealing with the leaders of these armed factions during stability operations, it may appear that there are three general courses of action. The first is to aggressively locate and target the leaders and their forces militarily. The second is to ignore them completely, and the third is to meet and work with them diplomatically. Depending on METT-TC, commanders may reduce some of the risks involved in these first three courses of action by combining the last two to create a fourth approach. Nevertheless, the technique chosen must clearly support political and military objectives and lead to the desired end state.

3-59. In some cases, the leaders of armed factions pose a direct threat to Army forces, the urban populace, and the overall accomplishment of the mission. Therefore, it may become necessary to target them and their forces for capture or destruction. Yet, commanders should generally avoid framing the success of a major operation on the death or capture of one or more individuals. Even with increased ISR capabilities, commanders may find an individual manhunt difficult or impossible to achieve. Instead, it may be better to seek to isolate warlords and their forces militarily, while simultaneously seeking to marginalize their influence through a combination of political, economic, and informational means. This does not necessarily mean that commanders must abandon the search if the death or capture the faction leader will contribute to achieving the mission. However, commanders must maintain the capability to strike with speed and precision if the threat leader's location is discovered in the clutter of the complex urban environment.

3-60. In some situations, commanders may attempt to marginalize warlords by ignoring them. However, refusal to acknowledge warlords may increase the threat to Army forces and NGOs. Their militias may attack Army forces to achieve recognition or simply due to misunderstanding or inherent friction between armed forces. On the other hand, dealing with them may provide legitimacy to the exclusion of other elements of the urban population such as professional groups (for example, doctors or teachers), religious leaders, and traditional clan or tribal chiefs—which may have a greater claim to legitimacy and better form the foundation for a reconstituted urban society.

3-61. A compromise between the last two options may offer the best chance for success. In this fourth course of action, commanders can generally recognize these warlords and work with them diplomatically to decrease the risk of armed conflict and civilian casualties; however, this recognition can be kept at staff levels to avoid bestowing any perceived legitimacy on them. Instead, commanders themselves may visibly meet the other elements of society that have a more legitimate claim to political, social, or economic leadership. Inevitably, commanders may need to meet with warlords. In those circumstances, clan or tribal elders, and others who represent traditional authority, should attend the meetings. Moreover, commanders should ask for and give deliberate consideration to their opinions above those of the leaders of the armed factions.

This page intentionally left blank.

Chapter 4

Understanding the Urban Environment's Effects on Warfighting Functions and Tactics

War is, above all things, an art, employing science in all its branches as its servant, but depending first and chiefly upon the skill of the artisan. It has its own rules, but not one of them is rigid and invariable. As new implements are devised, new methods result in its mechanical execution; but over and above all its mechanical appliances, it rests upon the complex factors of human nature, which cannot be reduced to formulas and rules.

Captain Francis V. Greene, 1883

Commanders of major operations should understand the potential effects that the urban environment may have on warfighting functions. They should also understand the possible effects that the urban environment may have on lower-level tactics to properly plan, prepare, and execute major operations that may include UO. Otherwise, commanders may ask their subordinates to achieve effects, accomplish objectives, or adhere to a timetable that is unsupportable due to the constraints imposed by the urban environment. Commanders and their staffs must do more than simply understand the impossible; rather, they must apply the art and science of warfighting to the urban environment and determine what it will take to make it possible.

WARFIGHTING FUNCTIONS

4-1. Understanding the potential effects of the urban environment on warfighting functions permits the urban commander to better visualize his operational environment. With this appreciation, he can conduct a more thorough assessment and thereby determine the most efficient and effective means of employing Army forces. The staff should be intimately familiar with effects in their area of expertise and use that knowledge to understand the problem and develop creative and innovative solutions to achieve their commander's intent.

WARFIGHTING FUNCTIONS

- Intelligence
- Movement and maneuver
- Fire support
- Protection
- Sustainment
- Command and control

INTELLIGENCE

4-2. The intelligence function facilitates understanding of the threat and the environment. The urban environment affects this critical function in many ways. Impacts of the environment on the intelligence function include degraded reconnaissance capability; more difficult IPB process; and an increased importance of credible HUMINT (including the contribution of local civilian liaisons), and an established intelligence reach capability. The Army forces' response to these effects can result in timely, accurate, and actionable intelligence that permits the effective application of other warfighting functions to the mission within the urban environment (see FM 2-0).

Degraded Reconnaissance and Surveillance Capability

4-3. The physical environment creates a major challenge to the intelligence function. The man-made construction in the urban areas provides nearly complete cover and concealment for threats. Although improving many sensor capabilities cannot penetrate the subsurface facilities and much of the space within supersurface areas. The mass of buildings can also defuse electronic signatures. Tall buildings shield movement within urban canyons from aerial observation except from directly overhead. Urban threats may be less technology dependent and may thwart some signals intelligence efforts simply by turning off their radios and using messengers. Threat forces will likely use elements of the civilian telecommunications infrastructure for C2. These systems may include traditional landline phones, cellular telephones, and computer-to-computer or Internet data communications. Most urban telecommunications systems use buried fiber or cables or employ modern digital signaling technology. Such systems are difficult to intercept and exploit at the tactical level.

4-4. From the above, it is evident that these characteristics make it more difficult for the intelligence function to use electronic means to determine threat dispositions and, in offensive and defensive UO, identify decisive points leading to centers of gravity. While the environment limits some typical collection methods, all enemy electronic and human activity creates some form of observable signature and exposes the enemy to potential collection. Seeking ways to take advantage of these vulnerabilities will provide the commander an information advantage over his opponent.

Challenging IPB Process

4-5. The sheer complexity of the environment also challenges the intelligence function. The intelligence function applies the IPB process to the urban environment in accordance with Army doctrine (see Appendix B). With more data points for the IPB process to identify, evaluate, and monitor, this application becomes more demanding. The human and societal aspects of the environment and the physical complexity primarily cause this difference. Relationships between aspects of the environment, built on an immense infrastructure of formal and informal systems connecting the population to the urban area, are usually less familiar to analysts. Thus, the urban environment often requires more specifically-focused intelligence resources to plan, prepare for, execute, and assess operations than in other environments.

4-6. Compounding the challenges is the relative incongruity of all urban environments. No two urban areas are alike physically, in population, or in infrastructure. Thus, experience in one urban area with a particular population and pattern of infrastructure does not readily transfer to another urban area. Any experience in UO is valuable and normally serves as a starting point for analysis, but the intelligence function cannot assume (and treat as fact) that patterns of behavior and the relationships in one urban area mirror another urban area. The opposite is as likely to hold true. The intelligence function will have to study each urban area individually to determine how it works and understand its complex relationships.

4-7. Each characteristic of the urban environment—terrain, society, and infrastructure—is dynamic and can change radically in response to UO or external influences. Civilian populations pose a special challenge to commanders conducting UO. Civilians react to, interact with, and influence to varying degrees Army forces. Commanders must know and account for the potential influence these populations may have on their operations. Intelligence analysts must revisit or continuously monitor the critical points looking for changes, relationships, and patterns.

4-8. The actions of Army forces will affect, positively or negatively, their relationship with the urban population and, hence, mission success. NGOs may deliberately or inadvertently influence civilians. The intelligence function can monitor and predict the reactions of the civil population. However, accurate predictive analysis of a large population requires specific training and extensive cultural and regional expertise.

Increased Importance of Human Intelligence

> ***HUMINT*** is the collection of foreign information—by a trained HUMINT collector—from people and multimedia to identify elements, intentions, composition, strength, dispositions, tactics, equipment, personnel, and capabilities. It uses human sources as a tool, and a variety of collection methods, both passively and actively, to collect information. FM 2-0

4-9. The intelligence function adjusts to the degradation of its technical intelligence gathering systems by increasing emphasis on HUMINT in UO. HUMINT operations may be the primary and most productive intelligence source in UO. In urban offensive and defensive operations, HUMINT gathers information from refugees, immigrants and former citizens (especially previous civil administrators), civilian contractors, and military personnel who have operated in the area. Credible intelligence of this type can help meet requirements, provide more detail, and alleviate some of the need to physically penetrate the urban area with reconnaissance forces. In many urban operations where HUMINT is the primary source of intelligence, acting on single-source reporting is a constant pitfall. Yet, situations may arise where commanders must weigh the consequences of inaction against any potential negative consequences resulting from acting on uncorroborated, single-source information. (See also the Human Capabilities discussion under the Urban ISR portion of Chapter 5.)

4-10. In urban stability operations, HUMINT identifies threats and monitors the intentions and attitudes of the population. A chief source of information contributing to the development of accurate HUMINT, particularly at the tactical level, is reconnaissance forces—especially small-unit dismounted patrols. Urban reconnaissance forces and patrols should be thoroughly and routinely debriefed by unit intelligence personnel to obtain information that aids in developing a clearer picture of the threat and the urban environment. Reliable and trustworthy HUMINT is particularly important in foreign internal defense, counterterrorism, and support to counterdrug operations. Leaders must organize intelligence resources appropriately, and learn and apply valuable techniques, such as pattern and link analysis (see FM 34-3). Additionally, Soldiers, as part of reconnaissance and patrolling training, should be taught to handle captured documents, weapons, material, and equipment as legal evidence much like military and civilian police. Proper "evidence" handling is often a critical intelligence concern in counterterrorism and counterinsurgency operations.

Developing Local Liaisons

4-11. Whenever Soldiers encounter the urban populace, the resulting interaction may become an important source of information the commander can use to answer questions about the threat and the urban environment. While military intelligence units are the primary collectors and processors of HUMINT, commanders are not likely to have enough trained HUMINT Soldiers to satisfy their requirements—particularly in a larger urban environment and during longer-term stability operations. Therefore, commanders may need to cultivate and establish local civilian associations to provide relevant information for decision making and to support the overall HUMINT effort.

4-12. Urban liaisons can be developed through positive civil-military interaction with the urban populace. Critical information may be acquired through interface with the urban leadership (both formal and informal), administration officials, business owners, host-nation support workers, inhabitants along a unit's patrol route, pedestrians at a checkpoint, civilian detainees, or any other human source willing to volunteer information to Army forces or who respond positively to tactical questioning. (Noncombatants are never coerced to provide information.) Commanders may also direct unit leaders to conduct liaisons with specific local leaders and key members of the community to obtain command directed information. Critical information may also come from other U.S. and coalition forces and intelligence organizations operating near or within the commander's AO. To this end, commanders should ensure that collectors operating in an urban area coordinate and deconflict activities and, if possible, outbrief subordinate, geographically-responsible commanders with any relevant information that may affect their current operations. Any relevant information obtained incident to civilian liaison activities should be routinely provided to intelligence staffs not only to gain assistance in verifying the credibility of the information but to share the information with all affected echelons and units.

4-13. NGOs operating in urban areas can also be especially beneficial resources for credible and relevant information about the urban environment. (However, they are generally not a good source for information about the threat since providing such information can violate their neutrality thereby making it difficult for them to achieve their humanitarian aid objectives.) During the 1999 fighting in Kosovo, for example, the Red Cross provided the most accurate figures regarding the number of Kosovar refugees, helping U.S. and other coalition forces to estimate the appropriate level of support required to handle their needs. In addition to a developed understanding of the current needs of the local urban populace, NGOs may also have—

- A network of influential associations.
- Historical archives.
- Extensive understanding of the urban infrastructure.
- Key knowledge of political and economic influences.
- A keen awareness of significant changes in the urban environment.
- Insight into the current security situation.
- Up-to-date web sites and maps.

4-14. While productive civilian associations may become long term, they should not be confused with HUMINT source operations. Only trained HUMINT personnel can recruit and task sources to seek out threat information. Information obtained from these societal connections is normally incidental to other civil-military relationships. For example, as part of infrastructure repair in an urban stability operation, a commander may be instrumental in obtaining a generator for a local hospital. Within the context of this relationship, the commander may develop a rapport with one or more of the hospital's administrators or health practitioners. These civilians may be inclined to provide valuable information about the threat and the urban environment—often on a continuing basis. In any civil-military relationship, however, commanders ensure that the information provided is not tied to promises of assistance or that such assistance is in any way perceived as a means to purchase civilian loyalty.

4-15. Commanders also understand that repeated interaction with any one individual may put that individual and his family in danger from threat forces. Before this potential danger becomes a reality, they should refer their civilian connections to trained HUMINT personnel for protection and continued exploitation. In addition to civilian protection considerations, commanders may also deem it necessary to turn their civilian associations over to trained HUMINT collectors anytime during the relationship if they consider the information that the contact is providing (or may provide) is credible, relevant, and—

- Provides essential threat information on a repetitive basis.
- Helps answer higher-level CCIR.
- Affects operations in another AO.
- Requires interrogation or monetary compensation to obtain.

However, turning a liaison over to trained HUMINT teams does not necessarily preclude maintaining a continued, albeit a more guarded, relationship with the individual.

4-16. In developing these civilian liaisons essential to understanding the urban environment, commanders must avoid the distinct possibility of conducting unofficial source operations by non-HUMINT Soldiers. While prohibited by regulatory guidance (see Defense Intelligence Agency Manual 58-11 for requirements and restrictions for source operations), such actions also run the additional risks of—

- Obtaining unevaluated information that cannot be crosschecked with and verified by other sources of information.
- Creating inequities that result from illegally rewarding contracts which can undermine HUMINT Soldiers who are constrained by intelligence contingency fund regulations.
- Disrupting ongoing HUMINT operations when different sources are seen to be treated differently by non-HUMINT Soldiers vice HUMINT Soldiers.
- Providing non-HUMINT and HUMINT Soldiers with the same information potentially leading to a false confirmation of information.

- Increasing the likelihood that untrained Soldiers may fall victim to a threat deception and misinformation.

Established Intelligence Reach

4-17. Understanding the complex urban environment, particularly the infrastructure and the society, will require more sources of information beyond a unit's organic intelligence capabilities. Therefore, commanders will have to make extensive use of intelligence reach to access information and conduct collaboration and information sharing with other units, organizations, and individual subject matter experts. Before deployment (and throughout the operation), units should establish a comprehensive directory of intelligence reach resources. These resources may include national, joint, Army, foreign, commercial, and university research programs. (Prior to deployment for OPERATION IRAQI FREEDOM, some units established contacts within the local community outside their bases such as police, fire department, and government officials that expanded their reach once in theater—particularly for information regarding civilian infrastructure and urban administration.) Once deployed, intelligence reach includes effective information sharing and collaboration among adjacent units, sister services, coalition partners, and other governmental and nongovernmental agencies operating in the area. Effective information sharing and collaboration requires common network analysis software and databases to be used among all Army forces and, if possible, other governmental agencies.

MOVEMENT AND MANEUVER

4-18. Army maneuver forces—infantry, armor, cavalry, and attack aviation—move to achieve a position of advantage. Entire urban areas or specific threat forces located within may be isolated from affecting other operations and then bypassed. However, when the situation requires entering the urban area to accomplish the mission—such as when bypassed urban threat forces interdict critical lines of communications (LOCs), the environment will significantly affect the Army's ability to move and maneuver (see figure 4-1). These negative effects include canalization, compartmentalization, and increased vulnerability. However, tactics and techniques equip Army forces to overcome these challenges and maneuver successfully. One tactic, effective combined arms task organization, includes an increased dismounted maneuver capability combined with armor and combat engineers, continuous operations, and technological enhancements.

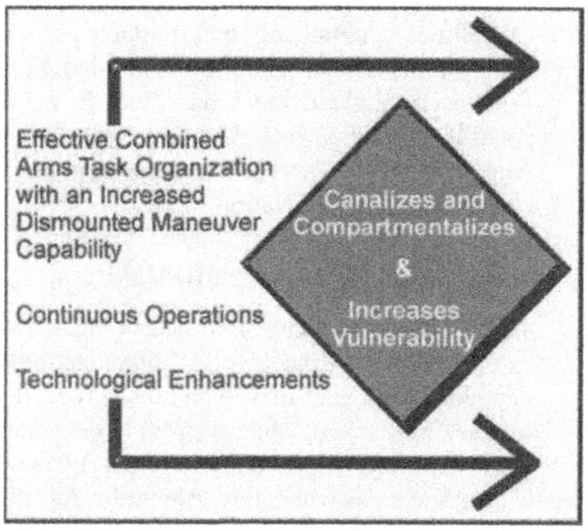

Figure 4-1. Urban maneuver challenges and means to overcome them

4-19. In all UO, mobility operations may allow civilian traffic and commerce to resume, letting the urban area return to some semblance of normalcy (often a critical objective). In stability operations, mobility often focuses on keeping lines of communications open and reducing the threat of explosive hazards to Soldiers and civilians. In civil support operations, mobility may focus on removing storm debris or reducing obstacles caused by destroyed property.

Canalization and Compartmentalization

4-20. The urban terrain will often canalize and compartment forces—and their fires—moving and maneuvering through it. Buildings pose obstacles to both mounted and dismounted movement, forcing units to be canalized along streets. The buildings also block movement between streets, thus compartmenting units. Fires are canalized into open and unmasked areas where vision is unobstructed, producing concentrated fire zones and areas, especially at road intersections and in front of defended positions. Hence, changing directions, repositioning committed forces, reinforcing forces in contact, bypassing threats, and maneuvering to the threat flank become extremely difficult. Units often breach obstacles to help solve this problem. Using helicopters to quickly move forces, both forward into contact and to rear areas as part of repositioning, also permits Army forces to overcome some terrain constraints.

4-21. The canalized and compartmented effects can slow movement and maneuver considerably. However, slowed movement also results from the physically demanding soldier tasks required in an urban environment. Soldiers operate dismounted across rubble and hard surfaces. Operating in three dimensions, they constantly move up the supersurface areas of building interiors and down into basements, cellars, and other subsurface areas. They breach many obstacles and use upper-body strength, ropes, and ladders to scale heights. The inability to see into the next room, floor, or building magnifies stress. The resulting fatigue slows the overall rate of Army force movement and maneuver.

Increased Vulnerability

4-22. The urban environment increases the vulnerability of Army forces executing movement and maneuver in offensive, defensive, and stability or civil support operations. Both the physical terrain and the urban population provide threat cover and concealment. Air movement and maneuver is vulnerable for many of the same reasons. In offensive or defensive operations, enemy forces can remain undetected in buildings and in position to ambush Army forces. Consistent with METT-TC, Army forces should clear buildings along maneuver routes prior to mounted movement along those axes. Failure to clear routes (and effectively mark cleared portions) may expose mounted movement to ambush at close range. Movement back across streets and obstacles may be difficult particularly if the element of surprise was essential in the initial crossing or breach. The same buildings also provide cover and concealment to enemy air defense capabilities—particularly man-portable air defense systems that can be fired from multiple positions hidden amongst the clutter of fires, lights, smoke and dust, and easily concealed and transported in civilian vehicles throughout the urban area. In all operations, but especially stability operations, civilians can conceal threat elements. The threat can then initiate offensive operations against Army forces from close range and where ROE will hamper applying combat power. Thus, maneuver through a dense population can be a high-risk operation.

Combined Arms Task Organization

4-23. Effective combined arms task organization ensures that forces are task organized with infantry—the essential building block for all organizations conducting UO. Infantry protects mounted elements as the combined arms unit moves and maneuvers through the urban area. (In some urban situations, mechanized infantry may not be able to provide dismounted support beyond support to its own vehicles—tanks may require the support of additional light infantry). The infantry destroys the enemy in buildings, bunkers, and subsurface areas where they cannot be defeated by mounted forces and prevents infiltration of threat forces back into hard-won urban terrain. Field artillery aids in dismounted and mounted (to include air) maneuver by suppressing known and suspected enemy positions with precision fires. Attack aviation make best use of their standoff capabilities (see later fire support discussion) and aircraft speed to conduct running and diving fires; in UO, hovering fire is generally avoided. Armored elements protect Soldiers from small arms fire and destroy or suppress enemy positions with precise, direct fire. Carefully protected artillery may also be used in this direct fire role. Armored forces and attack helicopters also can facilitate maneuver through shock action that can have a psychological effect, particularly against less well-trained threats and, in discrete instances, hostile crowds. (Although, commanders consider that the "intimidation" value of any method erodes quickly with its repetitive use.)

4-24. Combined arms also ensure that combat engineers support dismounted maneuver by assisting in covered and concealed maneuver through buildings and off exposed streets. In addition to combat engineers, explosive ordnance disposal teams, military police, chemical personnel, and other with essential expertise to conduct mobility missions significantly reduces mobility and maneuver challenges (see FM 3-34.2). Urban buildings are often obstacles to movement and mobility. Combat engineers, trained and equipped for UO, can turn these obstacles into an advantage by breaching them with "mouse holes" made by explosives, sledgehammers, bulldozers or armored vehicles, or high-strength (diamond or carbide-tipped) cutting devices. These breaches permit dismounted movement through buildings under both cover and concealment.

4-25. Combat engineers must also be trained and equipped to facilitate mounted mobility in the urban environment. Generally, buildings restrict mounted movement to the compartmented and canalized streets. Threats can block streets with roadblocks ranging from sophisticated log and concrete cribs reinforced with antitank and antipersonnel mines to the expedient use of cars, buses, and trucks to create obstacles. Combat engineers must be capable of breaching these obstacles to maintain the coherence of the combined arms team (mounted and dismounted). Combat engineers should be forward, often task organized down to platoon level, and have the expertise and equipment to rapidly reduce point obstacles. It even may be necessary that every armored vehicle (or section of two vehicles) be task organized with an associated engineer squad and vehicle. Because of the increased density and hardness of many urban building and construction materials, heavy engineer equipment (such as the D9 bulldozer) will be in great demand to accomplish mobility (and countermobility, and survivability) functions in an urban environment. However, commanders will need to consider increased protection requirements and the availability of equipment transport to move these slower-moving engineer assets around the urban battlefield.

4-26. A major difference of UO combined arms is in proportion and organization. Although based an accurate METT-TC assessment, UO will often require an increased proportion of dismounted infantry and engineer capabilities; armor may not be required in the same high numbers. As significant, genuine combined arms urban operations are required at lower tactical levels where small, well trained and led units will dominate. Commonly, company level will require true combined arms capability and may include combat engineers, military intelligence, reconnaissance, and artillery. Combined arms teams can then form at platoon and squad levels. Because of this, larger units will need more CA, military intelligence, and combat engineers than habitually attached for combat in more open or less restrictive terrain.

4-27. In determining the appropriate task organization, commanders consider a suitable span of control for subordinate commanders. They also consider the potential of dissipating a unit's combat power, capabilities, and synergy by breaking a unit up into smaller units in an attempt to ensure subordinate maneuver units have a complete combined arms capability. For example, an additional engineer battalion may be task organized to a brigade combat team. In turn, the brigade combat team may task organize this battalion into engineer companies under the control of their subordinate maneuver battalions. If this type of organization continues, maneuver companies may end up with an engineer platoon and, in the end; maneuver platoons each having an engineer squad. Ultimately, a combined arms capability may have been established at lower tactical levels but the parent maneuver unit (in this example, the brigade combat team) may have lost the ability to conduct larger engineer operations without having to re-task organize and potentially disrupt current operations and established relationships. As a guide, urban commanders may consider task organizing to create combined arms organizations at lower tactical levels when operations are predominately offensive or defensive, and bringing those assets back under their own control when the operation transitions to predominately stability or civil support operations.

Continuous Operations and Technology Enhancements

4-28. Two other means to improve Army forces' ability to move and maneuver in urban terrain is through continuous operations and the leveraging of technology, such as the Army's night operations capability. Historically, urban battles have been fought primarily during daylight because of technological limitations and fatigue. By utilizing night vision technologies, accurate situational understanding, a common operational picture (COP), training, and rotated units, Army forces can defeat threats who use the same soldiers in day and night operations and who are less well-equipped and adept at night operations. Night operations are also a means of mitigating the air defense threat against air maneuver. Continuous operations through night maneuver with fresh forces are challenging, but it can overcome many advantages that a stationary force has against maneuver in the urban environment. However, commanders should also consider that streetlights, fires, background illumination (as well as dark building interiors without ambient light), the increased heat absorption of many urban structures, and the skillful use of searchlights by threat forces may limit the effectiveness of night vision devices and make thermal imagery identification difficult.

Countermobility

4-29. Countermobility capabilities in urban terrain are also an essential consideration in all UO. In defensive operations, commanders use countermobility capability to control where the enemy moves in the urban area. Repositioning defensive forces in the urban area can be difficult and obstacles are essential to limiting the enemy's maneuver options. During offensive operations, countermobility protects exposed flanks and air assaulting forces from counterattack. In stability and civil support operations, countermobility operations may take the form of constructing barriers to assist in populace and resources control at critical urban locations.

FIRE SUPPORT

4-30. The fire support function includes the collective and coordinated use of several means to attack targets in the urban area (see Appendix D for joint capabilities). These means include target acquisition data, indirect fire weapons, rotary- and fixed-wing aircraft, offensive information operations (IO), and other lethal and nonlethal means. The urban environment affects these components of the fires function and their employment.

Target Acquisition

4-31. Target acquisition in an urban environment faces several challenges. First, forces have difficulty penetrating the urban environment's increased cover and concealment using sensors and reconnaissance. Acquiring targeting information and tracking targets throughout the depth of the urban area may prove challenging. Moving personnel or vehicular targets are normally easiest to acquire. However, the cover and concealment provided by urban terrain gives moving targets short exposure times requiring firing systems to act rapidly on targeting data. Targeting of opposing indirect fire units by acquisition radar may work more effectively in urban terrain because of the necessary high angles of indirect fire. The urban environment presents similar difficulties for battle damage assessment.

4-32. Targeting challenges are met by innovatively integrating reconnaissance capabilities. These capabilities include SOF, cavalry, unmanned aircraft systems, and aerial observers as well as the standard reconnaissance assets. More artillery systems may need to be used to ensure the responsiveness (rather than the weight) of fires. Positioning numerous artillery systems reduces the dead space (as discussed below) and permits units to establish more direct sensor-to-shooter links.

The Targeting Process

4-33. Heightened concerns for collateral damage will require that commanders pay particular attention to their targeting process. This process ensures that all available combat power, both lethal and nonlethal (including offensive IO), is effectively integrated and synchronized to accomplish the mission. Commanders ensure that techniques and procedures are in place, rehearsed, and understood by all members of their staffs. Additionally, the C2 system must be responsive and agile; otherwise, an elusive and adaptable threat will likely disappear before units can employ the appropriate weapon systems. In an urban area, even 10-digit grid coordinates may not be sufficient to accurately identify targets as buildings may be connected to each other—often throughout the entire block. Target locations, in addition to grid coordinates, may need to routinely include the street address, number of stories, shape, color, or any other distinguishing characteristic essential for ground and air forces to achieve targeting precision. A common urban reference system with graphics, reference points, and other control measures adequate for both ground and air forces may also help facilitate identification of targets and facilitate the rapid clearance of fires.

4-34. Greater concerns exist for the safety and health (environmental matters) of the urban populace and the protection of critical infrastructure and cultural structures. Hence, CA and staff judge advocates (see Chapters 5 and 10 respectively) will play a greater role for the expert advice they can provide regarding these elements of the urban environment. Nonetheless, all members of the staff ensure that operations minimize collateral damage. That responsibility does not end with identifying potential collateral damage; the goal, as always, is successful mission accomplishment. Again, staffs are guided by the commander's intent and work to develop courses of action that incorporate collateral damage concerns (short- and long-term) yet accomplish the mission. This requires a keen understanding of the legal issues and both friendly and enemy weapon systems' effects in an urban environment.

Urban Effects on Fire Support Systems

4-35. Both the physical and human components of the urban area affect how units use fire support weapon systems. The physical aspects of the urban environment, such as the heights and concentration of buildings, may cause significant masking and dead space (see figure 4-2). Intervening buildings that stand three or more stories tall hinder close indirect fire support. Tall buildings can potentially mask several blocks of area along the gun-target line of artillery. (For low-angle artillery fire, dead space is about five times the height of the building behind which the target sits.) The potential for collateral damage to adjacent buildings may also prevent engagement with artillery. Such damage might cause noncombatant and friendly troop casualties and unintentional (and unwanted) rubbling. Commanders can offset these effects by carefully placing artillery positions, repositioning artillery as targets change, and using mortars. Mortars have a steep angle of fall and short minimum ranges as a high-angle alternative to field artillery fire. (In comparison to artillery, dead space for mortar fire is only about one-half the height of the building.) For fixed-winged aircraft, precision munitions and weapons with low explosive yields and near-vertical impact angles resulting in bomb burial can significantly reduce collateral damage. Collateral damage concerns may also cause commanders to—

- Maintain approval authority for some sensitive or protected targets (churches or mosques, for example) at higher echelons of command.
- Restrict attacks to certain times of day.
- Give warning prior to an attack so that noncombatants can evacuate the area.
- Incorporate indigenous forces into the operation.
- Abort an attack unless the required level of precision effects can be achieved.
- Prepare specific branches and sequels to the IO plan to inform the populace why the collateral damage was justified. These plans may include filming or otherwise documenting the operation to thwart threat propaganda and claims of excessive collateral damage.
- Develop and rehearse detailed staff battle drills that address clearance of fires in an urban environment.

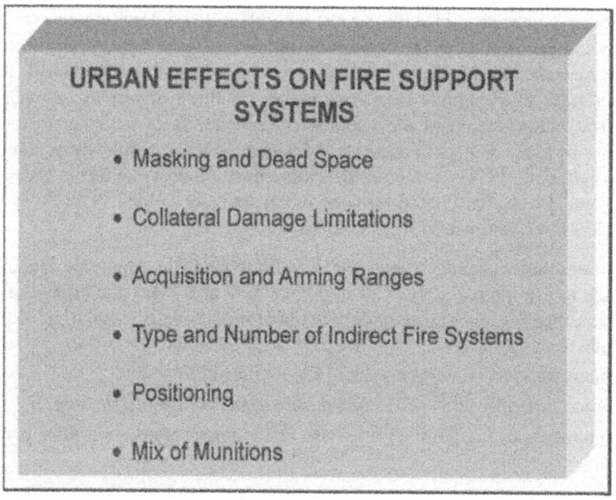

Figure 4-2. Urban effects on fire support systems

4-36. Vertical structures interrupt line of sight (LOS) and create corridors of visibility along street axes. The result is thereby shortened acquisition and arming ranges for supporting fires from attack helicopters and subsequently affected engagement techniques and delivery options. Pilots maintain LOS long enough to acquire targets, achieve weapons delivery solutions, and fly to those parameters. For example, tube-launched, optically tracked, wire-guided heavy antitank missile systems require 65 meters to arm. Similarly, the Hellfire missile requires at least 500 meters to reliably arm and stabilize on the intended target. Thus, attack helicopters firing from longer ranges actually improve the probability of a hit. Poor weather and heavy smoke and dust rising from urban fires and explosions may hinder target identification, laser designation, and guidance for rotary- and fixed-winged aircraft. Poor air-to-ground communications may also hinder effective use of airpower. The close proximity of friendly units and noncombatants requires units to agree on, thoroughly disseminate, and rehearse clear techniques and procedures for marking target and friendly locations. The ability for ground units to "talk-on" aircraft using a common reference system described earlier helps expedite aerial target acquisition (and helps mitigate potential fratricide). FM 3-06.1 details other aviation TTP in an urban environment.

4-37. The urban environment also affects the type and number of indirect fire weapon systems employed. Commanders may prefer high-angle fire because of its ability to fire in close proximity to friendly occupied buildings. Tactically, commanders may consider reinforcing units in UO with mortar platoons from reserve units. This will increase the number of systems available to support maneuver units. Unguided Multiple Launch Rocket Systems (MLRSs) may be of limited use in urban areas due to their exceptional destructive capabilities and the potential for collateral damage. However, commanders may use unguided MLRSs to effectively isolate the urban area from outside influence. Commanders may also employ field artillery systems as independent sections, particularly self-propelled systems, in the direct-fire role; decreasing volume and increasing precision of artillery fire helps minimize collateral damage. While discretely applying the effects of high-explosive and concrete-piercing munitions, these self-propelled systems take advantage of the mobility and limited protection of their armored vehicles.

4-38. The urban area may also affect the positioning of artillery. Sufficient space may not exist to place battery or platoon positions with the proper unmasked gun line. This may mandate moving and positioning artillery in sections while still massing fires on specific targets. Commanders must protect artillery systems, particularly when organized into small sections. Threats to artillery include raids and snipers. Therefore, maneuver and firing units will have to place increased emphasis on securing their positions and other appropriate force protection measures.

4-39. The mix of munitions used by indirect fire systems will change somewhat in urban areas. Units will likely request more precision-guided munitions (PGM) for artillery systems to target small enemy positions, such as snipers or machine guns, while limiting collateral damage. Currently, only conventional tube artillery, not mortars, has this capability. However, large expanses of polished, flat reflective surfaces common in urban areas may degrade laser designation for these munitions (as well as attack helicopter PGM). The vertical nature amplifies the geometrical constraints of many precision munitions. Remote designators need to be close enough to accurately designate but far enough away not to be acquired by the PGM during its flight path. PGMs based on the global positioning system (for instance, guided MLRS or the Air Force's joint direct attack munitions) or other optically guided PGMs may be more effective if urban terrain hinders laser designation.

4-40. The urban environment greatly affects the use of nonprecision munitions. Building height may cause variable time fuses to arm prematurely. Tall buildings may also mask the effects of illumination rounds. Units may choose not to use dual-purpose conventional munitions if (similar considerations apply to Air Force cluster bombs)—

- The enemy has several building floors for overhead protection.
- Dismounted friendly units need rapid access to the area being fired on.
- Large numbers of civilians will operate in the target areas soon after combat operations have ceased.

4-41. Depending on the building construction, commanders may prohibit or limit illumination, smoke, and other munitions because of fire hazards. In particular instances, they may specifically use them for that effect. Structure fires in an urban area are difficult to control and may affect friendly units. Conventional high-explosive munitions may work best against concrete, steel, stone, and other reinforced structures. When not used in the direct-fire role, a greater mass of indirect fire is often required to achieve desired effects. Commanders balance firepower and collateral damage since the rubbling caused by massive indirect fires may adversely affect a unit's ability to maneuver and provide a threat with additional cover and concealment.

4-42. Nonlethal weapons, munitions, and devices can help commanders maintain the desired balance of force protection, mission accomplishment, and safety of noncombatants by expanding the number of options available when deadly force may be problematic. As additional nonlethal capabilities are developed, they are routinely considered for their applicability to UO. In determining their use and employment, commanders, in addition to any previous experience at using these weapons, munitions, and devices, consider—

- **Risk**. The use of nonlethal weapons in situations where lethal force is more appropriate may drastically increase the risk to Army forces.
- **Threat Perspective**. A threat may interpret the use of nonlethal weapons as a reluctance to use force and embolden him to adopt courses of action that he would not otherwise use.
- **Legal Concerns**. Laws or international agreements may restrict or prohibit their use (see Chapter 10).
- **Environmental Concerns**. Environmental interests may also limit their use.
- **Public Opinion**. The apparent suffering caused by nonlethal weapons, especially when there are no combat casualties with which to contrast it, may arouse adverse public opinion.

PROTECTION

4-43. The protection function includes those tasks and systems that preserve the force so that commanders can apply maximum combat power. Preserving the force includes enhancing survivability and properly planned and executed air and missile defense as well as defensive IO (see IO discussion in Chapter 5) and chemical, biological, radiological, nuclear, and high-yield explosive (CBRNE) counterproliferation and consequence management activities (see CBRNE discussion in Chapter 9).

Survivability

4-44. Survivability in the urban environment is a significant force multiplier. Properly positioned Army forces can take advantage of the increased survivability afforded by the physical terrain. Even a limited engineer effort can significantly enhance the combat power of small Army forces. In stability operations, properly planned and constructed survivability positions can enable small groups of Soldiers to withstand the assaults of large mobs, sniping, and indirect fire. Well-protected support bases are often critically essential to minimizing casualties during long-term stability operations and can become a key engineer task.

4-45. While executing major combat operations or campaigns, in particular defensive operations, well planned and resourced engineer efforts can enhance the survivability characteristics of the urban area. These efforts, though still requiring significant time and materiel, can establish defensive strong points more quickly and with greater protection than can be done in more open terrain. Skillfully integrating the strong point into the urban defense greatly increases the overall effectiveness of the defense disproportionately to the number of forces actually occupying the strong point (see Chapter 8).

4-46. Commanders increase survivability by ensuring that all Soldiers have necessary protective equipment and are trained and disciplined in their use. In addition to standard equipment such as helmets, gloves, boots, and chemical protective overgarments, commanders should ensure, as necessary, availability of other protective equipment and material such as—

- Body armor.
- Goggles or ballistic eye protection.
- Knee and elbow protectors.
- Riot control equipment such as batons, face masks, and shields.

- Barrier material such as preformed concrete barriers, wire, sandbags, and fencing material.
- Up-armored or hardened vehicles.
- Fire extinguishers and other fire-fighting equipment.
- Immunizations.

4-47. The Army's urban survivability operations can become complex if the Army is tasked to support survivability operations for civilians. Such operations can range from constructing civil defense shelters or evacuating the population to assisting the population in preparing for or reacting to the use of weapons of mass destruction. However, Army forces are not organized or equipped to support a major urban area's requirements as well as its own mission needs. Normally, Army forces can render this type of support only as a focused mission using a unique, specially equipped task organization.

Air and Missile Defense

4-48. The air and missile defense protects the force from air surveillance and air and missile attack. This system uses—

- The careful massing of air and missile defense combat power at points critical to the urban operation.
- The proper mix of air defense weapon and sensor systems.
- Matched (or greater) mobility to the supported force.
- The integration of the air defense plan into the overall urban operation.
- The integration of Army systems with those of joint and multinational forces.

4-49. Properly planned and executed air and missile defense prevents air threats from interdicting friendly forces and frees the commander to synchronize maneuver and other elements of firepower. Even in a major combat operation or campaign, the enemy will likely have limited air and missile capabilities and so seek to achieve the greatest payoff for the use of these systems. Attacking Army forces and facilities promises the greatest likelihood of achieving results, making urban areas the most likely targets for air and missile attack.

Rotary- and Fixed-Winged Aircraft

4-50. Enemy rotary-wing aircraft can be used in various roles to include air assault, fire support, and combat service support. Some threats may use unmanned aircraft systems to obtain intelligence and target acquisition data on friendly forces. Increased air mobility limitations and targeting difficulties may cause enemy fixed-wing aircraft to target key logistics, C2 nodes, and troop concentrations outside the urban area, simultaneously attacking key infrastructure both in and out of the urban area.

Increased Missile Threat

4-51. The intermediate range missile capability of potential threats has increased to be the most likely air threat to an urban area. Urban areas, particularly friendly or allied, make the most attractive targets because of the sometimes-limited accuracy of these systems. By firing missiles at an urban area, a threat seeks three possible objectives:

- Inflict casualties and materiel damage on military forces.
- Inflict casualties and materiel damage on the urban population.
- Undermine the confidence or trust of the civil population (particularly if allied) in the ability of Army forces to protect them.

4-52. If facing a missile threat, commanders conducting UO work closely with civil authorities (as well as joint and multinational forces) to integrate the Army warning system with civil defense mechanisms. Similarly, Army forces may support urban agencies reacting to a missile attack with medical and medical evacuation support, survivor recovery and assistance in damaged areas, and crowd control augmentation of local police forces. Before such an attack, Army engineers might assist and advise urban officials on how to construct shelters.

Increased Security of Assets

4-53. When defending against an air or missile threat in a neutral or hostile urban environment, air defense assets are concerned with security. Separating air defense locations from high population and traffic centers, as well as augmenting these positions with defending forces, can prevent or defeat threat efforts to neutralize them. Additionally, increased density of UO means increased concentration of all friendly and enemy systems engaged in air and counter-air operations. This density may increase friend and foe identification challenges, air space management challenges, and the overall risk in the conduct of air operations. Finally, limited air defense assets, difficulties in providing mutual support between systems, potential mobility limitations, and other effects of the urban environment increase the need for (and effectiveness of) a combined arms approach to air defense (see FM 44-8).

SUSTAINMENT

4-54. The sustainment function incorporates support activities and technical service specialties, to include maximizing available urban infrastructure and contracted logistics support. It provides the physical means with which forces operate. Properly conducted, the sustainment function ensures freedom of action, extends operational reach, and prolongs endurance. Commanders conducting sustainment to support full spectrum operations must understand the diverse logistic requirements of units conducting UO. They must also understand how the environment (to include the population) can impact sustainment support. These requirements range from minimal to extensive, requiring Army forces to potentially provide or coordinate all life support essentials to a large urban population.

4-55. Commanders and staffs consider and plan for Army sustainment operations that are based in a major urban area. These operations are located in major urban areas to exploit air- and seaports, maintenance and storage facilities, transportation networks, host-nation contracting opportunities, and labor support. These operations are also UO. The commander gains additional factors to consider from basing the sustainment operation in an urban environment. See Chapter 10 for a detailed discussion of urban sustainment considerations.

COMMAND AND CONTROL

> *Fighting in a city is much more involved than fighting in the field. Here the "big chiefs" have practically no influence on the officers and squad leaders commanding the units and subunits.*
>
> Soviet General Vasili Chuikov
> during the 1942-43 Battle for Stalingrad

4-56. The command and control function is the related tasks and systems that support the commander in exercising authority and direction. The urban environment influences both the commander and his C2 system (which includes INFOSYS). The leader's ability to physically see the battlefield, his interaction with the human component of the environment, and his intellectual flexibility in the face of change all impact the mission. The C2 system faces difficulties placed on the tactical Internet and system hardware by the urban environment, by the increased volume of information, and by requirements to support the dynamic decision making necessary to execute successful UO.

Unity of Command

4-57. Although severely challenged, the principle of unity of command remains essential to UO. However, the number of tasks and the size of the urban area often require that Army forces operate noncontiguously. Noncontiguous operations stress the C2 system and challenge the commander's ability to unify the actions of his subordinates, apply the full force of his combat power, and achieve success. To apply this crucial principle in an urban environment requires centralized planning, mission orders, and highly decentralized execution. The method of C2 that best supports UO is mission command (see FM 6-0). Mission command permits subordinates to be innovative and operate independently according to clear orders and intent as well as clearly articulated ROE. These orders and ROE guide subordinates to make the right decision when facing—

- A determined, resolute, and adaptive threat.
- A complex, multidimensional battlefield.
- Intermittent or complete loss of communications.
- Numerous potentially hostile civilians close to military operations.
- The constant critique of the media and military pundits.

4-58. Decentralized execution allows commanders to focus on the overall situation—a situation that requires constant assessment and coordination with other forces and agencies—instead of the numerous details of lower-level tactical situations. Fundamentally, this concept of C2 requires commanders who can accept risk and trust in the initiative, judgment, and tactical and technical competence of their subordinate leaders. Many times, it requires commanders to exercise a degree of patience as subordinate commanders and leaders apply mental agility to novel situations.

Political and Media Impact

4-59. Commanders of a major operation consider how the need to maintain a heightened awareness of the political situation may affect their exercise of C2. A magnified political awareness and media sensitivity may create a desire to micromanage and rely solely on detailed command. Reliance on this method may create tactical leaders afraid to act decisively and with speed and determination—waiting instead for expected guidance from a higher-level commander. Threats may capitalize on this hesitation by conducting operations faster than Army forces can react. Mission orders that express the overarching political objectives and the impact of inappropriate actions, combined with training and trust, will decrease the need for detailed command. Leaders must reduce a complex political concept to its simplest form, particularly at the small-unit level. Even a basic understanding will help curtail potentially damaging political actions and enable subordinates to make the often instantaneous decisions required in UO—decisions that support military and political objectives.

Commander's Visualization

> *I heard small-arms fire and RPG explosions and felt shrapnel hit the vehicle.... Land navigation at this time was impossible; every time I tried to look out, I was thrown in a different direction.... At this time, I was totally disoriented and had not realized we were on our own.*
>
> Captain Mark Hollis
> "Platoon Under Fire"

4-60. Leaders at all levels need to see the battlefield to lead Soldiers, make effective decisions, and give direction. Sensors and other surveillance and reconnaissance assets alone cannot provide all the information regarding the urban environment that commanders will need. The focus of lead elements narrows rapidly once in contact with a hostile force limiting their assessment to the local area. Therefore, tactical commanders will not be able to observe operations from long, stand-off ranges. Their personal observation remains as critical in urban areas as elsewhere and helps to preclude commanders from demanding their subordinates accomplish a task or advance at a rate inconsistent with the immediate situation. In urban offensive and defensive operations, seeing the battlefield requires that commanders move themselves and their command posts forward to positions that may be more exposed to risk. Thus, commanders modify their C2 system capabilities to make them smaller, reduce their signature, and increase their mobility. Because of the greater threat to C2, security efforts may be more intense.

4-61. In stability operations, commanders often intervene personally to reassure the urban population and community and faction leaders about the intentions of Army forces. In these type operations, threats may attack leaders to gain the greatest payoff with the least expenditure of resources. Commanders carefully evaluate risk and potential benefits of such exposure. These risks however, cannot stop them from seeing the battlefield, personally intervening in situations as appropriate, and leading their Soldiers.

4-62. Commander's visualization also requires having an accurate understanding of friendly and enemy locations, detailed maps, other appropriate intelligence products, and INFOSYS that accurately depict the urban environment and help establish a COP. The reliability of these items is as important to planning major operations as it is to tactical-level operations. The commander of the major operation ensure that subordinate tactical-level commanders have the necessary products to achieve accurate situational understanding and dominate the urban environment as subordinate commands often lack the personnel or assets to develop these products. Frequently, satellite or aerial imagery is requested to compensate for the drastic changes that can occur due to UO, natural disasters, and outdated or imprecise maps. (Even maps developed and maintained by urban area's administrative activities may not be up-to-date. Extensive and continually expanding shantytowns, for example, may not be mapped at all. Maps may have even been purposefully distorted or critical detail intentionally omitted. The systems used to transliterate some languages such as Arabic and Chinese to Anglicized alphabets often result in the same location being spelled several—and frequently considerably different—ways. Maps may also assign names to features that are completely different than those used by locals to refer to them.)

4-63. Other critical intelligence products needed in the COP include overlays or gridded reference graphics. (Whenever possible, gridded reference graphics should conform to standard military grid reference system formats to reduce the probability of error when entering target coordinates into targeting systems that use global positioning systems.) These products should be developed and distributed to all participants prior to the UO. Overall, their focus should be on ease of reference and usefulness for all forces—ground and air (see Appendix B). Overlays and graphics can also portray important societal information or urban infrastructure, such as—

- Religious, ethnic, racial, or other significant and identifiable social divisions.
- Locations of police, fire, and emergency medical services and their boundaries or zones of coverage.
- Protected structures such as places of worship, hospitals, or other historical and culturally significant buildings or locations.
- Underground subway, tunnel, sewer, or water systems.
- Bridges, elevated roadways, and rail lines.
- Electrical generation (to include nuclear) and gas storage and production facilities and their distribution lines.
- Water and sewage treatment facilities.
- Telephone exchanges and television and radio stations.
- Toxic industrial material locations.

Mental Flexibility

4-64. Commanders conducting UO must remain mentally flexible. Situations can change rapidly because of the complexity of the human dimension. Typical of the change is a stability operation that suddenly requires the use of force. Commanders must be capable of quickly adjusting their mental focus from a noncombat to combat situation. Equally important is dealing with populations during combat operations. Consequently, commanders must also be capable of rapidly adjusting plans and orders for sudden stability or civil support tasks that emerge during or soon after a combat mission. In developing their vision, commanders must consider the second- and third-order effects of UO.

Information Systems

4-65. The urban environment also challenges INFOSYS that support the commander, especially communications. Urban structures, materials, densities, and configurations (such as urban canyons) and power constraints associated with man-portable radios significantly degrade frequency modulation (FM) communications. This causes problems at brigade-level and below where commanders rely heavily on constant FM radio contact with subordinates. Tactical communication problems might also cause an inability to maintain a COP, to give orders and guidance, to request support, or to coordinate and synchronize elements of the combined arms team. Communication problems in urban areas can prevent the achievement of information superiority and contribute directly to mission failure. In UO, allocating critical or high-value communication assets will be significant and essential to the main effort.

4-66. In an urban environment, units and staffs properly prepare for and mitigate the communication problems in urban areas (see figure 4-3). Adequate communications, in most cases, are ensured by—

- Training in and use of retransmission and relay sites and equipment, which may include unmanned aircraft systems (UAS).
- Airborne command posts, satellite communications, high-frequency radios, and other redundant communication platforms and systems.
- Careful positioning of commanders, command posts, and antennas to take advantage of urban terrain characteristics.
- Detailed communications analysis for movement from one AO to another due to the likely density of units operating in the urban environment.

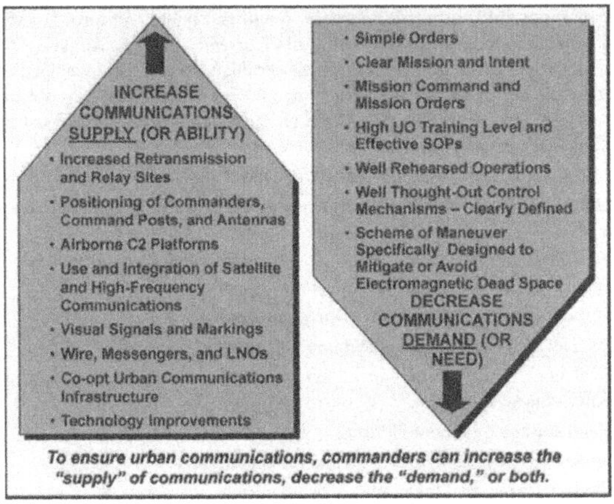

Figure 4-3. Methods to overcome urban communications challenges

4-67. Standing operating procedures (SOPs) for visual markings (both day and night) may assist in command and control. These SOPs indicate unit locations and other essential information. They coordinate with units across common boundaries. Given adequate consideration to limitations on multinational capabilities, these SOPs may assist in command and control and preclude fratricide incidents resulting from loss of FM communications. However, visual signals, including pyrotechnics, are less effective in buildings and enclosed spaces.

4-68. In defensive, stability, or civil support operations, positions often do not change as frequently as in offensive operations. Urban commanders then rely more on military wire (properly camouflaged amongst the civilian communications infrastructure), commercial communications, and messengers. Even in combat, some if not all of the urban area's organic communications structure remains intact for Army use. For example, every building may have one or more telephone distribution boxes that can control hundreds of individual telephone lines. Setting up wire communications using these points is relatively simple but, like all wire communications, is susceptible to wire-tapping. Cellular telephones can usually work well in urban areas; however, locating and destroying the repeater stations or other land-based elements of the cellular telephone system (or the effects of natural disasters) can easily disable them. (Cell phones may be a critical and only means to rapidly communicate with key civilian organizations and important community leaders.) Consequently, the C2 system may use these alternatives to FM communications but with proper operations and physical security procedures in place.

**Example of Simple Communications Innovation:
Israel's Six-Day War – 1967**

In the 1967 battle for Jerusalem, the Israeli Defense Force placed flags on top of cleared buildings so that aircraft providing close air support could monitor the Israeli forward line of troops. They also used a spotlight during the night to mark specific buildings as close air support targets.

4-69. Units will likely use multiple means to communicate throughout the urban area. Hence, commanders emphasize proper operations security procedures (OPSEC) despite the level of security provided by the communications system. This emphasis helps to lessen the probability that Soldiers will inadvertently compromise essential information as they switch from one mode of communications to another (for example, from secure FM radio to unsecured cellular telephones or from classified to unclassified Internet domains).

4-70. Command posts above brigade-level ensure that they can communicate in an urban area without significant disruption. In stability and civil support operations, immediate and reliable communications between tactical and strategic levels may be necessary. Higher commanders anticipate that although the urban area does not significantly challenge their INFOSYS, the area may severely challenge systems at the lower tactical levels. For this reason, information flow from lower to higher may take longer. If the situation is not acceptable, the higher headquarters takes steps to mitigate it, such as increasing the number of liaison officers operating with units engaged in decisive operations. In some instances, the scheme of maneuver may be specifically designed to account for communications interference, propagation characteristics, and electromagnetic dead space. (However, this will require more time, resources, and a detailed urban communications IPB.)

4-71. Finally, urban areas can overload the INFOSYS with information. UO across the spectrum of operations can generate large volumes of information when crises threaten. This sheer volume can easily overwhelm UO commanders and command posts, and the information conduit connecting the two. Training prepares command posts to handle this volume of information and to filter the critical from the merely informative. Staffs must create products (visual or textual) that help their commanders understand the urban environment, not just present them information to know.

KEY TACTICAL CONSIDERATIONS

4-72. Commanders and planners of major UO must thoroughly understand the tactical urban battle. Especially, they must understand the effects of the environment on men, equipment, and systems. The complexity of urban environment changes and often compresses many tactical factors typically considered in the planning process. Figure 4-4 shows some of these compressed factors, however, commanders and their staffs should carefully review FM 3-06.11 for doctrine to support tactical urban operations.

Figure 4-4. Compressed tactical factors

TIME

4-73. The time available to think and act is compressed in urban combat operations. The tactical engagements that comprise battles and major UO are often quick and decisive; therefore, higher-level commanders require the ability to conduct battle command on the move so that their decision making remains correspondingly fast. The impact of decisions (or lack of) and the outcome of battle can occur in mere minutes. Often the amount of information and the number of decisions can overwhelm the overall ability of INFOSYS to respond. Commanders have little time to influence tactical actions with resources kept in reserve. Reserves and fire support assets are close to the point of decision so that they can respond in time to make a difference. The terrain causes C2 challenges that further inhibit commanders from responding quickly to changes in the situation. Small unit leaders receive training that emphasizes understanding the commander's intent so that they can recognize tactical opportunities and can act quickly to take advantage of them.

DISTANCES AND DENSITY

4-74. Distances in UO are compressed to correspond to the density of threat forces and noncombatants. In open terrain, squads, platoons, and companies may be able to control or influence thousands of meters of space. In UO, large buildings can absorb the efforts of several companies or battalions. Crowds of thousands can assemble in areas of a few hundred meters requiring correspondingly large forces for control. Maximum engagement ranges, as influenced by the urban terrain, are usually closer. Units may require field artillery for direct fire at targets ranging fewer than a hundred meters. Commanders and staffs understand the telescoping nature of the battlefield, the density of threat forces, and the density of noncombatants. In addition to the actual conduct of urban tactical operations, these factors will directly affect training, planning, force deployment, and strength.

4-75. Time-distance considerations are especially important throughout planning cycles. Though distances may be short, the physical nature of the environment can drastically change the planning factors for unit movements. The advance of a battalion may be measured in hundreds of meters per day. Thus, all time and distance calculations that relate to sequencing of forces, synchronizing combat power and other capacities, and making decisions require reevaluation based on the urban conditions.

4-76. Although it may appear counterintuitive, airspace above the urban area may be also be severely compressed as multiple fixed-wing, rotary-wing, and indirect fires (including multinational assets) compete for the same space. Increasing this density is the use of airspace by civilian aircraft and the proliferation of tactical UAS. Due to the potential for a high volume of air traffic, commanders and planers pay close attention to the integration and deconfliction of airspace over urban areas. Commanders should consider specific techniques and procedures for UAS which may include—

- Defining airspace control areas.
- Regulating flight times.
- Creating altitude restrictions.
- Including UAS on the air tasking order.

COMBAT POWER

4-77. The urban terrain can also compress combat power. This terrain increases the utility and effects of some weapons and systems, increasing overall combat power. One system that dramatically demonstrates this effect is the sniper. In open terrain, snipers slightly influence operations. In UO, snipers—well concealed, positioned, and protected—can take on significance disproportionate to their combat capability in other situations (see FM 23-10).

4-78. The density of ground combat power in a given size area is also increased because of the effect of the terrain on ranges. The complex terrain precludes standoff engagement from extended ranges by dispersed forces. Commanders often position weapon systems closer together and at shorter ranges to mass effects on the same target. Thus, commanders may position armored vehicles, which typically position themselves Hundreds of meters from friendly troops and other vehicles, within a few meters of each other to provide mutual support. Targets, which in open terrain are engaged at thousands of meters, are engaged in tens of meters on the urban battlefield.

4-79. The dense clutter of the urban environment also affects target acquisition. Systems, such as radar optimized for open terrain, will not be able to acquire targets as effectively. Decreased acquisition capability equates to diminished combat power. It may also require increasing the density of acquisition systems to compensate for reduced capability.

4-80. Finally, the density of combat power may also increase the vulnerability of Army forces. Many Army systems are protected from enemy systems at longer ranges. The number of enemy systems that can threaten Army forces at a short range increases dramatically. Lack of dispersal will make it more likely that multiple Army systems can be targeted by a single enemy threat.

LEVELS OF WAR

4-81. The levels of war are also compressed in the urban area. The tactical actions of individuals and small units can directly influence operational and even national and strategic objectives. Conversely, the decisions of the President can directly affect the conduct of tactical operations. UO have short cause and effect links between the tactical, operational, and strategic levels of operations. Because of the close media scrutiny of UO, the President can sometimes observe the actions of platoons in real time. For example, the media may film a platoon applying nonlethal force for crowd control. The President can view that film on the nightly news before the platoon even disengages from the action, much less reports formally through the various levels of command. If appropriate, the President can decide and direct the strategic and operational commanders to adjust ROE before the platoon has reported. Therefore, commanders at all levels must understand the urban environment's potential compressive effects on the levels of war. A major impact of these effects can be a lower tolerance for tactical errors and a greater need for detailed planning and precision in execution and weapons' effects (lethal and nonlethal).

DECISION MAKING

4-82. The nature of the urban environment compresses the time available to make decisions and increases the number of decisions to make. This is particularly true at the lower tactical levels. Units observing an urban AO face more potential unknowns than in other situations. A large structure presents many more potential firing positions that are observed than simpler terrain. Movement in one of those windows forces the soldier or unit to quickly make a decision regarding the nature of the target—deciding whether it is a threat or a noncombatant. Incorporating combatant and noncombatant discriminatory considerations into all live-fire training will improve Soldiers' ability to make these critical judgment decisions. Overall, lack of understanding regarding the urban environment requires commanders to rely more on analytic decision making, while greater understanding facilitates adaptability and allows commanders to make rapid and intuitive decisions while still accounting for second, third, and higher-order effects (see Figure 4-5 and FM 6-0). Realistically, however, all decisions made during urban operations will likely require a combination of analytic and intuitive decision-making abilities.

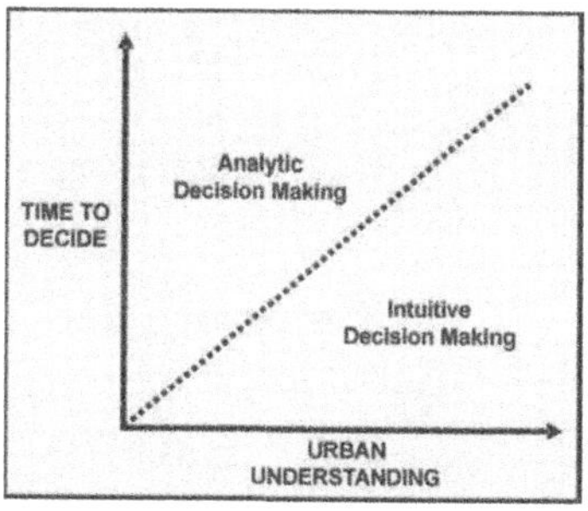

Figure 4-5. Urban understanding and decision making

This page intentionally left blank.

Chapter 5
Contemplating Urban Operations

We based all our further calculations on the most unfavorable assumptions: the inevitability of heavy and prolonged fighting in the streets of Berlin, the possibility of German counter-attacks from outside the ring of encirclement from the west and south-west, restoration of the enemy's defence to the west of Berlin and the consequent need to continue the offensive.

General of the Army, S. M. Shtemenko
describing operational level planning for taking Berlin
The Soviet General Staff at War

In any potential situation and in any area, Army commanders will likely need to assess and understand the relevance and impact of one or more urban areas on their operations. They will also need to determine whether full spectrum urban operations (UO) will be essential to mission accomplishment. UO may be the commander's sole focus or only one of several tasks nested in an even larger operation. Although UO potentially can be conducted as a single battle, engagement, or strike, they will more often be conducted as a major operation requiring joint resources. Such actions result from the increasing sizes of urban areas. Army commanders of a major urban operation must then ensure that UO clearly support the operational objectives of the joint force commander (JFC), requesting and appropriately integrating critical joint resources. Whether the urban operation is the major operation itself or one of many tasks in a larger operation, Army commanders must understand and thoroughly shape the conditions so subordinate tactical commanders can engage and dominate in this complex environment.

A ***major operation*** is a series of tactical actions (battles, engagements, strikes) conducted by various combat forces of a single or several services, coordinated in time and place, to accomplish operational, and sometimes strategic objectives in an operational area.

DETERMINING THE NECESSITY OF URBAN OPERATIONS

5-1. Early in planning, commanders of a major operation must address the necessity and feasibility of conducting operations in urban areas located throughout their areas of operations (AOs). Chapter 1 discussed strategic and operational considerations that compel forces to operate in urban areas. These reasons include the location and intent of the threat force; critical infrastructure or capabilities that are operationally or strategically valuable; the geographic location of an urban area; and the area's political, economic, or cultural significance. Additionally, humanitarian concerns may require the control of an urban area or necessitate operations within it. Several considerations exist, however, that may make UO unnecessary, unwarranted, or prohibitively inefficient. When determining whether to operate in an urban environment, commanders must consider the operational (and accidental) risks and balance them with mission benefits. The factors shown in Figure 5-1 highlight some measures to evaluate the risks associated with UO.

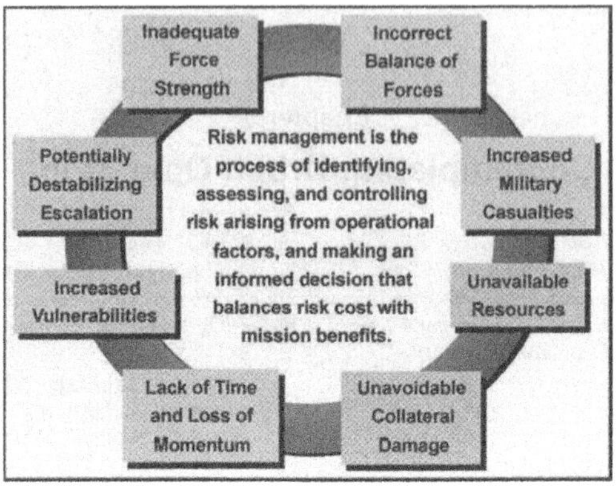

Figure 5-1. Risk management and the associated risks with urban operations

Inadequate Force Strength

5-2. When facing prospective UO, commanders must consider if they have the necessary troops available to conduct the operation properly and within acceptable risk. Under normal circumstances, large urban areas require many forces merely to establish control. The New York City police department has over forty thousand officers and hundreds of support staff simply to conduct peacetime law enforcement. Major UO, particularly those that are opposed, will often require a significant number of forces. Offensive missions, for example, can require three to five times greater troop density than for similar missions in open terrain. If commanders lack sufficient force to conduct effective operations, they may postpone or consider not initiating those operations until they have the necessary strength. Commanders should also add to their analysis the requirements for troop strength elsewhere in the AO. Additionally, commanders must consider the number (and type) of forces required to make the transition from urban offensive and defensive to stability or civil support operations when determining overall force requirements—not just the number of forces required to realize objectives for major combat operations.

Incorrect Balance of Forces

5-3. Along with force strength, commanders must consider the type and balance of forces available. This consideration includes an assessment of their level of training in urban operations. Generally, UO put a premium on well-trained, dismounted infantry units. The operational or tactical necessity to clear threat forces from the dense urban environment, hold hard-won terrain, and interact with the urban population greatly increase dismounted requirements. Therefore, Army forces conducting UO are often force tailored to include a larger infantry component. In addition, special operations forces (SOF) are invaluable in UO. SOF include psychological operations (PSYOP) and civil affairs (CA) forces. They should always be considered as part of the task organization.

5-4. UO must include combined arms to ensure tactical success in combat. Although masses of heavy forces may not be required, successful UO require the complementary effects of all Army forces. Even if an urban operation is unlikely to involve offensive and defensive operations, armor, combat engineers, and field artillery may be essential to mobility and force protection. In urban stability and civil support operations, successful mission accomplishment requires more robust CA organizations. They are also valuable in urban offensive and defensive operations. While commanders may have sufficient combat and combat support forces, they may lack enough sustainment forces to provide the support necessary to maintain the tempo. Again, commanders without balanced types of forces, to include their proficiency in operating in urban environments, should consider alternatives to UO or delaying UO until proper force types are trained and available in sufficient numbers.

Increased Military Casualties

5-5. Casualties in UO are more likely than in operations in other environments. In urban offense and defense, friendly and threat forces often engage at close range with little space to maneuver. The urban terrain provides numerous advantages to the urban defender; higher casualties occur among troops on the offensive, where frontal assaults may be the only tactical option. Conversely, defenders with limited ability to withdraw can also suffer high casualties when isolated and attacked. Casualties can be more difficult to prevent in urban stability operations because of the dense complex terrain, the close proximity of the urban population, and the possible difficulty in distinguishing friend from foe. The potential for high casualties and the subsequent need for casualty evacuation under difficult circumstances make the positioning and availability of adequate medical resources another important consideration. Additionally, high-intensity urban combat and the potential for increased stress casualties may require additional units to allow for adequate unit rotations so that Soldiers receive the rest they require.

5-6. Though casualties occur in all operations, commanders should recognize the likelihood of more casualties during large-scale or high-intensity UO. During the battle for Hue in 1968, for example, many company-size units suffered more than 60 percent casualties in only a few days of offensive operations. Commanders conducting urban stability operations must know the casualty risk and how it relates to national and strategic objectives. While a lower risk normally exists in stability operations than in offensive and defensive operations, just one casualty may adversely impact the success of the stability mission. A realistic understanding of the risk and the nature of casualties resulting from UO critically affect the decision-making process. If commanders assess the casualty risk as high, they must ensure that their higher headquarters understands their assessment and that the objectives sought within the urban area are commensurate with the anticipated risk.

Unavailable Resources

5-7. Offensive and defensive operations in an urban environment put a premium on certain types of munitions and equipment. Forces may want to use vast amounts of precision munitions in the urban environment. At the tactical level, they will likely use more munitions than during operations in other environments. These munitions include—

- Grenades (fragmentation, concussion, stun, riot control, and smoke).
- Mortar ammunition (due to its rate of fire, responsiveness, and high-angle fire characteristic).
- Explosives.
- Small arms.

5-8. Soldiers need access to special equipment necessary to execute small-unit tactics effectively. In urban stability and civil support operations, this equipment may include antiriot gear, such as batons, protective clothing, and other nonlethal crowd-control devices. In urban offensive and defensive operations, special equipment can include sniper rifles, scaling ladders, knee and elbow pads, and breaching equipment such as door busters, bolt cutters, and sledgehammers. Soldiers can conduct UO with standard clothing and military equipment. However, failure to equip them with the right types and quantities of munitions and special equipment will make mission success more difficult and costly. When commanders consider whether to conduct UO, they must evaluate the ability of logistics to provide the necessary resources (see Chapter 10). Considerations must include the ability to supply all Soldiers regardless of branch or military occupational specialty with urban-specific equipment.

Unavoidable Collateral Damage

5-9. UO require an expanded view of risk assessment. When considering risk to Army, joint, and multinational forces, commanders must also analyze the risk to the area's population and infrastructure. This comprehensive analysis includes the second- and third-order effects of significant civil casualties and infrastructure damage. Collateral damage can influence world and domestic opinion of military operations and thus directly affect ongoing operations. It also influences the postconflict physical environment and attitudes of the population. Negative impressions of the civilian population caused by collateral damage can take generations to overcome. Destroying an urban area to save it is not a viable course of action for Army commanders. The density of civilian populations in urban areas and the multidimensional nature of the environment make it more likely that even accurate attacks with precision weapons will injure noncombatants. Unavoidable collateral damage of sufficient magnitude may justify avoiding UO, which, though it may be tactically successful, may run counter to national and strategic objectives.

Lack of Time and Loss of Momentum

5-10. Commanders conducting major operations must analyze the time required to conduct UO successfully. UO can be time consuming and can require larger quantities of resources. The density of the environment, the need for additional time to conduct a thorough reconnaissance, the additional stress and physical exertion imposed on Army forces operating in urban areas, and the potential requirements to care for the needs of the urban population consume time and slow momentum. Commanders cannot permit UO conducted as a shaping operation to divert resources from the decisive operation. Nor can they allow UO to interrupt critical time lines, unnecessarily slow tempo, or delay the overall operation. Threat forces may conduct UO with the primary purpose of causing these effects. Commanders must recognize that time generally works against political and military objectives and, hence, they must develop plans and operations to avoid or minimize UO that might unacceptably delay or disrupt the larger, decisive operation. Once commanders achieve major combat objectives, however, they will likely need to shift resources and focus on the urban areas that they previously isolated and bypassed.

Increased Vulnerabilities

5-11. Commanders must weigh the potential for increased vulnerabilities when executing UO. The density of the environment makes protection (safety, field discipline, force protection, and especially fratricide avoidance) much more difficult. Forces operating in a large urban area increase their risk of isolation and defeat in detail. Joint capabilities, such as air power, may work less effectively to support a close urban battle than in some other environments. Thus, responding to unexpected situations or augmenting disadvantageous force ratios when applying joint capabilities is significantly more difficult. Although organized, trained, and equipped for success in any environment, the Army vulnerability to weapons of mass destruction (WMD) increases when forces concentrate to conduct UO. Commanders may consider not committing forces or limiting the size of a force committed to an urban area because of increased vulnerability to (and likelihood of) attack by WMD.

5-12. Fratricide avoidance is a matter of concern for commanders in all operations. The complex urban terrain and density of participating forces coupled with typical battlefield effects—smoke, dust, burning fires—and weather effects—fog, snow, rain, and clouds—immensely increase the potential for urban fratricide. Additionally, safety requirements make it difficult to fully replicate training conditions that allow leaders to become more aware of the conditions that contribute to fratricide. The effects of fratricide can be devastating to UO and spread deeply within the Army force. Critical effects include—

- Needless loss of combat power.
- Decreased confidence in leadership, weapons, and equipment. These lead to a loss in initiative and aggressiveness, failure to use supporting combat systems, and hesitation to conduct limited visibility operations.
- Disrupted operations and decreased tempo.
- General degradation of cohesion and morale.

5-13. Therefore, commanders should increase fratricide awareness and emphasis on prevention measures during UO. Causes can be procedural, technical, or a combination of the two and include—

- Combat identification failures due to poor situational understanding, lack of communication, failure to effectively coordinate, and short engagement ranges coupled with the need for quick reaction.

- Location errors involving either the target or enemy forces due to poor situational understanding.
- Inappropriate command and control and fire support coordinating measures; a failure to receive, understand, or adhere to these measures.
- Imprecise weapons and munitions effects such as, an antitank round that penetrates several walls before exploding near friendly forces.

Exacerbating these difficulties will be the likelihood of Army forces conducting operations with (or within proximity of) SOF, coalition forces, and indigenous security forces including local police. (During OPERATION IRAQI FREEDOM, the head of the Australian Defence Forces SOF in Iraq made it a point to carefully coordinate with U.S. Army forces—down to the company level—whenever his personnel would be in an American AO. He felt the danger of fratricide was a much greater risk than the loss of operational security.)

Potentially Destabilizing Escalation

5-14. In the urban environment, Army forces cannot avoid close contact with enemy forces and civilians that may potentially become hostiles. In urban stability and civil support operations, commanders should consider the chance of this contact escalating into confrontation and violence, which may become destabilizing. This consideration may delay, limit, or altogether preclude UO using Army forces.

Consider Alternatives and Risk Reduction Measures

5-15. Since UO are often high risk, commanders should consider courses of action that provide alternatives. When the objective of an urban operation is a facility, commanders should consider replicating that facility outside of the urban area. For example, a critical requirement for an airfield to sustain operations may lead commanders to consider UO to seize or secure one located in an urban area. However, if adequate resources exist (especially time and adequate general engineering support), Army forces may build an airfield outside of the urban area and eliminate the need to conduct the urban operation. Similarly, logistics over-the-shore operations may be an alternative to seizing a port facility. In some situations, the objective of UO may be to protect a political organization such as a government. Relocating the government, its institutions, and its personnel to a safer area may be possible. Commanders can also design an operation to avoid an urban area. For example, if an urban area dominates a particular avenue of approach, use a different avenue of approach. Using a different avenue of approach differs from isolating and bypassing because the entire operation specifically makes the urban area irrelevant.

5-16. If commanders execute UO, they must assess potential hazards, and then they develop controls to either eliminate or reduce the risks to Army forces. The first means to offset risk is always to ensure a thorough understanding of the urban environment and its effects on operations by all members of the force. Other measures to mitigate risk may include—

- Detailed planning to include thorough intelligence preparation of the battlefield and the development of appropriate branches and sequels.
- Integrated, accurate, and timely intelligence, surveillance, and reconnaissance (ISR).
- Clear missions and intent, which includes a well-articulated end state that looks beyond the cessation of combat operations.
- Sufficient reserves and rotation of forces.
- Vigilant physical security precautions to include increased use of barriers and other defenses, particularly when urban areas are used as support areas.
- Operative communications and other information systems (INFOSYS).
- Effective populace and resources control measures.
- Comprehensive and flexible rules of engagement (ROE) continuously reviewed to ensure they remain adequate for the situation.
- Sufficient control measures (which often include a common urban reference system) and standard marking and identification techniques that adequately consider limited visibility concerns for both air and ground forces. Measures should allow commanders to satisfactorily

control UO and minimize fratricide without unreasonably restricting subordinate commanders' ability to accomplish assigned missions. Commanders must ensure that all subordinate units thoroughly disseminate any approved nonstandard reference systems.

- Proper targeting procedures (including effective fire support coordinating measures and a streamlined legal review of targets), positive identification of targets, and controlled clearance of fires. The goal is achievement of precise (yet rapid) effects with both lethal and nonlethal means. In close air support, positive air-to-ground communications are essential to coordinate and authenticate markings.
- Well-synchronized information operations (IO) that begin before introducing Army forces into the urban environment and well through transition. Commanders should emphasize vigilant operations security (OPSEC) particularly when operating closely with the media, nongovernmental organizations (NGOs), and elements of the civilian population.
- Active and effective integrating, synchronizing, and coordinating among all forces, agencies, and organizations involved in the operation. Commanders should allow adequate planning and rehearsal time for subordinates.
- Responsive, sustainable, and flexible urban sustainment.
- Forces well trained in joint, interagency, multinational, and combined arms UO.
- The creation of adaptable, learning organizations. This requires thorough after-action analyses conducted during actual operations as well as after training exercises. In addition to official Army sites (such as the Center for Army Lessons Learned and the Battle Command Knowledge System), commanders must create a unit-level system to allow hard-won, lessons learned and tactics developed to be immediately passed on to other units and soldiers—even in the midst of an operation. This system may be technology-based, procedural, or both. (During OPERATION IRAQI FREEDOM, the 1st Cavalry Division developed an effective web-based knowledge network—called CAVNET—that allowed them to actively capture and share lessons learned among subordinate units.)

INTEGRATION INTO LAND OPERATIONS

5-17. The commander of the major operation, after determining that urban operations are required, will then integrate the urban operation into his overall operation. He does this by articulating his intent and concept for the urban operation to his subordinates. The commander of the major operation must also set the conditions for successful tactical urban operations by his subordinates. He should define ROE, focus ISR efforts, task organize his capabilities, ensure information superiority, design the operational framework, and initiate and sustain effective coordination with other agencies and organizations (see FM 6-0).

CONCEPT OF THE OPERATION

5-18. The commander's concept of the operation should address all operationally important urban areas in his AO. It should also articulate his vision of the urban operation through directions to his staff and subordinates. Subordinate commanders should address urban areas that the higher commander does not specifically address. The commander's concept should discuss each urban area in terms of task and purpose (see FM 5-0). The commander should also describes his vision of the situation's end state in terms of—

- The threat.
- The urban environment (terrain, society, and infrastructure).
- Friendly forces.
- The conditions necessary to transition control of urban areas within his AO to another agency or back to legitimate civilian control.

RULES OF ENGAGEMENT

5-19. National- or joint-level command authorities may develop urban-specific ROE. If not, Army commanders, as part of their assessment, must determine if urban-specific ROE are required for their situation and provide supplemental ROE. However, commanders must forward any conflicts or incongruities to their higher headquarters for immediate resolution.

5-20. Developing effective ROE relies on thoroughly understanding the national and strategic environment and objectives. It also relies on understanding how to conduct urban operations at the tactical level including weapons effects. For example, broad ROE may result in significant collateral damage and civilian casualties. Even in a major operation or campaign, significant collateral damage caused during UO can make postcombat operations difficult. Such damage may even change national and international public opinion or threaten the achievement of national and strategic objectives. In contrast, restrictive ROE can hamper tactical operations causing mission failure, higher friendly casualties, or both. ROE are often part of essential elements of friendly information (EEFI), protected to reduce the potential for threat exploitation. Even in a limited urban operation, ROE will frequently need to change as circumstances warrant. Therefore, commanders should plan ROE "branches" for anticipated changes in the operational environment.

5-21. In urban operations, ROE are flexible, detailed, and understandable. They should preclude the indiscriminate use of deadly force while allowing soldiers latitude to finish the mission and defend themselves. ROE should recognize that the urban area is not homogenous and may vary according to the key elements of the threat and environment: terrain, society, and infrastructure. To be effective, ROE are consistent throughout the force (an increased challenge in multinational urban operations), and soldiers are thoroughly trained and familiar with them.

Enemy Considerations

5-22. The nature of an urban enemy affects ROE as well. Commanders must consider the type of enemy weapon systems, the degree of defensive preparation, the ability to target enemy vulnerabilities with precision systems, and the ability to distinguish combatant from noncombatant.

Terrain Considerations

5-23. ROE may vary according to the terrain or physical attributes of an urban area. Physical factors may drive the ROE to preclude certain types of munitions. For example, if the construction of a portion of the area is sensitive to fire, then ROE may preclude using incendiary munitions in that area. The ROE may lift this prohibition when units move into areas of masonry construction. Toxic industrial chemicals or radiological contaminants in an industrial area may also affect ROE.

Societal Considerations

5-24. The societal or human dimension of the urban environment will often affect ROE the most. Commanders must base the ROE development on a thorough understanding of the civilian population and threat. They evaluate the loyalty of the population, its dynamic involvement in activities that affects the operation, and its size and physical location. A population that is present and supports Army forces will likely elicit more restrictive ROE than a hostile population actively supporting forces opposing the Army forces. A neutral population, not actively involved in behavior affecting Army forces, supports consideration of more restrictive ROE. In all cases, ROE conforms to the law of war. However, ROE may be much more restrictive than the law of war requires.

5-25. The location of the population also affects ROE. The evacuation or consolidation of noncombatants into controlled, safe areas may result in less restrictive ROE. A U.S. or allied population that remains in the urban area conducting routine business in and amongst Army forces during noncombat UO will normally require the most stringent ROE.

Infrastructure Considerations

5-26. Commanders must consider the urban infrastructure when developing ROE. Urban infrastructure vital to current or future Army operations may dictate that commanders adjust ROE to ensure that critical elements of the infrastructure remain intact during the conduct of operations. If Army forces conduct an urban operation to capture port facilities, for example, the ROE should address damage to the key facilities that are the objective of the operation.

RESOURCE ALLOCATION

5-27. Commanders of a major operation must ensure that subordinate tactical commanders have the resources necessary to conduct UO effectively. They assign appropriate forces to subordinate commanders tasked to conduct UO; support them with Army forces at the operational level; and request and coordinate their support by joint resources.

Task Organization

5-28. Task organizing subordinate units for urban operations will depend largely on the nature of the operation. Some units, however, are always part of the task organization to ensure the success of UO. Infantry, CA, aviation, military police, PSYOP, military intelligence, and engineers are units required for all urban operations across full spectrum operations. Other type forces—such as armor, artillery, and chemical—have essential roles in specific types of urban operations but may be less applicable across the spectrum of operations. Commanders and staffs of a major operation must understand their mission, the particular urban environment in which they operate, and the general effects of the environment across the warfighting functions to allocate the appropriate forces to their tactical commanders. See Chapter 4 for details.

Operational-Level Support

5-29. Commanders of a major operation should also support the tactical commander with forces remaining under their direct control. These forces can include Army SOF, such as CA, PSYOP, and Special Forces, ground and air cavalry, aviation, logistics, engineers, and communications support. These forces may not be under operational control of the supported command, but their efforts are synchronized and coordinated.

Coordinating and Requesting Joint Support

5-30. Commanders of a major operation will often provide forces to the JFC as well as receive assets. They will also coordinate for and integrate joint assets to support the tactical battle. These assets will usually include air support, such as close air support, tactical airlift, and aerial reconnaissance and surveillance. Intelligence support comes in the form of reach to strategic and national intelligence capabilities and to space-based systems. This reach to space assets provides reliable, robust long-range communications, environmental monitoring, ISR, positioning and navigation, and warning of enemy missile launch. Special operations capabilities can assist the tactical mission with special reconnaissance and direct action (including special operations aviation) against high-payoff targets. Joint resources also provide the Army forces augmentation by Marine ground forces. In coastal areas, Navy forces and Coast Guard elements assist Army forces with security, sealift, and fire support. Additionally, the potential agility, reach, and power projection provided by Navy forces and platforms, combined with their self-sustaining capability, can play a vital role in supporting UO. This is especially true when UO are conducted in the littorals, require forcible entry operations, and when host nation infrastructure is severely taxed or destroyed. Commanders of a major operation will coordinate with the JFC regarding available joint resources and their allocation. They then ensure that their efforts coordinate with and complement those of tactical Army forces in the urban area. Appendix D discusses the potential contribution of joint capabilities to Army UO.

URBAN ISR

5-31. Commanders at all levels require accurate and timely information to conduct assessments for successful urban operations. This is critical to planning and execution. Senior commanders have a large role in coordinating the urban ISR effort. National strategic sources (as well as open sources) provide most of the information that commanders and staffs require on the characteristics of the human dimension, the physical properties of the terrain, and the infrastructure. The general characteristics of these aspects of the urban environment do not change drastically over time, with one exception. Military operations or natural disasters can change physical characteristics radically. Analysts can obtain crucial information through diligent research of intelligence databases and open sources. However, the disposition and composition of the urban threat is time sensitive and not likely to be discovered through this type of investigation. Due to the effects of the urban environment, deceptive efforts may influence the threat more easily. The urban population is dynamic and critical aspects of this dimension must be updated or confirmed as a prelude to urban operations. Surveillance and reconnaissance (including geospatial assets) provide relevant information regarding threat dispositions, composition and the state of the population, and the specifics of the urban terrain. Successful urban operations depend on the successful conduct of urban reconnaissance. (See also the earlier discussion of effects on the intelligence and command and control warfighting functions in Chapter 4. While counterintelligence (CI) is an essential ISR component, it will be covered later in this chapter under IO.)

Challenges

5-32. The most significant challenge to urban ISR is physical. The physical organization and complexity of the urban terrain, both man-made and natural, challenges national strategic, operational, and tactical ISR capabilities. Commanders should understand the challenges when planning and allocating time and resources to their ISR efforts. They acknowledge that subordinate commanders will face similar challenges. Therefore, commanders consider subordinate capabilities, limitations, and needs when planning, requesting, allocating, and prioritizing ISR assets and capabilities.

Imagery Capabilities

5-33. A significant national, strategic, and operational ISR capability is imagery. Even at the tactical level, units employ unmanned aircraft systems (UAS) and vehicle and dismounted electro-optics, commercial cameras (still and video), and co-opt existing civilian surveillance systems to gain situational understanding. However, the structures of the urban area significantly degrade the information that imagery acquires and may make it susceptible to physical deception measures.

5-34. Imagery, however, is an excellent source regarding the arrangement and nature of many other physical aspects. It can be used to create updated mapping products and provide significant detail of major portions of the infrastructure. Imagery can also reveal what may be happening in structures through detailed study of patterns and other exterior indicators. Yet, the bulk of a skillful threat's forces, well positioned and concealed inside or underneath structures in the urban area, are largely immune from rapid detection by overhead imaging systems. The volume of civilian movement in an urban area will itself provide a degree of camouflage and increase the difficulty of employing pattern analysis. The success in 1999 of the Yugoslavian army concealing heavy forces when confronting NATO indicates the limits of these assets to penetrate an urban area. This leaves a requirement to supplement overhead imagery in the urban environment with a reliable vehicle-mounted or handheld imaging system. For example, thermal sights on a vehicle patrolling an urban street during limited visibility may note the hot engine of a vehicle on the side of the road, possibly indicating suspicious activity. Simply providing patrols with digital cameras or video cameras can also help provide needed information and fill in information gaps.

5-35. The requirement for high altitude and low oblique imagery in more open terrain has been supplemented by a growing need for detailed, close-up, and nearly vertical overhead shots of specific portions of the urban area. This forces collection managers and system operators to move expensive and scarce imagery collection systems closer to the target or to employ flight profiles that place them at greater risk from threat air defense systems. While gun-camera images from aircraft can provide a stand-off reconnaissance platform, UAS imagery may be one of the fastest, least risky methods by which commanders can conduct reconnaissance to update and verify current maps of the urban area and determine routes, obstacles such as damaged and destroyed buildings, and intact and destroyed bridges and other significant infrastructure. Increased UAS proliferation will require greater attention to airspace command and control procedures as well as the movement and protection of ground control stations.

5-36. As the media will likely be present in the urban environment, commanders may consider devoting intelligence resources (to include linguists) to monitoring the various local, national and international sources, particularly the medium of television. The media coverage of threat forces may provide important intelligence and relevant information of immediate tactical value. Commanders, however, should always be cognizant that threats may recognize that Army forces will be using this as a source of intelligence and, feeling no compulsion to be truthful, may use it for their own deception means.

Electronic Capabilities

5-37. The physical attributes of the urban area also diminish the effectiveness of electronic ISR capabilities. Buildings and other structures significantly disrupt radio communications in an urban area. Buildings not only make tactical radio communications difficult for the user, they also make them difficult to locate, intercept, and jam. The range and clarity of frequency modulation (FM) signals significantly diminish when antennas are located inside buildings or when buildings block line of sight between the source and receiving station. To mitigate this effect on signals intelligence means, detection capabilities often move closer to the transmission source. Thus, the threat's vulnerability to compromise by means of his FM and other wireless communications in an urban environment is much less than in many other environments.

5-38. Without losing tactical surprise and increasing risk, units cannot effectively use many electronic detection and surveillance capabilities until urban combat is imminent or perhaps already begun. Ground surveillance radars will have limited uses in urban areas because of the lack of wide-open spaces in which they best operate, but can be used along the periphery of urban areas to detect infiltration and exfiltration of threat forces. However, remote systems such as the remotely monitored battlefield sensor system and the platoon early warning device may have great utility in monitoring many of the numerous avenues of approach that cannot be kept under human surveillance due to manpower constraints. Among many urban applications, remote sensors and early warning devices can be used to monitor subsurface passageways, the entrances and exits to buildings, fire escapes, the perimeters of sustainment bases, and covert foot trails between urban and rural areas.

Human Capabilities

5-39. The limits on imagery and electronic ISR capabilities place a premium on human-based visual reconnaissance. Commanders have three types of human reconnaissance assets to augment electronic reconnaissance resources: special reconnaissance, conventional combat reconnaissance, and human intelligence (HUMINT) gathered by military intelligence from individuals and multimedia. The urban environment poses several challenges to these capabilities.

5-40. The urban area challenges special reconnaissance in several ways. First is the access to the urban area. Although avenues of approach may be numerous, concealed avenues of approach into a defended urban area may be limited and thoroughly covered. Air access is also more difficult because aircraft are detected more easily, airspace is smaller, drop and landing zones are limited or not secure, and more air defense systems probably exist. Still, special reconnaissance efforts to penetrate the urban area can be successful using unconventional techniques including high-altitude low-opening parachutes or underwater penetration.

5-41. Special reconnaissance then faces a second challenge: moving in and identifying targets in the urban area. Stealth movement in an occupied urban area is exceptionally difficult. Repositioning to new or alternate positions is also dangerous. The soldiers' ability to conceal themselves among the civil population can mitigate some of these challenges but includes inherent risks of a different nature. Also difficult is establishing observation positions that provide a field of view of several targets.

5-42. Lastly, special reconnaissance may face navigational and reporting challenges. Special reconnaissance's ability to locate themselves and communicate critical locations and routes are challenged by—

- Differences in language and numbering systems. Road signs, if they exist at all, may be removed or deliberately altered to confuse.
- Irregular street patterns.
- Outdated maps.
- Intervening structures that impede communications and global positioning systems.
- Changes to the landscape due to the effects of UO or natural disasters.
- Featureless shantytowns.

5-43. Conventional reconnaissance faces many of the same challenges as special reconnaissance. Conventional reconnaissance also may lack the advantage of surprise and the special equipment and training that provides special reconnaissance stealth capability. Conventional reconnaissance is not likely to operate undetected by the civilian population. Given the constraints discussed above on other sources, conventional reconnaissance units will likely begin their mission with much less information than they would have on threat dispositions in a less complex environment. Commanders may choose to have their reconnaissance elements fight for information in the urban area. While this high-risk option is more favorable under fluid conditions, it can be used at any time. It requires careful planning, rehearsal, and formulation of information requirements.

5-44. As previously indicated, human intelligence will be one of the most valuable sources for information regarding the situation inside an urban area. HUMINT may take advantage of the proximity and large numbers of potential informants to gather information about threat activities and capabilities. It is especially valuable because it can address all elements of the environment. HUMINT sources can describe political and religious nuances that may be significant to commanders. Such information is useful for insights regarding the human dimension but extremely difficult to obtain from other means. This intelligence also can describe the infrastructure uncovering essential details of how the infrastructure functions. Obtaining good HUMINT requires skilled interrogators and linguists—often in large numbers. Commanders must know and account for some of the possible shortcomings of HUMINT:

- It is susceptible to the influence of the threat; the threat can threaten and influence the source as well as local interpreters. Commander may need to consider that people may not want to talk through a local interpreter because they are providing information they might not want others to know; anonymity may be essential. On the other hand, a local interpreter may provide community access. Yet, those trusted by the locals may not be truthful with Army forces. Conversely, those truthful with Army forces may not be trusted by the indigenous population particularly if these interpreters come from another country, region, or demographic group. In such cases, commanders may need more than one interpreter.
- It is limited by the accuracy of the source and interpreter's perceptions. A local interpreter may have difficulty understanding and translating military topics or the intricacies of urban infrastructure. Some local translators may want to please both participants, so they may shade the conversation, telling each side what they want to hear. Finally, accuracy may be affected by the collector's own perceptions if not firmly grounded in an understanding of social, cultural, and other local conditions.
- Multiple echelons collecting from a given area may inadvertently collect from the same sources and thereby give the false impression of independent sources confirming the same data. To help mitigate this possibility, commanders should ensure that all HUMINT sources are properly recorded on intelligence source registers.
- It may not be timely. The process of identifying and cultivating a source (particularly in an environment where most civilians support threat forces), gathering information, analyzing the information, and providing the intelligence to commanders can be extremely time consuming.
- Some informants may come from unscrupulous or sordid elements of the urban society and may also have their own agenda. They may attempt to use protection afforded them by their relationship with Army forces to conduct activities (even atrocities) that will compromise

political and military objectives. In some cases, a local interpreter's own agenda may influence his translation. (For example, a local schoolteacher used as a translator may report to the collector that the source believes that repairing schools in their urban area is critical, when actually the source may have said that potable water is most needed.)

Conducting Urban ISR

5-45. To be successful, ISR efforts (national to tactical level) are exceptionally comprehensive and synchronized. Moreover, success necessitates the integration of all ISR sources into operational and tactical planning. This requires that ISR assets deploy and execute early, diversify, properly focus, and integrate under a comprehensive ISR plan. Successful ISR also requires flexibility to adapt to the operational and tactical needs of the commander (see Figure 5-2). Commanders must ensure that the appropriate intelligence architecture (including robust links to joint and multinational elements) is established and tested prior to execution of ISR effort.

Early Deployment

5-46. One of the first requirements for effective urban ISR is the early deployment and employment of assets. The complex urban terrain presents a significant challenge. Commanders should consider that ISR assets will normally take longer to gather data amid the complexity.

5-47. Limited national, strategic, and operational imagery intelligence (IMINT) and signals intelligence (SIGINT) capabilities should be considered and quickly requested. If they are approved, they are tasked and deployed or repositioned to begin urban ISR operations. This takes time. Fortunately, spacing the ISR effort over time permits the analysis of the information or data as it is received. Such time also permits subsequently refining the ISR effort before all assets are committed.

5-48. SOF or conventional units will require significantly more time to execute reconnaissance missions and maintain an acceptable survivability rate. Urban reconnaissance operations require additional time for stealthy insertion into the urban area. IMINT and SIGINT capabilities are used to identify possible locations of high-value targets and corresponding observation positions; this helps minimize time-consuming and high-risk repositioning in the urban area. Again, reconnaissance units may require extensive time to observe from observation positions for indicators of threat activity and disposition and identify patterns.

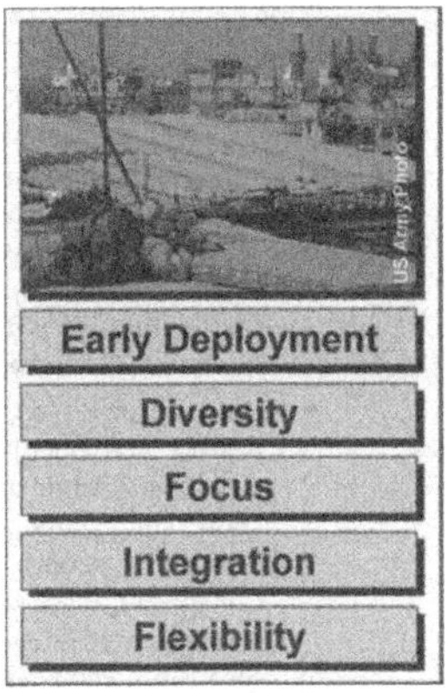

Figure 5-2. Urban ISR considerations

5-49. As conventional combat forces prepare to commit to the urban area, conventional reconnaissance precedes their actions. Conventional reconnaissance will often be a slow and methodical effort. Such forces may need time to reconnoiter the interior of structures for snipers and other small threat teams. They may also need time to deploy and destroy snipers and small delaying elements and to breach harassing obstacles. If necessary, they may need time to mass the combat power necessary to fight through enemy security and continue the reconnaissance.

Diversity

5-50. No single ISR capability can solve the riddle of the urban defense. In order to gain an accurate common operational picture of the complex urban terrain necessary to focus combat power on decisive points, commanders must employ diverse ISR capabilities. These capabilities will each contribute pieces of relevant information to permit identifying operational objectives and leveraging tactical combat power to achieve those objectives quickly. Higher-level commanders should understand that tactical reconnaissance capabilities alone often cannot provide all the tactical information required for success at lower echelons.

5-51. Using diverse capabilities challenges the threat's ability to conduct deception and otherwise defeat the friendly ISR effort. A threat that focuses on minimizing his vulnerability to overhead imagery (for example, satellites) may increase his reliance on communications and thus his vulnerability to SIGINT. At the same time, he may decrease his ability to detect the actions of ground reconnaissance units. Conversely, a threat that actively campaigns to detect ground reconnaissance may make himself more vulnerable to SIGINT and IMINT. Moreover, as Army forces enter urban areas, threat activity and susceptibility to SIGNINT often increases.

5-52. Diverse capabilities also facilitate the tactical ISR effort. Tactical reconnaissance units often consist of small dismounted teams and small combined arms teams with a dismounted element and an armor-protected mounted element. Combat engineers with their technical expertise and breaching capability are essential to the combined arms reconnaissance effort. The teams' movements are synchronized and coordinated with other assets, such as UAS and air cavalry reconnaissance. These teams use several movement techniques including infiltration, with the primary objective of conducting zone reconnaissance along key axes that support brigade and battalion actions against decisive points. To accomplish this mission, reconnaissance reconnoiters the proposed routes and alternate approaches. This supports deception and contingency planning. (Reconnaissance of alternate routes and objectives also applies to aerial reconnaissance; helicopters and UAS are not invisible. They can alert a threat to impending operations.) Infiltration of dismounted reconnaissance is made easier when a threat focuses on combined arms reconnaissance teams. Aerial reconnaissance, such as air cavalry and UAS, provides early warning of threat elements to ground reconnaissance, identifies obstacles and ambush sites, and helps select the routes for ground reconnaissance. Air elements may also reduce the mobility of counterreconnaissance forces.

Focus

5-53. Another key to successful ISR is the ability to focus the assets on commander's critical information requirements (CCIR). This focus begins with mission analysis and the commander's initial planning guidance. It is incrementally refined throughout planning and execution as each ISR effort provides information and permits more specific focus in subsequent efforts. The size and complexity of the urban environment require that the ISR effort center strictly on decisive points or centers of gravity (COGs). Therefore, the overall ISR effort will have two major focuses. The first is to uncover and confirm information on the decisive points and COG. The second is to determine the approaches (physical, psychological, or both) leading to the decisive points. The first focus will likely drive ISR in support of major operations. The second focus will likely provide the impetus for tactical ISR efforts. For example, special operations reconnaissance might focus on a major command center that controls the entire urban area and that is one of the higher-echelon command's CCIR. Tactical reconnaissance might focus on the nature of the defense along a particular avenue of approach to the objective.

Integration

5-54. Another important aspect of urban ISR is integration. All reconnaissance capabilities provide both distinctive information as well as information that confirms and adds to that coming from other sources. Essential to urban ISR is the link between all of these sources, either directly or through an integrating headquarters.

5-55. ISR operations must be vertically and horizontally linked. Vertical links ensure that ISR operations among the various levels of command are complementary and that the information flow between these levels is rapid. Horizontal links ensure that forces operating in close proximity (particularly adjacent units), where areas of interest overlap, can rapidly share results of their individual ISR efforts. Together, this helps ensure that all Army forces share a common operational picture and permits the greatest flexibility and survivability of ISR resources.

5-56. ISR operations also are integrated into the planning system, especially the targeting process. As part of targeting, positioned reconnaissance and surveillance elements may become the trigger and terminal control for applying precision fires when appropriate and after considering the risks of compromise of the position or platform. ISR operations and plans must also be synchronized with the air tasking order, as many surveillance platforms are coordinated on the same timeline.

Flexibility

5-57. The urban ISR effort must be more flexible than in other operations. This flexibility permits the ISR effort to meet unforeseen circumstances and to deal with the challenges of the urban environment. As indicated previously, the urban environment is particularly difficult to penetrate. The practical effects of this characteristic are that—

- The initial ISR effort may not be as successful as in other operations.
- More intelligence requirements may be discovered later while executing ISR operations than otherwise.
- The threat may be more successful in active counterreconnaissance because of the concealment advantages of the urban environment (hiding in structures as well as among the urban population).

Therefore, tactical and operational commanders should consider requesting greater than usual ISR support from higher headquarters. Higher headquarters should be proactive in augmenting units conducting urban operations with additional ISR assets. Additionally, ISR assets remaining under the control of the higher headquarters must respond more quickly to the CCIR of supported commanders. Sequencing reconnaissance missions over time provides flexibility by creating uncommitted reconnaissance assets.

5-58. Time sequencing of ISR assets is essential to flexibility. It makes ISR assets more survivable and allows the intelligence cycle to mature the CCIR. It also creates a ready ISR capability to augment committed forces in critical areas if required or diverts them around centers of threat resistance. If not required, the original ISR tasks can be executed as envisioned in planning. Cueing allows a high-value ISR asset to be able to respond to multiple targets based on an ongoing assessment of the overall reconnaissance effort and the changing CCIR. Redundancy permits the effort to overcome line of sight restrictions, the destruction of an ISR asset, and the ability to combine ISR resources to create combat power if required. Maximizing the ISR effort requires applying all available ISR assets to support the urban operation. Additionally, assets—such as air defense artillery and field artillery radars and engineer reconnaissance teams—are integrated into the ISR effort. In urban operations, units will also commit infantry and armor elements (plus their organic reconnaissance elements) into the tactical reconnaissance effort. These units increase the dismount capability and the ability of reconnaissance elements to fight for information and fight through security zones.

Information Operations

5-59. Information operations are an integral part of all Army operations and a critical component in creating and maintaining information superiority. The information environment is the sum of individuals, organizations, or systems that collect, process, and disseminate information; it also includes the information itself. In UO, the information environment is extremely dense due to the proliferation of INFOSYS and widespread access to those systems. In urban operations, commanders must consider how the urban environment, particularly the human component, uniquely relates to executing IO. They must also anticipate the threat's information operation campaign and preclude successful operations on its part. Overall in regard to the urban society, commanders must convince the urban inhabitants of the inevitable success of UO, the legitimacy of Army actions, and the beneficial effects that Army success will eventually generate.

5-60. Because urban operations are likely to be joint, interagency, and multinational, commanders must guard against various forms of information fratricide as a multitude of actors plan and conduct IO in the urban area. For example, the lead governmental agency's key IO objective may be to establish the legitimacy of newly formed civilian authorities. In an attempt to develop close relationships with the civilian populace, Army commanders may continue to work closely with traditional, informal leaders to the exclusion of the new authority. These actions, while they are often conducted out of practical and immediate necessity, may run counter to the lead agency's goal. Overall, Army commanders must nest their IO campaign objectives and themes within those of the lead agency, aggressively coordinate with other governmental agencies and coalition partners, and synchronize activities down to the tactical level to prevent working at odds and avoid information fratricide. However, the process must not be so centralized and rigid that subordinate units lose the flexibility necessary to develop their own products and themes essential to addressing their own AO and the particular situation that confronts them.

5-61. IO are executed using core and supporting elements and related activities (see Figure 5-3 and FM 3-13). The elements of IO are employed in either an offensive or defensive role. Many aspects of IO are not affected differently in an urban environment from any other environment. The following sections outline some IO considerations unique to urban operations.

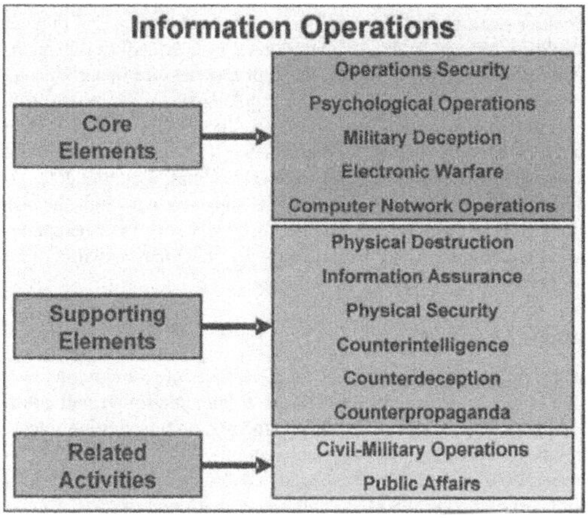

Figure 5-3. IO elements and related activities

Operations Security

5-62. In the urban environment, Army forces can leverage existing urban infrastructure, including the communications and information infrastructure, to enhance Army operations. The danger in integrating these systems is violating OPSEC. Commands ensure that Army forces use only approved systems and proper safeguards exist. Commands also supervise subordinate units for inadvertent breaches of OPSEC policies when using existing urban systems. Finally, established OPSEC procedures must be reasonable and practical. Overly complicated and stringent procedures increase the likelihood that unintentional OPSEC violations may occur as Army forces seek to rapidly accomplish the mission.

5-63. Of particular concern are computers (see also the discussion of Computer Network Operations below). During longer-term UO, commanders will constantly be upgrading and improving the quality of life for their soldiers through the establishment of well-resourced forward operating and sustainment bases. This will likely include the provision of unclassified access to the Internet and e-mail allowing Soldiers to communicate with friends and families around the world. E-mail addresses, facsimile numbers, cell phone numbers, photographs, and other sensitive (but unclassified) information can be valuable sources of threat information. Soldiers must be periodically trained and reminded of everyday procedures to combat potential threat exploitation of these unclassified sources (such as not opening e-mails from unknown sources as they may harbor Trojan programs and viruses).

5-64. To compound concerns, commanders will also establish protected systems to disseminate tactical lessons learned to individual soldiers. This creates the potential that Soldiers can inadvertently release classified information about ongoing operations to threat intelligence systems. In addition to OPSEC training and persistent awareness, commanders must establish procedures to shutdown unclassified Internet and e-mail access immediately, both before and after critical events, to mitigate potential OPSEC violations. (Commanders may also need to randomly shutdown Internet and e-mail access as part of their overall OPSEC measures to preclude such information blackouts serving as a potential warning of impending operations.) Technical surveillance countermeasures—the identification of technical collection activities conducted by threat intelligence entities—will be an important CI service to identify this and other technical OPSEC concerns.

5-65. The close proximity of Army operations to a civil population, particularly in stability and civil support operations, makes Army activities themselves an additional OPSEC concern. Hostile civilians or other threats integrated into the urban population may have more chances to observe Army activities closely. Such observations can provide insight into tactics, techniques, and procedures (TTP) and expose operational vulnerabilities. However, threats may coerce even friendly civilians to provide a threat with EEFI. Therefore, commanders in an urban environment must ensure that civilians cannot observe critical TTP. Any observable patterns and TTP vary and are supplemented with deception efforts. Physical security is increasingly important in urban areas to control civilians' access. Although many urban operations require close coordination with NGOs, commanders should screen information provided to these organizations to protect EEFI. Release of EEFI to NGOs is controlled and done with full recognition and understanding of potential consequences—the benefits must far outweigh the risks involved. (Even the best-intentioned NGO might inadvertently compromise security. For example, by revealing that a commander has closed a particular route the next day, the NGO may unwittingly reveal the objective of a pending offensive operation.)

Psychological Operations

5-66. PSYOP aim to influence the behavior and attitude of foreign audiences, both military and noncombatant, in the urban environment. PSYOP are a force multiplier and contribute in many ways to mission success (see FM 3-05.30). Effective PSYOP focuses on transmitting selected messages to specific individuals and groups in order to influence their actions. Their ability to influence the attitudes and disposition of the urban population cannot be overstated. While the complexity of the societal component of the urban environment can make PSYOP challenging, the urban society also offers many options and resources. Potentially, PSYOP (with other political and economic actions) may help limit or preclude the use of military force in urban areas. In some circumstances, UO may be relevant to the major operation only in terms of their psychological effect.

5-67. The positive influence created by PSYOP is often useful in developing an effective HUMINT capability particularly in an urban area where many civilians actively or passively support the threat. Persuading and influencing a few to support friendly forces may pay great dividends. These few supporters may allow Army forces to penetrate the urban area and obtain essential information. Such information can apply to threat capabilities, threat intentions, and even the urban environment itself.

5-68. PSYOP, combined with other elements of offensive IO, aid in isolation of a threat—a critical shaping action for any urban operation. For example, commanders may use PSYOP to inform civilians about new food distribution points located away from urban combat operations. This action supports the UO fundamental of separating combatants from noncombatants and helps to further isolate the threat (both physically and psychologically) from the civilian populace. Aside from projecting a positive image of friendly forces over threat forces, PSYOP can also isolate the threat by identifying and exploiting ethnic, cultural, religious, and economic differences between the elements of the civilian populace and threat forces as well as the differences among supportive and unsupportive civilian factions. The complexity of the urban environment enables quick changes in opinion or attitude. Commanders must continually evaluate the results of PSYOP for mission relevance.

Military Deception

5-69. Urban operations present numerous challenges to tactical commanders; however, higher-level commanders may help to mitigate some challenges through effective military deception Commanders can use military deception efforts designed to mislead threat decision makers as to friendly force disposition, capabilities, vulnerabilities, and intentions. Military deception actions may allow commanders to achieve tactical surprise or improve relative combat power at a selected location. For example, allowing the threat to observe certain activities on a selected avenue of approach may cause the threat to shift his forces (and effort) to the area perceived to be threatened. (This movement may also aid in determining the overall disposition of threat forces and intentions.) Repositioned forces or effort to activities or locations that are not decisive to the achievement of friendly objectives, combined with other IO designed to overwhelm threat information and intelligence systems, may create the force and tempo differential necessary to achieve success. Commanders should tailor urban deception plans to the specific urban area, paying close attention to the societal characteristics of the target population.

Electronic Warfare

5-70. Electronic warfare (EW) includes all actions that use electromagnetic or directed energy weapons to control the electromagnetic spectrum or to attack a threat. Conducting EW in urban areas seeks to achieve much the same results as in other environments. A major consideration in urban areas is collateral effects on portions of the urban infrastructure that rely on the electromagnetic spectrum for service. Thus, precision is a major factor in planning for EW operations. For example, EW attacking a threat's television broadcasts avoids affecting the television broadcasts of neutral or friendly television. Likewise, EW attacking military communications in a large urban area avoids adversely affecting the area's police and other emergency service communications. Urban offensive and defensive operations will have the least restrictions on EW operations while urban stability may have significant constraints on using EW capabilities.

Computer Network Operations

5-71. Computer network operations (CNO) include computer network attack (CNA), computer network defense (CND), and computer network exploitation (CNE). CNO are not applicable to units at corps and below. Echelons above corps (EAC) units will conduct CNA and CNE. If tactical units require either of these network supports, they will request it of EAC units.

Computer Network Defense

5-72. In urban operations, CND will require extreme measures to protect and defend the computers and networks from disruption, denial, degradation, or destruction. The nature of the urban environment and configuration of computer networks provides the threat with many opportunities to interdict local area networks (LANs) unless monitored by military forces. LANs controlled by military forces are normally more secure than the civilian infrastructure. Commanders should prepare for opportunities by the threat to insert misinformation.

Computer Network Attack

5-73. Considerations regarding the execution of CNA in urban operations are similar to those of EW: CNAs that do not discriminate can disrupt vital civilian systems. However, possible adverse effects on the civilian infrastructure can be much larger—potentially on a global scale. In the short term, CNAs may serve to enhance immediate combat operations but have a debilitating effect on the efficiency of follow-on urban stability operations. Because of these far-reaching effects, tactical units do not execute CNA. CNA is requested of EAC units. EAC units will receive all requests from lower echelons, carefully consider second- and third-order effects of CNA, and work to ensure its precise application.

Computer Network Exploitation

5-74. CNE consists of enabling operations and intelligence collection to gather data from target or adversary automated INFOSYS or networks. Tactical units do not have the capability for CNE. CNE contributes to intelligence collection at EAC. In UO, CNE will be centrally controlled.

Physical Destruction

5-75. Physical destruction includes those actions—including direct and indirect fires from air, land, sea, space, and Special Forces—taken with, to augment, or supplement IO actions. Like many other IO elements, major concerns with employing physical destruction in UO are precision and follow-on effects. Thus, commanders using physical destruction to support IO must adhere to the same constraints as all other urban fires.

Information Assurance

5-76. Information assurance in UO takes on an added dimension. As with other operations, availability of information means timely, reliable access to data and services by authorized users. In UO, the timeliness of information may be restricted because structures block the transmission waves. The need for retransmission facilities may overwhelm the signal community. The reliability can be questioned because of the blockage between units and communications nodes. Unauthorized users may intercept the communications and input misinformation or disinformation. Commanders must protect the integrity of all information from unauthorized changes, including destruction. INFOSYS with integrity operate correctly, consistently, and accurately. The authentication of information may be accomplished by sophisticated electronic means. However, it is more likely that communications-electronics operating instructions authentication tables will authenticate the information. Commanders should consider the confidential nature of all information in UO and establish procedures to protect the information from unauthorized disclosure. Of additional concern, the density of the infrastructure in urban areas may inhibit receipt by the intended individual or unit. Therefore, commanders must develop alternative means to determine if the message was received.

Physical Security

5-77. Physical security consists of those actions and measures to safeguard personnel; to prevent unauthorized access to equipment, installations, material, and documents; and to safeguard them against espionage, sabotage, damage, and theft. The joint, interagency, and multinational nature of urban operations increases the need for physical security measures, especially with organizations that do not emphasize physical and operational security within their own organizations. The potential for locating sustainment bases in or near urban areas and the use of indigenous personnel on these facilities to accomplish a multitude of tasks mandates that commanders carefully consider the effectiveness of their physical security measures. In some stability operations where Army forces must recruit and train security or police forces, and then conduct joint operations and patrols with them, commanders must examine how best to establish close working relations while maintaining essential physical (and operations) security.

Counterintelligence

5-78. The urban environment, particularly in stability operations, is ideal for espionage, other intelligence activities, sabotage, or assassination. Threats can approach, conduct reconnaissance, and escape under the concealment offered by the urban environment. The dense environment and electromagnetic clutter serves to hide technological collection means. To counter threat intelligence collection activities, the Army employs CI forces. They provide analysis of the threat's human, imagery, signals, and measurement and signature intelligence capabilities to develop countermeasures against them (see FM 34-60).

5-79. In conjunction with military police, engineers, medical personnel, and other as required, CI help create urban threat and vulnerability assessments necessary in developing comprehensive force protection measures and evaluating their effectiveness. CI personnel can also supplement the screening of linguists and other critical host-nation personnel employed by Army forces. As Soldiers operate in close proximity to the urban society, threats will attempt to use three, time-tested tools to influence them: money, sex, and drugs. Periodic CI refresher training remains crucial to decrease the effectiveness of these threat tactics.

Counterdeception

5-80. In UO, threat forces can easily accomplish deception operations. The force that controls the area above and below ground will have freedom of movement. Deception aimed at friendly commanders will cause them to deploy combat power at the wrong place and the wrong time. Counterdeception by friendly commanders will identify and exploit threat attempts to mislead friendly forces. Counterdeception is difficult. Cultures of certain rhetoric and actions are more predisposed to deception than others. Knowing a threat's previous deception methods is important. Dismissing tactical indicators because they conflict with preconceptions may allow a hostile deception operation that plays on the preconception to succeed.

Counterpropaganda

Four hostile newspapers are more to be feared than a thousand bayonets.

Napoleon Bonaparte

5-81. Because propaganda is aimed at both combatants and noncombatants, UO are especially concerned with its use. Propaganda can rapidly and dramatically affect the attitudes of the urban population and will probably occur (or increase significantly) after urban operations have begun. Thus, it can create situations in the human dimension of the environment quite different from those discovered in the pre-operations assessment. Counterpropaganda is, therefore, essential to urban operations. To negate, deflect, or destroy the threat's propaganda capability, counterpropaganda requires—

- Monitoring and evaluating the effectiveness the threat's propaganda efforts.
- Determining innovative methods using all Army force capabilities, especially PSYOP and PA units.

5-82. In particular, humanitarian assistance, when coupled with an effective information operations campaign, can be a powerful counterpropaganda tool to sway public support. For example, media coverage of Army forces directly providing or coordinating NGO humanitarian efforts to feed, shelter, and provide medical care to the urban population can help counter threat efforts to discredit Army forces as an evil occupying force. This type of coverage can be used to turn opinion against the threat and could also prove an effective tool in gaining favorable international support.

Public Affairs

5-83. A closely related activity to IO is public affairs (PA). PA influences urban operations by transmitting information through the media to internal (urban Army forces as well as the urban civilian populace) and external audiences. At higher levels of command, PA can help maintain popular national support for the urban operation by clarifying the links between strategic goals and operational objectives. At both the operational and tactical levels, it links Army units, the urban inhabitants, the U.S. and international public, and the media. The media will often focus on the negative actions that are ongoing in an AO; hence, the urban commander's challenge will be to publicize his unit's good news stories. Commanders should not wait for media inquiries and interview requests. Instead, they should aggressively plan public affairs to get the Army's story out to the media and, ultimately, the public.

5-84. PA can help determine potential media issues that may influence planned UO. It can also aid commanders in understanding the impact of UO on the environment (particularly its citizens) and other agencies and organizations operating in the urban area. PA also helps to counter rumors, uncertainty, fear, loneliness, confusion, and other factors that cause stress (to both soldiers and civilians) and undermine effective UO. If the populace does not understand the mission or the Army forces' capabilities, false expectations may be created that Army forces may not be able to meet. PA can help prepare the American public for the possibility of high casualty rates. Overall, PA supports urban commanders in their goals to achieve information superiority and preserve public support.

5-85. The density of information sources and reporters in UO ensures that all Army activities will be subject to media and public scrutiny. Many reporters will congregate in cities for their own comfort and take advantage of established communications networks. Urban areas are densely populated and, together with Army forces and NGOs operating there, will present the greatest number of human-interest stories.

The local urban or host-nation media, however, will often have their own agendas developed over a longer period of time. This local media may also have a greater influence over the urban population than the international media. The indigenous media may not follow international norms. Commanders are responsible for understanding the media (particularly the local media), its role, and its potential for negative or positive influence.

5-86. Commanders cannot allow themselves to be intimidated by the media. Media sources are going to get and report their story one way or another. Therefore, commanders should support open and independent reporting and grant access to their units as early and as far forward as the situation permits. Successful units often go beyond simply accommodating the media to actually integrating them into their units. In the future, embedding of media may become the standard for reporting Army operations. Other considerations for improving urban media relations include:

- Encouraging the media to report on all ongoing aspects of the unit's urban operations—not exclusively on combat operations. This may create a unit responsibility to protect journalists that participate in covering suggested activities.
- Hosting media luncheons or other events for key journalists. This provides commanders the ability to get the whole story out to the press from their unit's perspective. To gain media interest and unit credibility, this will often require key leadership participation.
- Identifying Soldiers that speak the local and international journalists' native language. Even if journalists speak English as a second language, they may overlook or misunderstand important points due to language subtleties. Additionally, people tend to be friendlier to those that speak their own language.
- Identifying selected Soldiers at lower tactical levels to act as PA representatives for their unit as an additional duty. However, commanders must provide them with appropriate training and resources to accomplish specific PA tasks.
- Writing articles for local newspapers that emphasize the unit's good news. Depending on the situation, subordinate units may be tasked to write stories about the good accomplishments in their AO. Commanders, with appropriate OPSEC considerations, may hire locals to assist in this effort. However, commanders must take extra precautions to ensure that these efforts are not perceived as covert propaganda or as undermining free and independent news reporting.

5-87. PA does not distort, direct, or manipulate information. Its effectiveness stems directly from establishing and maintaining credibility with the urban population and media. PA protects the overall integrity of the Army as an institution by ensuring that the Army is recognized as the most reliable source for official information among all other competing sources. Commanders must synchronize PA with the core and supporting elements of IO (particularly PSYOP and counterpropaganda) to ensure that all Army sources send only one message. Urban commanders must plan for the media and integrate PA into their decision-making and (through IO) targeting processes. Similar to lethal targeting, commanders should consider target audiences and coverage areas for the various new agencies and mediums. They also pay attention to print media circulation statistics and television and radio broadcast times. For example, an urban populace may tend to read morning newspapers vice late editions and watch or listen to nightly news instead of morning or daytime programming. Additionally, low literacy rates among the urban population may indicate that television and radio are better mediums to use to influence the urban population than print media such as newspapers, magazines, leaflets, and posters.

5-88. The PA principles listed in Figure 5-4 and addressed in FM 46-1 summarize PA. They serve as useful guides toward planning and executing PA operations regardless of the environment. However, the principles of "practice security at the source," and "truth is paramount" particularly apply to the urban environment. The compartmented nature of most UO impede commanders' and their PA officers' ability to be at all places where the media will likely be. Friendly inaccuracies, intentional or not, will help fuel the threat information campaign. Therefore, all soldiers are trained, provided with clear and understandable PA guidance, and prepared to accurately communicate to the civilian media. The keys at all levels are understanding, prepared acceptance, and truthfulness tempered with an essential concern for OPSEC.

- Truth is Paramount
- If News is Out, It is Out
- Public Affairs Must be Deployed Early
- Not All News is Good News
- Practice Security at the Source
- Media are Not the Enemy
- Telling Our Story is Good for the Army
- Soldiers and Families Come First

Figure 5-4. Public affairs principles

Civil-Military Operations

5-89. Civil-military operations (CMO) are critical and will be an essential, implied task of any urban operation. They are included here as a closely related activity of IO. Commanders use CMO to establish, maintain, influence, or exploit relations to achieve operational objectives. These relations are among military forces, civilian authorities and organizations (both governmental and nongovernmental), and the civilian population. CMO range from support to combat operations to assisting in establishing political, economic, and social stability. In some instances, commanders may need to be prepared to assume temporary responsibility for the functions and capacities of the urban government. Like PA, effective CMO is based on establishing and maintaining credibility and trust with the urban populace and civilian organizations operating in the urban environment. Because of the potential importance of CMO, commanders may consider placing an assistant commander, executive officer, or some other senior leader to monitor, guide, liaise with other units and organizations, and, overall, ensure that the appropriate command emphasis is placed on these activities.

Civil Affairs

5-90. Civil affairs (CA) units and teams will be critical during UO. While any military force can conduct CMO, CA units are specifically organized, trained, and equipped to conduct activities in support of CMO. They have experience in planning and conducting CMO, a regional focus (which includes enhanced cultural awareness and language training), civilian technical expertise that parallel common urban government and administration functions, and experience in negotiation and mediation. Such experience ensures relevant support to commanders conducting urban operations. (As it is likely that there will not be enough CA forces to cover all requirements, commanders should survey their Soldiers to determine the whether any possess civilian expertise that may be beneficial to CMO.) CA units organize their skills into six functional specialty areas (see Figure 5-5). Commanders should use these skills, often unfamiliar to most military personnel, to—

- Develop their situational understanding of the urban environment (particularly the infrastructure and society).
- Plan CMO to support UO.
- Achieve many of the fundamentals of UO.

Note: To ensure optimal integration, commanders must establish clear command and support relationships between CA and maneuver forces.

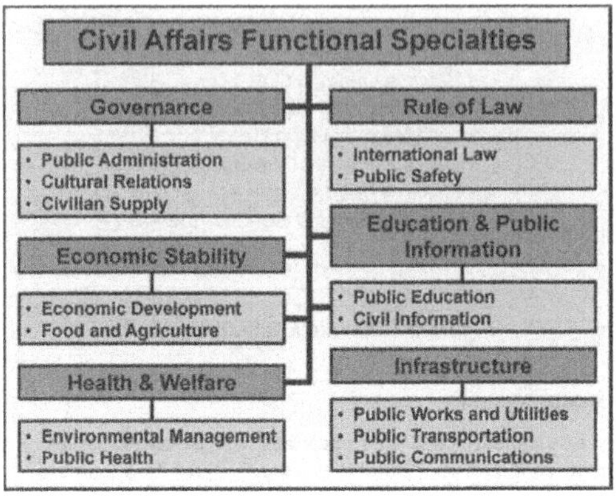

Figure 5-5. Civil affairs functional specialties

5-91. In addition to providing essential information for understanding the urban environment, CA personnel and activities help shape the battlefield, engage a civil problem, consolidate gains, and transition to a legitimate civil authority. Specifically, CA teams and CMO help urban commanders—

- Minimize civilian interference with UO and the impact of urban operations on the populace and infrastructure. CA personnel can help establish and run a civil-military operations center (CMOC), discussed later in this chapter, to coordinate UO with civilian agencies (both governmental and nongovernmental), other services, and multinational partners.
- Provide advice and assistance to restore or rehabilitate portions of the infrastructure.
- Plan, supervise, and execute necessary populace and resources control measures (in close coordination with military police units) until no longer required or the urban operation is completed.
- When requested or when military necessity or legitimate directives require, establish all or portions of the urban civil administration.
- Determine available supplies and services in the urban area and if necessary assist in negotiating their acquisition. CA also help commanders assess the capability, dependability, and willingness of urban sources to provide and sustain identified needs as well as to calculate the impact of using them on other aspects of the urban environment.
- In conjunction with the SJA, fulfill the Army's responsibilities toward the urban population under international, host-nation, and U.S. law.
- Plan and conduct the transition of control for the urban area or operation to another military or civilian governmental or nongovernmental organization or agency.

5-92. As part of CMO, stability or civil support operations will be critical as commanders begin to restore essential services to meet the emergency needs of a potentially devastated urban population. Besides repairing or restoring vital services and other critical infrastructure, these and other projects may also stimulate the urban area's economy and provide work for the unemployed. (Employed workers are less likely to be co-opted by terrorists, insurgents, and criminal gangs.) As such, they can be a critical stabilizing tool for commanders as they attempt to transition control and responsibility to other agencies or to legitimate civilian control. These projects may include—

- Providing clean water.
- Restoring waste management functions (less-extensive sewer repairs and trash collection and disposal activities).
- Improving medical care.
- Rehabilitating schools.
- Assisting in the establishment of local governments.
- Improving security and emergency services (police, fire, and rescue).
- Providing electrical generators for hospitals, clinics, and factories.
- Restoring irrigation to surrounding farmland.
- Repairing important cultural infrastructure.

5-93. As part of the initial planning process, CA units conduct an *area assessment*, which can provide commanders with essential information about the environment (see FM 41-10). Commanders should integrate this initial assessment into the overall urban-focused IPB process. To help analyze civil considerations in any environment, commanders and staffs can consider many characteristics such as areas, structures, capabilities, organizations, people, and events (ASCOPE). These characteristics easily align with an urban area's three main characteristics of terrain, society, and infrastructure; and, like them, they are overlapped and interdependent (see FM 6-0). Of these, the category of events is a critical addition to urban characteristics previously categorized and discussed. Significant spikes (and lulls) to threat and civilian activity can be expected to occur in conjunction with important religious, historic, and political events (for example elections). Therefore, commanders should determine, track, and pay particular attention to these events when planning operations and managing transitions (see also the Key Event Chart discussion in Appendix B, paragraph B-43).

5-94. Similar to PA operations, CMO are related to information operations. The nature of CMO and the need for CA personnel to develop and maintain close relationships with the urban population put CA personnel in a favorable position to collect information. CA personnel work daily with civilians, their equipment, and their records that may be prime sources of information. If used correctly, CA personnel can complement the intelligence collection process necessary to understand the dynamic societal component of the urban environment and detect significant changes. However, CA personnel are not, and cannot appear as, intelligence agents; otherwise, it will undermine their ability to interact with the civilian community. Examples of information available to CA units include government documents, libraries, and archives; files of newspapers and periodicals; industrial and commercial records; and technical equipment, blueprints, and plans.

5-95. Overall, CA expertise is critical to a commander's understanding of the complexities of the infrastructure and societal components of the urban area. These components (together with the terrain or physical component of the urban area) interconnect. CA forces help identify and understand the relationships and interactions between these urban components. From this understanding, commanders can anticipate how specific military actions affect the urban environment and the subsequent reactions. CA personnel help commanders predict and consider the second- and third-order effects and reactions as well as the long-term consequences. Understanding these long-term consequences helps in achieving stability and ensuring a smoother and speedier transition of the urban area back to civilian responsibility and control. Oppositely, an unconsidered short-term solution may, in the end, turn out to be the solution requiring the greatest amount of time. Poor solutions may exacerbate the situation. Others may contribute little to conflict resolution and waste precious resources (particularly time) until a more reasoned approach is formulated—one that should have been the first solution.

Civil-Military Reconstruction Efforts

5-96. Reconstruction efforts and activities, an integral part of CMO, should complement NGO efforts and are conducted in concert with larger urban area and nationwide projects often headed by the U.S. Army Corps of Engineers (USACE) and the U.S. Agency for International Development (USAID). For example, extensive, large-scale sewer projects or major repairs to the urban area's electrical infrastructure would be left to the USACE and USAID. When planning and conducting urban reconstruction, commanders should—

- **Keep expectations realistic.** With so much to do, it is easy for Army forces to become overcommitted. Commanders must focus on the achievable, build off successes, and create constructive forward process.
- **Seek opportunities.** While commanders think long-term, they should look for catalytic investment of resources. Minimal resources provided at the right time can generate significant activities resulting in increased momentum toward the desired end state.
- **Align resources with local needs.** In analyzing the merit of a project, commanders must check to determine if it is supported and desired by the urban populace. Local participation in the process reinforces ownership and acceptance and eventual transition of responsibility.
- **Balance speed and quality.** Commanders must balance the achievement of rapid response with quality construction and repairs.
- **Ensure efforts do not undermine locals.** Army reconstruction efforts should not undermine the growth and legitimacy of local institutions and authorities.
- **Integrate in the targeting process.** Commanders should routinely include approval of civil-military reconstruction products as part of the information operation's contribution to the overall targeting process.

INTEGRATION OF CONVENTIONAL AND SPECIAL OPERATIONS FORCES

5-97. One important Army and joint resource that commanders of a major operation can use to influence urban operations is SOF. Several types of these forces exist (including the CA forces discussed above), each with unique and complementary capabilities. They can be extremely valuable in UO for their ability to execute discrete missions with a higher degree of precision than conventional forces, to provide information, and to enhance cultural understanding. However, the challenges of using SOF include command and control; integration; coordination with conventional forces that will normally command, control, and conduct the bulk of UO tasks; and the imprudent inclination to use SOF forces for conventional purposes. The density and complexity of UO make close coordination and synchronization of conventional forces and SOF essential to mission success. The nature of the environment dictates that both forces will work in close proximity to each other; the separation in space and time between SOF and conventional forces will often be much less in urban areas than in other environments. Overall, the nature of the environment demands a synergistic combination of capabilities to achieve effects on the threat and mission success.

5-98. Successfully integrating SOF occurs with proper integration into, or coordination with, the command structure of the force conducting the UO. SOF within a theater (less CA and tactical PSYOP) ordinarily falls under joint command and control. Therefore, the commander of the major operation responsible for an urban area, if he is not a JFC, will have to coordinate through the JFC to integrate SOF capabilities into the UO. Examples of critical coordination elements include:

- Boundaries.
- No-fire areas. (If assigned, SOF liaison officers (LNOs) must be routinely included in targeting meetings.)
- Coordination points.
- The exchange of key intelligence and information.
- Requirements to support personnel recovery contingencies.

Conventional and Special Forces Integration

On 20 April 2003 during OPERATION IRAQI FREEDOM (OIF), an Army Special Forces (SF) operational detachment alpha (ODA) was patrolling through villages and secondary roads east-southeast of An Najaf. A local Iraqi approached the team and volunteered the location of a senior Ba'ath Party official in the nearby town of Ghamas. Unfortunately, he did not provide an exact location but rather an approximate location and the name of the family dwelling. The ODA coordinated with their company commander and nearby elements of the 101st Airborne Division to conduct a raid on the residence at dawn the next day. Joint planning, preparation, and rehearsals among the units occurred throughout the night.

At 0430, the raid convoy departed its base en route to Ghamas. The small task force consisted of an assault team, a security team, a command and control element, and three blocking forces composed of Soldiers from a scout platoon and antiarmor company of the 101st. At 0515, the convoy hit the release point outside Ghamas. The 101st vehicles moved to their blocking positions at three bridges surrounding the town. The remainder of the task force moved into the town. An interpreter quickly found a local guide who knew where the house was located. He took the SF team to a walled two-story dwelling with a courtyard. The security team isolated the objective and provided overwatch while the assault team forcibly seized the house and apprehended three adult males. Among them was Abd Hamden, the target of the raid and a senior Ba'ath Party official from Baghdad.

After a tactical interrogation, one of the men provided the location of a Fedayeen major nearby. The SF and conventional force raided this house minutes later but only found the major's relatives. As the Soldiers left the town with their captives, they were cheered by the locals throughout the town. Although only one small example, the extensive coordination between SOF and conventional forces during urban operations was a tremendous success during OIF.

SYNCHRONIZATION OF ACTIVITIES WITH OTHER AGENCIES

5-99. The population density of the urban environment, its economic and political importance, and its life-supporting infrastructure attracts many types of organizations. These organizations include—

- Other U.S. governmental agencies.
- International governmental organizations.
- Allied and neutral national governments.
- Allied and coalition forces.
- Local governmental agencies and politicians.
- NGOs.

5-100. Even in a major operation or campaign, many organizations operate in the area as long as possible before combat or as soon as possible after combat. Therefore, coordination with these organizations sharing the urban AO will be essential to achieve synchronization; however, effective synchronization is challenging, time consuming, and manpower intensive. The staffs of larger headquarters (divisions or higher) normally have the breadth of resources and experience to best conduct the necessary synchronization. They can effectively use or manage the organizations interested in the urban area and mitigate their potential adverse effects on UO. By taking on as much of the synchronization requirements as possible, the operational headquarters permits its tactical subordinates to remain focused on accomplishing their tactical missions. The higher headquarters should assume as much of the burden of synchronization as possible. However, the density of the urban environment will often require that smaller tactical units coordinate and synchronize their activities with other agencies and the local civilian leadership (formal and informal) simply because of their physical presence in the units' AOs. In urban stability and civil support operations, mission accomplishment will require effective civil-military coordination and synchronization activities and measures at all levels as either a specified or implied task.

Civil-Military Operations Centers

5-101. To coordinate activities among the varied agencies and organizations operating in an urban area and the local population, urban commanders can establish a CMOC. The CMOC synchronizes Army activities and resources with the efforts and resources of all others involved (see FM 41-10). This can be particularly important in stability and civil support operations where combat operations are not the dominant characteristic of the major operation. CMOCs can be established at all levels of command. Hence, more than one CMOC may exist in an AO, particularly large urban areas. CMOCs may be organized in various ways and include representatives from as many agencies as required to facilitate the flow of information among all concerned parties. Commanders must still ensure that force protection and OPSEC requirements are not compromised. Effective CMOCs can serve as clearinghouses for the receipt and validation of all civilian requests for support, can aid in prioritizing efforts and eliminating redundancy, can decrease the potential for inappropriate displays of wealth by one or more of the participating organizations, and, most importantly, can reduce wasting the urban commander's scarce resources.

Liaison Officers

5-102. LNOs—sufficiently experienced and adequately trained in liaison duties and functions—are necessary to deal with the other agencies that have interests in the urban area. Army LNOs work with the lead agency or other organizations (including local civilian agencies such as police) that the commander has identified as critical to mission success. Together they work to rapidly establish unity of effort and maintain coordination, often before a CMOC is established. The additional coordination afforded by the physical presence of LNOs within these organizations may be required even after the CMOC is fully functional. When commanders lack enough LNOs to meet requirements, they should prioritize and often assign a single LNO to several organizations. That LNO will then share his time and presence to those organizations based on the situation and his commander's guidance.

Commander's Personal Involvement

5-103. Overall, establishing a close relationship with other agencies and the urban civilian population will often be a major, positive factor in successful mission accomplishment, particularly in urban stability operations. Despite internally established command and control relationships, commanders that develop a direct and personal relationship with the leaders and staff of other agencies can often avoid conflict, win support, foster trust, and help eliminate the "us versus them" mentality that can frustrate cooperation among Army forces and civilian organizations.

Chapter 6

Foundations for Urban Operations

Utilities such as electricity and water are as much weapons of war as rifles, artillery pieces or fighter aircraft. . . . In the case of Manila, where there was a noncombatant, civilian population of one million in place, it was the attacker's aim to capture the utilities which the defender planned to destroy.

The Battle for Manila

Commanders conducting major urban operations (UO) use their ability to visualize how doctrine and military capabilities are applied within the context of the urban environment. An operational framework is the basic foundation for this visualization. In turn, this visualization forms the basis of operational design and decision making. To accurately visualize, describe, and direct the conduct of UO, commanders and their staffs must understand the basic fundamentals applicable to most UO.

URBAN OPERATIONAL FRAMEWORK

6-1. Army leaders who have an urban area in their area of operations (AO) or are assigned missions in an urban area follow an urban operational framework. They identify the portion of the urban area essential to mission success, shape the area, precisely mass the effects of combat power to rapidly dominate the area, protect and strengthen initial gains without losing momentum, and then transition control of the area to another agency. This framework divides into five essential components: understand, shape, engage, consolidate, and transition. These five components provide a means for conceptualizing the application of Army combat power and capabilities in the urban environment.

6-2. The urban operational framework assists commanders in visualizing urban operations. This framework is simply an aid to the commander. It is not sequential, nor is it a planner's tool for phasing an operation. Commanders should combine the urban operational framework with—

- The principles of war.
- The tenets of Army operations.
- The components of operational design.
- Considerations for stability operations and civil support operations.
- Sustainment characteristics.
- Running estimates.
- Commander's critical information requirements (CCIR).
- Each commander's experience.

The framework contributes to the visualizing, describing, and directing aspects of leadership that make commanders the catalysts of the operations process (see Figure 6-1). In the same manner, the urban operational framework contributes to the overall operations process (see FM 3-0).

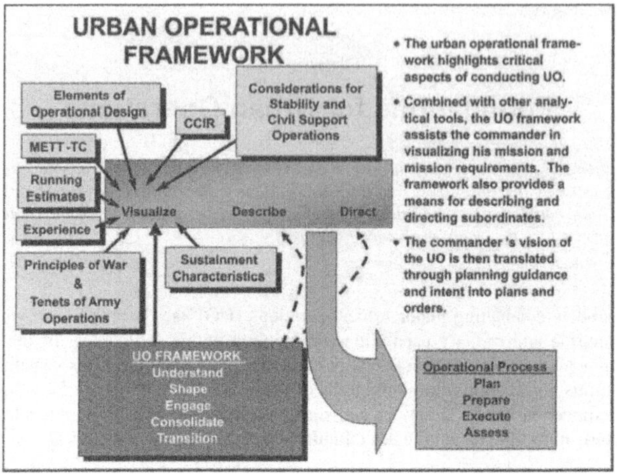

Figure 6-1. The urban operational framework and battle command

UNDERSTAND

6-3. Understanding requires continuous assessment—throughout planning, preparation, and execution—of the current situation and progress of an operation, and the evaluation of it against measures of effectiveness to make decisions and adjustments. Commanders use visualization, staff officers use running estimates, and all use the intelligence preparation of the battlefield (IPB) process to assess and understand the urban environment. Commanders and staffs begin to understand by observing and then collecting information about the situation. They observe and learn about the urban environment (terrain, society, and infrastructure), and other factors of METT-TC—mission, enemy, weather, troops and support available, and time available. They use intelligence, surveillance, and reconnaissance means; information systems (INFOSYS); and reports from other headquarters, services, organizations, and agencies. Then they orient themselves to the situation and achieve situational understanding based on a common operational picture (COP) and continuously updated CCIR. Largely, the ability to rapidly and accurately achieve a holistic understanding of the urban environment contributes to the commanders' abilities to seize, retain, and exploit the initiative during UO.

Disproportionately Critical

6-4. The Army operations process requires continuous assessment and understanding; it precedes and guides every activity. In UO, however, in-depth understanding is disproportionately critical for several reasons. First, each urban environment is unique. Other environments can be studied and their characteristics quantified in a general manner with accuracy. This is fundamentally not true of different urban areas. The characteristics and experience in one urban area often have limited value and application to an urban area elsewhere. This characteristic sets UO apart from operations in other environments.

Extremely Dynamic

6-5. The urban environment is also extremely dynamic. Either deliberate destruction or collateral damage can quickly alter physical aspects of the urban environment. The human aspect is even more dynamic and potentially volatile. A friendly civil population, for example, can become hostile almost instantaneously.

These dynamics (combined with initial difficulty of understanding and describing this unique environment) make it difficult for commanders and staffs to initially develop and maintain a COP and establish situational understanding. Furthermore, public reaction to media coverage of the urban operation and political changes influence national security strategy and objectives. Such changes can affect the basic nature of an operation, especially after it has commenced. Anticipating these potential effects and developing appropriate branches and sequels based on an accurate understanding often determines how quickly commanders can achieve the desired end state.

Risk Assessment

6-6. As in any environment, UO pose both tactical and accident risks. However, the level of uncertainty, ambiguity, and friction can often be higher than that of many other environments. Such challenges increase the probability and severity of a potential loss due to the presence of the enemy, a hostile civilian group, or some other hazardous condition within the urban environment (see Determining the Necessity of Urban Operations in Chapter 5). Therefore, commanders must—

- Identify, assess, and understand hazards that may be encountered in executing their missions.
- Develop and implement clear and practical control measures to eliminate unnecessary risk.
- Continuously supervise and assess to ensure measures are properly understood, executed, and remain appropriate as the situation changes.

6-7. Risk decisions are commanders' business. However, staffs, subordinate leaders, and even individual Soldiers must also understand the risk management process and continuously look for hazards at their level or within their area of expertise. Any risks identified (with recommended risk reduction measures) must be quickly elevated to the appropriate level within the chain of command (see FM 3-100.12 and FM 5-19).

Complex and Resource Intensive

6-8. The urban environment is the most complex of all the environments in which the Army conducts operations. It is often comprised of a diverse civil population and complex, ill-defined physical components. A sophisticated structure of functional, social, cultural, economic, political, and informational institutions unites it. Thus, the analysis to understand the environment is also complex and time and resource intensive. The nuances of the urban environment can take years to uncover. Hence, constant analysis of the environment requires greater command attention and resources. Accurately understanding the environment is a prerequisite to shaping it, and both understanding and shaping activities are crucial to effectively engage the elements of the urban environment critical to success.

SHAPE

6-9. Shaping operations, part of all Army operations, are essential to successful UO. They set the conditions for decisive operations at the tactical level in the urban area. Rapid action, minimum friendly casualties, and acceptable collateral damage distinguish this success when the AO is properly shaped. Failure to adequately shape the urban AO creates unacceptable risk. The commander of a major urban operation has several resources with which to begin shaping the AO. Important capabilities include—

- Fires.
- Information operations.
- Special operations capabilities.
- The maneuver of major subordinate units.

6-10. Critical urban shaping operations may include actions taken to achieve or prevent isolation, understand the environment, maintain freedom of action, protect the force, develop cooperative relationships with the urban population, and train Army forces for sustained UO.

Isolation

6-11. Isolation of an urban environment is often the most critical component of shaping operations. Commanders who's AO includes operationally significant urban areas often conduct many shaping operations to isolate, or prevent isolation of, those areas from other parts of the AO. Likewise, commanders operating in the urban area focus on isolating decisive points and objectives in the urban area or averting isolation of points that are critical to maintaining their own freedom of action. Isolation is usually the key shaping action that affects UO. It applies across full spectrum operations. Most successful UO have effectively isolated the urban area. Failure to do so often contributed to difficult or failed UO. In fact, the relationship between successful isolation and successful UO is so great that the threat often opposes external isolation actions more strongly than operations executed in the urban area (or critical areas within). In some situations, the success of isolation efforts has been decisive. This occurs when the isolation of the urban area compels a defending enemy to withdraw or to surrender before beginning or completing decisive operations. In UO that are opposed, Army forces attempt to isolate the threat three ways: physically, electronically, and, as a resultant combination of these first two, psychologically (see Figure 6-2).

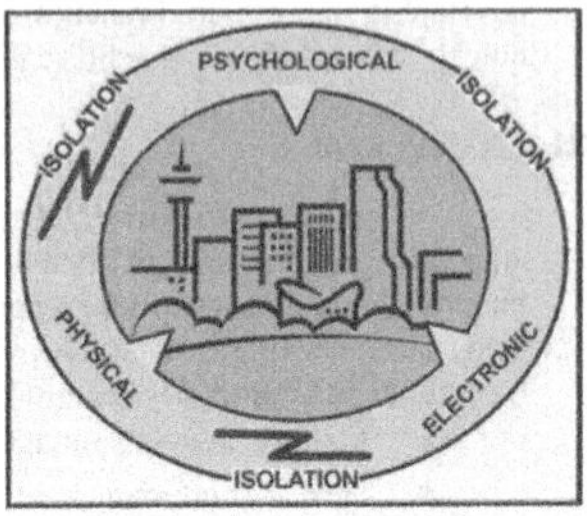

Figure 6-2. Urban isolation

Physical Isolation

6-12. In offensive UO, physical isolation keeps the threat from receiving information, supplies, and reinforcement while preventing him from withdrawing or breaking out. Conversely, a defending Army force attempts to avoid its own physical isolation. Simultaneously, this force conducts operations to isolate the threat outside, as they enter, or at selected locations in the urban area. Physical isolation can occur at all levels. In many situations, the commander of a major combat operation may attempt to isolate the entire urban area and all enemy forces defending or attacking it. At the tactical level, forces isolate and attack individual decisive points often using a cordon technique. In stability operations, physical isolation may be more subtly focused on isolating less obvious decisive points, such as a hostile civilian group's individual leaders. In many operations, isolation may be temporary and synchronized to facilitate a decisive operation elsewhere. To effectively isolate an urban area, air, space, and sea forces are often necessary additions to the capabilities of ground forces.

Electronic Isolation

6-13. Electronic isolation is achieved through offensive information operations (IO). Electronic warfare (particularly two of its components: electronic warfare support and electronic attack) and computer network attack are critical to electronic isolation (see FM 3-13 and the Information Operations discussion in Chapter 5). At the operational level, offensive IO aims to quickly and effectively control the information flow into and out of an urban area. This isolation separates the threat's command and control (C2) system in the urban area from its operational and strategic leadership outside the urban area. Offensive IO also focuses on preventing the threat from communicating with civilians through television, radio, telephone, and computer systems. At the tactical level, IO aim to isolate the threat's combat capability from its C2 and leadership within the urban area, thus preventing unity of effort within the urban area. Defensive IO can prevent isolation of friendly forces defending in an urban area.

Psychological Isolation

6-14. Psychological isolation is a function of physical actions, electronic warfare, and other forms of IO, especially military deception and psychological operations. Psychological isolation denies the threat political and military allies. It separates the enemy or hostile civilian group from the friendly population, nongovernmental organizations (NGOs) operating in the urban area, and from political leaders who may consider supporting Army forces. Psychological isolation destroys the morale of individual enemy soldiers or hostile civilians. It creates a feeling of isolation and hopelessness in the mind of the threat. It undermines the threat leadership's ability to think, plan, decide, and act. As such, it undermines the confidence of the threat in their leadership. On the other hand, IO, as well as the disciplined conduct of Army personnel, can help to forge legitimacy for Army operations. In stability operations, psychologically isolating the threat can result in the friendly urban population and NGOs positively supporting Army operations.

Other Shaping Actions

6-15. Other shaping actions can include the proper sequencing and deployment of forces, reconnaissance operations, and force protection. These actions contribute equally to the success of any urban operation. Commanders must understand how the urban environment affects their ability to accomplish these shaping actions. However, civil-military operations (CMO), another closely related activity of IO, are important to shaping the urban operational environment for decisive operations. The specific civil-military task can vary greatly and may include affecting a cooperative relationship with the civil political system, protecting portions of the civil population or infrastructure, or establishing refugee camps or safe areas for noncombatants. This is most true in stability and civil support operations. Successful CMO also can contribute to the psychological isolation of the threat. (See Civil-Military Operations in Chapter 5 for a more detailed discussion.)

Training and Education

6-16. Finally, Army commanders must consider that critical shaping actions often occur prior to the urban operation in the form of professional education, and home-station and in-theater training. Commanders can enhance training through joint, interagency, multinational, and combined arms exercises and effective rehearsals.

ENGAGE

6-17. In UO, Army forces engage by appropriately applying their full range of capabilities against decisive points leading to centers of gravity. Successful engagement takes advantage of the Army force's superior training, leadership, and, within the constraints of the environment, equipment and technology. Successful engagement also requires the establishment of necessary levels of control and influence over all or portions of the urban environment until responsibilities can be transferred to other legitimate military or civilian control. Engagement may range from the overwhelming, yet precise, application of combat power to defeat a threat, to large-scale humanitarian assistance operations, to unobtrusive advice and assistance to the urban security forces. It may require brief engagement in a relatively small portion of an urban area or with small segments of the population to long-term engagement throughout multiple, large urban areas and with vast sectors of the population. The commander's ability to engage in a specific urban operation depends, as always, on the situation and the assigned mission.

Offense: Attack Decisive Points

6-18. In offensive UO, forces successfully engage by striking at the enemy's center of gravity using multiple offensive actions from unexpected directions and throughout all dimensions. Army forces aim to engage identifiable decisive points. Successful efforts against decisive points lead to effects on the center of gravity. The center of gravity will differ in each offensive situation. It may be an individual enemy leader, the enemy's combat power, the enemy's communications capability, or a physical structure of cultural, political, or economic significance.

Defense: Deny Vital Functions and Critical Infrastructure

6-19. In defensive UO, engage translates into denying the enemy control of the vital functions and critical infrastructure of the urban area. Forces achieve this by leveraging the defensive advantages of the urban terrain, defending essential areas in depth, using economy of force in nonessential areas, controlling the enemy direction of attack with natural and man-made obstacles, and retaining the initiative through counterattacks.

Stability and Civil Support: Apply Innovation and Imagination

6-20. The ability to engage in urban stability operations hinges on the type of stability operation commanders execute. In a noncombatant evacuation operation, forces limit engagement to finite geographic areas and times. In contrast, a peace operation may require engagement of a large urban area for an extended time. In this operation, engage is defined as using the array of Army capabilities to create specific conditions among the belligerents. Thus, the techniques used for engagement in stability operations vary according to the situation and as situations mature during long-term operations.

6-21. In urban civil support operations, engagement is accomplished by innovative and subtle application of Army capabilities. Since Army forces usually support other agencies that lead the operation, effective engagement results from carefully and discretely applying Army capabilities to the tasks assigned by the lead agency. In a humanitarian relief situation, Army forces may be tasked to transport supplies in the urban area. Successful engagement in this activity then becomes the goal of Army forces and may be achieved by providing, managing, and protecting transportation assets and clearing and maintaining sufficient lines of communications (LOCs).

CONSOLIDATE

6-22. Army forces consolidate to protect and strengthen initial gains and ensure retention of the initiative. Consolidation includes actions taken to eliminate or neutralize isolated or bypassed threat forces (including the processing of enemy prisoners and civilian detainees) to increase security and protect LOCs. It includes the rapid repositioning and, as necessary, reorganization of forces and intelligence, reconnaissance, and surveillance (ISR) assets. Consolidation may also include activities in support of the civilian population such as relocating displaced civilians, reestablishing law and order, humanitarian assistance and relief operations, and restoration of key urban infrastructure. A significant consolidation effort may be the reduction or elimination numerous explosive hazards resulting from previous urban combat operations. During consolidation, commanders often coordinate with NGOs to establish unity of effort in the execution of stability or civil support operations. Consolidation begins with the initiation of the urban operation and follows each critical activity.

TRANSITION

6-23. When planning UO, commanders ensure that they plan, prepare for, and manage transitions. Transitions are movements from one phase of an operation to another or a change in responsibility for all or portions of the urban environment from one unit, organization, or agency to another. They involve significant changes in the type of operation, concept of the operation, mission, situation, task organization, forces, resource allocation and support arrangements, or C2. Transitions occur in all operations, but in UO they occur with greater frequency and intensity, are more complex, and often involve agencies other than U.S. military organizations. For example, a successful attack may transition to a defend mission that includes not only defense tasks but also stability tasks. All operations will likely include a transition of responsibility for some aspect of the urban environment to (or back to) a legitimate civilian authority. Unless planned and executed effectively, transitions can reduce the tempo of the urban operation, slow its momentum, and cede the initiative to the threat.

Mental and Physical Preparation

6-24. Transitions occur as conditions warrant. They can be carefully planned and controlled, or they can be quick and dramatic, such as the swift transformation of a stability operation into offense or defense—and back again. Units prepare mentally and physically to address rapid transitions. Accordingly, plans include branches and sequels that address anticipated or possible transition points. When the dominant type of operation changes from an offense to stability or civil support, the types of units originally conducting the UO may no longer be appropriate. A large mobile reserve may permit increased flexibility to react to unplanned transition requirements. Operations or units in one part of an urban area may transition before operations or units in a different part of the same urban area. This will require commanders to exercise careful command and control and execute various types of operations and associated tasks simultaneously.

Transition to Legitimate Civilian Authorities or Agencies

6-25. Combat operations are not an end to themselves. They are inevitably a means of transition, an activity for moving from an unsatisfactory state of affairs to an improved end (as determined by national objectives). In UO, a distinct aspect of transition is the requirement to quickly and efficiently transition the major portions of Army responsibilities to civil agencies. Some tasks to which units will transition are not traditional combat tasks but rather stability tasks more closely associated with CMO. In stability or civil support operations this is often a near-term critical mission objective. In these operations, commanders aim to alleviate the circumstances requiring Army forces and ensure that other civilian agencies assume the functions provided by Army forces. Following combat operations, civilian agencies are assisted and encouraged to quickly resume specific support activities—such as providing sanitary services, food services, law enforcement, and health services—because of their high demand on Army resources.

6-26. Transitioning responsibility for the urban environment (or selected portions) to legitimate civilian agencies or authorities requires commanders to make an accurate assessment and develop an in-depth understanding of the civilian organization's ability to accept and handle the responsibility. Commanders consider not only resources available but whether the organization has the appropriate leadership and technical expertise as well as a vision in consonance with U.S. objectives. Commanders should avoid forcing a transition only to alleviate the burden on Army forces. On the other hand, they may need to encourage acceptance of responsibility while continuing to provide "behind-the-scenes" advice, assistance, and resources. Transitioning responsibility prematurely may alleviate the commander's immediate resource concerns but may, in the longer term, negatively impact the achievement of strategic objectives.

Clearly Visualize and Describe the End State

6-27. Army UO conclude when Army forces depart and have no further mission requirements in the urban area. At the outset, commanders visualize and describe the intended end state of a unit's execution of UO. Commanders then clarify and update this visualization as the political or strategic situation is refined or changes. Importantly, commanders must create a command climate and procedures that allow their staffs and subordinate commanders the ability to ask questions that will enable all to share in an accurate perception of the commander's visualization. This enables subordinate units to identify likely transitions and ensures that current operational planning takes into account second- and third-order effects. As long as an active Army AO contains an urban area, some type of urban operation will exist. After urban combat successfully ends, combat forces may move on. Support forces conducting sustaining operations may then occupy the area and continue to conduct a different form of UO.

Applying the Urban Operational Framework: Panama – 1989

The U.S. conducted OPERATION JUST CAUSE in December 1989 to remove the illegal ruler of Panama, Manuel Noriega, and to restore that country to a democracy. It also conducted the operation to ensure the safety of a substantial number of U.S. personnel as well as the security of U.S. interests in Panama. The major focus of JUST CAUSE was in Panama City, the country's capital. Most operations occurred in this large urban area, one of the numerous smaller urban areas, or the urban-like military bases. These bases increased the AO and were directly linked to operations in the capital city. This successful operation illustrates how commanders can apply the urban operational framework to visualize, describe, and direct the critical aspects of urban operations.

Understand

The synchronization achieved during the operation may have obscured the challenges faced in the initial assessment process in Panama. However, it was not as simple as it may have seemed. Using the framework of the urban environment, U.S. forces required details of the physical characteristics of the environment, the infrastructure, and the human dimension including the capabilities of the Panamanian military.

Because Army forces had a long history in Panama, commanders clearly understood the physical challenges and layout of critical urban areas (see Figure 6-3), particularly Panama City. They also understood how the infrastructure in each urban area functioned and which parts was a key element to success. Examples of key portions of the infrastructure included the Madden Dam, which controlled the water flow through the Panama Canal, and the Cerro Azul television tower, which was the main Panamanian broadcast tower.

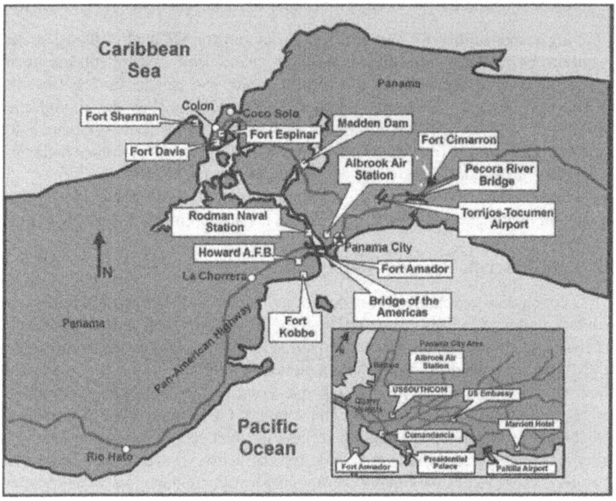

Figure 6-3. Panama

Collecting information and developing intelligence on the human elements of the urban environment was critical to operational success and a challenge. Because the Panamanian Defense Force (PDF) had traditionally been an ally of the United States, Army forces did not have a systemic database that adequately depicted their order of battle and their true capabilities. Additionally, much of the situation in Panama was colored in political terms making it more difficult for traditional military sources to evaluate the status of PDF forces. For example, Army planners needed to know if PDF military units (when faced with a formidable U.S. force) would fight at all for Noriega and if they did fight, how hard and long would they fight. The answers depended largely on their political loyalty to Noriega and on the individual loyalty of the unit officers to the Panamanian president. Thus, Army commanders needed to understand the military characteristics of PDF units and their political affiliations and tendencies.

Because transition from combat to noncombat tasks would be critical to achieving all objectives, particularly the restoration of democracy, Army forces also needed an accurate assessment of the political opposition to Noriega—including that opposition's capabilities and vulnerabilities. Again, Army forces were required to make assessments outside those needed solely for combat operations. Ultimately, assessing the political opposition's vulnerabilities led to assigning Army units to protect them throughout the operation so that they could serve as a foundation for a new democratic government.

Finally, the commander's assessment included an evaluation (often subjective) of the attitudes and disposition of the Panamanian people. Human intelligence (HUMINT) was the primary source of information on the population. Army forces had good access to the population because of their close proximity and historical ties to Panama. Many Soldiers were married to Panamanians, and the Army had total access to local media and to prominent individuals.

National imagery and special operations forces (SOF) also contributed to the ability of Army forces to assess the urban environments of JUST CAUSE. All units executing operations had detailed satellite photos of objective areas. Additionally, key objectives were placed under SOF surveillance well in advance. This surveillance revealed unit readiness, vulnerabilities, detailed disposition, and other patterns critical to mission success. The combination of the two capabilities allowed units to plan and achieve the synchronization necessary for such a complex urban operation.

Shape

During OPERATION JUST CAUSE, commanders conducted numerous shaping operations to establish the conditions for the decisive operations. Many operations were designed to control information, such as an assault on the Azul television tower identified during the assessment of the infrastructure. Planners designed many shaping operations to isolate various garrisons and PDF units. An example of tactical isolation was the plan for the Pacora River Bridge to prevent reinforcements from reaching the garrison at Torrijos-Tocumen Airport.

Operational isolation was achieved through the Ranger Regiment's and 82nd Airborne Division's assault on targets at Rio Hato in the west and Fort Cimarron in the east. These actions in conjunction with the securing of Maddam Dam had the primary objective of isolating Panama City. They were also the largest of the major actions occurring during OPERATION JUST CAUSE. The airborne assault was also the largest airborne operation conducted by U.S. forces since World War II. This large-scale shaping operation demonstrates that shaping operations are critical to mission success and can be more resource intensive than the actual operations that achieve domination.

Engage

U.S. Army forces achieved successful engagement in OPERATION JUST CAUSE by establishing unchallenged military control over Panama City and eliminating Noriega's capability to challenge that control. Toward this end, the operation attacked two decisive points. The first was the assault on the PDF headquarters located in Panama City: the Comandancia. The second was the operation undertaken to locate and seize Noriega himself.

Three battalions of task force (TF) Bayonet (5-87th Infantry, 1-508 Infantry [Airborne], and 4-6th Infantry [Mechanized]) executed the attack on the Comandancia and Fort Amador. They were also tasked to protect the American Embassy in downtown Panama City. To execute these missions, they moved from various staging areas located throughout the city to their assigned objectives using air assault, mounted, and dismounted approaches. The ground movement through the city proved to be the most difficult and hazardous part of the mission due to the vulnerability of the troops in their armored personnel carriers and trucks. The dismounted movement was slower than the mounted movement but allowed the Soldiers greater cover and concealment.

The strongest opposition to TF Bayonet occurred at the Comandancia. Elements of three PDF companies and two public order companies held out for three hours. The troops moving to Comandancia were subject to a large volume of sniper fire, and in the assault, unidentified indirect fire caused significant casualties among the mechanized forces. TF Bayonet forces, supported by airborne armored reconnaissance vehicles and Hellfire missiles from Apache helicopters, captured the

> Comandancia. Commanders noted in particular the precision of the supporting fires from attack helicopters. The assault by fire from supporting AC 130 gunships severely damaged much of the reinforced Comandancia building.
>
> Simultaneously, SOF attacked several targets where Noriega might be located. These initial attacks were unsuccessful. However, many subsequent actions neutralized Noriega's influence and eventually resulted in his apprehension on 3 January 1990. These actions included the well-organized and relentless manhunt conducted by SOF units, the isolation of Panama City itself, population control efforts, sophisticated IO, and cooperation with other U.S. agencies.
>
> Consolidate and Transition
>
> OPERATION JUST CAUSE demonstrated the vital need for a thought-out plan that adequately addresses consolidation and the transition from combat to noncombat before commanders initiate operations. Normally in complex UO, commanders cannot leave the details of transition until after the operation has begun without unacceptable risk to overall mission accomplishment. The follow-on stability operation, OPERATION BLIND LOGIC (later renamed OPERATION PROMOTE LIBERTY) began 24 hours after the initial assault and thus both operations were occurring simultaneously. This simultaneity of different types of operations is typical in major operations conducted in a large urban area. The stability operation involved more time than the combat operation and continued well after the close of OPERATION JUST CAUSE and after most of the major combat units had redeployed. It involved significant resources without the same level of risk to U.S. forces as the combat operations.
>
> Civil affairs (CA) were a dominant part of the transition from combat to stability operations. The 96th Civil Affairs battalion was central to this operation. CA forces established a civil police force, emergency food distribution, property protection, production and distribution of newspapers, cleanup of the city, and building support for a new civil government. Most tasks were coordinated through Army CA forces and executed by other Army forces under the supervision of CA.
>
> IO were also a major aspect of affecting a stable transition and successful postcombat operations. These operations built support for the U.S. operation among the population. They emphasized that the U.S. conflict was with Noriega and not the Panamanian people and that the U.S. forces would depart as soon as a new Panamanian government could take over.
>
> Other U.S. agencies played critical roles in stability operations in Panama. The U.S. Drug Enforcement Agency and Justice Department were important to the negotiations that led to Noriega's capture. The U.S. State Department helped to negotiate for Noriega and develop military policies and plans during stability operations. The American Embassy advised commanders regarding the large diplomatic community that existed in Panama City.

FUNDAMENTALS OF URBAN OPERATIONS

6-28. UO often differ from one operation to the next. However, some fundamentals apply to UO regardless of the mission, geographical location, or level of command. Some of these fundamentals are not exclusive to urban environments. Yet, they are particularly relevant to an environment dominated by man-made structures and a dense noncombatant population (see figure 6-4). Appendix A provides an historical example of how many of these fundamentals applied to an actual conflict situation. Vitally, these fundamentals help to ensure that every action taken by a commander operating in an urban environment contributes to the desired end-state of the major operation.

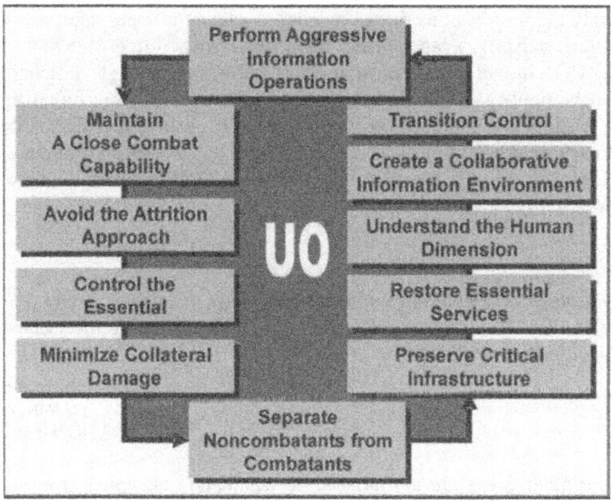

Figure 6-4. Fundamentals of urban operations

PERFORM AGGRESSIVE INFORMATION OPERATIONS

6-29. Information operations aimed at influencing non-Army sources of information are critical in UO across the spectrum of operations. Because of the density of noncombatants and information sources, the media, the public, allies, coalition partners, neutral nations, and strategic leadership will likely scrutinize how Army forces participate in UO. The proliferation of cell phones, video cameras, Internet capability, and media outlets ensure close observation of the activities of Army forces. With information sources rapidly expanding, public information of Army operations will often be available faster than the internal military INFOSYS can process it. Army forces should aggressively integrate IO into every facet and at all levels of the operation to manage perceptions and mitigate unintended consequences. Under media scrutiny, the actions of one Soldier may have significant strategic implications. IO aim to make the information accurate; placed in the proper context of the Army's mission; and available to all interested parties: the public, the media, and other agencies. Ensuring that the urban population understands how they and their urban area fit within the commander's vision, intent, and end state will be a critical task.

MAINTAIN A CLOSE COMBAT CAPABILITY

6-30. Close combat is required in all offensive and defensive UO. This core capability must also be present and visible in urban stability operations and may be required in urban civil support operations. Close combat in any urban operation is resource intensive, requires properly trained and equipped forces, and has the potential for high casualties. However, the ability to decisively close with and destroy enemy forces as a combined arms team remains essential. Potential threats respect strength and, instead, seek to attack ill-prepared or complacent Army forces—not necessarily forces that are better protected or armored, but those less likely or capable of inflicting casualties against them. In stability and civil support operations, a lack of respect and fear of Army forces can hinder recovery as much as the ill-advised use of forces. Hence, all Soldiers, regardless of branch or military occupational specialty, must be properly equipped and trained to fight in an urban environment. This ability allows Army forces to deter aggression, compel compliance, morally and physically dominate a threat and destroy his means to resist, and terminate urban conflicts on the Army commander's terms.

Avoid the Attrition Approach

6-31. Previous Army doctrine was inclined towards a systematic linear approach to urban combat. This approach emphasized standoff weapons and firepower. Army force structure does not support this approach towards UO. It can result in significant collateral damage, a lengthy operation, and an inconsistency with the political situation and strategic objectives. Enemy forces that defend urban areas want Army forces to adopt this approach because of the likely costs in resources. Commanders should only consider this approach to urban combat as an exception and justified by unique circumstances. Instead, commanders should seek to achieve precise, intended effects against multiple decisive points that overwhelm a threat's ability to react effectively.

Control the Essential

I'm talking about attacking those things from which the regime draws power but being very careful about it so that we don't get large bodies of young Americans caught up in house-to-house Berlin, World War II-type scenario.

Lt. Gen. William C. Wallace
Commander, V Corps
OPERATION IRAQI FREEDOM

6-32. Many modern urban areas are too large to be completely occupied or even effectively controlled without an enormous force. Therefore, Army forces focus their efforts on controlling only the essentials to mission accomplishment. At a minimum, this requires control of key terrain. Key terrain is terrain whose possession or control provides a marked advantage to one side or another. In the urban environment, commanders determine key terrain based on its functional, political, economic, or social significance. A power station or a place of worship may be key terrain.

6-33. All principles of war can apply to UO. The principle of mass and the principle of economy of force (in addition to the principle of unity of command discussed in the previous chapter) are particularly important in guiding UO and providing mission focus. Army forces mass combat power only to control those requirements essential for mission success. This permits conservation of combat power. It also implies economy of force and associated risk in those areas where Army forces choose not to exercise control.

Minimize Collateral Damage

6-34. Forces should integrate precision fires, IO, and nonlethal tactical systems consistent with mission accomplishment while decreasing the potential for collateral damage. Commanders develop unique rules of engagement (ROE) for each urban operation and provide necessary firepower constraints. IO and nonlethal systems may compensate for some restrictions, especially in stability or civil support operations. Commanders continually assess the short- and long-term effects of operations and firepower on the population, infrastructure, subsequent missions, and national and strategic objectives. They must also consider what, if any, provisions should be made to amend or address potential collateral damage. Overall, commanders must balance restraint and precision with speed and overwhelming combat power. By avoiding unnecessary harm to all elements of the urban environment, commanders retain the moral high ground and help sustain legitimacy for their operations. Minimization of collateral damage allows civilians to continue to provide for their own needs or the rapid return of the urban area to civilian self-sufficiency.

Separate Noncombatants from Combatants

6-35. Promptly separating noncombatants from combatants (psychologically and physically) may make the operation more efficient and diminish some of the threat's potential asymmetrical advantages. This separation also may reduce restrictions on the use of firepower, enhance force protection, and strip the threat from its popular support base. This important task becomes more difficult when the threat is an unconventional force that can mix with civilians.

6-36. In more recent operations, threats have sought to integrate their military capabilities as closely as possible into the civilian population and infrastructure. In these conditions, commanders must increase their efforts to discriminate between the two. Soldiers managing violence in this setting require the highest level of individual and organizational discipline and judgment. Soldiers will require the mental agility to separate their aggression toward threats from the noncombatant civilian population. The training, effort, and command emphasis in this area is as important as fully successful results. Such efforts strongly impact national and international perceptions of the operation.

PRESERVE CRITICAL INFRASTRUCTURE

6-37. Commanders should analyze the urban area to identify critical infrastructure. They should attempt to preserve and protect the critical elements for postcombat sustainment operations, stability or civil support operations, or the overall health and well-being of the indigenous population. Urban areas remain in the AO after combat operations have ceased. Postcombat UO are unavoidable. Different from simply avoiding collateral damage, Army forces may have to initiate actions to prevent an enemy or a hostile civilian group from removing or destroying critical infrastructure and assets. This may include cultural infrastructure such as religious and historical places. In some cases, preserving the infrastructure and the urban society's sources of economic and cultural wealth may be the assigned objective of the urban operation.

RESTORE ESSENTIAL SERVICES

6-38. Army forces plan to restore essential services that may fail to function before or during an operation. Essential services include power, food, water, sewage, medical care, and security and law enforcement. When planning for and conducting Army UO, units can use nonlethal and less destructive munitions and capabilities to keep potentially vital infrastructure intact. Initially, Army forces may be the only force able to restore or provide essential services and commanders must plan accordingly. Failure to do so can result in serious health problems for the civilians, which can affect the health of Army forces and negatively impact overall mission success. Army forces seek to transfer responsibility for providing essential services to other agencies, NGOs, or the local government as quickly and effectively as possible. Despite the potential causes for the failure or destruction of essential services, commanders also seek to ensure that civilians continually perceive Army restoration activities as assistance rather than an Army requirement. Otherwise, the civilian populace may be slow to accept or resume responsibility for their urban area.

UNDERSTAND THE HUMAN DIMENSION

> *Do not try to do too much with your own hands. Better the Arabs do it tolerably than that you do it perfectly. It is their war, and you are to help them, not to win it for them. Actually, also, under the very odd conditions of Arabia, your practical work will not be as good as, perhaps, you think it is. It may take them longer and it may not be as good as you think, but if it is theirs, it will be better.*
>
> T. E. Lawrence
> "Twenty-Seven Articles"

6-39. Commanders carefully consider and manage the perceptions, allegiance, and morale of the civilians. Their assessment of the environment needs to accurately identify the attitudes of the people toward Army forces. Operational guidance to subordinates—including ROE, protection, sustainment operations, and fraternization—is based on this assessment. Commanders expect and consider the demographic variance in the attitudes of an urban population. They cannot inadvertently apply Western cultural norms to a non-Western urban population. Commanders can only make reliable assessments based on a thorough understanding and appreciation of the local society and their culture. Developing this ability also requires the additional ability to effectively share information among all echelons of command.

6-40. Sound policies, proper discipline, adequate consideration for local culture, and rapid engagement of local urban leaders will positively affect the attitudes of the population toward Army forces. Additionally, well-conceived and executed IO will enhance the position of Army forces relative to the urban population. Even during high-intensity urban combat, heightened awareness of and sensitivity toward the civilians can lead to a better postcombat situation than if civil considerations were unobserved or diminished in importance. An improved postcombat situation enhances transition. As the environment of conflict becomes more complex, the human dimension (and associated moral aspects) takes on greater importance and may have the greatest potential for affecting the successful outcome of UO. Therefore, the human aspect and development of cultural acuity creates a discrete planning factor.

Create a Collaborative Information Environment

6-41. The complexity of the urban environment, particularly the human dimension, requires rapid information sharing—providing and receiving—from the national level to the tactical level, among Army headquarters at each echelon, with sister services and coalition partners, and with participating governmental and (at appropriate times) nongovernmental agencies. The analysis of urban information into the relevant intelligence necessary to refine and deepen a commander's understanding of the urban environment and its infrastructure of systems also demands collaboration among the various information sources and consumers. Therefore, commanders seek to establish streamlined procedures, develop commonality among databases, and make best use of existing information systems to disseminate and receive the necessary intelligence and relevant information for subordinates and partner organizations and agencies to exercise effective leadership and decision making and establish synergistic unity of effort in this multifaceted environment.

Transition Control

6-42. Because UO are resource intensive, commanders should plan to end them as quickly as possible, yet consistently with successful mission accomplishment. (Depending on the factors of METT-TC, successful transition may take a few days or many years to achieve.) The end state of all UO transfers control of the urban area to another agency or returns it to legitimate civilian control and responsibility. Rapid transition releases Army resources for use elsewhere and improves the civilian morale and disposition toward Army forces. This requires the successful completion of the Army force mission and a thorough transition plan. The transition plan may include returning control of the urban area to another agency a portion at a time as conditions permit. A successful transition plan should consider early alignment of military capabilities with existing urban governmental and administrative organizations, agencies, structures, and districts. Transition planning must be conducted before the onset of operations and continually adjusted as the situation develops.

Chapter 7
Urban Offensive Operations

... Capture Suez City "provided it does not become a Stalingrad situation."

Order to the Adan Armored Division
prior to its 1973 attack on Suez City
On the Banks of the Suez

Offensive urban operations (UO) are one of the most challenging operations that military forces can undertake. Campaigns and wars have sometimes hinged on their success or failure. Costly in resources, even when successful, they are not lightly entered into. Once engaged, they are executed rapidly and decisively. For reasons already discussed, threat forces defending in UO may gain advantages from the environment while Army force capabilities may diminish. Despite the challenges, Army forces conduct successful urban offensive operations by combining the Army's existing offensive doctrine with a thorough understanding of the environment.

PURPOSE OF URBAN OFFENSIVE OPERATIONS

7-1. Like all offensive operations, urban offensive operations are designed to impose the will of commanders on the threat. The urban offense often aims to destroy, defeat, or neutralize a threat force. However, the purpose may be to achieve some effect relating to the population or infrastructure of the urban area. Army forces may conduct offensive operations to secure a port or a communications center, to eliminate a threat to a friendly government or the urban population, or to deny the threat use of urban infrastructure. No matter the purpose, commanders must use a combined arms approach for successful urban offensive operations.

OFFENSIVE CHARACTERISTICS APPLIED TO URBAN OPERATIONS

7-2. All offensive operations, including those in urban areas, contain the characteristics of surprise, concentration, tempo, and audacity (see FM 3-0). Commanders consider and incorporate these characteristics in their offensive UO plans.

SURPRISE

7-3. Army forces can achieve offensive surprise at two levels: operational and tactical. In urban offensive operations, operational surprise can be decisive. The goal is to attack the urban area before the threat expects it, from a direction he doesn't expect, or in a manner he doesn't expect. In UO, this requires an attack against urban areas that the threat believes will provide sanctuary from the technological advantages of Army forces. Usually, urban areas that meet this criterion are not easily accessible. Army forces launch an attack against these urban complexes in different ways: through a vertical assault using airborne or air assault forces, through an amphibious assault, or through a penetration followed by a rapid and deep advance. All three attacks aim to achieve surprise and to deny the threat time to prepare and establish a defense. Surprise in a major urban operation prevents a threat from falling back to occupy prepared positions in and around an urban area.

7-4. At lower tactical levels, forces achieve surprise by attacking asymmetrically. An asymmetric method attacks the threat so he cannot respond effectively. This may be achieved by using special operations forces

(SOF) against a threat prepared for a conventional attack, by attacking decisively with heavy forces when the threat expects an effort by light forces or SOF, or by leveraging Army forces' extensive information operations (IO) capability. Offensive IO—primarily using IO elements of deception, electronic warfare, and operations security (OPSEC)—can help achieve surprise at all levels (see Chapter 4). Attacking at night surprises the threat and maximizes the Army forces' training, command and control (C2), and technological advantages. Attacking from unexpected or multiple directions achieves surprise by leveraging Army information systems (INFOSYS) and superior synchronization of combat power and capabilities.

CONCENTRATION

7-5. In UO, the attacking force creates a major advantage by concentrating the effects of combat power at the point and time of its choosing. The area and its compartmented effects naturally disperse and dissipate combat capability. Often commanders must position considerable amounts of stay-behind forces to protect hard-fought gains. The environment also hinders repositioning forces rapidly. Such effects can work equally against defending and attacking forces. However, in a well-prepared defense, the defender often has the advantage of interior lines. The defender can reinforce or reposition forces more quickly using covered and concealed routes (such as, sewers, tunnels, or prepared holes made in walls). Successful UO need synchronized air and ground maneuver with overwhelming effects from fires at decisive points on the urban battlefield. To achieve proper synchronization and precise effects, commanders should consider the unique time and distance relationships set by the environment.

TEMPO

7-6. Tempo is the rate of military action; it differs between UO and operations in more open terrain. Often, the primary purpose of the threat's urban defense is to take advantage of that differential to disrupt the rapidity and overall tempo of the Army's major operation. Additionally, the complexity and the potential for increased risk in an urban environment may invoke a cautious and methodical response on the part of commanders and their staffs thereby exacerbating tempo differences. While careful and meticulous preparation and planning are required in UO, commanders conducting major operations that include urban areas must strive to maintain an active tempo in offensive operations through synchronization of combat power and anticipation of threat reactions. A high tempo allows Army forces to achieve surprise and quickly gain positions of advantage. Controlling operational tempo and not allowing the different tempo of UO to adversely affect other operations is a challenge that commanders of major operations must overcome.

Operational Context of Urban Operations: Brittany Ports – 1944

The plan for the invasion of Normandy, France, in June 1944 was meticulously developed. The plan not only addressed the invasion itself, but also contained detailed planning for the campaign to follow. A major concern of the detailed campaign planning was logistics. To address this critical concern, and specifically the problem of ports to supply the allied armies once ashore, the pre-invasion planning called for the major ports of the French province of Brittany—Brest, Lorient, and Saint Nazaire—to be objectives of General Patton's Third Army, once it was activated.

Early August 1944, almost two months after the successful Normandy invasion, the operational situation significantly differed from that envisioned by the D-Day planners. General Montgomery's Twenty-first Army Group was still fighting in the Bocage of Normandy. In contrast, General Bradley's Twelfth Army Group had just achieved a major breakthrough at Saint Lo, secured the Cotentin Peninsula, and reached the city of Avranches. Here was a decision point. Bradley and Eisenhower had to decide whether to adhere to the original plan and turn west with Patton's forces to secure the peninsula or to take advantage of the breakout at Saint Lo and turn east to exploit the disruption of the German defenses.

Ultimately they reached a compromise. General Middleton's VIII Corps was tasked to secure the peninsula, and the bulk of Patton's Army, three Army corps, was turned northeast to exploit the operational collapse of the main German defenses. (See figure 7-1.)

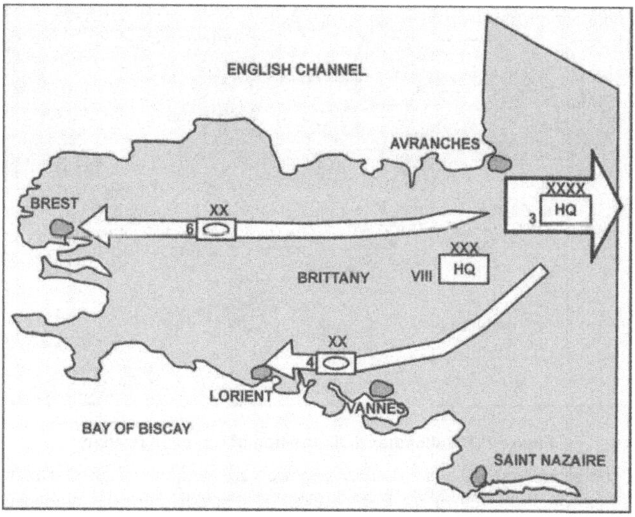

Figure 7-1. Initial attack in Brittany

Middleton's corps sprinted into the peninsula with the 4th and 6th Armored Divisions leading the way. However, poor communications, disagreements between commands, and contradictory orders caused the corps to hesitate before pushing the two divisions to continue to exploit toward the ports. The result: the 6th Armored Division missed an opportunity to seize Brest against light resistance by one day. The 4th Armored Division, after capturing the smaller port of Vannes, was also frustrated on the approaches to Lorient. The American reaction to the inability to rapidly seize the ports demonstrated an understanding of changing circumstances. The 6th Armored Division turned the attack at Brest to the 8th Infantry Division and then relieved the 4th Armored Division at Lorient. The 4th Armored was moved to rejoin the rest of Third Army exploiting to the east and north. Ultimately Brest fell to VIII Corps on 19 September after a 43-day siege by three infantry divisions. The victory yielded 36,000 German prisoners of war (POWs). However, the German defense and demolitions of the port left the port without an impact on the logistic situation of the allies. Brest cost the U.S. Army almost 10,000 casualties and the commitment of significant supplies. The experience convinced commanders to surround and bypass the other major Brittany ports. Lorient and Saint Nazaire remained under German control, deep in allied territory, until the war ended ten months later (see figure 7-2).

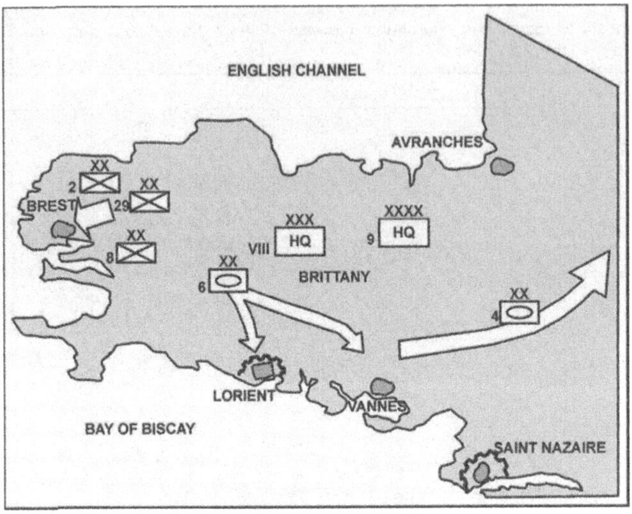

Figure 7-2. Subsequent disposition of forces in Brittany

The operational lessons of the Brittany campaign are numerous. First, commanders are responsible to continually assess assumptions and decisions made during planning based on the changing circumstances of the battlefield. This includes the planning decision to conduct urban offensive operations. When the allies arrived at the Brittany Peninsula, the focus of the operational maneuver was no longer securing logistics facilities but exploiting the breakthrough at Saint Lo and the disintegrating the German defense. The bulk of Third Army then was turned to the north and east rather than west into the peninsula.

The Brest experience also demonstrates that the costs of urban offensive operations are continually assessed against the operational value of the objective. This lesson was applied to the cities of Lorient and Saint Nazaire. The cities were never seized from the Germans because their logistic value failed to warrant the required resources to take them. German retention of the ports had no major adverse effect on the overall campaign.

Another lesson is that commanders cannot allow urban operations to disrupt the tempo of other offensive operations. One German goal of defending the ports was to disrupt the rapid tempo of the U.S. exploitation. They failed to achieve this goal because General Bradley continued the exploitation with the bulk of Third Army and executed the original plan with only a single corps.

Finally, commanders cannot allow emotion to color their decision to conduct or continue UO. The failure of 6th Armored Division to seize Brest rapidly caused some commanders to believe that Brest had to be captured because the prestige of the Army was committed to the battle. Costs of the continuing combat operations to seize Brest were significant. These resources might have been better committed elsewhere in the theater.

7-7. Tactical tempo is also important in urban combat. Because of the complex terrain, defending forces can rapidly occupy and defend from a position of strength. Once Army forces initiate tactical offensive operations, they cannot allow the threat to set the tempo of the operation. Instead, attacking forces seek to maintain a high tempo of operations. However, the tactical tempo of urban operations differs from operations in other terrain. Not necessarily slow, it requires a careful balance of preparation, speed, and security. In terms of unit fatigue, resource consumption, and contact with the threat, the tempo of most urban offensive operations may be rated as very high. On the other hand, in distances traveled and time consumed to achieve objectives, the tempo of many urban offensive operations might be rated as slow. The urban battlefield's density concentrates activity and consumes resources in a relatively small area. The lack of terrain seized or secured is not to be construed to mean a low tempo in the battle. Although the tempo may seem excruciatingly slow at higher levels of command and exceeding fast at lower, tactical levels, in reality, the natural tempo of urban operations is not faster or slower than other types of operations, merely different. Creating and operating at a tempo faster than an opponent can maintain, however, can favor forces which are better led, trained, prepared, and resourced.

7-8. A high tactical tempo in urban offensive operations challenges logisticians to provide for the increased consumption of munitions and degrades soldiers' physical capabilities. Commanders must anticipate these challenges and develop the means and abilities to overcome them. In the past, these challenges forced commanders to conduct urban offensives cyclically. They used night and other periods of limited visibility to resupply, rest, and refit forces. The environment influenced the tempo of their operations. This type of "battle rhythm" resulted in the forces spending each new day attacking a rested threat that was in a well-prepared position.

7-9. Army forces must maintain the tempo. Offensive operations continue even during darkness. Moreover, Army forces increase the tempo of operations at night to leverage the limited visibility capabilities, increased situational understanding, training, and INFOSYS that give an advantage to Army forces in all environments. To overcome the physical impact of the environment on soldiers, commanders can retain a large reserve to rotate, continuing offensive operations at night. The force that fights in daylight becomes the reserve, rests, and conducts sustaining operations while another force fights at night. Army forces can then maintain the tempo of operations and leverage technological advantages in urban offensive combat.

7-10. Tempo in UO does not necessarily mean speed. Offensive operations balance speed, security, and adequate firepower. Commanders must plan for the complex tactical environment and the requirements to secure flanks and airspace as the operation progresses. Mission orders allow subordinate units to make the most of tactical advantages and fleeting opportunities.

AUDACITY

7-11. Audacity is a simple plan of action, boldly executed. Superb execution and calculated risk exemplify it. In an urban attack, a thorough understanding of the physical terrain can mitigate risk. The terrain's complexity can be studied to reveal advantages to the attacker. Audacity can also be embodied in an operation by inventively integrating and coordinating the direct action tasks of SOF throughout the operation. Combining SOF actions with conventional attacks can asymmetrically unhinge a threat's defensive plan. Well trained Soldiers confident in their ability to execute urban offensive operations foster audacity.

URBAN OFFENSIVE OPERATIONS AND BATTLEFIELD ORGANIZATION

7-12. Urban offensive operations, like all operations, are arranged using the overall battlefield organization of decisive, shaping, and sustaining operations. Each operation is essential to the success of an urban offensive, and usually two or more of these operations occur simultaneously. Decisive operations are attacks that conclusively determine the outcome of UO. These attacks strike at a series of decisive points and directly lead to neutralizing the threat's center of gravity. Shaping operations in urban offensive operations create the conditions for decisive operations. In UO, much of the shaping effort focuses on isolation, which is critical in both major operations and tactical battles and engagements. Sustaining operations in urban offensive operations ensure freedom of action. They occur throughout the area of operations (AO) and for the duration of the operation.

Decisive Operations

7-13. A tactical commander fights decisive urban combat, whereas commanders conducting a larger major operation influence urban combat by setting the conditions for tactical success. Higher commanders may directly influence urban offensive operations by operational maneuver, by coordinating joint fires, by closely coordinating conventional forces, or with SOF.

7-14. Tactical urban offensive operations quickly devolve into small-unit tactics of squads, platoons, and companies seizing their objectives. The compartmented effect of the terrain and the obstacles to command and control of small units, especially once they enter close combat inside buildings or underground, often restricts the higher commander's ability to influence operations. Commanders can influence the actions of subordinates by clearly identifying the decisive points leading to the center of gravity; using mission orders (as discussed in Chapter 5); developing effective task organizations; synchronizing their decisive, shaping, and sustaining operations; and managing transitions.

7-15. Like all operations, successful decisive operations in UO depend on identifying the decisive points so the forces can destroy or neutralize the threat's center of gravity. Seizing a key structure or system that makes the threat's defense untenable; interdicting a key resupply route that effectively isolates the threat force from his primary source of support; or isolating the threat so that his force can no longer influence friendly activity may be more effective than his outright destruction.

7-16. Commanders must select the right subordinate force for the mission and balance it with appropriate attachments. Higher commanders do not direct how to organize the small tactical combined arms teams, but they must ensure that subordinates have the proper balance of forces from which to form these teams. Successful urban offensive operations require small tactical combined arms teams. Urban offensive operations require abundant infantry as the base of this force. However, successful urban combat requires a combined arms approach adjusted for the conditions of the environment. In urban offensive operations, it will normally not be a question of whether include armor and mechanized elements but, rather, how best to accomplish this task organization. Precision-capable artillery systems generally support urban operations better than unguided rocket artillery.

7-17. Divisions entering urban combat may require additional resources. These resources include military intelligence support in the form of linguists, human intelligence (HUMINT) specialists, and unmanned aircraft systems (UAS). Engineering assets will be at a premium; the task organization of a task force executing the decisive operation may require a one-to-one ratio of engineer units to combat units. Corps and higher engineering support may be necessary to meet these requirements and to repair vital and specialized infrastructure. A tailored and dedicated support battalion or group may need to assist in providing anticipated support to a displaced and stressed civil population. Finally, divisional CA units may require augmentation to deal with NGOs and civilian government issues.

7-18. Successfully conducting decisive operations in the urban environment requires properly synchronizing the application of all available combat power. Army forces have a major advantage in the command and control of operations. Commanders can use this advantage to attack numerous decisive points simultaneously or in rapid succession. They also use it to attack each individual decisive point from as many directions and with as many different complementary capabilities as possible. Commanders must completely understand urban environmental effects on warfighting functions to envision and execute the bold and imaginative operations required. Significantly, these operations require that C2 systems account for the mitigating effects of the environment as execution occurs.

7-19. Properly synchronized actions considerably enhance the relative value of the combat power applied at the decisive points. They present to the threat more requirements than he has resources with which to respond. Synchronized IO and multiple maneuver actions paralyze the threat's decision-making capacity with information overload combined with attacks on his C2 systems. Additionally, well-synchronized actions limit the time the threat has to make decisions and forces him into bad decisions. In the urban environment, these effects are enhanced because C2 systems are already strained, poor decisions are harder to retrieve, and units that do not react effectively are more easily isolated and destroyed.

Shaping Operations

7-20. Shaping operations that support the urban attack separate into those focused on isolating the threat and all others. Army forces isolate the threat to ensure successful urban offensive operations. Depending on the threat reaction to isolation efforts and the nature of the threat center of gravity, this task may become decisive. Other shaping operations include those common to all offensive operations and others unique to urban operations. Unique urban shaping operations may include securing a foothold in a well-fortified defensive sector, securing key infrastructure, or protecting noncombatants. Because of the nature of UO, shaping operations may consume a much larger proportion of the force than during other operations and may take place both inside and outside the urban area (see Applying the Urban Operational Framework: Panama in Chapter 5). By successfully isolating a threat force, the force needed to conduct the decisive operation may be relatively small.

Sustaining Operations

7-21. Commanders conducting urban offensive operations must ensure security of the sustaining operation and bases; in many situations, protecting lines of communications (LOCs) and sustaining operations may be the greatest vulnerability of the attacking force. Those supporting an urban offensive are tailored to the urban environment and are well forward. Ideally, the supporting forces closely follow the combat forces and move within or just outside the urban area as soon as they secure an area. Operating in the urban area during offensive operations allows the sustaining operation to take advantage of the defensive attributes of the environment for security purposes.

7-22. Counterattacks against sustaining operations may take the form of special operations activities aimed at LOCs leading to or within the urban area. Choke points—such as bridges, tunnels, and mountain passes—are vulnerable to these attacks and may require combat forces to protect them. Threat forces attack the LOC to blunt the Army's combat power advantage in the urban area.

7-23. Attacks against the LOC into the urban area may also attempt to isolate the attacking Army forces from its sustainment base. Isolated forces in an urban area are greatly disadvantaged. Commanders must plan and aggressively execute strong measures to protect their LOC, even if it requires reduced combat power to execute their offensive operation.

7-24. Sustaining operations anticipate the volume and unique logistics requirements of urban operations. Specialized individual equipment—such as grappling hooks, ladders, and pads—is identified and provided to troops in quantity before they are needed. Forces stockpile and distribute their attacking units' special munitions requirements including small arms, explosives, and grenades of all types, precision artillery munitions, and mortar ammunition. Forces also supply transport to move the resources rapidly forward, both to and through the urban environment. Sustaining operations cannot rely on "operational pauses" to execute their tasks. Commanders must plan to continuously supply resources and capabilities to the most forward combatants as offensive operations advance.

7-25. Sustaining operations also anticipate the growth of sustainment requirements as Army forces secure and take responsibility for large portions of the urban area. The success of Army urban offensive operations will often uncover the civil population in former threat occupied areas. It may attract the civil population from sections of the urban area where the Army is not operating to areas occupied by Army forces. Rural populations may migrate to the urban area as the result of successful Army offensive operations.

7-26. Army forces may be required to take initial responsibility to provide for the urban population. This consideration is integrated into logistics planning and organization from the start of the planning process. To be successful and efficient in such a situation, sustainment planning includes Army civil affairs (CA) specialists and local government representatives. It also integrates and consults with the international community and nongovernmental organizations (NGOs) that might augment or supplement Army sustainment capabilities.

FORMS AND TYPES OF URBAN OFFENSE

7-27. Traditional forms of offensive maneuver include envelopment, turning movement, infiltration, penetration, and frontal attack. Traditional types of offensive operations are movement to contact, attack, exploitation, and pursuit. These traditional forms and types listed apply to urban combat. Some have greater application to an urban environment than others do. Moreover, success will belong to commanders who imaginatively combine and sequence these forms and types throughout the depth, breadth, and height of the urban battlefield. This is true at the lowest tactical level and in major operations.

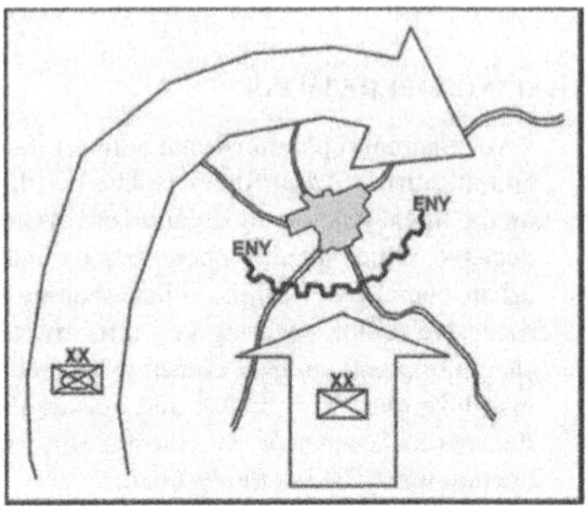

Figure 7-3. Envelopment isolates an urban area

FORMS OF OFFENSIVE MANEUVER

Envelopment

7-28. The envelopment is the ideal maneuver for isolating threat elements in the urban area or isolating the area itself. A deep envelopment effectively isolates the defending forces and sets the conditions for attacking the urban area from the flank or rear. Yet, enveloping an objective or threat force in the urban area is often harder since achieving speed of maneuver in the environment is so difficult (see figure 7-3). Vertical envelopment, however, works effectively if Army fires can effectively suppress or neutralize the threat air defense.

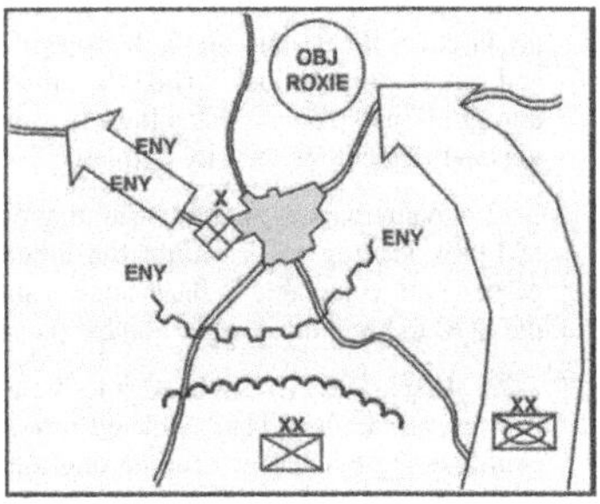

Figure 7-4. Turning movement

Turning Movement

7-29. Turning movements can also be extremely effective in major operations (see figure 7-4). By controlling key LOCs into the urban area, Army forces can force the threat to abandon the urban area entirely. These movements may also force the threat to fight in the open to regain control of LOCs.

Infiltration

7-30. Infiltration secures key objectives in the urban area while avoiding unnecessary combat with threat defensive forces on conditions favorable to them (see figure 7-5). This technique seeks to avoid the threat's defense using stealthy, clandestine movement through all dimensions of the urban area to occupy positions of

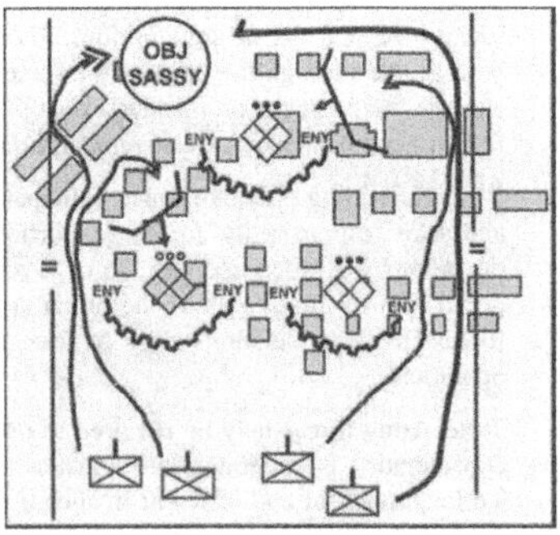

Figure 7-5. Infiltration

advantage in the threat's rear (or elsewhere).It depends on the careful selection of objectives that threaten the integrity of the threat's defense and a superior common operational picture (COP). Well-planned and resourced deception operations may potentially play a critical role in masking the movement of infiltrating forces. The difficulty of infiltration attacks increases with the size and number of units involved. It is also more difficult when Army forces face a hostile civilian population. Under such circumstances, infiltration by conventional forces may be impossible. Armored forces are generally inappropriate for infiltration operations. However, they may infiltrate large urban areas if the threat is not established in strength and had insufficient time to prepare defenses.

Penetration

7-31. Penetration is often the most useful form of attack against a prepared and comprehensive urban defense (see figure 7-6). It focuses on successfully attacking a decisive point or on segmenting or fragmenting the defense—thereby weakening it and allowing for piecemeal destruction. (The decisive point may be a relatively weak or undefended area that allows Army forces to establish a foothold for attacks on the remainder of the urban area.) Ideally in urban combat, multiple penetrations in all dimensions are focused at the same decisive point or on several decisive points simultaneously. In urban combat, the flanks of a penetration attack are secure, and resources are positioned to exploit the penetration once achieved. Although always a combined arms team, rapid penetrations are enhanced by the potential speed, firepower, and shock action of armored and mechanized forces.

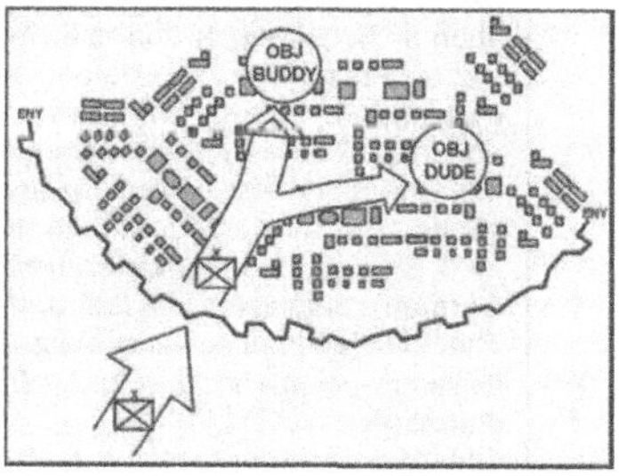

Figure 7-6. Penetration

7-32. Importantly, commanders must consider required actions and resources that must be applied following a successful (or unsuccessful) penetration. A penetration may result in the rapid collapse and defeat of the threat defense and complete capitulation. On the other hand, success may cause threat forces to withdraw but leave significant stay-behind forces (or disperse into the urban population as an insurgent-type force) which may necessitate methodical room-to-room clearance operations by significant dismounted forces. Additionally, securing portions (or all) of an urban area requires occupation by Army forces to prevent re-infiltration of threat forces thereby further increasing manpower requirements. Based on the factors of the situation, commanders may conclude that over time methodical clearance operations conducted from the outset, while frontloading time and resource requirements, will be less costly than a penetration followed by systematic clearing operations.

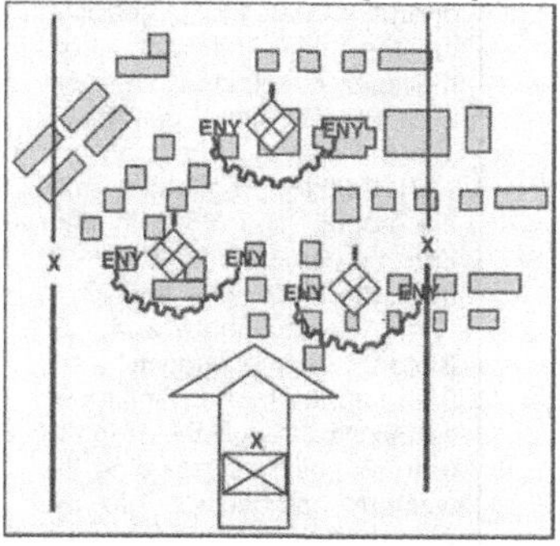

Figure 7-7. Frontal attack

Frontal Attack

7-33. For the commander of a major operation, the frontal attack is generally the least favorable form of maneuver against an urban area unless the threat is at an obvious disadvantage in organization; numbers, training, weapons capabilities, and overall combat power (see figure 7-7). Frontal attacks require many resources to execute properly, risk dispersing combat power into nonessential portions of the area, and risk exposing more of the force than necessary to threat fires. In urban offensive combat, forces most effectively use the frontal attack at the lowest tactical level once they set conditions to ensure that they have achieved overwhelming combat power. Then the force of the frontal attack overwhelms the threat with speed and coordinated and synchronized combat power at the point of attack. The assigned frontage for units conducting an attack on an urban area depends upon the size and type of the buildings and the anticipated threat disposition. Generally, a company attacks on a one- to two-block front and a battalion on a two- to four-block front (based on city blocks averaging 175 meters in width).

Forms of Attack in the Urban Offense: Metz – 1944

In November 1944, the U.S. Third Army launched its final effort to take the French city of Metz from the defending Germans. This was the Army's third attempt. The first attempt had been a surprise, mounted attack. This was followed by a series of piecemeal infantry assaults on the surrounding fortresses. Finally, a deliberate effort was made to take the city in a coordinated effort by XX Corps.

The initial failures stemmed from a shortage of resources on the U.S. side, to include fuel and units—especially infantry. This added to the ad-hoc nature of the first two efforts. The third effort, though more deliberate, was still constrained by resources. XX Corps could only muster three nearly full-strength infantry divisions to attack the German's defending with four under-strength divisions.

The third attempt to take Metz demonstrates how a corps operates with multiple divisions using various forms of attack to achieve its objective in urban offensive operations.

The opening phase of the Metz battle had attacks by the 90th and 5th Infantry Divisions to envelop the city from the north and south (see figure 7-8). This isolated the city and ensured the garrison could not escape nor be reinforced. The garrison was under orders from Hitler not to conduct a breakout and to resist to the last man; thus, German forces strongly counterattacked both wings of the envelopment to prevent isolation.

The second phase of the operation was the penetration of the city defenses from multiple directions. The 5th Infantry Division penetrated into the city with the 10th and 11th Infantry Regiments from the south. The 95th Infantry Division penetrated into the city from the north with the 377th Infantry Regiment and TF Bacon. Simultaneously, the 95th Infantry Division infiltrated battalions through the string of fortress positions guarding the western approaches into the city, isolated, and bypassed these positions with its other two regiments.

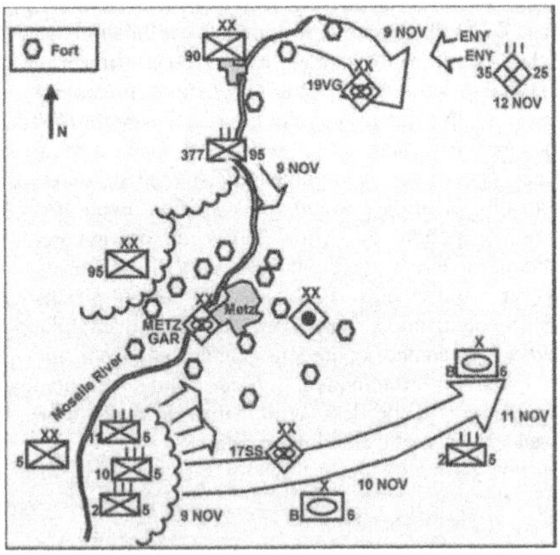

Figure 7-8. Metz envelopment

The final reduction of the defense was a series of battalion frontal attacks, which took place against the last remaining strongholds within the city. Even in these final engagements, however, the infantry battalions isolated, bypassed, and then attacked from the flanks and rear whenever possible (see figure 7-9).

The city was declared secured on 19 November. However, at that point more than a half-dozen of the fortresses had yet to be reduced. The 95th Infantry Division, after a four-day rest, quickly moved forward to rejoin the still rapidly advancing corps forward elements. Elements of the 5th Infantry Division remained in siege posture around the remaining strong points, the last of which surrendered on 19 December 1944 when it ran out of food. U.S. forces made no efforts to attempt to assault these bypassed fortresses although extensive psychological operations (PSYOP) were used.

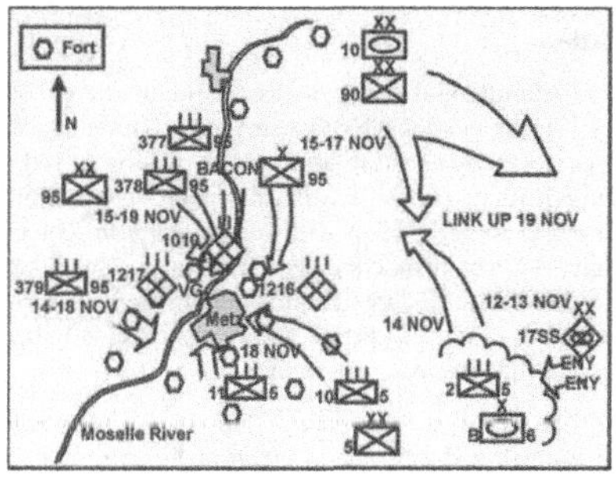

Figure 7-9. Metz final assault

TYPES OF OFFENSIVE OPERATIONS

Movement to Contact

7-34. In an urban area where the threat situation is vague, Army forces will often conduct a movement to contact to establish or regain threat contact and develop the situation. A movement to contact in an urban area occurs as both sides try to establish their influence or control over a contested urban area. The situation determines whether the movement to contact or its specific technique, the search and attack, is appropriate. A conventional force-oriented movement to contact will likely take place when friendly and threat conventional forces attempt to establish control simultaneously. Initially, neither side is defensive. The friendly force aims to quickly locate and fix the threat while establishing control of the urban area and its key infrastructure. The search and attack technique works well when a smaller threat has established a noncontiguous defense in an urban area. This operation is characterized by the friendly point defense of key infrastructure, robust reconnaissance, and rapidly concentrated combat power to fix and defeat or destroy threat resistance once located.

7-35. A meeting engagement often results from the movement to contact. It occurs when a moving force that is partially deployed for battle collides with and engages a threat at an unexpected time and place. In a meeting engagement in an urban area, the unit that reacts most quickly and decisively will likely win. Rapid and accurate decision making depends heavily on understanding the nature of the urban area and its impact on operations. Thus, in a meeting engagement, commanders should quickly assess the impact and role of all components of the urban environment (terrain, infrastructure, and society) on the operation. To this end, responsive reconnaissance is important. This permits accurate decision making regarding where to attack, where to defend, and how to allocate resources. Situational understanding enhanced by robust digital INFOSYS that provide an accurate COP also facilitates the rapid reaction of Army units and a synchronized response. This reaction and response allow Army forces to seize the initiative and dominate the threat.

Attack

7-36. The attack is the most common and likely offensive operation that Army forces will conduct in an urban environment. Commanders conducting major operations and commanders of large tactical units usually execute deliberate attacks. In the urban environment, units larger than battalion-size rarely conduct hasty attacks. Hasty attacks are common below company level as units use their initiative to take advantage of tactical opportunities. However, larger units will conduct hasty attacks when threat defenses are disrupted or unprepared, to take advantage of an unexpected situation, and to prevent the threat from establishing or re-establishing a coherent defense.

Exploitation

7-37. Exploitation follows a successful attack to disrupt the threat in depth. Commanders of major operations should consider focusing exploitation attacks on urban areas. A threat defeated in an attack will attempt to rally units, reinforce with reserves, and reorganize his defense. With its information and communications capability, transportation network, and defensive attributes, the urban area is the natural focal point to reestablish a disrupted defense. By establishing urban centers as the objectives of the exploitation, commanders deny the threat the sanctuary needed to reorganize and reestablish his defense. The exploitation focuses on the urban area as well as on the remnants of the threat. A successful exploitation to seize an urban area works efficiently because the attack preempts the defense and denies the threat the full advantages of urban terrain.

7-38. Commanders conducting exploitation must acknowledge the vulnerability of their forces to counterattack and ambush in urban areas. An urban area provides ideal cover and concealment to hide threat reserves, reinforcements, or reorganized forces. Constrictions of routes into and through the urban area make exploitation forces a potentially dense target and limit maneuver options. Robust and well-coordinated reconnaissance, tactical dispersal, and use of advance guard security forces protect against this threat (see Defensive Combat Power: Suez City vignette in Chapter 8).

Pursuit

7-39. The pursuit is designed to destroy threat forces attempting to escape. It focuses on the threat and not on urban areas. When conducting a pursuit, Army forces move through undefended urban areas and, if possible, bypass those in which threat forces successfully take refuge. The threat will likely attempt to use urban areas to disrupt the pursuit and permit the threat main body to escape. Commanders can prevent escape by denying the threat the time to establish forces in urban areas that cannot be bypassed. The agility of Army aviation forces for attack, reconnaissance, and transportation is essential to execute a successful pursuit around and through urban areas.

URBAN OFFENSIVE CONSIDERATIONS

7-40. The urban operational framework (understand, shape, engage, consolidate, and transition) provides a structure for developing considerations unique to urban offensive operations. The considerations vary depending on the situation and scale of the operation. Some considerations applicable to major operations that include an urban area will also be considerations at the tactical level focused in the urban area. However, no set rules exist. All urban operations are unique. Issues addressed at the operational level in one situation may be addressed in a new situation only at the tactical level. Under the right circumstances, a consideration may become an operational issue, a tactical issue, or a combination of the two. The following identifies some planning and execution issues that commanders conducting major operations should address.

UNDERSTAND

7-41. The first requirement, and a continuing requirement throughout the conduct of urban operations, is the assessment and understanding of the situation. Commanders should base this understanding on detailed information regarding the particular urban area. Since the threat will likely dominate or control most of the urban area during the planning phase of offensive operations, achieving an accurate understanding of the urban environment will be difficult. A comprehensive intelligence, surveillance, and reconnaissance (ISR) effort in support of a rigorous intelligence preparation of the battlefield (IPB) process overcomes this obstacle.

Integrated Intelligence, Surveillance, and Reconnaissance

7-42. The commander of a major operation that includes an urban area, unlike his subordinate commanders, can target reconnaissance deep into his AO and area of interest. This begins the application of ISR resources against the urban area that may lead to decisive ground operations. This ISR effort and the understanding it supports continue as long as the urban area remains in the AO. Commanders of major operations initially direct ISR assets on those information requirements that support determining whether or not to conduct urban offensive operations. Once decided, ISR resources shift to support the planning and execution of the operation in the urban area.

7-43. The first resources that a senior commander can use are national and strategic sensors. He requests them through the appropriate joint force commander. The commander aggressively pursues full use of these systems to begin building an initial database for analyzing the significant aspects of the terrain; key infrastructure considerations; the status and disposition of the population; and the size, type, disposition, and intentions of threat forces in the area.

7-44. Simultaneously, multiple intelligence sources contribute to the database. The sources collect, process, store, display, and disseminate the relevant information on large urban areas through open and classified resources. These information sources include—

- Historical research.
- Travel brochures that include cultural information and recent maps.
- Classified debriefings of diplomats, businesses, DOD personnel, and allies.
- Military maps and special geospatial products of the urban area.
- Previous intelligence assessments of the country, government, and population.
- Reachback to appropriate economic, political, cultural, and infrastructure subject matter experts not in the commanders AO.

The gathering and analysis of HUMINT plays a critical part of this process and assists commanders in understanding ethnic, cultural, religious, economic, political, and other societal and infrastructural facets of the environment.

7-45. As the intelligence and the national reconnaissance and surveillance efforts progress, commanders will insert, if available and feasible, Special Forces reconnaissance assets into the urban environment. These elements will seek to confirm or deny the information received from imagery intelligence (IMINT), signals intelligence (SIGINT), and HUMINT sources. Among many factors, using SOF depends on their availability, the particular urban area, the area's ethnic composition, and the relationship between the urban population and the threat. Other joint operational reconnaissance and surveillance assets that higher-echelon commanders may have available might include the Joint Surveillance Target Attack Radar System, Guard Rail targeting aircraft, UAS, and space-based systems.

7-46. The commander's staff will use all sources of information—IMINT and SIGINT sensors, HUMINT, historical research, and reconnaissance—to refine his ability to understand the urban environment. Digitally linking subordinate commanders with information sources helps to develop a COP essential to their situational understanding of the urban environment. The IPB process guides this assessment. As operations progress, additional reconnaissance and surveillance assets may become available. These may include UAS, long-range reconnaissance and surveillance units, counterfire radar, and air and ground cavalry. As these assets are employed, they are linked into the net of sources sharing information and further refine a common situational understanding of the environment.

Focused Assessment Efforts

7-47. In urban offensive operations, the tactical commander's assessment focuses on defeating the threat in the urban area within the constraints of the environment. Toward this end, identifying and assessing decisive points to attack is a commander's priority assessment task. Some unique aspects of the urban environment also require the focus of the commander's assessment efforts. These include the character of the urban defense, collateral damage considerations, and the effects of the environment on friendly and threat courses of action.

Character of the Urban Defense

7-48. To be both efficient and effective, Army urban offensive operations focus on what is decisive. Decisive points for an urban attack depend primarily on the mission within the urban area. They can vary widely in composition and size. Since commanders only focus on the essential, they may determine the decisive point to be a single building or a limited sector of an urban area. It could be an entire system within the urban infrastructure such as communications and information, or a limited subsystem of the transportation and distribution infrastructure such as a single airfield. Sometimes what is decisive in the urban area is the threat military capability, but even this large an objective, when carefully analyzed, may not require destruction of all threat forces or control of the entire urban area. Decisive points relate directly to the threat's center of gravity and to mission success. Some decisive points related to the urban threat's center of gravity may be physically located outside the urban area.

7-49. To gain specifics on threat dispositions within the urban area requires reconnaissance capability to see into the depths of the area and the intelligence capability to determine the threat's likely defensive course of action. With this information, commanders can determine decisive points and apply Army combat power discretely against them. Effective urban offensive operations require detailed situational understanding of an area of interest that extends well beyond the perimeter of the urban area.

7-50. Commanders see throughout the depth of the urban area using several actions (see figure 7-10). First, they evaluate sensor data and imagery. This guides targeting of special reconnaissance. Simultaneously, HUMINT is conducted using any persons who might know the urban area and threat. This includes civilians (allies, aides, neutrals, obstacles, and hostiles) and POWs. Finally, tactical conventional reconnaissance assets including reconnaissance forces, aviation, artillery radar, signals intelligence, and UAS are directed at the urban area. All these sources and data are linked through digital INFOSYS to provide commanders and their subordinates with improved situational understanding and a COP.

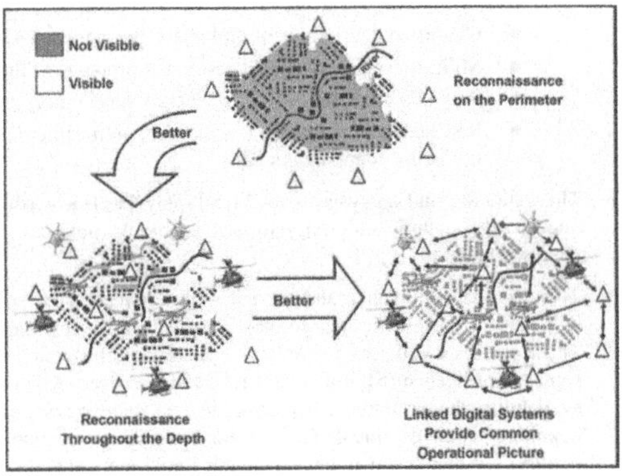

Figure 7-10. Required urban reconnaissance capabilities

Collateral Damage Considerations

7-51. Commanders must also assess the collateral damage risks that their operations may include. This assessment helps to initially determine the viability of a course of action. However, commanders should reassess their courses of action at frequent intervals in urban offensive operations based on known information to determine if the original evaluations remain valid. This reassessment minimizes potential collateral damage from a change in mission or a change in a course of action. Many aspects of the environment can change during mission execution.

The Environment's Effects on Courses of Action

7-52. The urban environment's unique aspects can significantly impact the course of action chosen by Army forces and the threat. Commanders must assess these effects in planning, but they must also verify and monitor these effects as forces execute offensive missions. In particular, commanders will want to confirm the civilian population's locations, beliefs, and actions and to monitor any changes. They will need to validate terrain considerations and monitor the effects of any changes due to rubble and other damage to structures. In urban terrain, dead space, cover, and concealment can only be identified physically and may change considerably as operations are executed.

SHAPE

7-53. Commanders of major operations have a primary contribution to urban operations: the planning and conduct of effective shaping operations that set the conditions for subordinate tactical success. In urban operations, isolation will be a critical condition. Effective isolation will require persistent, continuous surveillance and reconnaissance, innovative use of fires and maneuver (including effective force allocation decisions), and well-established sensor-to-shooter links. These efforts—combined and synchronized with SOF direct actions, IO that minimize noncombatant influences, and necessary shaping attacks (particularly the seizure of a foothold)—establish the conditions necessary for the subsequent offensive domination of the area.

Isolation is Essential

7-54. One key to success in the history of urban operations has been the effective isolation of the threat force (see figure 7-11). This applies today and equally well to major urban offensive operations as it does to smaller-unit attacks. This isolation not only denies access to the urban area from outside but also contains threat forces within. In a modern metropolis or megalopolis, this can appear a daunting task. Operational isolation requires dominating all physical and electronic contact between the threat in the urban area and supporting threat forces outside the urban area. This does not necessarily require physically encircling the urban area, but it does require that Army forces be able to exert control over the area's entire perimeter, as well as decisive points within. For a sprawling urban area, successful isolation may require the commitment of a large amount of resources.

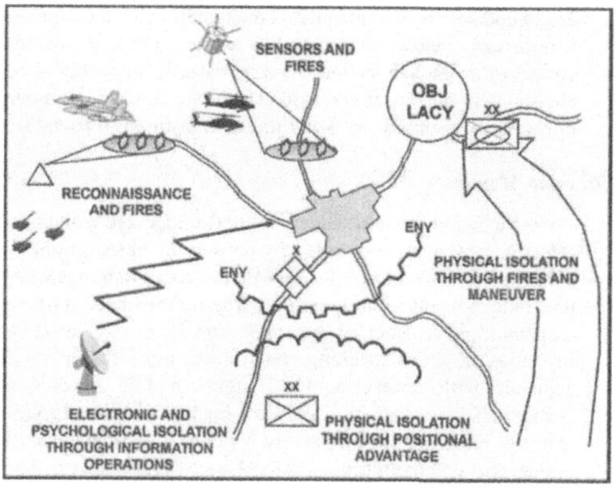

Figure 7-11. Shaping through isolation

7-55. Successful isolation of the urban area depends as much on the nature of the threat as it does on any other factor. A conventional threat in a large urban area may be much easier to isolate than an insurgent threat in a much smaller urban area. The forces needed in the former situation may be less than those needed in the latter. The more the characteristics of the threat are conventional in nature generally the easier it will be to isolate him using standard combat methods and equipment. Isolating a more unconventional force requires many of the same techniques as used against conventional forces. However, it also requires a much greater ability to simultaneously conduct offensive IO, to integrate CA units and civil-military operations (CMO), and to work with allies, nongovernmental organizations, and local authorities. Fundamentally, isolating a less conventional threat puts increased emphasis on separating combatants from noncombatants.

Offensive Isolation Objectives

7-56. Isolation seeks to achieve two primary objectives with respect to defeating a threat's urban defense:

- Weaken the overall coherence of his defense.
- Limit or manipulate his maneuver options.

Isolating the threat in the urban area from external support, as well as isolating him from sources of support within the urban area, weakens his overall defense. The defense is weakened through a combination of attrition (the threat cannot replace his losses) and the diversion of his combat power from the defense to operations to counter the isolation effort. Isolation can also prevent the threat from shifting his forces to reinforce decisive points in the urban area or to conduct counterattacks.

7-57. Commanders may choose not to isolate the urban area completely—or at least make it appear so to threat forces. Instead, they may afford the threat an apparent means of escape, create the conditions for its use through effective fire and maneuver against the defenders, and then destroy the threat through various ambush methods. While friendly forces may be able to move undetected to appropriate ambush sites, it is more likely that this technique will necessitate rapidly mobile air and ground forces moving along carefully chosen routes through the urban area. Commanders must consider maintaining the ability to complete the isolation of the urban area to prevent reinforcement and escape of urban threat forces—particularly if the ambush attack does not achieve desired effects.

Persistent Surveillance

7-58. Persistent surveillance of the urban area is essential to all types of actions used to isolate an urban area and as complete as resources will allow. Surveillance of the urban area relies on either reconnaissance forces or sensors continuously observing or monitoring urban avenues of approach. This network of ISR assets updates the commander's situational understanding and provides the means to quickly identify and, if necessary, attack threat elements as they move. However, particularly with sensors, commanders must consider that each detection is not necessarily an enemy to be attacked. Noncombatant activity clutters the environment making it easier for threats to disguise themselves and increases the burden (and the number of resources required) on Army forces to distinguish friend from foe.

Fires and Maneuver

7-59. Fires and maneuver may be used to achieve isolation, either singly or in combination. (As always, effective obstacles, monitored by sensors or observation, are integral to any isolation technique.) First, attacking forces can pre-position themselves along avenues of approach to deny entry and exit through positional advantage. Relying primarily on this method of isolation, particularly around a large urban area with multiple avenues of approach, can be overly resource intensive. Instead, the pairing of fires and maneuver provides attacking commanders more flexibility and allows them to isolate several avenues of approach with fewer resources. Highly mobile attack helicopters, operating outside threat-controlled portions of the urban area, are ideal for this purpose. Inside threat-controlled areas, it is more difficult to identify, eliminate, or effectively suppress the air defense threat. The threats may have numerous man-portable air defense weapons and enhanced effects of small arms used for air defense. Therefore, the risk to using this equipment may outweigh the potential benefits. However, mobile ground units—such as an air assault (subject to the same air defense threat considerations as attack aviation), armored, or mechanized forces—can also rapidly move to attack and destroy a threat moving in or out of an urban area. Potential disadvantages of the combined, fires and maneuver, option are that—

- Critical assets, on standby and dedicated to isolation efforts, may be unavailable for other missions.
- The attacking force may not locate the threat in time to complete its mission (an inherent risk to any attack).

7-60. Another alternative relies on indirect or joint fires alone to destroy the threat force. Its disadvantage is that fires alone often cannot completely destroy or stop a determined force from moving into or out of an urban area. Although targets and avenues of approach will require continual surveillance, it is usually a less resource-intensive option than those that include maneuver. It also does not normally require fires assets to remain on standby to accomplish the mission. However, fires must be able to reliably and quickly respond. For Army field artillery units and naval gunfire, the units must also be in range, which requires careful positioning. A skilled threat can avoid interdiction fires by using the geometry of the area to identify gaps due to obstructing terrain or the firing unit's range limitations. It can also use concealment and weather to avoid observation. However, effective sensor-to-shooter links throughout the urban battlefield will reduce the threat's ability to hide (see figure 7-12). A resolute threat may risk significant losses to fires to prevent isolation or may attempt to use noncombatants as a shield. Ultimately, commanders should use innovative combinations of all techniques discussed. Some units will physically block key avenues of approach. Surveillance will monitor less important routes and avenues. Artillery fires, joint fires, and maneuver units will then respond to the results of surveillance depending on the circumstances.

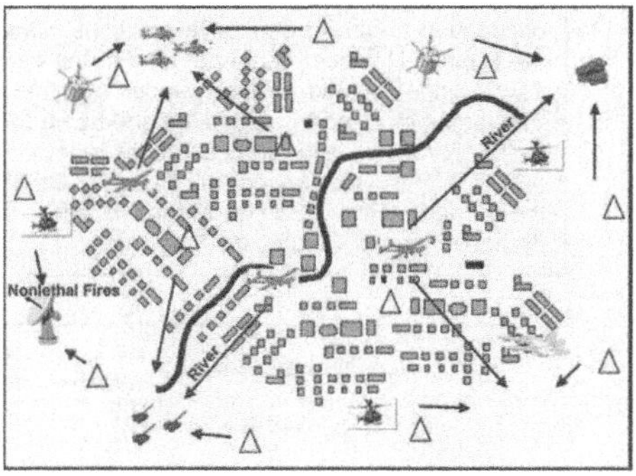

Figure 7-12. Critical sensor-to-shooter links

Threat Reactions

7-61. The reaction of the threat to the effects of isolation will depend on his mission, morale, force structure, and overall campaign plan. The threat may recognize isolation actions early and withdraw from the urban area before isolation is completed instead of risking destruction. On the other hand, the threat, based on a different or flawed assessment (perhaps a perception shaped by the Army force commander), may choose to—

- Continue to defend (or hide) and conduct local ambushes and counterattacks.
- Attack to break into the urban area or infiltrate forces and supplies in.
- Attack to break out of the urban area or infiltrate forces out.
- Execute any combination of the above (see figure 7-13).

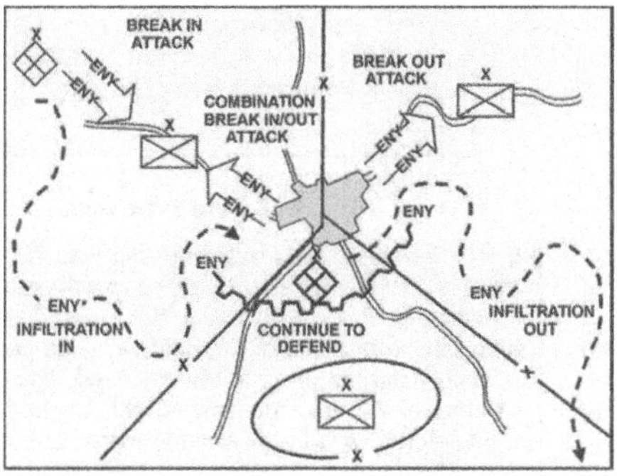

Figure 7-13. Reactions to isolation

7-62. Attacking commanders must consider how the threat leadership's subsequent actions will affect the continuance of overall offensive operations. They deliberate many considerations, to include—

- The allocation of more forces to the shaping operations to isolate the urban area.
- The allocation of more combat power to achieve rapid penetration and seizure of objectives to take advantage of developing threat dispositions in the urban area.

> **Isolating an Urban Area: Hue, Vietnam – 1968**
>
> On 31 January 1968, the 4th and 6th North Vietnamese Army (NVA) regiments and attached NVA and Vietcong (VC) sapper battalions attempted to seize control of Hue from the South Vietnamese Army's (ARVN) 1st Division as part of the North Vietnamese Tet Offensive. The attack, which was launched with complete surprise, successfully established temporary control over most of the city and facilitated the

occupation of strong defensive positions within the city's ancient fortress known as the Citadel. However, it did not lead to the capture of the headquarters of the ARVN 1st Division located in the Citadel nor the U.S. military assistance command—Vietnam compound located in the southeast part of the city.

Both ARVN and U.S. forces swiftly reacted to the NVA attack. Within 24 hours, ARVN infantry and cavalry units counterattacked to recapture the Citadel. South of the Perfume River, the 1st Marine Division began attacking to clear the southern portion of the city (see figure 7-14).

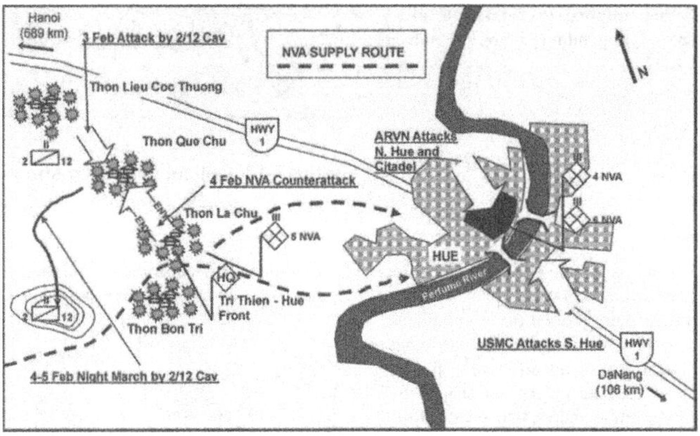

Figure 7-14. Initial attack to isolate Hue

On 2 February, the first Army unit was committed to Hue. The 2-12th Cavalry (an infantry battalion) was ordered to attack southeast along Highway 1 and clear enemy interdicting that route. On 3 February, 2-12th Cavalry began their attack moving along the south side of the highway. The plan was to clear a series of small hamlets in succession en route to the city itself. The first hamlet, Thon Lieu Coc Thuong, was cleared easily. Moving toward the second hamlet, Thon Que Chu, the battalion encountered a strong enemy force that was well entrenched. Air and artillery supported the battalion as it attacked and captured the hamlet against strong resistance. The battalion dug in for the night and prepared to resume the attack against the third hamlet, Thon La Chu, on 4 February.

Unknown to the 2-12th Cavalry, Thon La Chu was the headquarters of the Tri Thien-Hue Front, and it was defended by the NVA's 5th Infantry Regiment. Also important, the hamlet sat astride the NVA's primary supply route to the regiments fighting in Hue. The NVA was determined that 2-12th Cavalry's attack would not succeed; the NVA was determined to destroy the threat to its LOCs.

As the 2-12th Cavalry prepared to resume the attack on 4 February, the NVA fiercely counterattacked with all three battalions of the 5th NVA Regiment. It soon became apparent that 2-12th Cavalry could not continue to attack. As the day continued and the NVA pressure steadily increased, the battalion began to lose its position in Thon Que Chu. To avoid destruction, the battalion broke contact.

As darkness fell on 4 February, 2-12th Cavalry finally broke contact with the NVA. However, instead of retreating north back to its start line, the battalion moved west and then south to good defensive high ground deeper in the NVA AO. Eleven hours later the battalion was set in its new defensive position.

At dawn on 5 February, 2-12th Cavalry was established 5 kilometers west of Hue. The battalion soon observed enemy forces and supplies moving toward Hue. From its high ground position, the battalion directed artillery and air strikes against the NVA forces. By its bold move to bypass the 5th NVA Regiment, the battalion now held excellent position to direct fires on the primary NVA supply line into Hue. These fires were the first step toward isolating the NVA in Hue.

The fires controlled by the 2-12th Cavalry shut down the NVA LOCs into Hue during the daytime. However, under the cover of darkness supplies and reinforcements were still entering the city (see figure 7-15). It became apparent that the isolation of the NVA in Hue would require the capture of Thong La Chu. The problem facing American forces was concentrating combat power against the NVA. All U.S. units at this time were actively engaged against the numerous NVA attacks that constituted the NVA's 31 January Tet Offensive.

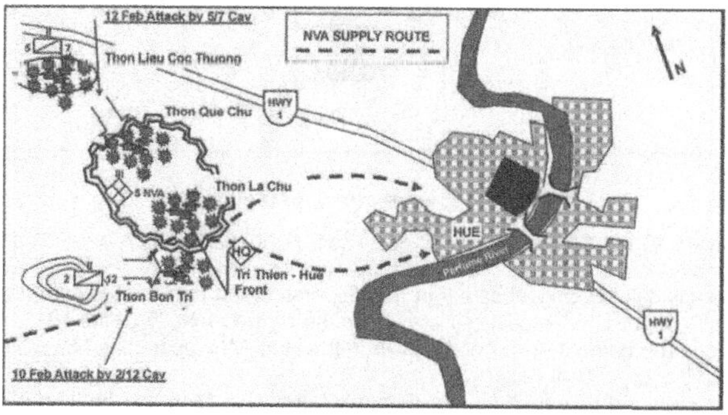

Figure 7-15. Subsequent attack to isolate Hue

The first additional American unit was not available until 12 February when the 5 7th Cavalry attacked Thong Que Chu much like the 2-12th Cavalry had attacked previously. The 5-7th Cavalry had even less success against the alert 5th NVA. The 5-7th Cavalry was forced to occupy defensive positions in Thon Lieu Coc Thuong and await the build up of combat power before it could continue to attack. In the interval, 2-12th Cavalry had moved off the high ground and captured the hamlet of Thon Bon Tri, south of the 5th NVA Regiment.

On 21 February, the 1st Cavalry Division had moved enough resources to the area to launch an effective attack to isolate Hue (see figure 7-16). In addition to the 5-7th and 2-12th Cavalry, the 1-7th Cavalry arrived in the AO and the 2-501st Airborne Infantry of the 101st Airborne Division was attached. On 21 February, after a combined artillery, air, and naval gunfire bombardment, the four battalions attacked the Thon La Chu stronghold. Elements of the 5th NVA Regiment were either destroyed in place or fled northeast. The next day resistance in Hue was noticeably lighter. U.S. Marine and ARVN units began the last phase of fighting to recapture the Imperial Palace. On 26 February, the North Vietnamese flag was removed from the Citadel and the ARVN I Corps declared the city secured.

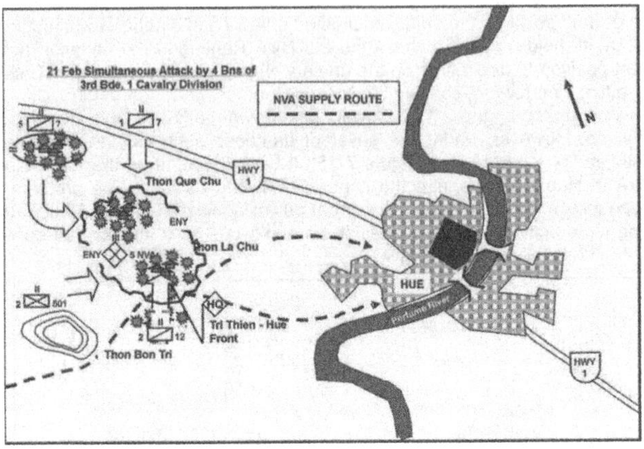

Figure 7-16. Final attack to isolate Hue

> The actions of the 1st Cavalry Division forces northwest of Hue demonstrated the importance and the difficulty of isolating an enemy fighting in an urban area. Isolating Hue was difficult not only because of the dispersion and surprise with which the Tet Offensive caught U.S. forces, but also because of the tenacity of the NVA. At least one-third of the combat power of the NVA in the Hue AO was focused on maintaining access to the city.
>
> Although only one aspect of this urban battle, Hue's isolation had an immediate and important, if not decisive, impact on operations. It not only resulted in restriction and then elimination of supplies and reinforcements, but it also immediately impacted the conduct of the defending NVA forces. Isolation caused an immediate drop in NVA morale and changed the nature of the defense. Once the enemy was isolated from external support and retreat, the objective of the NVA in the city changed from defending to avoiding destruction and attempting to infiltrate out of the city.

Civilian Reactions

7-63. Commanders must also consider the potential effects on and reactions and perceptions of the population living in the urban areas that they choose to isolate and bypass—either as a direct effect or as a response of the threat force being isolated. Isolation to reduce the threat's ability to sustain itself will likely have similar (and worse) effects on the civilian population remaining in the isolated area. (If food and water are in short supply, threat forces may take from noncombatants to satisfy their needs, leaving civilians to starve.) Isolation may also create a collapse of civilian authority within an urban area as it becomes apparent that the military arm of their government is suffering defeat. Due to their isolation, elements of the population may completely usurp the governmental and administrative functions of the former regime and establish their own local control, or the population may lapse into lawlessness. Returning later, Army commanders may find that these self-governing residents are proud of their accomplishments and, in some instances, less willing to allow Army forces to assume control since they may be perceived as having done nothing to earn that privilege. Alternatively, as witnessed in some urban areas during OPERATION IRAQI FREEDOM in 2003, a power vacuum may lead to intra-urban conflicts among rival factions coupled with general public disorder, looting, and destruction of the infrastructure (see also the discussion of Competing Power Structures in Chapter 3).

Direct Action by Special Operations Forces

7-64. Although SOF in urban offensive operations will likely conduct essential reconnaissance, they also have a direct action capability to shape the offensive operation (see figure 7-17). Special Forces and Rangers can use direct action capabilities to attack targets to help isolate the urban area or to directly support decisive actions subsequently or simultaneously executed by conventional forces. Successful attacks against urban infrastructure, such as transportation or communications centers, further the area's physical and electronic isolation. Direct action against command centers, logistics bases, and air defense assets can contribute to the success of conventional attacks by destroying or disrupting key threat capabilities. Direct action can also secure key targets such as airports, power stations, and television stations necessary for subsequent operations. Direct action by Special Forces and Rangers in these operations can help achieve precision and reduce potential damage to the target or noncombatant casualties.

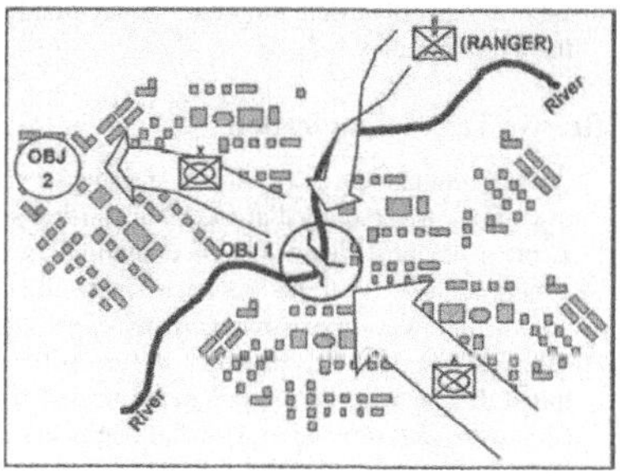

Figure 7-17. Coordination of SOF and conventional capabilities

Information Operations

7-65. Regardless of how Army forces physically isolate the urban area, they combine physical isolation with offensive IO to electronically isolate the threat and undermine his morale. Electronic isolation will cut off communications between forces in the urban area from their higher command to deny both from knowing the other's status. IO combined with isolation may persuade the threat's higher command or leadership that its forces located in the urban area are defeated. Thus, the command or leadership's intentions to break through to the besieged threat forces may be affected. PSYOP can undermine the morale of the threat in the area and reinforce electronic isolation and perceptions of abandonment. IO can be used to reduce any loyalty the civil population may have to the threat. IO can also ensure that civilians have the information that minimizes their exposure to combat and, as a result, overall noncombatant casualties. In addition, IO aim to deceive the threat regarding the time and place of Army force operations and intentions.

Detailed Leader Reconnaissance

7-66. Army commanders must clearly see the urban environment to understand the challenges facing their brigades, battalions, companies, platoons, and squads. Urban terrain can be deceptive until viewed from the soldier's perspective. Commanders are responsible to intimately know the conditions to allocate resources effectively to subordinate units. Often, particularly at battalion level and above, commanders will not be able to command and control dispersed forces from positions forward, but be forced by the terrain to rely on semifixed command posts. Detailed leader reconnaissance of the AO by commanders, their staff, and their subordinates before the mission can compensate for this challenge. This reconnaissance will give commanders a personal feel for the challenges of the terrain and will facilitate more accurate planning and better decision making during operations.

Mission Orders

7-67. Before contact, commanders can mitigate some terrain challenges to effective C2 using mission orders. Subordinates have mission orders to take advantage of opportunities before C2 systems can adversely impact the environment. To see the battle and provide effective and timely direction, tactical leaders will follow closely behind units as they assault buildings, floors, and rooms. Thus, only the most mobile INFOSYS can accompany tactical leaders into combat, and they will suffer the degrading effects of the environment. Mission orders permit rapid and decisive execution without commanders intervening at battalion level and above. Higher-level commanders can facilitate mission orders through their subordinates by articulating their desired end state, clearly stating their intent, and building flexibility into the overall plan.

Effective Task Organization

7-68. Commanders can shape urban offensive operations through effective and innovative task organization. Combined arms, often starting with an infantry base, are essential to success and may be the Army's asymmetric means of defeating an urban threat. Urban attacks will quickly break down into noncontiguous firefights between small units. To achieve the tactical agility for mission success in this nonlinear environment, many Army capabilities are task organized down to the company, platoon, and squad levels. Infantry provides the decisive capability to enter buildings and other structures to ensure threat destruction. Tanks, gun systems, and fighting vehicles provide additional speed and mobility, direct firepower, and protection. Combat engineers provide specialized breaching and reconnaissance capability. Field artillery provides the indirect (and if necessary, direct) firepower. Such mobility and firepower create the conditions necessary for the dismounted infantry to close with and destroy a covered threat in an urban defense. When a threat skillfully uses the urban area to limit ground maneuver, vertical envelopment or aerial attack using precision-guided munitions from Army aviation may circumvent his defenses and achieve necessary effects. Generally, ground systems used within the urban area will not be able to operate independently from dismounted infantry. The infantry will be required to protect armor and mechanized systems from close antiarmor weapons, particularly when those weapons are in well-prepared positions located throughout the urban area but especially on rooftops and in basements.

7-69. In urban offensive operations, direct fire support can be critical. Armor vehicle munitions types do not always achieve decisive effects against some urban structures. In some cases, field artillery high explosive munitions work better than armor for direct fire support of infantry. Large caliber (105 or 155mm) high explosives directly fired at a structure often produce a more severe shock effect than tank and fighting vehicle cannon and machine guns produce. Artillery is also able to achieve higher elevation than armor and engage threats located at greater heights.

7-70. However, commanders must view artillery as not just a weapon but a weapon system. As such, artillery should normally be placed under tactical control (TACON) of maneuver commanders, such as a platoon of three guns TACON to a company or a battery to a battalion, not just one gun to a company or other maneuver unit. Self-propelled artillery has some of the mobility characteristics of armor; however, it provides minimal ballistic protection from fragmentation for the crew. Although these systems seem formidable, they provide less crew protection than a Bradley fighting vehicle, for example, and contain large amounts of onboard ammunition and propellant. They are susceptible to catastrophic destruction by heavy automatic weapons, light cannon, and antitank fire. Therefore, infantry units must carefully secure and protect these systems (even more so than armored vehicles) when employed in urban offensive operations, particularly when forward in the direct fire role.

Creative Task Organization: Using Artillery in the Direct Fire Role

Task organizing artillery to permit its use in a direct fire role demonstrates the innovative task organization required for urban operations. The following provides three historical examples of task organizing and using field artillery for a direct fire role.

In 1944, U.S. Army units of the 1st Infantry Division were assigned to attack and seize the German city of Aachen. The city's internal defense included bunkers designed to serve as air raid shelters. These positions, buildings of stone, seemed impervious to direct fire tank weapons, demolitions, and small arms. To reduce the positions, the 1st Infantry Division relied on the artillery's direct fire.

Field artillery used this way had physical and psychological effects on the defenders. The 26th Regimental Combat Team's history of the battle describes the German reaction to the artillery pieces:

The chief shock to the defenders, Colonel Wilck (Aachen defense commander) said, came from the self-propelled 155s and tanks. The colonel spoke with considerable consternation of the 155mm self-propelled rifles. A shell from one of the guns, he said, pierced three houses completely before exploding and wrecking a fourth.

The 26th Infantry Regiment also described how the artillery, one piece attached to each assaulting infantry battalion, helped the infantry to penetrate buildings.

With solid blocks of buildings comprising most of the city, there wasn't any easy way to get at the Germans in the buildings. The eight-inch gun solved the problem. Beginning on the eastern outskirts the gun would plow a round into the side of the built up block of buildings at about ground level. One shell would usually open an entrance into the first tier of floors, i.e. the first building. Then several more shells were fired through the first hole. Thus a tunnel would be rapidly made all the way to the next cross street. Soldiers could then rush the newly formed entrance, clear the upper floors with hand grenades and rifles and then move on to the next building to repeat the process. When a block or square, was thus completely cleared of Germans—soldiers, skulkers, or even snipers—the next square was treated in the same way, working forward square by square, right and left, thereby avoiding nearly all exposure in the streets.

In 1982, Israeli forces invaded southern Lebanon to destroy base camps of the Palestine Liberation Organization (PLO). This operation involved significant fighting in urban areas including major operations in Beirut. Artillery, firing in a direct fire role, played a major part of the tactical solution. Artillery was particularly effective in the 33-day siege of Beirut. During this siege, Israeli forces used artillery in its traditional role as well as in the direct fire role.

The Israeli army was committed to a policy of disproportionate response during the Beirut siege. When fired on with small arms, crew-served weapons, tanks, or indirect artillery, the Israeli forces responded with intense, high-caliber direct and indirect fire from tanks and artillery positioned around the city. Many firing positions were on heights to the south and southwest that dominated much of the city. These positions had almost unrestricted fields of view. Israeli artillery fired from these positions directly into high-rise buildings concealing PLO gunners and snipers. The artillery, using direct fire, destroyed entire floors, collapsed floors on top of each other, and completely removed some upper floors. Such a response, as in Aachen in 1944, had as much a psychological impact as it did a physical impact on the PLO defenders.

In the early hours of 20 December 1989, the United States launched OPERATION JUST CAUSE. One of this operation's objectives was removing the Panamanian dictator, Manuel Noriega. U.S. forces carefully planned using all fires before the operation to minimize casualties and collateral damage. Part of this detailed fire planning called for applying artillery in a direct fire role.

The Panamanian Defense Force's (PDF) 5th Rifle Company based at Fort Amador was one of the key objectives of U.S. forces at the start of hostilities. This unit was high priority because it was the closest PDF unit to Noriega's headquarters, the Comandancia. U.S. forces expected the 500-man company to react first to

OPERATION JUST CAUSE by reinforcing the defense of the Comandancia. It also posed a threat to U.S. military dependents housed at Fort Amador.

To quickly neutralize this force, the United States assembled a three-company force composed of A, B, and headquarters elements of 1-508th Infantry (Airborne), supported by 105mm towed howitzers of 320th Field Artillery and M113 armored personnel carriers. The howitzers and the personnel carriers were covertly pre-positioned at the fort. At approximately 0100, helicopters transported the two airborne rifle companies into position. The howitzers then suppressed any personnel in the PDF-controlled buildings on Fort Amador while demonstrating the firepower of the U.S. task force. They used direct fire into the PDF barracks. The impact of the 105mm high explosives and .50-caliber fire from the M113s convinced the PDF infantry to give up after token resistance. Following the direct fire, U.S. infantry assaulted and cleared the dozen PDF buildings, finding that most occupants had fled or surrendered. For more details of OPERATION JUST CAUSE, see Applying the Urban Operational Framework: Panama in Chapter 5.

The three examples cited indicate the importance of the situation-dependent and innovative task organization of artillery and its use in the direct fire role. Using artillery helps overcome some challenges of offensive operations in the urban environment, and it has an important psychological effect on a defending threat. Such task organization takes a traditional tool of a higher-level tactical commander and uses it to directly influence the company-level battle. This philosophy of task organization can be applied to other types of forces—not just artillery. PSYOP teams, interpreters, CA specialists, armor, engineers, and reconnaissance teams may require task organization different from traditional organization. The compartmented urban environment drives the requirements for these assets lower in the tactical scheme than in open operations. Consequently, commanders must understand and account for more of these assets for UO than for operations in less restrictive environments.

7-71. Army aviation may also be inventively task organized. It can support urban operations with lift, attack, and reconnaissance capabilities. Tactical commanders down to company may use all these capabilities to positively influence ground close combat. Army attack and reconnaissance aircraft can provide flank security for attacking ground forces. Attack aircraft may also provide direct fire support to individual platoons or squads. Lift may move entire battalions as part of brigade operations, or it may move single squads to a position of advantage (such as a roof) as part of a small unit assault. Army aviation can assist with C2 by providing airborne retransmission capability, airborne command posts, and the confirmed status and position of friendly forces. However, Army aviation is a limited and high-value asset; commanders should review its use in innovative task organizations. It is particularly vulnerable to urban air defense threats unless used over terrain secured by ground forces. From these positions, aircraft can use enhanced sensors to conduct reconnaissance and use precision weapons with standoff capability.

Shaping Attacks

7-72. In a large urban area, the defending threat cannot be strong everywhere. Shaping operations can also take the form of attacks against vulnerable positions to force the threat to maneuver and redeploy his forces in the urban area. This prevents him from merely defending from prepared positions. Forcing the threat to move negates many of the defensive advantages of urban terrain, confirms his dispositions, exposes vulnerable flanks, and permits target acquisition and engagement with precision standoff fires.

7-73. A critical shaping operation in urban offensive operations is usually an initial attack to seize a foothold. Once Army forces establish this foothold, they accrue some of the defensive advantages of urban terrain. From this protected location, Army forces continue offensive operations and have a position of advantage against neighboring threat defensive positions.

ENGAGE

7-74. Commanders may employ several methods to decisively engage elements of the urban area during offensive operations. These include—

- Rapid maneuver.
- Appropriate use of SOF.
- Precise application of fires and effects.
- Proper balance of speed and security.

7-75. None is unique to UO. Their effective execution, however, allows Army commanders to dominate in this challenging environment by effectively using resources with the least amount of collateral damage. Overall, decisive engagement results from urban offensive operations when forces achieve the objective of the assigned mission and establish preeminent control over the necessary terrain, population, and infrastructure. Largely, the Army commander's ability to engage is based on superior situational understanding and the correct application of unit strengths to the challenges found in the urban environment.

Rapid and Bold Maneuver

7-76. Commanders of major operations may have or create the opportunity to seize an urban area with rapid and bold maneuver. Such maneuver requires striking while the area remains relatively undefended—essentially preempting an effective defense. This opportunity occurs when the urban area is well to the rear of defending threat forces or before the onset of hostilities. Under such conditions, an attack requires striking deep behind threat forces or striking quickly with little time for the threat to make deliberate preparations. Attacks under such conditions may entail significant risk; however, the potential benefit of audacious offensive operations may be well worth possible losses. Three potential ways to accomplish such attacks (and their combinations) are:

- Airborne or air assault.
- Amphibious assault.
- Rapid penetration followed by an exceptionally aggressive exploitation, for example, a heavy force using shock, armor protection, and mobility.

7-77. Commanders should analyze all potential urban operations to seek an opportunity or advantage to apply rapid and bold operational maneuver to the task. Using operational maneuver to avoid urban combat against an established threat defense potentially marks a significant operational achievement and can have decisive strategic consequences. Just influencing the threat's morale can positively affect all future operations. However, commanders must evaluate the challenges of such a course of action. These challenges may include—

- Sustaining the operation.
- Avoiding isolation and piecemeal destruction.
- Successfully conducting shaping attacks.
- Achieving the necessary tactical, operational, and strategic surprise.

7-78. Commanders can also build on the shaping effects of isolating the urban area internally and externally by attacking urban decisive points from multiple directions. They can attack multiple decisive points either simultaneously or in a systematic, synchronized manner. This complicates the threat's situational understanding of the urban environment, further impedes his decision making, and allows Army commanders to dictate the tempo.

Bold Operational Maneuver to Seize an Urban Area: Inchon and Seoul, Korea – 1950

In August 1950, UN forces in Korea were desperately attempting to stave off defeat by establishing a viable defense in southern Korea. This defense, known as the Pusan Perimeter, was the focus of the world's attention. The commander of U.S. Forces Far East, General Douglas MacArthur, was not focused on the U.S. Eighth Army in Pusan. Instead, he focused on how to dramatically and decisively transition to the offense—the recapture of the South Korean capital of Seoul.

Seoul was more than just the South Korean political capital, although that was an important consideration. Seoul was the key to the Korean transportation system, the hub of the national communications system, and the center of the country's economy and culture. It was also strategically placed astride maneuver corridors to the north, south, and northeast. It had been Korea's capital since 1394 when King Taejo Yi Sung-Gye established it as the center of the Chosun dynasty. It had been captured by the surprise communist invasion of the south in June 1950, and the key to success in Korea was its recapture. The question was how.

The obvious military solution to the Korean situation in August 1950 was to build military strength within the Pusan Perimeter and then, when sufficiently strong, to breakout and counterattack north and recapture Seoul. The disadvantage of this course of action was that the counterattack would be through hundreds of miles of mountainous terrain, through several other major cities, and across numerous rivers. The North Korean People's Army (NKPA) would be able to fall back from one mountain defense and one river line to another and would have weeks or months to prepare the defenses of Seoul.

MacArthur's solution was to delay a conventional counterattack, boldly maneuver by sea deep into the flanks of the NKPA, land on the Korean west coast at Inchon, and rapidly seize Seoul before the enemy could react (see figure 7-18). This course of action had numerous disadvantages: achieving surprise; assembling a landing force trained in amphibious operations; few good landing sites; supplying a large force once landed; and needing to simultaneously continue to defend and prepare a counterattack with the Eighth Army from Pusan. MacArthur understood the challenges of the operation but believed that the potential value compensated for the risk.

The potential value of a rapid assault on Seoul through a deep amphibious operation was immense. The operation would trap the bulk of the NKPA in South Korea and facilitate its destruction. It would also capture Seoul before the NKPA could prepare a defense. The quick recapture would immeasurably impact the morale of UN forces and South Korea. MacArthur felt the payoff of success justified the risks and challenges of the operation.

On 15 September, elements of the 1st Marine Division landed in Inchon taking the NKPA completely by surprise. On D+1 they were joined by elements of the 7th Infantry Division, both divisions under the control of U.S. X Corps. Naval gunfire and Marine close air support flying from Navy aircraft carriers supported the landings. Seven days later, the 5th Marine Regiment had battled 25 miles to the outskirts of Seoul and began attacking the city from the north. Three days later, on 25 September, the 1st Marine Division and elements of the 7th Infantry Division had defeated the 10,000 defenders of the NKPA's 18th Rifle Division.

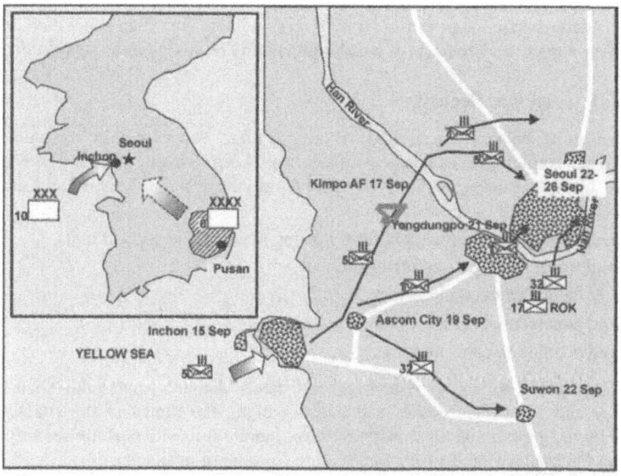

Figure 7-18. Inchon-Seoul campaign, September 1950

> Most of MacArthur's predictions for the Inchon-Seoul campaign were accurate. Superb staff work, excellent logistics operations, and unmatched training permitted the UN forces to assemble a trained landing force, land it over one of the most difficult shores in the history of military amphibious operations, and keep it supported. More importantly, the operation achieved total strategic and operational surprise. Seoul was recaptured quickly and, although resistance was fierce, the NKPA could not react fast enough to influence the outcome. In conjunction with the Eighth Army counterattack from Pusan, 75 percent of the NKPA was destroyed. The Inchon-Seoul campaign was an important turning point in the war, and had not the Chinese People's Army intervened two months later, it would have been the decisive campaign of the war.

Appropriate Use of Special Operations Forces

7-79. Sometimes Army forces can decisively engage (instead of simply shape) the urban area using the direct action capability of SOF. When the threat fails to develop a comprehensive defense and does not possess large, capable conventional forces, then Army forces can achieve operational surprise. Commanders, by synchronizing conventional and SOF effects, may actively control offensive operations to dominate the area. Importantly, however, conventional ground forces must be available to quickly assume the mission because SOF units acting as the primary striking force have limited logistic capability to sustain long-term operations. OPERATION JUST CAUSE offers several examples of this type of synchronization (see Applying the Urban Operational Framework: Panama in Chapter 6).

Precise Application of Fires and Effects

7-80. Precisely applied fires and the massed effects of combat power characterize successful urban attacks. The fires can be direct fire from combined heavy or light ground teams; direct or indirect fires from supporting Army aviation standing off from the target and any possible air defense threat; precision indirect fires from conventional tube artillery; or direct and indirect fires from supporting joint assets including Air Force and naval assets. All efforts strive to reduce collateral damage around the point of attack, consistent with mission success. Forces use fires to deny the threat the ability to maneuver in the urban area and to destroy the threat when he attempts to maneuver. When the threat exposes himself by moving, the environment no longer protects him, and fires can effectively engage him. Overall, precise fires and effects demonstrate the power of Army forces and help the urban population understand that only legitimate military targets are the focus of attacks, potentially building public support of Army UO.

Proper Balance of Speed and Security

7-81. Attacking units balance speed and security. Forces secure flanks as units advance, control dominating terrain (buildings), evacuate civilians, and keep the integrity and synchronization of the combined arms team. Obstacles are anticipated and rapidly breached. Commanders can choose avenues of approach to—

- Provide cover and concealment for following aviation and support units.
- Permit travel by all classes of vehicles.
- Easily defend from counterattack.
- Avoid nonessential centers of threat resistance.
- Avoid population concentrations.

7-82. Army aviation is a critical resource to protect flanks. Another important resource is engineers who seal off surface and subsurface entries and avenues along the flanks of the attack. Finally, as in all offensive operations, ground and air cavalry are ideal mobile forces to perform security in an economy of force role along flanks allowing decisive forces more freedom of maneuver.

CONSOLIDATE

7-83. Commanders, at all levels, must consolidate to strengthen their position during urban offensive operations without loss of momentum. They take the steps necessary to make permanent any temporary battlefield successes while maintaining relentless pressure on threat forces. Consolidation provides security and protection, facilitates reorganization, and allows forces to prepare for counterattack.

Repositioning of Forces

7-84. Following seizure of the objective, commanders normally consolidate by adjusting and repositioning forces. While UO are likely to be noncontiguous, commanders reposition ISR assets (including observation posts and reconnaissance patrols) to maintain contact with the threat, to establish contact with nearby friendly units and ensure that there are no exploitable gaps or seams, and to help maintain freedom of action. Physical occupation of the terrain (as well as continued reconnaissance) will provide commanders with a fuller understanding of the urban environment. With this enhanced understanding, commanders may need to adjust boundaries and other control measures to better adapt to the effects of urban terrain features such as canals, subway tunnels, raised roadways, and tall buildings. As necessary, commanders may also need to reposition communications assets and C2 facilities to enable subsequent operations.

7-85. Immediately after the conduct of successful urban operations, units must also be alert to the potential for rapid and violent counterattacks. Defenders may launch a quick counterattack to regain terrain before the offensive forces have consolidated and fully assumed the defensive advantages of the urban terrain. Delaying a counterattack in UO, even for a few minutes, permits the environment's advantages to shift to the successful attacker. Thus, attacking units anticipate this reaction and reposition forces such that they are prepared to defeat it.

7-86. As necessary, commanders should reposition armor and artillery (and other fire support assets) to account for the changing situation and battlefield geometry. Commanders should also consider consolidating and repositioning armored and artillery forces in positions—either inside or outside the urban area—to add significant combat power to a hasty defense, to defeat threat counterattacks, or to allow for quick resumption of the attack. If integrated into a hasty defense inside the urban area, these forces will require continued infantry protection. As a mobile counterattack force positioned inside the urban area, armored forces will also require careful selection of attack positions and counterattack routes. Damage to buildings and infrastructure may limit maneuver and the use of direct and indirect weapon systems.

Repositioning forces outside the urban area may contribute to strengthening or reestablishing the isolation of the urban area and will make better use of range and standoff capability, allow them to take full advantage of their speed and mobility, and decrease their need for additional protective support.

Eliminating Pockets of Resistance

7-87. Focusing on controlling the essential and attacking decisive points means that attacking units will have bypassed some elements of the threat's defense. Within the bounds of the initial plan and current situation, commanders should consider whether it is necessary to devote the time and resources to immediately clear and eliminate remaining threat forces and pockets of resistance with all or parts of the attacking force or leave the task to follow-on forces. In their deliberations, commanders should determine if remaining threat forces are capable of consolidating and mounting effective counterattacks before follow-on units can engage the remaining threat and take advantage of their dispersion and disunity. As another part of their deliberations, commanders should also consider if remaining threat forces will likely (and immediately) endanger—

- The urban inhabitants.
- Lines of communications.
- Critical resources within the urban area necessary to support the inhabitants or subsequent operations.

Reorganization

7-88. Reorganization includes all measures taken by the commander to maintain combat effectiveness or return to a specified level of combat capability. As necessary, these actions include—

- Redistributing or cross-leveling supplies and equipment until resupply can be accomplished.
- Replacing key personnel and combining units or crews to form mission-capable ones.
- Integrating replacement soldiers and systems into the unit and matching operational systems with available crews.
- Recovering, treating, and evacuating casualties, enemy prisoners of war, and damaged equipment.
- Conducting training and disseminating critical lessons learned.
- Other actions to reestablish unit cohesion.

TRANSITION

7-89. Effective transitioning allows commanders to continue full spectrum operations in the urban area and elsewhere in the AO without unnecessary delays. Commanders can accomplish this by thorough planning—including appropriate branches and sequels (revised as the situation changes)—that gives adequate consideration to post-offensive organizational, training, psychological, and civilian requirements. If properly prepared, commanders can anticipate rather than react to potential mission changes.

Early and Concurrent Transition Planning

7-90. Commanders can ensure smooth transitions of urban offensive operations by planning for post-offensive operations early. Based on the mission envisioned, they determine which subordinates and what type of force structure to use. Post-offensive missions, like all urban operations, encompass full spectrum operations. At the successful conclusion of offensive operations, Army forces transition to some type of stability or civil support operation. Commanders may leave the subordinate unit in place to execute the new mission, may reorganize the subordinate unit for the mission, or relieve the unit that just completed offensive operations with a new unit.

Changes to Task Organization

7-91. Commanders should consider the organization of forces following offensive operations. Hostile civilians may require significant combat forces or military police forces to maintain stability. On the other hand, friendly civilians may require a minimum of military police or combat forces, but significant logistic support. Commanders must carefully consider the urban situation before deciding how to use combat forces that recently participated in a high-intensity offensive operation.

Training and Psychological Considerations

7-92. Many Army combat tasks may not support follow-on stability or civil support operations without considerable modification (if at all). Often, noncombat skills—not normally part of a unit's mission essential task list such as negotiating or mediating skills—will be required. However, the greatest modification required applies to each soldier's mental outlook. Forces that transition directly from combat to stability operations may not be psychologically prepared for a rapid and drastic change of mission. Commanders cannot expect troops who have just completed high-intensity offensive operations to rapidly adjust and exercise the sensitivity and judgment required in most stability operations. This especially applies if the population is hostile to Army forces. If possible, combat forces assisting in stability operations, particularly in hostile civilian situations, should not have had recent experience in high-intensity urban operations and they should have trained for the mission.

Return to Civil Agencies

7-93. Commanders of major operations also have the critical role of transitioning aspects of the urban offensive operation to civil, allied, NGOs, and other agencies as appropriate. Transition planning is detailed and aims to quickly return as much civil control of the area as is feasible after the attack. Beyond local civil control, outside civil agencies and NGOs assume tasks as completely and as rapidly as possible. These organizations are consulted and integrated into the planning process as early as possible. Commanders should begin planning for transition simultaneously with planning for offensive operations. They consider the feasibility of relinquishing control of urban areas to civil government, law enforcement, or NGOs even before completing offensive operations. During the conduct of urban operations, these transition operations are closely synchronized with the execution of the attack.

Transition to a New Mission

7-94. In urban offensive operations, like other offensive missions, the change in mission after a successful urban attack may be to a hasty defense or a continuation of offensive operations outside the area. However, in urban offensive operations the mission will just as likely rapidly change to a civil support or stability mission. This is particularly true if the unit has had special training and is task organized for urban operations. Transition to stability or civil support operations is often accompanied by a transition in roles from supported to supporting.

7-95. Even more challenging than transition at the end of the mission is transition during the conduct of the mission. Soldiers may have a difficultly transitioning from stability or civil support to offense and defense, and back again multiple times during an urban offensive operation. Soldiers may apply the tactics, techniques, and procedures of urban offensive operations directly to the stability or civil support missions with potentially disastrous results. Commanders need to segregate missions in time and space, and, if sufficient forces exist, by unit. To this end, commanders should consider permanently designating specific units to conduct civil-military and humanitarian support tasks and avoid rapid mission changes that rotate units (particularly at company level and below) between violent and nonviolent tasks. However, commanders may not have that luxury and may need to rely heavily on preparatory training (including the inculcation of Army values) and strong unit leadership to mitigate potential difficulties.

Chapter 8
Urban Defensive Operations

Generally, a modern city magnifies the power of the defender and robs the attacker of his advantages in firepower and mobility. A city can ingest an invading army, paralyze it for weeks on end, and grind it down to a state of ineffectiveness.

"Military Operations on Urbanized Terrain:
The 2d Battalion, 26th Infantry, at Aachen, October 1944"

The skillful defense of an urban area can decisively affect a campaign. The urban area offers many advantages to defending forces. An adroit defender can use the advantages of the urban environment to negate combat power disparities, blunt the tempo of an attack, attrit threat forces, and sap the morale of attacking troops. The defender gains an opportunity to concentrate resources, reconstitute attrited units, and transition to the offense. A successful defense of an urban area can also deny the threat vital resources. Defense in the urban environment is an essential Army capability and can significantly affect the outcome of entire campaigns and the achievement of national objectives.

PURPOSE OF URBAN DEFENSIVE OPERATIONS

8-1. Army forces defend urban areas for various reasons: defeating a threat attack, buying time, economizing forces, protecting an ally's political institutions and economic infrastructure, protecting an urban population, shaping conditions for decisive offensive operations, and shaping conditions for executing stability or civil support operations. During force projection operations, urban areas may be used as initial lodgment areas that Army commanders may need to defend at the outset until they build sufficient combat power. Usually two or more of these purposes apply to the urban defense. Urban defensive operations provide commanders great opportunities to turn the environment's characteristics to the advantage of Army forces. Urban areas are ideal for defensive operations and greatly enhance the combat power of defending units.

DEFENSIVE CHARACTERISTICS APPLIED TO URBAN OPERATIONS

8-2. There are five general characteristics of the successful defense: preparation, security, disruption, massing effects, and flexibility. All apply to the successful urban defense and to the higher commander supporting a subordinate defending in the urban area.

PREPARATION

8-3. The urban area suits the defense since the area's physical characteristics naturally enhance the combat power of defending units. These characteristics include protection, obstacles, and concealment. Urban terrain provides superb defensive positions with minimum preparation. With deliberate preparation, urban defensive positions can rapidly become strong points.

8-4. One primary characteristic of urban terrain that enhances the defense is protection. With little or no advance preparation, buildings, subsurface structures, and walls protect Soldiers from direct and indirect fire, interdict indirect fire, limit observation, and limit engagement ranges (requiring Soldiers to be skilled at combat in close quarters and quick-fire techniques). Nearly all buildings provide some ballistic protection from direct and indirect fire. Mason and stone buildings with basements and cellars can protect Soldiers from most fires except the largest caliber or tonnage bomb. Minimal additional preparation turns them into formidable, defensive strong points.

8-5. Buildings in urban areas, because of their height and close proximity, also can protect Soldiers by masking them from indirect fire. The height of a building may interdict the flight path of an artillery round, rocket, missile, or bomb at a point short of the intended target. Masking protects static defending forces and protects forces moving along routes bordered with tall buildings that form urban "canyons". These protected routes can be used for sustainment, counterattacks, and maneuver.

8-6. Structurally significant buildings in an urban area can create major obstacles to maneuver. These obstacles immediately canalize maneuver into existing streets and routes without any preparation by the defense. These obstacles then become kill zones for well-positioned and sited defensive forces. Minimal obstacle construction as point obstacles blocking streets and routes can further restrict the maneuver options of the attacking force. Rubble from structures collapsing into streets after fires (intentional or unintentional) can also block routes.

8-7. Buildings also conceal the location, disposition, and intent of the defense. They limit visual observation to the external perimeter of the urban area. They degrade radar and electronic position identifiers and decrease the utility of overhead imagery. The physical aspect of the urban environment greatly enhances the defense by degrading the opposition's intelligence, surveillance, and reconnaissance (ISR) capabilities. Buildings can conceal static defensive positions and the maneuver of defensive forces in the urban area. Although the environment constrains defensive mobility in much the same manner as offensive mobility, the defender has the time and opportunity to conduct careful reconnaissance and select and prepare routes. This gives the defender the ability to move reserves, maneuver counterattack forces, and plan sustainment without observation. Careful preparation provides the defender a mobility advantage over attacking forces.

SECURITY

8-8. The urban area can be an advantage or a disadvantage to the security of defending forces. This largely depends on the nature of the human dimension of the environment. If the population is evacuated or allied with Army forces, then the environment may assist in the security of defending Army forces. However, if the population is present and hostile, then the environment may make security difficult.

8-9. The physical aspects of the urban environment, uninfluenced by the human dimension, may assist in the security of defending Army forces. The combat power of small security forces manning observation posts is greatly enhanced. Forces can more easily restrict and monitor avenues of approach for threat reconnaissance. Defending forces positioned mostly in structures are difficult to locate.

8-10. The physical aspects of the environment may also present some security challenges, primarily with observation. The compartmented terrain limits the field of observation from any one point. The defense may require more security forces to adequately observe the mounted and dismounted avenues to prevent infiltration. Threat forces that successfully infiltrate will be more difficult to locate. These forces will gain numerous hide positions for small reconnaissance units in complex terrain and the effect the terrain has in masking electronic signatures.

8-11. Friendly civilians in the urban area can help identify threat forces attempting to conduct reconnaissance. Civilian activity will also help to mask defense preparations. However, a hostile element of the population may pass intelligence information to the threat. They may assist threat reconnaissance to infiltrate the urban area or provide guides, manpower, or resource support for threat forces. Commanders should take measures to ensure strict control of hostile populations. If resources permit, commanders may consider removing potentially hostile civilians from the area.

DISRUPTION

8-12. The urban environment's attributes assist defending Army forces to disrupt the attacker. It does this through compartmentalization, inhibiting command and control (C2), and facilitating counterattacks.

8-13. The physical aspects of the urban area force the attacking threat into compartmented urban canyons that make mutual support between attacking threat columns difficult. Shifting resources from one portion of the threat attack to another is also difficult. Physically, the urban area disrupts tactical communications making synchronization of combat power difficult.

8-14. The urban terrain hinders the mobility capabilities of the defense. However, careful planning, preparation, and rehearsals can facilitate more rapid movement of larger forces. Defending forces can assemble counterattacks undetected, move them along covered and concealed routes, and achieve surprise at the point of the counterattack. Attacking forces, using the compartmented terrain, often leave forward elements in position to be isolated or expose long and vulnerable flanks to friendly counterattack and interdiction.

MASSING EFFECTS

8-15. The urban environment allows defenders to better protect their centers of gravity and decisive points. The restrictive terrain reduces the attacker's maneuver options. Defenders can position forces in protected and mutually supportive positions oriented on deadly engagement areas. Relatively few well-positioned defenders can generate significant combat power. Without the positional advantage and the corresponding protective effects of the terrain, attacking forces often mass numbers to achieve the necessary combat power.

8-16. Knowing the complex terrain permits defending forces to plan engagement areas that maximize the effects of their combat power. Defending forces can remove fences, walls, rooftops, and even entire buildings to facilitate fields of fire and unmask indirect fire flight paths. Forces carefully choose firing positions for indirect fire systems so that flight paths travel between buildings into engagement areas. By leveraging this knowledge of the terrain, numerically inferior defenders can synchronize devastating fires on offensive forces that are forced by terrain and reinforcing obstacles to mass in confined spaces where fires can have the greatest effect.

FLEXIBILITY

8-17. Defensive flexibility results from detailed planning and mental agility. Commanders develop defensive flexibility by ensuring that plans adequately address branches and sequels that include alternate and subsequent positions and emphasize counterattack options. The urban area facilitates defensive flexibility because the urban terrain can be quickly adapted for defensive operations with little or no preparation. The effect is similar to having multiple, prepared positions on nearly every possible approach. The urban area can also permit rapid, covered movement on interior lines. This permits swift movement to and occupation of strong defensive positions with little or no preparation. The defense also has more flexibility since defenders often know and better understand the urban terrain's effects on operations. Normally, defenders will not get lost as easily, will know complex lines of sight and masking effects, and will best understand the ballistic characteristics of individual structures.

8-18. Mental agility allows commanders to see that the best urban defense may actually be to defend outside of the area. Such a defense mitigates the danger to the urban population and potentially reduces collateral damage. It takes advantage of Army long-range engagement capabilities and denies the threat the opportunity to "hug" Army forces or noncombatants as protection from fires. This defense may be appropriate when Army forces have enough resources to defend more open terrain, when time permits deploying extensive obstacles and constructing protected positions, and when natural terrain such as river obstacles aids the defense.

URBAN DEFENSIVE OPERATIONS AND BATTLEFIELD ORGANIZATION

8-19. Urban defensive operations are organized within the overall battlefield organization of sustaining, shaping, and decisive operations. The success of urban defense depends on each operation, but commanders must synchronize these simultaneous operations as one action. Sustaining operations in defensive urban operations (UO) ensure freedom of action. Critically, urban sustaining operations ensure security of the lines of communications and establish effective movement control. Shaping operations in defensive UO create the conditions for decisive operations. Shaping operations vary greatly depending on the type of defense. For example, in a mobile defense the shaping operation may be the fixing force. In contrast, in an area defense the fixed defense may be the decisive operation. In the urban defense, decisive operations focus on accomplishing the commander's mission. The decisive operation may not defeat the threat's main effort, and it may not prevent threat occupation of large portions of the urban area if those tasks are not essential to mission accomplishment. For example, if the defense's objective is to protect a critical communications node, then, depending on the commander's overall intent, threat actions to secure an airfield elsewhere may not be important.

TYPES OF URBAN DEFENSE

8-20. Commanders can view urban area defensive operations two ways: as conducting a major defensive operation with an urban area in their area of operations (AO) and as defending entirely in an urban area.

AREA DEFENSE

8-21. At the operational level, an area defense may include both urban areas and open maneuver areas. The most common defense in an urban area and the most suitable for the characteristics of this distinct environment is the area defense. As a defensive operation, the area defense concentrates on denying threat forces access to designated terrain for a specific time rather than destroying the threat outright. Although an area defense in an urban area does not directly seek to destroy or defeat attacking threat forces, as an objective it does aim to force culmination of the threat's attack. The urban area defense often works effectively to exhaust threat resources and shape conditions for a transition to offensive operations. The urban area may also be used as a strong point to force threat movement in a different direction or to fix threat forces as part of a large, mobile defense taking place in the AO outside the urban area.

MOBILE DEFENSE

8-22. A mobile defense can operate in an urban area but only under specific conditions. It focuses on destroying or defeating the threat through a decisive attack by a striking force. It requires the defender to have greater mobility than the attacker. To shape a mobility advantage, the urban defender effectively uses the terrain and correctly task organizes his forces' mobility. The principles of applying the mobile defense in the urban area remain the same: a small fixing force stops the threat and limits any ability to maneuver while a striking force quickly maneuvers and counterattacks to destroy him.

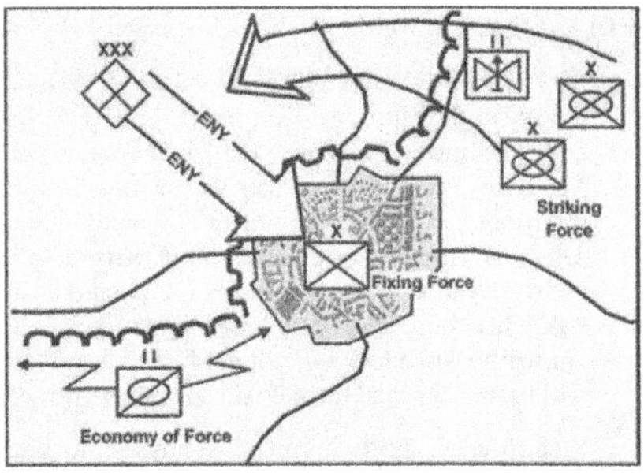

Figure 8-1. An urban area incorporated into a larger mobile defense

8-23. One key to executing a mobile defense in the urban area is to entice a threat force into the depths of the urban area where it begins to lose mobility options. A well-placed fixing force augmented with man-made obstacles and taking advantage of the naturally constrictive terrain can stop a much larger force. If the attacking force is largely mounted and armored, its mobility in the urban area may be reduced to less than that of dismounted infantry. In addition, if the attacking force's movement into the urban area is mounted and rapid, the commander's situational understanding also diminishes. Then the striking force, consisting of dismounted infantry forces, can execute the counterattack with surprise from multiple directions and dimensions (subsurface, surface, supersurface, and airspace). Man-portable antiarmor weapons—firing from flanks and top down and supported by precision indirect fires from both organic and joint systems—can rapidly destroy the threat.

8-24. From the perspective of commanders of the major operation, the urban environment can help defending forces achieve a mobility advantage over an attacker in a broader sense. Defending commanders can attempt to shape the battlefield so that the attacker commits significant resources into an urban area, where his maneuver capabilities are reduced (see figure 8-1). A disproportionately small defending force, which relies on the defensive combat power advantages of the urban environment, reduces and fixes the attacker's maneuver capabilities. Other defending forces mass outside the urban area then strike the threat with a combined mobility and firepower advantage.

Urban Defense in a Major Operation: Stalingrad – 1942 to 1943

The German and Soviet battle for Stalingrad in late 1942 and into early 1943 illustrates how a tactical urban area defense integrates into a larger mobile defense. Stalingrad was a battle fought on a huge scale: it involved army groups on both sides and thousands of square kilometers. Though the city was relatively small, it remained the focus of both German and Soviet offensive and defensive operations during the six-month battle.

In the summer of 1942, the Germans launched a strategic offensive in southern Russia. Its goal was the valuable oil fields of the Caucasus. German forces turning south into the Caucasus exposed a vulnerable flank to Soviet forces positioned between the Don and Volga Rivers. For the German Caucasus operation to succeed, it had to destroy Soviet forces between the Don and Volga, establish a good defensive line, and capture Stalingrad. This city would anchor the German defense and interdict the critical flow of supplies from the Caspian Sea via the Volga River into central Russia. Stalingrad, by virtue of its name, also had important political and cultural value to the Germans and Soviets.

The opening phases of the German offensive were successful; German forces—the 6th Army and 4th Panzer Army—entered the outskirts of Stalingrad in late August 1942 (see figure 8-2).

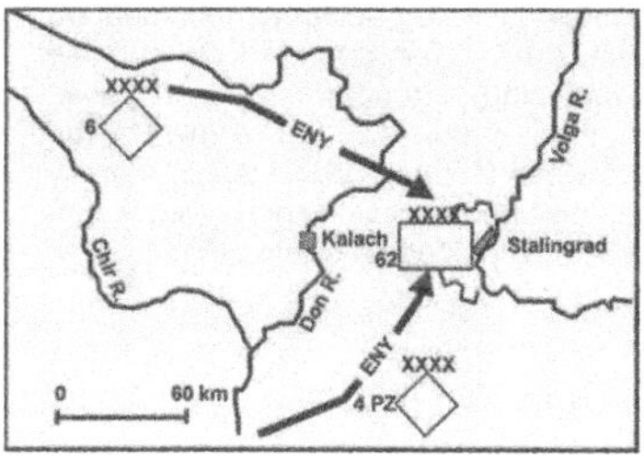

Figure 8-2. German attacks to seize Stalingrad

After a month of intense fighting, the Germans possessed nearly 90 percent of the city. At this point, the 6th Army commanded all German forces in and around Stalingrad. The Soviet 62nd Army's defense was reduced to a front only a few hundred meters deep and a couple of kilometers long on the banks of the Volga. The Soviet defenses hinged on fortress-like concrete industrial buildings and the fanatical bravery and tenacity of soldiers and civilians fighting in the city's remains (see figure 8-3 on page 8-6).

Beginning in mid-September, the Soviet command began looking at how to convert the defense of Stalingrad into an operational opportunity. During October and November, the 62nd Army held on to its toehold in Stalingrad. While maintaining the defense of the 62nd Army, the Soviets secretly began to build up strength on both flanks of the German 6th Army. The Germans increased their vulnerability by securing the German 6th Army's flanks with less capable Romanian, Hungarian, and Italian armies. Also, the 6th Army moved powerful German divisions into the city and rotated with German divisions that were exhausted by urban combat.

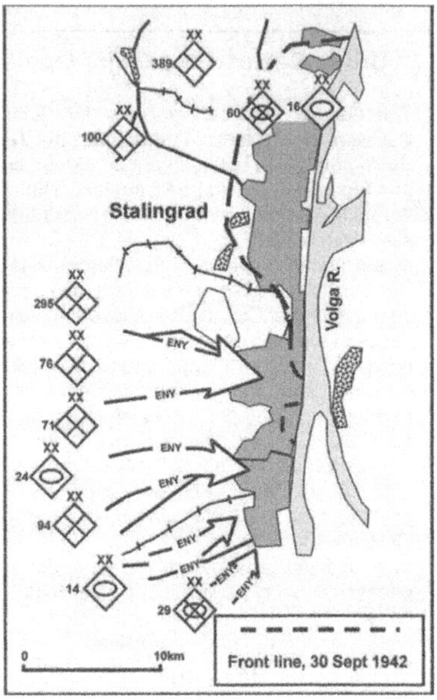

Figure 8-3. German attacks to seize Stalingrad, September 1942

On 19 November, the Soviets launched Operation Uranus that attacked two Romanian armies with seven Soviet armies. Simultaneously, the 8th Russian Army attacked to aid the 62nd Army in further fixing the German 6th Army. Within five days, the Soviet armies of the Don Front, Southwest Front, and Stalingrad Front met near the city of Kalach and sealed the fate of the German 6th Army's 300,000 troops in Stalingrad (see figure 8-4).

On the third day of the Soviet offensive, when encirclement seemed inevitable but not yet complete, the 6th Army commander asked permission to withdraw from the trap. The German high command denied permission believing that the Army could be supplied by air and then a renewed offensive could break through to the city. On 12 December, the German LVII Panzer Corps launched an offensive north to break through to Stalingrad. This offensive made progress until another Soviet offensive on 16 December forced its cancellation. This ended any hope of recovering Stalingrad and the 6th Army. On 31 January 1943, the 6th Army surrendered after sustaining losses of almost two-thirds of its strength. The Soviets took over 100,000 prisoners.

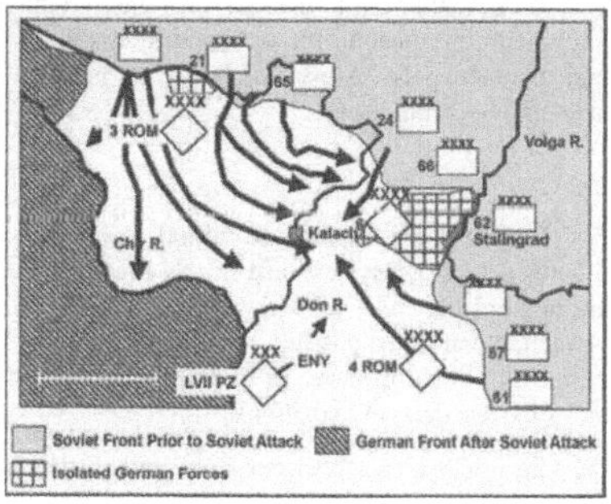

Figure 8-4. Soviet attacks trap German 6th Army

Many lessons emerge from the successful defense of Stalingrad. Tactically, the defense showed how using the terrain of a modern industrial city wisely could increase the combat power of an inferior, defending force and reduce the maneuver options of a mobile, modern attacking force. Another element in the Soviet's tactical success was the Germans' inability to isolate the defenders. The Germans never threatened the Soviet supply bases east of the Volga and, despite German air superiority, the Soviets continuously supplied and reinforced the 62nd Army across the Volga River. Also, Soviet artillery west of the river was able to fire in support of Soviet forces and was never threatened with ground attack.

At the operational level, the Soviets demonstrated a keen understanding of using an urban area within the context of a mobile defense. The 62nd Army's stubborn area defense of Stalingrad drew the bulk of the German combat power into the urban area where they were fixed by a smaller and quantitatively inferior defending force. This allowed the Soviets to build combat power outside the urban area. The Soviets set the conditions for a mobile defense by positioning powerful Soviet armor forces in open terrain outside the urban area against quantitatively inferior German allied forces. In OPERATION URANUS, the mobile defense's strike force destroyed the enemy outside the urban area and trapped the greater part of the best enemy formations inside the urban area. The trapped units were then subjected to dwindling resources and extensive psychological operations, further isolated into pockets, and defeated in detail.

RETROGRADE

8-25. A retrograde involves organized movement away from the threat. Retrograde operations include withdrawals, delays, and retirements. These defensive operations often occur in an urban environment. The urban environment enhances the defending force's ability to conduct retrograde operations successfully (see figure 8-5).

8-26. The cover and concealment afforded by the urban environment facilitates withdrawals where friendly forces attempt to break contact with the threat and move away. The environment also restricts threat reconnaissance, which is less able to detect friendly forces moving out of position, and presents excellent opportunities for deception actions. Finally, a small security force's ability to remain concealed until contact in the urban environment significantly slows threat attempts to regain contact once Army forces have broken contact and begun to move.

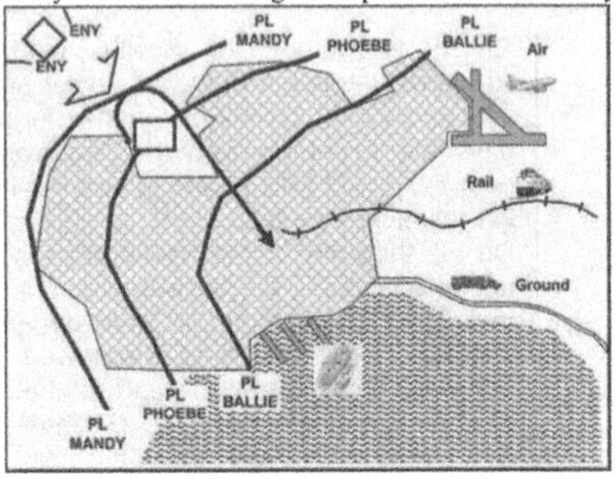

Figure 8-5. Retrograde through an urban area

8-27. The urban environment's natural cover and concealment, as well as the compartmented effects, facilitates delays. Delays can effectively draw the threat into the urban area for subsequent counterattack or as an integral part of a withdrawal under threat pressure. Delaying units can quickly displace from one covered and concealed position to another; the repositioning options are vast. Compartmented effects force the attacking threat to move on well-defined and easily interdicted routes and limit the threat's ability to flank or bypass delaying positions.

8-28. The urban area's transportation and distribution network facilitates retiring forces that are not in contact. Properly used, the urban transportation system can quickly move large forces and associated resources, using port facilities, airfields, railheads, and well-developed road networks.

URBAN DEFENSIVE CONSIDERATIONS

What is the position about London? I have a very clear view that we should fight every inch of it, and that it would devour quite a large invading army.

Winston Churchill
War in the Streets

8-29. The urban operational framework—understand, shape, engage, consolidate, and transition—provides structure to developing considerations for defensive operations. The considerations can vary depending on the level of war at which the operation is conducted, the type of defense, and the situation. Most issues discussed may, in the right circumstances, apply to both commanders conducting major UO and commanders at lower tactical levels of command.

UNDERSTAND

8-30. The commander defending in the urban area must assess many factors. His mission statement and guidance from higher commanders help him focus his assessment. If the mission is to deny a threat access to port facilities in an urban area, the commander's assessment will be focused much differently than if the mission is to deny the threat control over the entire urban area. The mission, enemy, terrain and weather, troops and support available, time available, civil considerations (METT TC) structure guides the commander's assessment. Of these, the impacts of the threat and environment—to include the terrain, weather, and civil considerations—are significant to the commander's understanding of urban defensive operations.

The Threat

8-31. In the urban defense, a key element is the commander's understanding of the threat. One of his primary concerns is to determine the attacker's general scheme, methodology, or concept. Overall, the attacker may take one of two approaches. The most obvious would be a direct approach aimed at seizing the objectives in the area by a frontal attack. A more sophisticated approach would be indirect and begin by isolating Army forces defending the urban area. Innumerable combinations of these two extremes exist, but the threat's intentions toward the urban area will favor one approach over another. The defending Army commander (whose AO includes but is not limited to the urban area) conducts defensive planning, particularly his allocation of forces, based on this initial assessment of threat intentions. This assessment determines whether the commander's primary concern is preventing isolation by defeating threat efforts outside the area or defeating a threat attacking the urban area directly. For the higher commander, this assessment determines how he allocates forces in and outside the urban area. For the commander in the urban area, this assessment clarifies threats to sustainment operations and helps shape how he arrays his forces.

The Environment's Defensive Characteristics

8-32. A second key assessment is the defensive qualities of the urban environment. This understanding, as in any defensive scenario, is based on mission requirements and on a systemic analysis of the terrain in terms of observation and fields of fire, avenues of approach, key terrain, obstacles, and cover and concealment (OAKOC). It is also based on potential chemical, biological, radiological, nuclear and fire hazards that may be present in the urban area. This understanding accounts for the unique characteristics of urban terrain, population, and infrastructure as discussed in Chapter 2.

8-33. Generally, units occupy less terrain in urban areas than more open areas. For example, an infantry company, which might occupy 1,500 to 2,000 meters in open terrain, is usually restricted to a frontage of 300 to 800 meters in urban areas. The density of building in the urban area, building sizes and heights, construction materials, rubble, and street patterns will dictate the actual frontage of units; however, for initial planning purposes, figure 8-6 provides approximate frontages and depths for units defending in an urban area.

UNIT	Frontage (Blocks*)	Depth (Blocks*)
Battalion	4 – 8	3 – 6
Company	3 – 4	2 – 3
Platoon	1 – 2	1
*Average block is 175 meters		

Figure 8-6. Approximate defensive frontages and depths

SHAPE

8-34. Commanders of a major operation shape the urban battle according to the type of defense they are attempting to conduct. If conducting an area defense or retrograde, they use shaping actions like those for any defensive action. Important shaping actions that apply to all defensive UO include—

- Preventing or defeating isolation.
- Separating attacking forces from supporting resources.
- Creating a mobility advantage.
- Applying economy of force measures.
- Effectively managing the urban population.
- Planning counterattacks.

Preventing or Defeating Isolation

8-35. Failure to prevent isolation of the urban area can rapidly lead to the failure of the entire urban defense. Its importance cannot be overstated. In planning the defense, commanders must anticipate that the threat will attempt to isolate the urban area. Defensive planning addresses in detail defeating threat attacks aimed at isolation of the urban area. Commanders may defeat this effort by allocating sufficient defending forces outside the urban area to prevent its isolation. Defensive information operations (IO) based on deception can also be used to mislead the threat regarding the defensive array in and outside the urban area. Such information can convince the threat that a direct attack against the urban area is the most favorable approach.

8-36. If the threat has successfully isolated the urban area, commanders of a major operation have several courses of actions. Three options are an exfiltration, a breakout attack by forces defending the urban area, or an attack by forces outside the urban area to relieve the siege. A fourth option combines the last two: counterattacks from both inside and outside the urban area to rupture the isolation (see breakout operations in FM 3-90). Time is critical to the success of either operation. Commanders should plan for both contingencies to ensure rapid execution if necessary. Delay permits threat forces surrounding the urban area to prepare defenses, reorganize their attacking force, retain the initiative, and continue offensive operations. The passage of time also reduces the resources of defending forces and their ability to breakout. Therefore, commanders and staff of a major operation must vigilantly avoid isolation when Army forces are defending urban areas in their AO.

Separating Attacking Forces from Supporting Resources

8-37. Commanders of the major operation primarily use fires and IO for separating in space and time threat forces attacking the urban area from echelons and resources in support. The purpose of this shaping action is the same as for any conventional area defense. It aims to allow the defending forces to defeat the threat piecemeal as they arrive in the urban area without support and already disrupted by deep fires and IO against information systems. This separation and disruption of the threat also sets the conditions for a mobile defense if commanders choose to execute that type of defense. These operations also prevent the threat commander from synchronizing and massing his combat power at the decisive point in the close battle.

8-38. If the urban area is part of a major mobile defense operation, the urban defense becomes the fixing force. Commanders can shape the defense to encourage the threat to attack into the urban area. They can lure the threat using a combination of techniques depending on the situation. They may make the urban area appear only lightly defended while other alternative courses of action appear strongly defended by friendly forces. Placing the bulk of the defending forces in concealed positions well within the urban area and positioning security forces on the periphery of the urban area portray a weak defense. In other situations, the opposite is true. If the urban area is an important objective to the threat, friendly forces can make the urban area appear heavily defended, thus ensuring that he commits sufficient combat power to the urban area to overwhelm the defense. Both cases have the same objective: to cause a major commitment of threat forces in the urban area. Once this commitment is made, the mobile defense striking force attacks and defeats the threat outside the urban area. This isolates the threat in the urban area and facilitates its destruction.

8-39. In the urban tactical battle, many shaping actions mirror those in all defensive operations. The size and complexity of the urban area prevent defending forces from being strong everywhere; shaping operations designed to engage the threat on terms advantageous to the defense have particular importance. Shaping actions include reconnaissance and security operations, passages of lines, and movement of reserve forces prior to their commitment. In addition, shaping operations critical to urban defense include mobility and countermobility operations, offensive IO, economy of force operations, and population management operations.

Creating a Mobility Advantage

8-40. In urban terrain, countermobility operations can greatly influence bringing the threat into the engagement areas of defending forces. Countermobility operations—based on understanding the urban transportation system, design, and construction characteristics—can be unusually effective (see Chapter 2). Demolitions can have important implications for creating impassable obstacles in urban canyons as well as for clearing fields of fire where necessary. Careful engineer planning can make the already constrictive terrain virtually impassable to mounted forces where appropriate, thus denying the threat combined arms capabilities. Countermobility operations in urban terrain drastically increase the defense's ability to shape the attacker's approach and to increase the combat power ratios in favor of the defense. As with all aspects of UO, countermobility considers collateral damage and the second- and third-order effects of obstacle construction.

8-41. Well-conceived mobility operations in urban terrain can provide defending forces mobility superiority over attacking forces. This is achieved by carefully selecting routes between primary, alternate, and subsequent positions, and for moving reserves and counterattack forces. These routes are reconnoitered, cleared, and marked before the operation. They maximize the cover and concealment characteristics of the terrain. Using demolitions, lanes, and innovative obstacles denies the defense of these same routes.

Applying Economy of Force Measures

8-42. Economy of force is extremely important to effective tactical urban defense. A megalopolis is too large and too easily accessible for defending forces to be strong everywhere. Economy of force enables the defending force to mass effects at decisive points. Forces used in an economy of force role execute security missions and take advantage of obstacles, mobility, and firepower to portray greater combat power than they actually possess. They prevent the threat from determining the actual disposition and strength of the friendly defense. If, contrary to expectations, they are strongly attacked, their mobility—stemming from a mounted maneuver capability, planning, and an intimate knowledge of the terrain—allows them to delay until reserves can meet the threat. Security forces in an economy of force role take position in parts of the urban area where the threat is less likely to attack.

Defensive Combat Power: Suez – 1973

At the end of October, the Israeli Army was in the midst of effective counterattack against the Egyptian Army. The Israelis had success attacking west across the Suez Canal. Their armored divisions were attempting to achieve several objectives, to include destroying Egyptian air defense sites and completing the encirclement of the Egyptian 3rd Army, which was trapped on the canal's east side.

To completely encircle the Egyptian 3rd Army, the Israelis had to seize all possible crossing sites to it from the canal's west bank and the Red Sea. Also, as international negotiations towards a cease-fire progressed, the Israeli government wanted to capture as much Egyptian territory as possible to improve their negotiating position after hostilities.

Consequently, the Israeli Adan Armored Division was tasked to seize the Egyptian Red Sea port of Suez on the morning of 24 October. A cease-fire was to begin at 0700, and the Israeli intent was to be decisively engaged in the city by that time and then consolidate their position as part of the cease-fire compliance.

The Adan Division plan to seize Suez was a two-part operation. Each of the division's armored brigades would have a role. The 460th Brigade would attack west of the city and complete the city's encirclement. Simultaneously, the 217th Brigade would attack in columns of battalions through the city to seize three key intersections in the city. This was in accordance with standard Israeli armored doctrine for fighting in an urban area. The 217th Brigade would seize its objectives through speed, firepower, and shock action. Once the objectives were seized, infantry and armored teams would continue attacking from the secured objectives to mop up and destroy pockets of resistance. The Israeli commanders expected to demoralize the defending Egyptians—two infantry battalions and one antitank company—by this rapid attack. The armored division commander was specifically advised by his commander to avoid a "Stalingrad" situation.

The attack got off to an ominous beginning as mist greatly inhibited a scheduled aerial bombardment in support of the attack. The 217th Brigade began its attack without infantry support and was quickly stopped by antitank missiles and antitank fire. Infantry was quickly integrated into the brigade and the attack resumed.

At the first objective, the Israelis encountered their first problems. A withering barrage of small arms, antitank missiles, and antitank fire hit the lead tank battalion, including direct fire from SU-23 anti-aircraft guns. Virtually all the officers and tank commanders in the tank battalion were killed or wounded, and several tanks were destroyed. Disabled vehicles blocked portions of the road, and vehicles that turned on to secondary roads were ambushed and destroyed. The battalion, however, successfully fought its way through the first brigade objective and on to the final brigade objective.

Hastily attached paratroop infantry in company strength were next in column following the tanks. They were traveling in buses and trucks. As the lead tank battalion took fire, the paratroopers dismounted, and attempted to secure adjacent buildings. The tank battalion's action of fighting through the objective caused the paratroopers to mount up and also attempt to move through the objective. Because of their soft skinned vehicles the paratroopers were unable to remain mounted and again dismounted, assaulted, and secured several buildings that they could defend. Once inside the buildings, the paratroopers found they were cut off, pinned down, and unable to evacuate their considerable casualties, which included the battalion commander. The paratroopers were on the initial brigade objective but were unable to maneuver and were taking casualties.

A second paratroop company also dismounted and quickly became stalled in house-to-house fighting. The brigade reconnaissance company in M113 personnel carriers brought up the rear of the brigade column and lost several vehicles and was also unable to advance.

By 1100 the Israeli attack culminated. Elements of the 217th Brigade were on all three of the brigade's objectives in the city. However, the armored battalion, which had achieved the deepest penetration, was without infantry support and under severe antitank fire. Both paratroop companies were isolated and pinned down. In addition, an attempt to link up with the paratroopers had failed. At the same time, the civilian population of the city began to react. They erected impromptu barriers, ambushed isolated Israeli troops, and carried supplies and information to Egyptian forces.

The Israeli division commander ordered the brigade to break contact and fight its way out of the city. The armored battalion was able to fight its way out in daylight. The paratroop companies were forced to wait until darkness and then infiltrated out of the city carrying their wounded with them. Israeli casualties totaled 88 killed and hundreds wounded in addition to 28 combat vehicles destroyed. Egyptian casualties were unknown but not believed to be significant.

The fight for Suez effectively demonstrates numerous urban defensive techniques. It also vividly demonstrates the significant effect on defensive combat power of the urban environment.

The Egyptian defense demonstrates how the compartmented urban terrain restricts the mobility and the massing of firepower of armored forces. Trapped in column on the road, the Israelis were unable to mass fire on particular targets nor effectively synchronize and coordinate their fires. The short-range engagement, also a characteristic of urban combat, reduced the Israeli armor protection and eliminated the Israeli armor's ability to keep out of small arms range. Thus, hand-held antiarmor weapons were more effective in an urban area. Additionally, Egyptian small arms and sniper fire critically affected Israeli C2 by successfully targeting leaders.

The Egyptian defenders effectively isolated the mounted Israelis by defending and planning engagement areas in depth. The Egyptians synchronized so that they engaged the entire Israeli force simultaneously. This forced the Israelis to fight in multiple directions. It also separated the Israeli infantry from the armor and prevented the formation of combined arms teams necessary for effective urban offensive operations.

Suez also demonstrated how civilians come to the advantage of the defense. After the battle was joined, the population—by threatening isolated pockets of Israelis and building barricades—helped prevent the Israelis from reorganizing while in contact and hindered the Israelis breaking contact. The population was also a valuable source of intelligence for the Egyptians and precluded covert Israeli movement in daytime.

Suez shows the ability of a well-placed defense in depth to fix a superior force in an urban area. Despite the Israeli commander's caution to avoid a "Stalingrad," the Israeli division, brigade, and battalion commanders were quickly trapped and unable to easily break contact. Even a successful defense on the perimeter of the city would not have been nearly as effective, as the Israelis would have easily broken contact once the strength of the defense was recognized.

Another key to the success of the Egyptian defense was the Israelis' inadequate reconnaissance. While the Israelis knew the approximate size of the defending forces, they had no idea of the Egyptian dispositions. In this case, time prevented adequate reconnaissance. Key to a successful defense is adequate security to obscure defense dispositions, which permits surprise and shock effect.

The Suez defense was a decisive defeat of elite Israeli forces by regular infantry units inferior in training, morale, and numbers. Total disaster was averted only because of the professionalism of the Israeli armored forces and paratroopers that permitted them to continue to fight and eventually exfiltrate the urban trap. The Israeli forces thus escaped total destruction. Suez strongly demonstrates how the enhancing effects of the urban environment on defensive combat power are significant enough to permit inferior regular forces to defeat elite formations. Since

> the 1973 Suez battle, U.S. forces in Mogadishu, Somalia, and Russian forces in Grozny, Chechnya have faced similar urban defensive ambushes.

Effectively Managing the Urban Population

8-43. Another way to shape the urban defensive battle is population management. In most cases, defending force commanders are in the urban area before combat. This time gives them the chance to manage civilians. Consequently, they can better manage and protect the population (a legal requirement) and gain more freedom of action for their forces.

8-44. Managing the civilians during the defense is a function of the size, disposition, and needs of the population and the resources available to the commander. Requesting higher support or coordinating with nongovernmental organizations and the local civil leadership for support may make up shortages of resources. Resources devoted to population management are carefully weighed against availability, military mission requirements, and possible collateral damage affecting tactical, operational, or strategic success. It may prove impractical to evacuate an urban area's population; still, commanders should attempt to create and move most civilians to protected areas. Moving the population allows defending forces to more liberally apply fires, emplace obstacles, and relieve combat units and support units of requirements to continue life support for civilians while executing combat operations. Overall, effective civil-military operations can turn a friendly (or a neutral) population into an effective force multiplier providing support to every warfighting function.

Planning Counterattacks

8-45. Counterattacks are also an important tool in shaping the battlefield for defensive success. Counterattacks as a shaping tool have two applications: retaining the initiative and separating forces. However, opportunity for effective counterattacks will be brief and, therefore, timing will be critical. If conducted too soon, the counterattack may expend resources required later; if conducted too late, it may not be effective. Commanders should understand the effect of the urban environment on time-distance relationships; otherwise, the timing of the attack may be upset and the operation desynchronized. Additionally, commanders should develop plans beyond the counterattack to exploit potential success.

ENGAGE

8-46. Engaging the urban area in a defensive operation requires decisively defeating the threat's attacks. Defensive forces use the terrain to their advantage, employ precision supporting fires, and use direct fire from protected positions aligned against carefully selected avenues of approaches and kill zones. The combat power of the defense augmented by shaping actions and the characteristics of urban terrain force culmination of the threat attack. Like urban offensive operations, effective engagement in urban defensive operations typically results from successful actions at the tactical level of war. These actions include—

- Performing aggressive ISR.
- Creating depth.
- Executing an effective obstacle plan.
- Conducting coordinated counterattacks.

Performing Aggressive ISR

8-47. ISR efforts of the defender are focused initially on identifying relevant information about the location and nature of the threat's main effort. Once identified, the defender's ISR focus shifts to assessing the rate at which the threat attack moves to its culminating point. Indicators of culmination may be physical fatigue of Soldiers, a breakdown in C2 capability, difficulty providing logistics support, or the increasing time required to reorganize small units to attack. When that culmination is achieved, friendly forces counterattack before the threat has a chance to transition to a hasty defense.

Creating Depth

8-48. Depth in the defense is the key to forcing the threat to culminate. The urban defense cannot allow itself to be penetrated nor permit forward elements to be destroyed. The defense is designed with the greatest depth possible. Defending forces weaken the threat to the fullest extent possible by attack from each position but not permit themselves to be destroyed by fires or close assault. Instead, as threat combat power builds up against individual positions, the use of mission orders permits subordinate leaders to disengage on their own initiative and move on preplanned routes to subsequent positions. Positions are designed to be mutually supporting—withdrawing from one position to a subsequent one while supporting positions cover by fires. The attacker is constantly forced to deploy and reorganize without being able to achieve decisive effects against the defender.

Executing an Effective Obstacle Plan

8-49. Obstacles in the urban defense are designed to break up the threat's combined arms capability. Separating dismounted forces from mounted forces disrupts the cohesion of the attacker and reduces his combat power. It also exposes his individual elements to the effects of asymmetric counterattack. The leading threat dismounted force can be effectively counterattacked by a friendly combined arms element while the threat armored force remains vulnerable to antiarmor attack by dismounted forces.

Conducting Coordinated Counterattacks

8-50. The counterattack is one of the key actions of the urban defense. However, commanders do not counterattack unless there is a reasonable chance of success. As the attacker moves into the depth of the urban area, his forces become fatigued, attrited, and increasingly disorganized. He likely also creates an increasingly long and exposed flank. At all levels, forces defending in urban terrain look for opportunities to counterattack. As the offensive force reaches the culmination point where it can no longer continue to attack with the available forces, the defensive commander executes a planned and coordinated counterattack. The counterattack aims to regain the initiative and to make the threat fight in multiple directions. Infiltration using superior knowledge of the terrain (including supersurface and subsurface capabilities) and the skillful use of stay-behind forces permits attacking the threat throughout the depth of his formations. Small-scale counterattacks focus on C2 and sustainment capabilities. These counterattacks can set the conditions for a deliberate attack leading to the ultimate destruction of the attacking threat force.

CONSOLIDATE

8-51. Consolidation is as important to urban defensive operations as to offensive operations. Many of the same consolidation considerations for the urban offense apply equally well to the defense. Commanders should reinforce or reposition maneuver forces and fire support assets on the urban battlefield based on weaknesses uncovered during rehearsals and opportunities discovered during actual execution. While maximizing the many advantages of the urban defense, the commander of the urban defense aggressively seeks ways of attriting and weakening enemy forces before they enter the urban area and close combat is initiated. Commanders should combine the static and mobile elements of the defense to strengthen their positions in relation to the enemy while seeking every opportunity to transition to urban offensive operations. As in urban offensive operations, commanders should conduct any necessary reorganization actions that they were unable to accomplish during execution.

TRANSITION

8-52. Transitions in urban defensive operations occur at all levels. As with offensive operations, commanders of major operations should address which units are assigned to continue to operate in the area after defensive operations have ceased. In defensive UO, this task is not as challenging as an occupation mission during urban offensive operations. The psychology of troops defending an urban area differs from those attacking into it. Defending forces become accustomed to the environment, having experience in the environment before combat. In terms of training, it is easier for follow-on missions to be assigned to a unit that has successfully defended the urban area. This course of action takes advantage of the defending unit's experience in the area and its relationships with other agencies—agencies that were operating alongside the units before and possibly during the defense. In defensive operations, regardless of the civilians' attitudes, policies regarding that population are established before the successful defense, and the command likely has experience executing operations with civil authorities and other agencies. Thus, these relationships are neither new nor as significant an issue as in offensive operations. Therefore, commanders must be prepared to execute various stability or civil support operations or use a successful defense to springboard into more decisive offensive operations elsewhere in the commanders' AO.

Transition Emphasis to Stability or Civil Support Operations

8-53. At the end of a successful urban defense, operational commanders should generally expect civil authority, control, and jurisdiction to increase. Additionally, the civil population may be anxious to return. Defensive combat will require virtually complete military control of the urban area; however, after the successful defense, a rapid transition will occur from military control to civil or joint military and civil control afterward. Although the full spectrum of Army operations will often be conducted simultaneously in UO, commanders will transition emphasis to stability or civil support operations. Important transition tasks will include demilitarizing munitions, clearing obstacles, and searching for isolated threat pockets of resistance. Conclusion of the defensive operations will also require transition to joint civil-military tasks, such as evaluating structures for safety, restoring essential services, and possibly creating joint law enforcement. Commanders of major operations, using a civil-military operations center and G9, should anticipate these requirements and begin early preparations to ensure a smooth, successful transition.

Transition to Offensive Operations

8-54. Units that have successfully defended the urban area may then transition to offensive operations. A rapid transition to offensive operations will require identification, preparation, and training of units designated to assume missions as the defending units leave the urban area. This preparation emphasizes continuity of policies and relationships already established. A relief in place occurs. The new occupying units provide not only a continuity of policy, but also a continuity of attitude toward the urban area, its population, and its institutions.

Chapter 9

Urban Stability and Civil Support Operations

The fundamental shared aims between stability and civil support operations and any operation conducted in an urban environment are the increased significance and influence of the civil population and nonmilitary organizations. Often, no military victory is to be achieved. The center of gravity for these operations normally cannot be attacked through military means alone; the Army (and the military in general) is often but one tool supporting a larger, civil-focused effort. When conducted, military defeat of an armed adversary is only a phase in a larger campaign, the overall objectives of which will be to bring about a more satisfactory political, social, diplomatic, economic, and military situation than existed prior to the conflict. Without a tightly coordinated civil-military effort, overall success will be difficult or impossible to achieve. Commanders who can understand and cope with the complexities of stability or civil support operations gain insights that directly apply to executing any urban operation. Urban stability or civil support operations may complement urban offensive and defensive operations, or may dominate the overall operation. Army forces may need to conduct offensive and defensive operations to defend themselves or destroy urban threats seeking to prevent the decisive stability or civil support mission. During hostilities, urban stability operations may keep armed conflict from spreading, encourage coalition and alliance partners, and secure the civilian population's support in unstable urban areas (and the surrounding rural areas under their influence). Following hostilities, urban stability operations may provide a secure environment for civil authorities to rebuild. Urban civil support operations can range from transporting, feeding, and sheltering the population made homeless as a result of man-made or natural disasters to conducting civil search and rescue and providing urgent medical care.

PURPOSE OF URBAN STABILITY AND CIVIL SUPPORT OPERATIONS

9-1. Army forces conduct stability and civil support operations to deter war, resolve conflict, promote peace, strengthen democratic processes, retain U.S. influence or access abroad, assist U.S. civil authorities, and support moral and legal imperatives. Stability operations promote and sustain regional and global stability. Civil support operations meet the urgent needs of designated U.S. civilians until civil authorities can accomplish these tasks without military assistance. Nearly every urban operation will involve some type or form of stability operation or civil support operation combined and sequenced with offensive and defensive operations.

CHARACTERISTICS OF URBAN STABILITY AND CIVIL SUPPORT OPERATIONS

9-2. Worldwide urbanization, migration trends from rural to urban areas, and more centralized populations in urban areas increase the chance that Army forces will conduct stability or civil support operations in or near urban areas. Simply put, many people live in urban areas, and their welfare will be the primary reason for conducting these operations. Urban areas that serve as economic and government centers (the ideal location for U.S. and allied embassies) are often the focal point for many threat activities. Therefore, Army forces may need to conduct stability operations in these foreign cities to counter those threats. Additionally, urban areas may contain the resources and infrastructure to support both types of operations, regardless of whether the overall focus is in urban or rural areas. Repairing or restoring the infrastructure may be a critical task in accomplishing either type of operation. Supported governmental and nongovernmental agencies are not as logistically self-sufficient as the Army. As such, these agencies will often need to center their operations in and around urban areas to use the area's infrastructure to support themselves and their objectives. These agencies may require military protection to accomplish their missions. Figure 9-1 lists some defining characteristics of these wide-ranging operations.

- Long or Short Duration
- Joint and Interagency
- Unilateral or Multinational
- Increased Civil-Military and Legal Considerations
- Greater Potential for Ambiguity
- Increased Constraints Necessitating More Restrictive ROE
- Amplified Need for Cultural and Political Sensitivity

Figure 9-1. Characteristics of stability and reconstruction and support operations

9-3. Stability and civil support operations are diverse, varied in duration, and unilateral or multinational. Stability operations are conducted overseas as part of a campaign or major operation, while civil support operations are conducted within the United States as part of homeland security. Like all urban operations (UO), they are usually joint. Unlike urban offensive and defensive operations, they are more often interagency operations and require more restrictive rules of engagement (ROE). The multiplicity of actors involved usually increases the scope and scale of required coordination and communication. In urban stability or civil support operations, adverse conditions arising from natural or man-made disasters or other endemic conditions—such as human suffering, disease, violations of human rights, or privation—will significantly modify the urban environment. Unresolved political issues and tenuous agreements, difficulties discriminating combatants from noncombatants or between parties of a dispute, and the absence of basic law and order all serve to complicate an already complex and uncertain environment. Civil-military and legal considerations take on added significance in all urban operations (see Civil-Military Operations and Legal Support in Chapters 5 and 10 respectively), but even more so in urban stability and civil support operations. Finally, recognizing, defining, and achieving the desired end state is often more difficult than in offensive and defensive operations.

9-4. Overall, commanders of major operations involving urban stability and civil support operations should not expect clear guidance. They must learn, adapt, and live with ambiguity. They cannot expect to operate in a political vacuum (even commanders at the tactical level); rather, they should expect to work alongside both governmental and nongovernmental leaders and organizations. Commanders conducting UO should not expect an easily identifiable enemy located across a clearly demarcated line. In fact in many peace operations, they and their Soldiers must resist the need to have an enemy—difficult at best when one side or another (or both) may be sniping at them. Commanders should also expect changing and additional missions and tasks, without being allowed to use every means at their disposal to carry out those missions. Many tasks required may be ones for which their units have never, or rarely, trained. Finally, commanders should expect to show restraint with a keen sensitivity to political considerations and to alien cultures, either or both of which they might find confusing or even repugnant.

URBAN STABILITY AND CIVIL SUPPORT OPERATIONS AND BATTLEFIELD ORGANIZATION

9-5. Each urban stability operation or civil support operation will be distinct from one another. These operations differ even more when applied to a specific urban area. Due to the complexity of the environment, commanders must carefully arrange their forces and operations according to purpose, time, and space to accomplish the mission. In most UO the terrain, the dense population (military and civilian), and the participating organizations will further complicate this arrangement.

9-6. The support and assistance that Army forces will provide during these operations is only temporary (although it may be of long duration) and this facet requires particular emphasis. Commanders should plan and execute both of these types of operations with that essential consideration always in mind. Eventually, the government and administration—either foreign or domestic—must secure and support their population by themselves. Therefore, commanders must envision and set the conditions that allow for the transition of control and responsibility to shift to legitimate civilian authorities. While commanders may only provide assistance and support based on specific and well planned civilian requests, more often commanders will need to determine requirements in collaboration with competent civilian authorities and agencies or, in some cases, with little or no initial civilian assistance at all. Ultimately, transition planning should occur as an integral part of the overall operational planning and include collaboration with appropriate civilian agencies and organization as early as possible to ensure a seamless transition to civilian control without major setbacks and loss of forward momentum.

STABILITY OPERATIONS

9-7. Stability operations establish, sustain, and exploit security and control over foreign areas, populations, and resources. Urban areas will be decisive to accomplishing many types of stability operations because urban areas are the centers of population, culture, economy, and government. Much of the support provided by Army forces will aim to assist local, regional, or national governments to restore essential services and infrastructure and reestablish civil order and authority. The location of civilian authorities in urban areas will, by necessity, be a dominating factor in accomplishing the mission. As importantly, many stability operations—enforcing peace in Bosnia for example—will require interacting with, influencing, controlling, or protecting all or parts of the civilian population. Assessing, understanding, and gaining the support of civilians in key economic, cultural, or political urban areas may influence surrounding regions (smaller urban areas and the rural countryside) and may be decisive to achieving overall stability objectives.

9-8. Stability operations may involve both coercive and cooperative actions. They are conducted in situations in which legitimate civil authority is unable to provide the necessary security and control for the urban population as a result of—

- Deliberate operations to influence a regime change.
- Offensive or defensive operations or natural disasters resulting in ineffective civil authorities, conditions beyond the capabilities of the foreign urban government, or both.

CIVIL SUPPORT OPERATIONS

9-9. Civil support operations are an integral part of homeland security which aim to protect U.S. citizens and infrastructure from conventional and unconventional threats and mitigate the effects of man-made and natural disasters. (The other part to homeland security is homeland defense.) Army forces conduct civil support operations by providing Army resources, expertise, and capabilities in support of a lead agency—which may be to Department of Defense—to address the consequences of man-made or natural accidents beyond the capabilities of local or regional U.S. civilian authorities. Such a situation could result from a chemical, biological, radiological, nuclear, and high-yield explosive (CBRNE) incident, hurricane, flood, or civil disturbance in a domestic urban area.

CBRNE Incidents

9-10. CBRNE incidents can be disastrous and are of particular concern for both stability and civil support UO. In urban areas, the potential for catastrophic loss of life and property is enormous. The Army categorizes CBRNE incidents separately from other natural and man-made disasters because it has specific expertise with these weapons. CBRNE incidents usually result from a military or terrorist threat (adding a law enforcement dimension to the disaster). Other urban CBRNE considerations include the following:

- Subways and other subsurface areas offer ideal areas for limited chemical or biological attacks. Nuclear attack (and high-yield explosives) can produce tragic results due to the effects of collapsing structures, flying debris, and fires. Dispersion patterns are affected by the urban terrain and are more difficult to predict and monitor. Large-scale incidents may produce hundreds of thousands of casualties, but even a limited attack may require evacuating and screening large numbers of civilians. Requirements for medical support, basic life support, and, if necessary, decontamination may quickly overwhelm the Army force's capabilities even with augmentation.
- Similar to natural disasters, panic and disorder may accompany the CBRNE event. Fleeing civilians may clog elements of the transportation and distribution infrastructure. Physical destruction may also affect other components of the infrastructure of critical and immediate concern, such as energy and administration and human services (water, sanitation, medical, fire fighting, and law enforcement). Because all elements of the infrastructure may be affected, the overall recovery time may be lengthened and the effects broadened to include much of the surrounding area. The effects of a single urban CBRNE event potentially could be felt nationally or globally.

Decisive Operations

9-11. In urban stability operations, decisive operations may take many years and include multiple actions before achieving the desired end state. This particularly applies to the strategic and operational levels. Oppositely, decisive operations involved in an urban civil support operation for mitigating or reducing disease, hunger, privation, and the effects of disasters normally achieve faster results. However, any operation that attacks the underlying cause and seeks to prevent or relieve such conditions will usually take longer. In urban areas, establishing law and order to protect critical infrastructure and the inhabitants from lawlessness and violence is often critical and also the decisive operation.

Shaping Operations

9-12. Shaping operations establish and maintain the conditions for executing decisive operations. In urban stability or civil support operations, shaping operations always include information operations (IO) that influence perceptions and maintain legitimacy. Often, various participants, and their potentially divergent missions and methods, are involved. Army commanders must coordinate their planning and efforts (early and continuously) to ensure that their decisive, shaping, or sustaining operations are not working against other agencies' efforts and operations—agencies that may have the lead role in the operation. Thus, a critical shaping operation may be to establish the coordination to help develop a common purpose and direction among agencies, particularly those that may experience continuous personnel turnover during the conduct of a lengthy operation. In some instances and with some organizations and agencies, particularly nongovernmental organizations (NGOs), genuine unity of effort may not be achievable; however, recognizing the differences in aims and goals will allow Army commanders to conduct operations with less friction. Commanders should actively request and include NGOs and appropriate governmental agencies in mission readiness exercises or any other training for stability or civil support operations.

SUSTAINING OPERATIONS

9-13. Sustaining operations enable decisive and shaping operations and include logistics, rear area and base security, movement control, terrain management, and infrastructure development. Sustainment bases, especially those located in urban areas, become an attractive target for hostile civilians; therefore, commanders must actively and aggressively protect these bases as well as lines of communications (see Chapter 10).

CONSIDERATIONS FOR URBAN STABILITY AND CIVIL SUPPORT OPERATIONS

In wars of intervention the essentials are to secure a general who is both a statesman and a soldier; to have clear stipulations with the allies as to the part to be taken by each in the principal operations; finally, to agree upon an objective point which shall be in harmony with the common interests.

Lieutenant General Antoine-Henri,
Baron de Jomini

9-14. The urban operational framework (understand, shape, engage, consolidate, and transition) provides a structure for developing considerations unique to urban stability and civil support operations. Many considerations presented in urban offensive and defensive operations apply to urban stability and civil support operations, particularly those that address how to understand the urban and overall operational environment. Because the situations in which stability and civil support operations normally occur share strong similarities with any urban environment, many of these considerations are closely linked to the urban fundamentals presented in Chapter 5. Taken together, commanders will often find them useful in conducting UO throughout full spectrum operations. Appendix C has a historical example of how to apply the urban operational framework to a stability operation.

UNDERSTAND

9-15. In urban stability or civil support operations, commanders must carefully assess and understand the political dimension of the operational environment, as well as their role and the media's part in managing information. These operations are inherently tied to the exercise of diplomatic power. All operations in urban areas are often the focus of the media and thus gain considerable public and political attention. Therefore, military objectives in urban stability or civil support operations are more directly linked with political objectives. The relationship between the levels of war—strategic, operational, and tactical—is often closer than in urban offensive and defensive operations. Military objectives must be carefully nested within political objectives. Commanders must ensure that the ways and means to accomplish their objectives, to include security and force protection measures, will hold up to media scrutiny and are appropriate for the situation and environment. All levels of command must understand the link between political and military objectives, to include a basic understanding at the Soldier level. One uncoordinated, undisciplined, or inappropriate action, even at the lowest level, could negate months or years of previous, disciplined effort. Commanders must balance security and force protection measures with mission accomplishment. Ineffective measures can put Soldiers at too great a risk and jeopardize the mission. Conversely, overly stringent measures may make it difficult for forces to interact with the population closely—essential in many of these operations. Finally, commanders will need a thorough assessment of the governmental and nongovernmental organizations and agencies that will be operating in or near urban areas that fall within their area of operations.

Political and Military Objectives

9-16. Commanders must translate political objectives into military objectives that are clear and achievable (clear tasks and purposes) and can lead to the desired end state. (In most stability operations, defeat of a threat is not the ultimate desired end state.) Political objectives may be vague making it difficult for commanders to conduct their mission analysis. This applies to tactical- and even operational-level commanders, unskilled at higher level, strategic political-military assessments. Each type of stability operation or civil support operation is distinct, often unfamiliar to the executing unit, and unique to the specific situation. These factors often make it difficult to confidently determine the specific tasks that will lead to mission success. Therefore, commanders must also establish measures of effectiveness that aid in understanding and evaluating progress and help gauge mission accomplishment.

Note: Commanders should consult the US Agency for International Development's (USAID) Field Operations Guide for Disaster Assessment and Response when conducting their assessments and developing measures of effectiveness for many urban relief operations.

9-17. These measures should be quantifiable (in some circumstances, a qualitative assessment may be most appropriate) and link cause with effect. They help determine the changes required and are essential to the assessment cycle required for urban stability or civil support operations. In a humanitarian relief operation to aid the starving, commanders could determine that the decisive effort is delivering safe food to the urban area. To judge success or effectiveness, they could determine that the appropriate measure is the number of food trucks dispatched daily to each distribution site; the more trucks, the more effective the efforts. However, this measure must correlate with the overarching measure of effectiveness: decline in the mortality rate. If no significant decrease in deaths due to starvation occurs, they may need to reassess and modify the tasks or measures of effectiveness. A better measure may be to track the amount of food consumed by those in need instead of simply counting the number of trucks dispatched. Areas around which measures of effectiveness can be formed for many stability or civil support operations (including the example above) and which will help return most societies to some degree of normalcy and self-sufficiency include:

- Restoring law and order.
- Decreasing morbidity and mortality rates.
- Securing safe food and water.
- Restoring critical infrastructure.
- Resettling the population.
- Reestablishing economic activity.

Commanders can often develop measures of effectiveness to address these areas in terms of establishing or restoring security and providing logistics. (Figure 9-2 provides example measures of effectiveness from the strategic to tactical levels for a possible stability operation.)

9-18. Political objectives are fluid and modified in response to new domestic and international events or circumstances. Thus, assessment is continuous, and commanders must adjust their own objectives and subsequent missions accordingly. In urban stability or civil support operations, commanders often develop military objectives that support or align with the objectives of another agency that has overall responsibility for the urban operation. In this supporting role, commanders may receive numerous requests for Soldier and materiel assistance from the supported agency and other supporting agencies operating in the urban area (to include elements of the urban population). With such unclear lines of authority and areas of responsibility, they must ensure that the tasks, missions, or requested Army resources fall clearly in the intended scope and purpose of the Army's participation in the operation. They must not develop or execute missions based on inadequate or false assumptions, misinterpreted intent, or well-meaning but erroneously interpreted laws or regulations by any organization, to include even the lead agency. When missions appear outside their scope, commanders must quickly relay their assessment to their higher headquarters for immediate resolution. The commander's goal is not to limit or slow military participation but to contribute as intended and in consonance with political objectives and the law.

Security and Force Protection Measures

9-19. Commanders must plan for and continually assess the security of their forces operating in an urban area as well as constantly review protection measures. Establishing a robust intelligence—particularly human intelligence (HUMINT)—network that can determine the intentions and capabilities of the threat and the urban populace is the basis for establishing force protection for Army forces operating in the urban environment. However, many such operations, particularly stability operations, require extra time to forge a lasting change. Over time, and particularly in peacetime when objectives center on helping others and avoiding violence, even the complex urban environment may seem benign. Without continued, aggressive command emphasis, Soldiers may become lulled into complacency. Also, during periods of transition or transfer of authority from one unit or organization to another, departing Soldiers' focus often shifts to redeployment activities and away from force protection concerns. It is usually then that Army forces are most vulnerable to terrorist and insurgent tactics, such as bombings, kidnappings, ambushes, raids, and other forms of urban violence.

Measures of Effectiveness: The Need to Measure Progress		
Strategic Level Criteria	**Operational Level Criteria**	**Tactical Level Criteria**
Accountable to the American people for defining and measuring progress towards defeating terrorism and meeting national security goals.	Accountable to the strategic level for measuring operational success and providing linkage to strategic goals.	Responsible to the operational level for measuring tactical success and providing linkage to operational goals
Examples: • Prevention of the insurgency from receiving aid or re-sources from other international groups. • A functioning national government. • Amount of international support and aid to reconstruction. • Number of nations contributing manpower to coalition forces.	Examples: • Host-nation security forces trained and equipped. • Denial of the merging of insurgent forces with terrorist groups. • Amount and distribution of— • Electricity. • Liquid propane gas. • Gasoline. • Functioning provincial governments.	Examples: • Number of insurgent forces in the AO neutralized. • Reduced attacks on coalition forces in the AO. • Reduced civilian-on-civilian violence in the AO. • Host-nation security force recruitment goals met. • Host-nation security force training goals met. • Number of reliable human intelligence walk-ins. • Amount of unexploded explosive ordnance cleared. • Functioning neighborhood and district advisory councils.

Figure 9-2. Example measures of effectiveness

9-20. Although force protection will not ensure successful urban stability or civil support operations, improper assessment and inadequate force protection measures can cause the operation to fail. In either operation, keeping a neutral attitude toward all elements of the urban population, while maintaining the appropriate defensive posture, enhances security. One threat principle discussed in Chapter 3 was that threats would seek to cause politically unacceptable casualties. An improper threat assessment and a lapse in security at the tactical level could result in casualties. That result could affect strategy by influencing domestic popular support and subsequently national leadership decisions and policy.

9-21. Emphasizing security and force protection measures does not mean isolating Soldiers from contact with the urban population. On the contrary, commanders should balance survivability with mobility according to the factors of METT-TC—mission, enemy, terrain and weather, troops and support available, time available, civil considerations. Survivability measures—such as sandbagging, hardening, or fortifying buildings and installations, particularly where large numbers of Soldiers are fed and billeted—will be necessary and require considerable command attention. On the other hand, mobility operations are essential in preserving freedom of action and denying a threat the opportunity to observe, plan, and attack urban forces. Continual Army presence in the urban area may provide the urban population a sense of security and allow Soldiers to develop a detailed knowledge of the "patterns of life" in their assigned area of operation (AO). Armed with this knowledge, they can detect the absence of the normal or the presence of the abnormal that might indicate a potential threat. Overall, mission degradation and increased risk to the force can result if force protection measures prevent Army forces from conducting prudent missions and establishing an active and capable presence.

Assessment of Security and Force Protection: Belfast, Northern Ireland

Since 1969, Belfast has significantly affected the British military campaign for stabilizing the area. British operations in Belfast illustrate the difficulty of balancing the security and protection of forces with maintaining the stabilizing presence necessary to uphold law and order, minimize violence, and control the urban population.

British successes in protecting Belfast's infrastructure and government facilities from terrorist attacks compelled various terrorist cells, especially the Irish Republican Army and the Provisional Irish Republican Army, to attack more military targets.

At the time, British soldiers and bases presented relatively unprotected targets to these factions, and attacks against them solidified their legitimacy as an "army." In response, British commanders implemented extreme security and force protection measures—from ballistic protection vests and helmets to fortress-like operational bases and large unit patrols. These protection measures successfully decreased the violence against British soldiers in Belfast. However, they also decreased the soldiers' interaction with the population and their ability to stabilize the city. The large patrols protected the soldiers, but also inhibited effective saturation of neighborhoods. These patrols, coupled with fortress-like bases and bulky protective clothing, created an "us-versus-them" mentality among civilians and soldiers.

As force protection increased and stabilizing effects decreased, the terrorists were provided more targets of opportunity among the civilians and infrastructure. British commanders reassessed the situation, identified this "see-saw" effect, and adapted to strike a better balance between force protection and effective presence patrols. For example, British forces switched to four-man patrols to enable greater mobility and wore berets instead of helmets to appear less aggressive.

While the above historical example illustrates a critical aspect of stability and civil support operations, it subtly reveals another important lesson for Army commanders. Many of our future coalition partners and assisting governmental and nongovernmental organizations will bring key insights to UO (particularly dealing with urban societies) based on their historical and operational experience. Army commanders and their staffs should seek out and remain open to others' unique understandings that are based on hard-earned and well-analyzed lessons learned.

Participating Organizations and Agencies

9-22. Across the spectrum of urban operations, but more so in these operations, numerous NGOs may be involved in relieving adverse humanitarian conditions. Dense populations and infrastructure make an urban area a likely headquarters location for them. (In 1994 during OPERATION UPHOLD DEMOCRACY, for example, over 400 civilian agencies and relief organizations were operating in Haiti.) Therefore, commanders should assess all significant NGOs and governmental agencies operating (or likely to operate) in or near the urban area to include their—

- Functions, purposes, or agendas.
- Known headquarters and operating locations.
- Leadership or senior points of contact (including telephone numbers).
- Communications capabilities.
- Potential as a source for critical information.
- Financial abilities and constraints.
- Logistic resources: transportation, energy and fuel, food and water, clothing and shelter, and emergency medical and health care services.
- Law enforcement, fire fighting, and search and rescue capabilities.
- Refugee services.
- Engineering and construction capabilities.
- Other unique capabilities or expertise.
- Previous military, multinational, and interagency coordination experience and training.
- Rapport with the urban population.
- Relationship with the media.
- Biases or prejudices (especially towards participating U.S. or coalition forces, other civilian organizations, or elements of the urban society).

Commanders can then seek to determine the resources and capabilities that these organizations may bring and the possible problem areas to include resources or assistance they will likely need or request from Army forces. These organizations will be critical to meeting the population's immediate needs and minimizing the effects of collateral damage or disaster. However, commanders should consider whether a close relationship with any of these organizations will compromise the organization's appearance of neutrality (particularly threat perceptions) and adversely affect their ability to assist the population.

SHAPE

9-23. Commanders conduct many activities to shape the conditions for successful decisive operations. In urban stability and civil support operations, two activities rise to the forefront of importance: aggressive IO and security operations.

Vigorous Information Operations

9-24. IO, particularly psychological operations (PSYOP) and the related activities of civil affairs (CA) and public affairs, are essential to shape the urban environment for the successful conduct of stability or civil support operations. Vigorous IO can influence the perceptions, decisions, and will of the threat, the urban population, and other groups in support of the commander's mission. IO objectives are translated to IO tasks that are then executed to create the commander's desired effects in shaping the battlefield. These operations can isolate an urban threat from his sources of support; neutralize hostile urban populations or gain the support of neutral populations; and mitigate the effects of threat IO, misinformation, rumors, confusion, and apprehension. Developing effective measures of effectiveness is essential to a good IO campaign strategy. One of the most valuable methods for obtaining data for use in this process is face-to-face encounters with targeted audiences by unit patrols and HUMINT, PSYOP, and CA teams. A valuable technique may be to conduct periodic, unbiased surveys or opinion polls of the civilian population to determine changes in their perceptions and attitudes.

Security Operations

Protecting Civilians and Critical Infrastructure

9-25. Security for NGOs and civilians may also be an important shaping operation, particularly for stability and civil support operations. Commanders may need to provide security to civil agencies and NGOs located near or operating in the urban area so that these agencies can focus their relief efforts directly to the emergency. Commanders may also need to protect the urban population and critical infrastructure to maintain law and order if the urban area's security or police forces are nonexistent or incapacitated, or the urban area's security situation has undergone drastic change (such as the result of a natural disaster) and security or police forces require additional augmentation. (See also the discussion of the legal aspects of Civilians Accompanying the Force in Chapter 10.)

Preserving Resources

9-26. Just as forces may be at risk during urban stability or civil support operations, so may their resources. In urban areas of great need, supplies and equipment are extremely valuable. Criminal elements, insurgent forces, and people in need may try to steal weapons, ammunition, food, construction material, medical supplies, and fuel. Protecting these resources may become a critical shaping operation. Otherwise, Army forces and supporting agencies may lack the resources to accomplish their primary objectives or overall mission.

Prioritize Resources and Efforts

9-27. During UO, commanders will always face limited resources with which to shape the battlefield, conduct their decisive operations, and accomplish their objectives. They must continually prioritize, allocate, and apply those resources to achieve the desired end state. To this end, they may develop an order of merit list for proposed projects and constantly update it over time. To some degree, the local urban population will usually need to be protected and sustained by Army forces. Hence, commanders must tailor their objectives and shape their operations to achieve the greatest good for the largest number. Commanders first apply the urban fundamental of preserving critical infrastructure to reduce the disruption to the residents' health and welfare. Second, they apply the urban fundamental of restoring essential services, which includes prioritizing their efforts to provide vital services for the greatest number of inhabitants possible. In operations that include efforts to alleviate human suffering, the criticism for any participating organization is likely to be there is not enough being done or Army forces are not being responsive enough. Therefore, commanders must develop clear measures of effectiveness not only to determine necessary improvements to operational plans but also to demonstrate their Soldiers' hard work and sacrifice and U.S. commitment to the operation.

Improve the Urban Economy

9-28. When conducting reconstruction and infrastructure repair, commanders should consider using such activities to simultaneously improve the urban economy. Hiring local civilians and organizations to do reconstruction work helps satisfies urban job requirements, may inspire critical elements of the urban society to assume responsibility for the success or failure of urban restoration efforts, and may potentially reduce threat influence by diminishing their civilian sources of aid. Hiring indigenous personnel for short-term projects does not replace the need for long-term economic planning and the development of stable jobs. The overall reconstruction effort must be guided by the commander's vision of the end state. However, the commander's guidance and intent should be broad and expansive enough to allow responsive decision making by subordinate commanders based on their analysis of the urban population's needs in their assigned AO.

ENGAGE

If there is any lesson to be derived from the work of the regular troops in San Francisco, it is that nothing can take the place of training and discipline, and that self-control and patience are as important as courage.

Brigadier General Frederick Funston
commenting on the Army's assistance following
the 1906 San Francisco earthquake and fire

9-29. The focus of the Army is warfighting. Therefore, when Army commanders conduct many urban stability and civil support operations, they must adjust their concept of success. Commanders will most often find themselves in a supporting role and less often responsible for conducting decisive operations. They must accept this supporting function and capitalize on the professional values they have instilled in each Soldier, particularly the sense of duty to do what needs to be done despite difficulty, danger, and personal hardship. Commanders must also put accomplishing the overall mission ahead of individual desires to take the lead—desire often fulfilled by being the supported rather than supporting commander. In many stability and civil support operations, success may be described as settlement and compromise rather than victory. Yet, the Army's professionalism and values—combined with inherent adaptability, aggressive coordination, perseverance, and reasonable restraint—will allow Army forces to engage purposefully and dominate during complex urban stability or civil support operations.

Adaptability

9-30. Adaptability is particularly critical to urban stability or civil support operations because these operations relentlessly present complex challenges to commanders for which no prescribed solutions exist. Commanders often lack the experience and training that provide the basis for creating the unique solutions required for these operations. Since the primary purpose for the Army is to fight and win the nation's wars, the challenge then is to adapt urban warfighting skills to the unique stability or civil support situation.

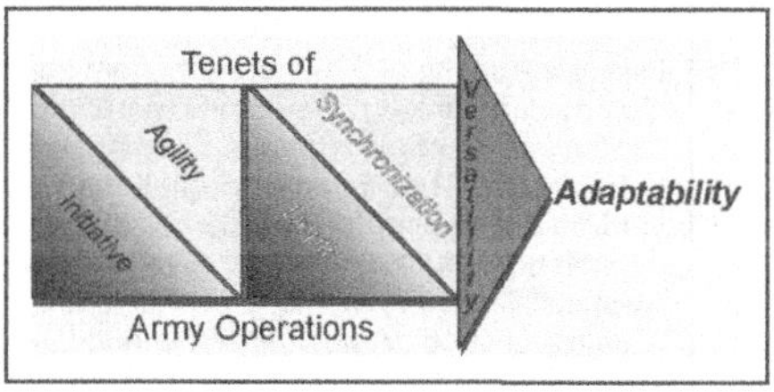

Figure 9-3. Adaptability

9-31. Doctrine (joint and Army) provides an inherent cohesion among the leaders of the Army and other services. Still, Army commanders conducting urban stability or civil support operations will often work with and support other agencies that have dissimilar purposes, methods, and professional languages. Army commanders must then capitalize on three of the five doctrinal tenets of Army operations: initiative, agility, and versatility (see figure 9-3 and FM 3-0). Commanders must bend as each situation and the urban environment demand without losing their orientation. They must thoroughly embrace the mission command philosophy of command and control addressed in FM 6-0 and Chapter 5 to encourage and allow subordinates to exercise creative and critical thinking required for planning and executing these UO. Commanders must also be alert to and recognize "good ideas," and effective tactics, techniques, and procedures, regardless of their source—other services, coalition partners, governmental and nongovernmental organizations, and even the threat—and adapt them for their own purposes in UO.

9-32. Adaptability also springs from detailed planning that carefully considers and realistically accounts for the extent of stability and civil support operations. Although no plan can account for every contingency and completely eliminate the unexpected, good plans—which include detailed civil considerations—provide platforms from which to adjust course more readily. Adequate planning allows commanders not only to react more effectively, but also to be forward-thinking and take actions that favorably steer the course of events.

Aggressive Coordination and Synchronization

9-33. In urban stability and civil support operations, the increased number of participants (both military and nonmilitary) and divergent missions and methods create a significant coordination and synchronization challenge. Significant potential for duplicated effort and working at cross-purposes exists. The success of UO often depends on establishing a successful working relationship with all groups operating in the urban area. The absence of unity of command among civil and military organizations does not prevent commanders from influencing other participants not under his direct command through persuasion, good leadership, and innovative ideas.

9-34. Commanders may consider establishing, as necessary, separate organizations for combat operations and for stability operations or civil support operations to increase coordination and enhance local, NGO, and international support. Further, aligning the unit or subordinate units with NGOs may contribute to establishing popular legitimacy for the operation and place greater pressure on threat forces. In some instances, commanders may consider organizing part of their staff around government, administrative, and infrastructure functions that mirror the urban area in which their forces are operating. Development of a mirror urban area organization may give greater legitimacy to the urban government or administration and ease transition of responsibility once the end state is achieved. Commanders must be mindful, however, that local groups seen as allying themselves with Army or coalition authorities will likely experience pressure to demonstrate their independence as established dates for redeployment or other critical events approach. In some instances that demonstration of independence may be violent.

Civil Support and Coordination with Civilian Authorities: Los Angeles – 1992

During the spring of 1992, Soldiers from the 40th Infantry Division, California National Guard were among the forces deployed to Los Angeles County to assist the California Highway Patrol, Los Angeles County Sheriffs, and civilian law enforcement. They worked to quell the riots that were sparked by the "not guilty" verdicts concerning four police officers who, following a lengthy high-speed chase through Los Angeles, were accused of brutally beating Rodney King.

Successful accomplishment of this civil support operation was attributed to the exercise of strong Army leadership and judgment at lower tactical levels, particularly among the unit's noncommissioned officers. An essential component of combat power, it was especially critical in executing noncontiguous and decentralized operations in the compartmented terrain of Los Angeles. As important, however, was the clear understanding that Army forces were to support civilian law enforcement—and not the other way around. The 40th Infantry Division aligned its area of operations with local law enforcement boundaries and relied heavily on police recommendations for the level at which Soldiers be armed (the need for magazines to be locked in weapons or rounds chambered).

One incident emphasized the need for coordination of command and control measures with civilian agencies even at the lowest tactical levels. To civilian law enforcement and Army forces, the command "Cover me" was interpreted the same: be prepared to shoot if necessary. However, when a police officer responding to a complaint of domestic abuse issued that command to an accompanying squad of Marines, they responded by immediately providing a supporting base of fire that narrowly missed some children at home. However, the Marines responded as they had been trained. This command meant something entirely different to them than for Army Soldiers and civilian law enforcement. Again, coordination at all levels is critical to the success of the operation (see also the vignette in Appendix B).

9-35. In the constraints imposed by METT-TC and operations security (OPSEC), commanders should seek to coordinate all tactical stability operations with other agencies and forces that share the urban environment. Importantly, they should seek to coordinate appropriate information and intelligence sharing with participating organizations. Commanders must strive to overcome difficulties, such as mutual suspicion, different values and motivations, and varying methods of organization and execution. Frequently, they must initiate cooperative efforts with participating civilian agencies and determine where their objectives and plans complement or conflict with those agencies. Commanders can then match Army force capabilities to the needs of the supported agencies. In situations leading to many urban civil support operations, confusion may initially make it difficult to ascertain specific priority requirements. Reconnaissance and liaison elements—heavily weighted with CA, engineers, and health support personnel—may need to be deployed immediately to determine what type of support Army forces provide. Overall, aggressive coordination will foster trust and make unity of effort possible in urban stability or civil support operations where unity of command is difficult or impossible to achieve.

Perseverance

9-36. The society is a major factor responsible for increasing the overall duration of urban operations. This particularly applies to urban stability operations where success often depends on changing people's fundamental beliefs and subsequent actions. Modifying behavior requires influence, sometimes with coercion or control, and perseverance. The urban population often must be convinced or persuaded to accept change. Necessary change may take as long as or longer than the evolution of the conflict. Decades of problems and their consequences cannot be immediately corrected. Frequently, the affected segments of the urban society must see that change is lasting and basic problems are being effectively addressed.

9-37. In most stability operations, success will not occur unless the host nation, not Army forces, ultimately prevails. The host urban administration must address the underlying problem or revise its policies toward the disaffected portions of the urban population. Otherwise, apparent successes will be short lived. The UO fundamental of understanding the human dimension is of paramount importance in applying this consideration. After all Army forces, particularly commanders and staff of major operations, understand the society's history and culture, they can begin to accurately identify the problem, understand root causes, quickly engage and assist key civilian leadership, and, overall, plan and execute successful Army UO.

Reasonable Restraint

9-38. Unlike offensive and defensive operations where commanders seek to apply overwhelming combat power at decisive points, restraint is more essential to success in urban stability and civil support operations. It involves employing combat power selectively, discriminately, and precisely (yet still at decisive points) in accordance with assigned missions and prescribed legal and policy limitations. Similar to the UO fundamentals of minimizing collateral damage and preserving critical infrastructure, restraint entails restrictions on using force. Commanders of major operations should issue or supplement ROE to guide the tactical application of combat power. Excessively or arbitrarily using force is never justified or tolerated by Army forces. Even unintentionally injuring or killing inhabitants and inadvertently destroying their property and infrastructure lessens legitimacy and the urban population's sympathy and support. Collateral damage may even cause some inhabitants to become hostile. In urban stability or civil support operations, even force against a violent opponent is minimized. Undue force often leads to commanders applying ever-increasing force to achieve the same results.

9-39. Although restraint is essential, Army forces, primarily during urban stability operations, must always be capable of decisive (albeit in some circumstances, limited) combat operations. This is in accordance with the UO fundamental of maintaining a close combat capability. This capability must be present and visible, yet displayed in a nonthreatening manner. A commander's intent normally includes demonstrating strength and resolve without provoking an unintended response. Army forces must be capable of moving quickly through the urban area and available on short notice. When necessary, Army forces must be prepared to apply combat power rapidly, forcefully, and decisively to prevent, end, or deter urban confrontations. Keeping this deterrent viable requires readiness, constant training, and rehearsals. It also requires active reconnaissance, superb OPSEC, a combined arms team, and timely and accurate intelligence.

Restraint: An Najaf, Iraq – 2003

On 31 March 2003, during OPERATION IRAQI FREEDOM, The 101st Airborne (Air Assault) Division transitioned from isolating An Najaf—among the holiest places for Shia Muslims in Iraq and the home of one of their leading holy men, the Grand Ayatollah Ali Hussein Sistani—to an attack to clear Iraqi fighters from the town. The decision to attack was made in response to increasing attacks against 101st forces situated near An Najaf and due to alarming reports that the Fedayeen were killing Iraqi family members to force adult males to fight coalition forces. Since An Najaf had an airfield, control of the town would also allow the 101st to obtain needed hardstands for their aircraft.

Ultimately, the division used two brigades to attack and clear the town. The 2nd Brigade Combat Team (BCT) attacked from the north and the 1st BCT—responsible for the initial penetration—attacked from the southwest. Using tanks, infantry fighting vehicles, and light infantry, the commander of the 1st BCT formed effective combined arms teams supported by artillery and air that successfully fought their way into the city.

On 3 April 2003, Soldiers of TF 2-327 Infantry turned a street corner to face a group of civilian men blocking their way and shouting in Arabic, "God is great." The crowd quickly grew into hundreds as they mistakenly thought the Soldiers were trying to capture Sistani and seize the Imam Ali Mosque. Someone in the crowd threw a rock at the Soldiers, which started a hailstorm of rocks; even the battalion commander was hit on the head, chest, and the corner of his sunglasses.

Based on information from a Free-Iraqi Fighter accompanying the battalion and his own observations during the attack, the commander of TF 2-327 believed that most of the people in An Najaf neither wanted to fight nor obstruct his unit's efforts. Consequently, he ordered his Soldiers to position themselves on one knee, smile, and point their weapons toward the ground. At this gesture, many of the Iraqis backed off and sat down, enabling the commander to identify the true troublemakers. He identified eight. In case these agitators were to produce weapons and start to shoot, the commander wanted to make sure that the remainder of the crowd would know from where the shooting would originate. Next, he ordered his Soldiers to withdraw to allow the tension to subside. With his own rifle pointed toward the ground, the battalion commander bowed to the crowd and led his Soldiers away. When tempers had calmed, the Grand Ayatollah Sistani issued a decree (fatwa) calling on the people of Najaf to welcome the American Soldiers.

The attack on An Najaf by the 101st provides an excellent example of well-led Soldiers capable of understanding cultural differences and possessing the discipline and mental flexibility to transition from the aggressive mindset required for high-intensity urban combat to the restraint essential for the initiation of stability operations. In his weekly radio address, President George W. Bush commented: "This gesture of respect helped defuse a dangerous situation and made our peaceful intentions clear."

CONSOLIDATE

9-40. As urban operations will often be full spectrum, many of the consolidation activities necessary to secure gains in urban offensive and defensive operations will be applicable to urban stability and civil support operations. However, the greatest obstacles to attaining strategic objectives will come after major urban combat operations. Therefore, emphasis will appropriately shift from actions to ensure the defeat of threat forces to those measures that address the needs of the urban population, manage their perceptions, and allow responsibility to shift from Army forces to legitimate indigenous civilian control (or the intermediate step to other military forces, governmental agencies, and organizations).

Continued Civilian and Infrastructure Protection

9-41. Following urban offensive or defensive operations, there will likely be a need to secure and protect the civilian population and much of the civilian infrastructure from the civilians themselves. After having minimized collateral damage and preserved critical infrastructure, commanders must implement measures to preclude looting and destruction of critical and essential infrastructure by the urban population and civilian-on-civilian violence. This may be as relatively simple as allowing the urban police force to return to work or may be as difficult as hiring, vetting, and training an indigenous police force. In the latter case commanders may need to determine—

- Number and operability of police stations.
- Responsibility for recruiting, hiring, training, and equipping the urban security or police force. The accountable unit, organization, or agency will need to consider a vetting process, suitable salaries and wages, and appropriate training standards.
- The appropriate responses toward those civilians who threaten, oppose, or harm the new police force.

9-42. Alternatively, civilian security firms (from inside or outside the urban area or country) can provide supplemental protection until indigenous police forces are fully functional. Further, the commander of the major urban operation may plan to manage expected instability primarily with Army forces. Often, this will require larger numbers of infantry, military police, and dismounted forces. Other populace and resource control measures such as curfews may also assist in civilian (including NGOs) and infrastructure protection. Previous shaping operations aimed at improving the local economy can also assist in this regard.

Resolute Legitimacy

9-43. Closely linked to the restraint described previously is legitimacy or the proper exercise of authority for reasonable purposes. Achieving or maintaining legitimacy during urban stability or civil support operations is essential in gaining and maintaining the support of the urban population. Commanders can ensure legitimacy by building consent among the population, projecting a credible force, and appropriately using that force. Perceptions play a key role in legitimacy, and skillful IO can shape perceptions. Commanders must send messages that are consistent with the actions of their forces. Generally, the urban population will accept violence for proper purposes if that force is used impartially. Perceptions that force is excessive or that certain groups are being favored over others can erode legitimacy and generate resentment, resistance, and, in some situations, violent acts of revenge.

9-44. In stability operations against an elusive insurgent threat, commanders may need to explain to urban residents why damage was necessary, apologize, or make near-instant restitution for some unsuccessful Army operations that may have been planned based on inaccurate or incomplete intelligence. (Soldiers may even make it a point to thank homeowners for allowing the search of their homes during cordon and search operations.) During urban operations, a single Soldier's misbehavior can significantly degrade a commander's ability to project an image of impartiality and legitimacy. Fortunately, disciplined Soldiers can contribute immeasurably to gaining and maintaining legitimacy, mitigating ill will, or otherwise winning the urban population's trust and confidence. In stability operations, the greater fight will often be to win the battle of perceptions and ideas instead of one to seize terrain and triumph over an enemy. Inconsistencies in message and behavior will provide threats with raw material for their propaganda and precipitate doubt in the minds of the urban populace who might otherwise support Army objectives.

TRANSITION

9-45. Commanders of major operations are the focal point for synchronizing tactical stability operations or civil support operations with strategic diplomatic and political issues. They are also the critical links between national intelligence resources and the tactical commander. Because strategic, diplomatic, and political changes can quickly transition the type of urban operation, they must keep subordinate tactical commanders abreast of changes in intelligence, policy, and higher decisions. The potential to rapidly transition to urban combat operations emphasizes the need to maintain the capability to conduct close, urban combat. Failure to recognize changes and transition points may lead to UO that do not support the attainment of the overall objective and needlessly use resources, particularly Soldiers' lives. Therefore, Army forces on the ground in an urban stability operation must be more aware of the strategic environment than the threat and the civilian population, each of whom will have their own means of monitoring the national and international situation.

Legitimate and Capable Civilian Control

I met immediately with the [special operations forces liaison officer] to discuss how we would establish a new government that would be able to make instant progress. We decided to select a city council with the following departments: commerce, public works, social services, health, emergency services, education, public relations, and agriculture. A city council director would lead the council, and each department would also have a director in charge of its respective services. A critical step in selecting the city council was ensuring that there was a balance on the council between Altun Kupri's three major ethnic groups. An ethnic imbalance on the council or perceived favoritism would have decreased the legitimacy of the Coalition forces' efforts in the city and served as fuel for the fire of competition for power between the Turkomen and Kurds.

Captain Jeffrey B. Van Sickle
"Stability Operations in Northern Iraq"

9-46. Commanders must maintain or enhance the credibility and legitimacy of the government and police of the urban area and, in the case of stability operations, of the host nation's military forces operating there. In accordance with the urban fundamental of transitioning control, urban commanders must conclude UO quickly and successfully, often to use assets elsewhere in their area of operations. This entails returning the control of the urban area back to civilian responsibility as soon as feasible. The host nation's military and the urban area's leadership and police are integrated into all aspects of the urban stability or civil support operations to maintain their legitimacy. They must be allowed (or influenced) to take the lead in developing and implementing solutions to their own problems. This requires commanders to transition from "leading from the front" to "leading from behind" in an advisory and assistance role. Effective transition to civilian control and responsibility requires commanders at all echelons who understand the basic operation of civil governments and the administration and management of key urban infrastructure.

9-47. If the host nation's leadership, military, and police are not up to the task, commanders can take steps to increase its capabilities through training, advice, and assistance by CA units or by other nongovernmental or governmental organizations and agencies. Sometimes, new leadership and a restructured police force may be required, particularly when corrupt and no longer trusted by the population. This candid assessment of the urban leadership's ability to govern, protect, and support itself is made early in the planning process. Only then can commanders ensure that resources and a well thought-out and coordinated plan (particularly with civilian organizations and agencies) are available for a speedy transition. IO will be paramount in these instances to ensure that the urban population sees the training and rebuilding process itself as legitimate. Throughout urban stability or civil support operations, commanders shape the conditions to successfully hand over all activities to urban civilian authorities.

Longer-Term Commitment

Maintain the Focus

9-48. Many stability operations often require perseverance and a longer-term U.S. commitment requiring operational endurance and the establishment of a sustainable battle rhythm. Lengthy operations will also require a transitional rotation of Army units into the area of operations to continue the mission. Considerations for these transitions are similar to a relief in place and battle handover (see FM 3-90) combined with considerations for deployment and redeployment. FM 41-10 contains comprehensive guidance on transition planning and coordination activities applicable to UO. In addition to any threat considerations, planning for urban transitions between units will often include emphasis and understanding of—

- Formal and informal civilian leadership and relationships.
- Government institutions and administrative functions.
- On-going reconstruction projects.
- The urban economy.
- Participating governmental and nongovernmental organizations and established relationships and cooperation activities, particularly information sharing.
- Significant key events affecting (or likely to affect) operations.
- Significant cultural lessons learned as a result of the outgoing unit's operations.

9-49. The commander of the major operation must ensure that the incoming unit also understands the political and strategic objectives behind the tasks that they will be required to accomplish. Otherwise, the new unit may begin to plan operations that are similar to those conducted by the previous unit without achieving the desired end state or accomplishing the mission.

Transition Trust

9-50. If units are successful in establishing and cultivating relationships with the urban populace and their leadership, and have built an effective level of trust, their anticipated departure from the AO as part of a unit rotation schedule may have detrimental effects on working relationships. As a result, local leaders may develop deep anxieties about the replacement force and their ability to "be as good" as the current command. Therefore, incumbent forces must take steps to identify successful and ongoing efforts and activities in which the new unit can continue to achieve success. In some instances, the current unit may delay a project or activity whose scheduled completion is close to the date of relief or the transfer of authority so as to allow the new unit immediate success. In other instances, projects or activities can be backward planned to ensure that the completion date is scheduled soon after the new unit arrives. The goal for the outgoing unit is to ensure that the new unit is readily accepted and that both units execute a seamless relief in place.

Modify Objectives to Match the Current Environment

9-51. While it is important that incoming commanders create a seamless transition with their outgoing counterparts, they must make full use of the opportunity to review the current political, strategic, and local urban environment to determine potential modifications to logical lines of operations and unit objectives. Otherwise, units may fall into the trap of executing a series of six-month to one-year rotations that do not significantly contribute to solidifying the conditions required to ultimately transition the urban environment back to legitimate civilian control. In operations that involve longer-term commitments, the potential exists for the same unit to conduct two (or more) rotations back into the same AO. Commanders must resist the desire to simply continue with previous lines of operations that, at the time, were deemed successful. The urban environment will change in response to Army actions and this change will necessitate subsequent operational adaptations.

Provide an End State Not an End Date

9-52. Though it may not be feasible, due to political constraints, commanders of major urban operations should fight the public desire to establish a fixed transition timetable. Instead, they should plan urban stability operations based on an attainable end state and not on a specified withdrawal or end date. Progress toward strategic (and operational) goals is susceptible to many changes and delay, particularly in coalitions which will likely be a part of most future urban operations. As such, an end date will only serve to unduly constrain commanders in the attainment of their urban operational objectives. Having set and announced a firm timetable, American and coalition publics will expect withdrawal to commence as scheduled and will not easily be convinced of the need to stay longer even if the situation changes drastically. In general, the public can understand and accept a logical and clearly articulated end state as long as commanders can demonstrate adequate progress based on well developed measures of effectiveness. As important, a timetable allows threats to adjust their plans to the anticipated withdrawal date. Instead of yielding to relentless diplomatic, informational, military, and economic pressure backed by firm political resolve, threats will likely exercise strategic and operational patience.

This page intentionally left blank.

Chapter 10
Urban Sustainment

Sustainment activities exist to enable the Army to initiate and sustain full spectrum operations. Inseparable from decisive and shaping operations, logistics is the central component of sustaining operations and provides the means for commanders to generate and maintain combat power and extend the operational reach of the force. In offensive and defensive operations, sustainment activities are not by themselves likely to be decisive or shaping, yet significantly contribute to those operations. In stability and civil support operations, when the critical objectives may be restoring the infrastructure and the welfare of civilians, sustainment activities will often be decisive. Their success will allow Army forces to dominate this complex environment. However, like all urban operations (UO), sustainment activities affect and are affected by the environment. The urban terrain, infrastructure, and existing resources, coupled with supportive civilians, may facilitate sustainment activities. In contrast, a poorly designed or damaged infrastructure and a hostile population may severely hamper sustainment activities. In the latter case, critical Army resources required elsewhere in the area of operations (AO) may be diverted to repair facilities and control and support the inhabitants of the urban areas.

URBAN SUSTAINMENT CHARACTERISTICS

10-1. Fundamental sustainment characteristics (see figure 10-1) guide prudent sustainment planning regardless of the environment. They provide commanders an excellent framework to analyze and help develop urban sustainment requirements, assess the impact of the environment on the provision of logistics, and gauge the effectiveness of urban sustainment activities.

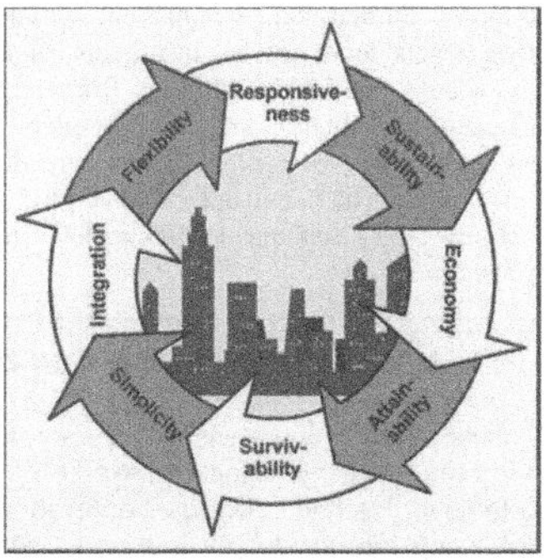

Figure 10-1. Sustainment characteristics

Responsiveness and Sustainability

10-2. UO require responsiveness and sustainability to establish and maintain the tempo necessary for success. Responsiveness—providing the right support in the right place at the right time—is the essential sustainment characteristic. It requires that sustainment commanders and planners accurately forecast urban operational requirements. Continuous urban operations will drain personnel, equipment, and supplies at rates vastly different than in other environments. Therefore, sustainability—the ability to maintain continuous support throughout all phases of the operation—will be a significant concern. Anticipation is critical to both responsiveness and sustainability. It requires that sustainment commanders and planners comprehend the potential effects that components of the urban environment (terrain, infrastructure, and society) may have on operations and sustainment, either benefiting or impeding UO. Effective urban operational and sustainment planning cannot be accomplished separately. Operational and sustainment planners, as well as sustainment operators, must be closely linked to aid in synchronizing and attaining responsiveness and sustainability. The key is often the right balance between efficiency and effectiveness.

Economy and Attainability

10-3. A thoughtful assessment and understanding of the urban environment can also help determine how specific urban areas can contribute to or frustrate the achievement of economy and attainability. Economy is providing the most efficient support at the least cost to accomplish the mission. Attainability means generating the minimum essential supplies and services necessary to begin operations. If available, obtaining support in the AO costs less than purchasing the supplies outside the area and then transporting them there. Critical resources may be available in urban areas to support the operation. However, relying on sources outside the established military sustainment system may create conflict with other sustainment characteristics. A strike by longshoremen, for example, may shut down port operations (at least temporarily) lowering responsiveness and sustainability.

Survivability

10-4. Survivability is being able to protect support functions from destruction or degradation. Commanders often choose to locate sustainment functions in an urban area because the buildings may better protect and conceal equipment, supplies, and people. Urban industrial areas and airports are frequently chosen as support areas because they offer this protection as well as sizeable warehouses, large parking areas, and materials handling equipment (MHE). Such areas facilitate the storage and movement of equipment and supplies. They may also provide readily available water, electricity, and other potentially useful urban resources and infrastructure. However, these areas may also contain toxic industrial materials (TIM) (see Chapter 2). These materials and chemicals in close proximity to support areas may unjustifiably increase the risk to survivability. Sustainment activities in any environment will always be targeted by threat forces. Furthermore, sustainment activities located in any type of confined urban area can offer lucrative targets for terrorists, insurgents, or even angry crowds and mobs. Therefore, no sustainment activity should be considered safe from attack during UO. (During OPERATION JUST CAUSE, Panamanian paramilitary forces and deserters even attacked marked ambulances). Although host-nation support may include assets to assist in defending sustainment units, bases, and lines of communications (LOCs), sustainment commanders must carefully consider if adequate protection measures can ensure survivability. The sustainment commander's greatest challenge to force protection may be complacency born of routine.

> **Base Security: Tan Son Nhut, Vietnam – 1968**
>
> Colonel Nam Truyen was the commander of the 9th Vietcong Division who planned and conducted the attack on the U.S. airbase at Tan Son Nhut during the 1968 Tet Offensive. He had previously entered the airbase during the 1967 Christmas cease-fire using forged identity papers to conduct his own personal and highly thorough reconnaissance.

> UO increase the likelihood that the indigenous population will be utilized as a workforce to supplement Army sustainment activities. Hence, strict operations security will be paramount. Commanders will need to carefully screen potential workers to determine their relative trustworthiness. Human intelligence (HUMINT) and counterintelligence (CI) assets will likely be required to assist in this effort. However, higher-level commanders should seek to take on as much of this burden as possible so that lower-level, tactical commanders can use their limited intelligence assets to accomplish other critical intelligence collection and analysis activities.
>
> All commanders with base protection responsibilities should implement random, changing force protection measures that focus on civilians—even those civilians who have, over time, earned a perceived measure of trust. (Civilians can be co-opted and coerced by threat forces.) As such, all civilian activity that seems out of the ordinary should be promptly investigated regardless of whether a specific force protection measure is in effect. Although leaders and Soldiers seek to establish mutual trust and respect between themselves and the local populace, commanders should normally err on the side of protecting their Soldiers.

SIMPLICITY

10-5. Simplicity is required in both planning and executing sustainment activities in this complex environment. Developing standard procedures among the Army, other services, and especially civilian governmental and nongovernmental agencies, liaison and open channels of communication, simple plans and orders, and extensive rehearsals contribute immeasurably to attaining this necessary characteristic.

INTEGRATION

10-6. The need for sustainment integration increases in UO due to its joint and multinational nature and greater numbers of other governmental and nongovernmental organizations (NGOs) and agencies operating in or near urban areas. More NGOs will likely exist because urban areas often contain most of a region's population. Most NGOs focus on people. Army forces and other military and nonmilitary groups must cooperate and coordinate their actions. Much of their coordination will revolve around sustainment activities. Cooperation and coordination will take advantage of each group's sustainment capabilities, help to avoid duplicated effort (contributing to economy), and create sustainment synergy. It will also help to curtail competition for the same urban resources and assist in developing a unified list of priorities. Such coordination will help ensure that other operations by one force or agency will not disrupt or destroy portions of the urban infrastructure critical to another's sustainment operations and the overall mission.

10-7. Success in UO requires combined arms and innovative task organizations. As such, UO will often result in the integration of heavy and light units necessitating a simultaneous integrated approach to sustainment. Gaining units must integrate forces into their maintenance systems and carefully consider potentially increased resource requirements. (For example, armor units consume greater amounts of Class III—bulk and petroleum—than a similar-sized light unit.) When integrating into a light unit, armored and mechanized units should bring as much of their critical parts and supplies as possible and immediately integrate into the gaining unit's logistical reporting system to ensure adequate resupply.

FLEXIBILITY

10-8. Lastly, commanders must develop flexibility. Although they and their staffs must thoroughly understand the urban environment essential to planning sustainment operations, they cannot possibly anticipate every eventuality. Urban commanders must possess the ability to exploit fleeting opportunities. Knowledge of the environment, particularly its infrastructure, can aid in developing innovative solutions to sustainment acquisition and distribution problems. Flexibility enables sustainment personnel to remain adaptive and responsive to the force commander's needs. Key to maintaining flexibility is the constant assessment of the sustainment situation and the readiness to modify or change procedures to adapt to the current conditions confronting Army forces operating within the urban environment.

10-9. Maneuver and sustainment commanders should consider and prioritize these characteristics as they visualize UO. Each characteristic does not affect every operation and urban area in the same way. The sustainment characteristics seldom exert equal influence, and their importance varies according to mission, enemy, terrain and weather, troops and support available, time available, civil considerations (METT-TC). Like the principles of war, commanders must not ignore the potential impact of sustainment characteristics and how their influence changes as the operation evolves (see FM 4-0).

LOGISTICS PREPARATION OF THE THEATER

[Sustainment] planning accounts for increased consumption, increased threats to lines of communications, and anticipated support to noncombatants...Urban operations place a premium on closely coordinated, combined arms teams and carefully protected [sustainment activities]. Urban operations are [logistic]-intensive, demanding large quantities of material and support for military forces and noncombatants displaced by operations.

FM 3-0

10-10. A thorough logistics preparation of the theater (LPT) is critical for an adaptable UO support plan. Sustainment planners conduct the LPT to understand the situation from a sustainment perspective and determine how best to support the maneuver commander's plan. However, decisions that impact the urban environment and the political situation require a combined operational and sustainment perspective. Sustainment planners must understand the urban environment, the fundamentals of UO, the proposed course of action, and the urban environment's effects on sustainment (as well as the other warfighting functions). Such knowledge allows the planners to develop a detailed estimate of support requirements. A thorough LPT helps commanders determine the most effective method of providing adequate, responsive support to meet support estimates while minimizing the sustainment footprint. Overall, it helps tie together UO requirements with acquisition and distribution. As with all operations, but particularly in a dynamic urban environment, this assessment process is continuous since requirements will change as the urban operation unfolds and matures.

SUPPORT TO INTELLIGENCE

10-11. The LPT resembles and runs parallel to the intelligence preparation of the battlefield (IPB). Products generated under IPB may be useful in the sustainment analysis. Conversely, the LPT may contribute to the IPB by identifying critical resources and infrastructure and assessing their potential to influence (positively or negatively) the operational plan. This information may warrant a course of action that includes offensive or defensive operations to seize, secure, or destroy those critical resources. In UO initially planned for other than sustainment reasons, the information may require altering the plan or imposing additional constraints to protect the identified resources. While these resources may or may not be critical to current operations, they are usually important to set or shape the conditions necessary for Army forces to consolidate and transition to subsequent missions or redeploy. This close relationship between IPB and LPT underscores the need to quickly and continuously involve sustainment personnel for their sustainment expertise and perspective in planning UO.

10-12. As sustainment units are among the most dispersed and omnipresent in any AO, sustainment Soldiers are potentially valuable intelligence collectors. Soldiers working in and around urban sustainment bases and executing frequent convoys through urban areas quickly become very familiar with the urban terrain and the routine associated with that terrain. As with all Soldiers, sustainment personnel must be incorporated into a comprehensive human intelligence collection system.

URBAN SUSTAINMENT INFORMATION

10-13. Figure 10-2 illustrates that a thorough analysis of the key components of urban areas in the commander's AO provides the data for an accurate LPT and subsequent UO sustainment plan (see Chapter 2 and Appendix B). Analyzing the urban terrain and infrastructure helps to determine—

- Geographic influences on consumption factors and on the provision of support (weather, climate, and topography).
- The availability of supplies, such as safe food, potable water, petroleum, electrical energy, barrier material, and compatible repair parts.
- The location of facilities, such as warehouses, cold-storage sites, water treatment facilities, manufacturing plants, hospitals, hotels for billeting, and waste treatment facilities.
- Transportation information, such as seaport and harbor facilities, airfields, rail and road networks, traffic flow patterns, choke points, and control problems.
- Locations and accessibility of maintenance facilities and equipment, and machine works for the possible fabrication of parts.
- The available general skills among the urban population, such as linguists, drivers, mechanics, MHE operators, longshoremen, and other vital trade skills.

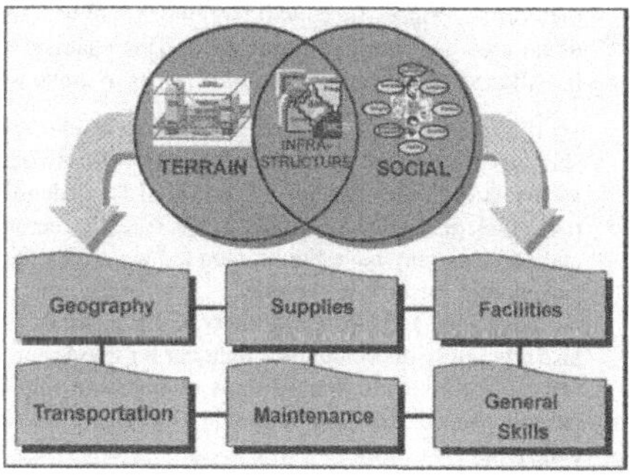

Figure 10-2. Urban environment and essential elements of sustainment information

POTENTIAL RESTRICTIONS

10-14. Commanders must be aware of restrictions that apply to the use of some non-U.S. resources. Security and requirements for U.S. national control dictate that only U.S. assets may perform certain services and functions. Therefore, some foreign urban-area capabilities, even if abundantly available, may not be used. These might include—

- Command and control of medical supply, service, maintenance, replacements, and communications.
- Triage of casualties for evacuation.
- Treatment of chemical, biological, radiological, nuclear, and nuclear (CBRN) casualties, as well as the decontamination of U.S. equipment, personnel, and remains.
- Identification and burial of deceased U.S. personnel.
- Veterinary subsistence inspection.
- Law and order operations over U.S. forces and U.S. military prisoner confinement operations, as well as accountability and security of enemy prisoners of war in U.S. custody.

URBAN SOCIETAL CONSIDERATIONS

10-15. As in all aspects of UO, the urban society is a critical element of the LPT analysis. Sustainment planners must assess whether they can acquire and use urban resources and infrastructure without overly disrupting the urban society and their environment. If the resources are only sufficient for the urban area's inhabitants (and dependent populations in outlying areas), and the facilities cannot increase production to accommodate the needs of Army forces, then commanders may not be able to rely on those resources to support their operations. In fact, the opposite may be true. The effects of UO on the inhabitants, particularly during offensive and defensive operations, may place increased burdens on the Army's resources. For example, commanders may require civilians to evacuate all or parts of an urban area to safeguard them from the effects of planned combat operations. This precautionary measure may require Army forces to support civilian evacuees with temporary housing, food, water, clothing, and medical and sanitary facilities. It will also require carefully planned evacuation routes to ensure that the movement does not interfere with sustainment of the fight.

10-16. Sustainment planners must take into account the potential effects that the provision of supplies to the urban population may have on the operational environment (provided by Army forces or not). For example, an embargo on petroleum products was lifted after Army forces entered Haiti in the 1990s and a less repressive regime assumed governmental control. Few civilian vehicles had moved on the streets of the Port-au-Prince, the capital city, when fuel was scarce and the movement of mounted forces through the urban area was virtually unimpeded. This changed almost overnight once fuel supplies became available, initially making it very difficult for forces to move where required (unless transported by helicopter).

10-17. Sustainment planners must also consider the urban society's ability to restore their own facilities and provide for themselves (if necessary with assistance from Army forces). Throughout this analysis, civil affairs (CA) units can advise and assist in identifying and assessing urban supply systems, services, personnel, resources, and facilities. Critically, commanders must understand that purchasing local goods and services may have the unintended negative consequence of financially sustaining the most disruptive and violent factions in the area. Army forces must seek to purchase urban products and services that will not contribute to prolonging the conflict or crisis. In many stability or civil support operations, they must also attempt to distribute the contracts for goods and services purchased locally as fairly as possible among urban factions and ethnic groups to maintain impartiality and legitimacy. As part of their coordination efforts, commanders should attempt to achieve the cooperation of relief agencies and other NGOs in this endeavor.

10-18. Finally, sustainment planners must also identify potential threats and increased protection requirements that the urban society (criminals, gangs, and riotous mobs) may present, particularly when support units and activities are positioned in urban areas. The disposition or allegiance of the urban population is also important to consider. The infrastructure of an urban area may exhibit great potential to support the sustainment efforts of Army forces, but if the population is hostile or unreliable, use of the facilities and resources may be unfeasible.

Urban Support Areas and Bases

10-19. A major influence on the operation plan and its subsequent execution is often the proper identification and preparation of support areas and bases. The LPT helps commanders determine the need, advantages, and disadvantages of using urban areas in the AO as bases from which to provide support and conduct distribution operations. Ideally, these areas should support reception, staging, onward movement, and integration operations. They should also allow easy sea and air access, offer adequate protection and storage space, increase the throughput of supplies and equipment, and be accessible to multiple LOCs. Consequently, commanders often establish support areas and bases near seaports and airports that are part of a larger urban area. However, threats recognize the Army's need for ports and airfields and may devote substantial resources and combat power to defend them. Army commanders also face the challenge of integrating airfield and port defensive operations with air and naval component commanders to deny threats the ability to conduct stand-off attacks against aircraft and ships. Only an integrated, joint approach can ensure persistent air and sea support– essential not only to sustainment but to firepower as well—is not degraded by the threat. Therefore, planners may determine during the LPT that the risks of seizing or establishing urban lodgment areas may be too high (see Chapter 4). Instead, they may recommend building an airfield, conducting logistics over-the-shore operations, or constructing sustainment bases in more isolated—but more easily secured—locations.

Overall Assessment

10-20. As shown above, the LPT process and analysis help to determine if urban areas in the AO—

- Are suitable as areas and bases for support.
- Can contribute sufficient quantities of and are a dependable source for resources for the overall operation.
- May additionally drain the supported commander's resources.

10-21. The results of this process often serve as a basis for reviewing requirements for civilian contract support and host-nation support and for developing sustainment input into time-phased force and deployment data. This chapter focuses on the effects urban areas may have on accomplishing sustainment functions and related activities, particularly when sustainment units and activities are in urban areas.

ACHIEVING FORCE AGILITY

10-22. To maintain the responsiveness necessary to support full spectrum operations against a wide array of threats and operational environments, sustainment forces must remain agile, possessing the mental and physical agility to transition within or between types of operation with minimal augmentation, no break in contact, and no significant additional training. Responsiveness, flexibility, and economy are key characteristics that enable sustainment forces to support a dynamic combat force; they represent the capability of sustainment forces to provide the resources to initiate, sustain, and extend UO. Agile sustainment forces are characterized by modular force designs, the ability to tailor sustainment organizations for the supporting mission, and the ability to conduct split-based operations. These sustainment aspects are critical to achieving the agility necessary for Army forces to succeed in UO.

MODULAR DESIGN

10-23. Sustainment units are structured as modular organizations, or must possess the ability to modularize in order to conduct decentralized sustainment operations required in UO. At the company level, modularity enables each major company sub-element to assume a cross-section of the company's total capabilities, allowing commanders to employ individual modules to provide a support function, while the rest of the unit remains operational. Modularity enhances responsiveness while enabling support forces to project capabilities as far forward as possible into urban areas, yet with the minimum forward footprint necessary to sustain operations.

FORCE TAILORING

10-24. Sustainment force tailoring refers to the process of determining and deploying the right mix of capabilities to support the force or mission. Sustainment commanders may task organize specific capabilities into functional elements in order to maximize effectiveness and efficiency, and to minimize the sustainment footprint. During UO, the sustainment commander can tailor the support element required to accomplish a specific mission or task, thereby mitigating the risk associated with deploying a larger, more robust capability package forward into the urban area.

SPLIT-BASED OPERATIONS

10-25. Split-based operations refer to performing certain sustainment administrative and management functions outside the area of operations. During UO, Soldiers and civilians can perform personnel, materiel, and distribution management functions without deploying forward into the urban area if the information systems are adequate. This is essential to minimizing risk to support forces and the sustainment footprint in the urban area, and still ensures that all support requirements are met.

SUSTAINMENT FUNCTIONS

- Supply and Field Services
- Maintenance Support
- Transportation Support
- Force Health Protection
- Explosive Ordnance Disposal Support
- Human Resources Support
- Financial Management Support
- Religious Support
- Legal Support to Operations
- General Engineering Support

Figure 10-3. Sustainment functions

SUSTAINMENT FUNCTIONS

10-26. Sustainment consists of eleven functions that must be carefully planned, managed, and synchronized to provide responsive and efficient support to maneuver commanders (see figure 10-3).

Similar to the components of the urban environment, particularly its infrastructure, they overlap and are interdependent. The success of one function depends on the success of several others. Like urban infrastructure, they have two components: a physical component (supplies, equipment, and facilities) and a human component (the Soldiers and civilians who execute these functions). Like city mayors, commanders must plan, manage, and synchronize these functions to provide responsive and efficient sustainment to UO.

10-27. Commanders and planners should consider two essential aspects when addressing these sustainment functions. One aspect looks outward and one looks inward. The first aspect is how these functions can best support full-spectrum UO—the outward analysis. The second aspect is how the urban area affects the conduct of sustainment functions, particularly when those functions are located or performed in an urban area—the inward analysis.

SUPPLY AND FIELD SERVICES

10-28. The supply function involves acquiring, managing, receiving, storing, protecting, maintaining, salvaging, and distributing all classes of supply (except Class VIII) required to equip and sustain Army (and joint) forces. For UO, commanders may need to decide early in the planning cycle whether to stockpile resources forward or to rely on distribution-based sustainment management to satisfy requirements. Some specialized items identified below may not be available through the normal military supply system and may take sustainment personnel much longer to obtain or fabricate. Operation planners must quickly identify the special equipment and increased supply requirements for UO to give sustainment personnel the time necessary to acquire them.

10-29. Field services are those essential services required to enhance a Soldier's quality of life. They consist of clothing exchange, laundry and shower support, textile repair, mortuary affairs, preparation for aerial delivery, food services, billeting, and sanitation. The urban commander determines the need and priority of each service after careful METT-TC analysis. Some facilities such as shower, laundry, and cold storage may be available in the urban area through theater support contracting or provided by external support contracts such as the logistics civilian augmentation program (LOGCAP). Additionally, requirements to care for the urban population will increase requirements for field services immensely. In some circumstances, most notably urban stability or civil support operations, adequately protected field service units or activities will be critical and may be the only support provided.

Increased Urban Supply Requirements

Water

10-30. Water production and distribution will become more difficult for sustainment forces during UO. (Many parts of an urban area can become virtual deserts if public water supplies are cut off or contaminated due to deliberate military action or natural disasters.) Consequently, sustainment managers will be called upon to find more efficient and effective means of moving and distributing potable water, especially when bulk storage of water within the urban area may not be possible or feasible. In sustainment bases that do not have access to urban or other water sources, the transportation and storage of non-potable water may be as great a concern as potable water. Unless adequate storage is available for both potable and non-potable water, units may be forced to rely solely on potable water for all bulk requirements to mitigate health risk. Due to the compartmented nature of many urban operations, planners may need to ensure that additional containers for water (and fuel) are available to support dispersed stock at small-unit level, and that man-portable systems are available in sufficient quantities to ensure timely replenishment.

Class II

10-31. Clothing and individual equipment exposed to the urban environment (concrete, glass, and steel) experience wear and damage at accelerated rates. Units will also require items such as rope, grappling hooks, crowbars, ladders, chain saws, elbow- and kneepads, special vehicle and personnel body armor, fire-fighting equipment, packboards, carpentry tools, and other specialized items to conduct UO. Planners must recognize this and plan accordingly to ensure operations are not interrupted due to equipment shortfalls.

Class III (Bulk)

10-32. With the potential exception of aviation fuel, bulk Class III requirements for UO generally decrease at the maneuver unit level. However, increased fuel requirements for engineer and power-generation equipment attached to or operating with forward units may offset these decreases. Generally, fuel sources within the urban area (refinement facilities, gas stations, garages, and airfields) should not be utilized unless tested and approved for use by a fuel laboratory.

Class IV

10-33. Requirements for construction and barrier material (and the power tools to cut it) during UO will likely far exceed the capability of the sustainment force to provide. Commanders must prioritize distribution, defensive projects, and construction and repair efforts in order to maximize the efficient use of available Class IV. Every effort should be made to obtain supplemental Class IV material from the local economy, as long as that effort does not adversely affect operations or the restoration and improvement of critical urban infrastructure. Of note, specialized, prefabricated road barriers will also be in great demand to block roads, create checkpoints, reinforce defensive positions, and protect headquarters and sustainment activities.

Class V

10-34. Urban combat operations will often increase requirements for Class V. Ammunition consumption rates for urban environments have been as much as five to ten times greater than operations in other environments. Urban offensive and defensive operations will often increase overall requirements for certain types of ammunition (crew-served and small arms ammunition, mortars, antitank rounds, grenades, mines, and demolitions) while others may typically decrease (tank, artillery, attack helicopter). Since forces operating on urban terrain may not be capable of storing ammunition, an established system of distribution is vital to ensuring a constant flow of ammunition necessary to sustain continuous operations and retain the initiative.

Class IX

10-35. Requirements for repair parts generally decrease in UO, with the notable exception of those systems that experience increased ammunition consumption. The heavy use of crew-served and small arms, mortar, and antitank systems in the urban area causes wear and failure at significantly increased rates. Additionally, wheeled-vehicle tires and armored-vehicle track pads and road arms, road wheels, center guides, wedge bolts, and track blocks may wear rapidly on urban streets, rubble, and debris. Based on the tempo of operations, it may be more feasible to replace whole sets of tracks rather than have crews replace individual track pads. Hence, commanders may consider resourcing all wheeled vehicles with spare tires and creating containerized and rapidly transportable sets of track and track parts for speedy delivery to urban sustainment bases. On the whole, planners should consider the use of repair parts packages specifically configured for UO that can be transported by tailored, modular sustainment elements that directly support operations within the urban environment. During long-term stability operations, outgoing units may leave large portions of their prescribed load list in theater to ensure that relieving units have adequate supplies to maintain the tempo of operations.

Storage and Distribution

10-36. UO present challenges to storage and distribution that may not be experienced outside of the urban environment. Forces operating within these areas may not be static and, therefore, cannot store equipment and resources beyond what they can transport organically. Even forces conducting operations from secure locations within the urban area may not be able to store materiel, necessitating a continuous, coordinated distribution system to ensure forces are adequately resourced. During UO, sustainment forces may need to utilize unit or throughput distribution techniques to provide the level of support required to sustain the force.

10-37. Urban commanders must also consider the benefits of stockpiling resources forward against relying on the Army's distribution-based sustainment system. Stockpiling brings supplies close to urban forces and helps ensure supplies are immediately available to support potentially increased consumption rates. However, this method may burden the support structure that moves, handles, and protects large quantities of resources often on a repetitive basis. The Army's normal distribution-based sustainment system reduces this burden significantly; however, available transportation assets (sea, air, and ground) impact delivery response times. These transportation assets often combine military assets overlaid on the host nation's (and urban area's) transportation and distribution infrastructure. Although the civilian infrastructure may initially support the Army's distribution system, later effects of UO, such as destruction of equipment and facilities or loss of civilian workers, may degrade the system. Moreover, Army forces may have to share these assets with other military, multinational force, or civilian organizations participating in the urban operation, as well as with the indigenous civilian population at large. This shared system may put at risk the timely delivery of critical supplies to Army forces.

10-38. Sustainment planners must understand the urban environment and its effects on the proposed method of distribution. They must also understand how urban operations (to include sustainment UO) may affect the urban environment. Storing bulk fuel or ammunition in or near an urban area, for example, may increase the risk of fire and explosive hazard to civilians and Army forces. With this awareness, planners present the commander with an estimate that considers both risks and benefits. Depending on the particular area and other METT-TC factors, they may recommend one method of distribution then transition later to another or a combination of methods. Preconfigured resupply loads delivered as close as possible to where they are needed is vital to responsiveness. Only an agile distribution-based sustainment system will allow Army forces to be responsive and operationally effective across full spectrum operations in an urban environment.

Food Preparation

10-39. The daily feeding standard for Soldiers operating in urban areas remains the same: three quality meals per Soldier. UO creates an increased energy demand on Soldiers, requiring a caloric intake of about 5,000 calories each day. Since producing and delivering prepared meals to forward elements may be difficult or even impossible due to the urban environment, the innovative use of meal supplements can be critical to meeting the dietary requirements of the force. Energy drinks and bars, rich in electrolytes and carbohydrates, are an essential supplement to prepared meals and should be provided in quantities that support or exceed the requirements of the forces operating in the urban area.

10-40. Food (and water) may be available in the urban area; however, local sources must be tested, carefully monitored, and medically approved before consumption. Garbage disposal may be an important consideration in the urban area. Improper trash disposal may leave a signature trail (particularly during urban defensive operations) that may degrade sanitary conditions leading to increased disease and nonbattle injuries (DNBIs). Commanders must understand that food operations, if not properly positioned and secured, can become a focal point for the urban population. Strict policy regarding distribution and control of any Class I supplies (including waste products) will be enforced. Black marketers will be attracted to Army food service activities as well (even during offensive and defensive operations).

Water Purification

Vigilant Monitoring

10-41. Urban areas will often have a ready source of water to support the urban inhabitants and its infrastructure. However, this water may not be suitable for U.S. and allied forces (though the urban population may have developed immunity toward its microorganisms). The higher concentration of TIM in urban areas compounds this problem, as ground water is highly susceptible to chemical contamination, even supplies located miles away from the source of contamination. U.S. Soldiers should be trained and cautioned against using water from an urban area (to include ice and bottled water) until preventive medicine and veterinary personnel can determine its quality. When water quality is unknown, commanders ensure use of tactical water purification equipment that will upgrade it to the Army's water quality standards. Even if initial testing indicates that urban water sources are safe for Army forces, personnel must continuously monitor the water quality. However, Army water purification, storage, and transportation requirements for UO can be greatly reduced if the existing urban water supply can be integrated into sustainment operations. An early assessment of the feasibility of this course of action is critical to sustainment planning.

Potentially Greater Requirements

10-42. Offensive and defensive UO are often intense and can produce more casualties, including civilians. Consequently, medical facilities, already consumers of large volumes of water, may require even more water. Water purification, particularly in the urban areas of developing nations and during urban disaster relief operations, will be a critical and constant concern for Army forces.

Possible Key Terrain

10-43. Sites that can control the water of the urban area may be key terrain, providing not only a resource for Army forces, but also a means to control the threat, the civilians, or both. These sites may be the sources of the water—the river, lake, reservoir, or storage tanks—or the means that process and transport the water—pipelines, pumping stations, or treatment facilities. Many sites may be outside the urban area, as many large urban areas draw water from distant sources. The seizure of a pumping station or pipeline may make it possible for commanders to control water supplies without expending resources required to enter the urban area. To preserve critical infrastructure, commanders may increase security to protect these locations from contamination or destruction. Engineers also may need to restore, maintain, or operate existing water facilities damaged by the threat or disaster and to drill new wells and construct new water facilities.

Mortuary Affairs

10-44. Mortuary affairs provide the necessary care and disposition of deceased personnel. It supports the Army across the spectrum of operations. It may directly and suddenly impact (positively or negatively) the morale of the Soldiers and the American public and may influence relations with the civilian population in the AO. It can also affect the health of Soldiers and the urban populace. Overall, commanders must carefully plan evacuation routes and temporary collection and internment sites, adhering to local customs and traditions to lessen potentially negative consequences.

10-45. Units are responsible for recovering the remains of their own fatalities and evacuating them to the closest mortuary affairs collection point, usually located at the nearest support area. Urban governmental authorities, family, and representatives of appropriate nongovernmental organization are responsible (and should always be relied on to the greatest extent possible) for recovering and handling indigenous civilian remains. However, because of the sheer density of noncombatants in UO, commanders may also find themselves responsible for civilian remains. Deaths of civilians under Army control, such as urban evacuees and refugees at Army-operated sites, often obligate the Army to care for their remains including medical certification and records of death. High-intensity urban combat may result in civilian deaths, and health concerns will require Army forces to deal with civilian remains expeditiously. Commanders should consult local religious leaders, the Staff Judge Advocate (SJA), CA personnel, and chaplains to verify that they are abiding by law and customs. Overall, commanders must ensure that forces treat all deceased, including civilians, with dignity and respect (another important aspect of adhering to the urban fundamental of understanding the human dimension).

Aerial Delivery

10-46. Aerial delivery is the movement by fixed- or rotary-wing aircraft and delivery by the use of parachute or sling load of Soldiers, supplies, and equipment. As a vital link in the distribution system, it contributes to flexibility and provides the capability of supplying the force even when urban ground LOCs are disrupted. Forces use aerial delivery to deliver supplies and equipment when not possible or timely by other means. During urban stability or civil support operations, it is used extensively to move supplies to meet the urgent needs of a population in crisis. In all UO, delivery aircraft are highly vulnerable to small arms, rockets, and air defense systems. A threat may further decrease an already limited number of urban drop zones (aircraft may be able to avoid air defense systems, but ground forces may not be able to secure the drop zone and retrieve the cargo). Equipment and supplies transported by helicopter sling-load lessen the latter disadvantage. There are usually more available sites to deposit sling loads, such as rooftops (engineers may need to determine if they have sufficient structural integrity), parking lots, and athletic fields. However, load instability during flight may restrict a helicopter's airspeed and maneuver capabilities making it more vulnerable to small arms and man-portable air defense systems. For these reasons, aerial delivery of supplies in UO may be much less efficient than in many other environments.

Maintenance Support

10-47. Maintenance entails keeping materiel in operating condition, returning it to service, or updating and upgrading its capability. It includes recovering and evacuating disabled equipment; replacing forward; performing preventive maintenance checks and services (PMCS); increasing battle damage assessment and repair (BDAR); and determining potential maintenance resources in the urban area.

Rapid Recovery Essential

10-48. Disabled vehicles produce obstacles to movement during UO, blocking narrow thoroughfares and creating vulnerable targets for threat forces. This makes rapid recovery operations essential to avoid clogging limited LOCs and mounted avenues with vehicle evacuation operations. Hence, maintenance personnel should be proficient in hastily securing unit maintenance collection sites near the damaged equipment, ensuring route security to and from the disabled vehicle, and methods of egress under fire. The increased potential for the task organization of armored units into smaller attachments during UO (often platoon-sized or smaller) will strain limited recovery assets and increase the need for sections to be well trained and resourced (to include adequate tow-bars and chains) for self-recovery operations. Units must maintain centralized and responsive control over these potentially critical recovery assets and position them as far forward as the situation allows.

Replace Forward

10-49. One of the guiding maintenance principles is to replace forward and fix rear. Maintenance activities, with a forward focus on system replacement, task and use the distribution and evacuation channels to push components and end items to the sustainment level for repair. However, the urban environment may make distribution and evacuation difficult or even impossible. Repairing equipment as far forward as possible will often be vital to maintaining the tempo necessary to succeed in UO. When recovery is required, equipment is moved only as far back as the point where repairs can be made. When selecting a maintenance site, commanders consider:

- Security and force protection measures.
- Sufficient area around equipment for lift or recovery vehicles.
- Availability of existing maintenance facilities or garages.

10-50. The unforgiving urban terrain will invariably increase damage to man-portable weapons and equipment, particularly electronic equipment sensitive to jarring. More frequent jarring can also translate into more frequent calibration requirements. Although by definition man-portable weapons and equipment are easier to transport, evacuating these systems may prove as difficult as evacuating vehicles and larger, heavier equipment. Therefore, field support maintenance support teams (MSTs) will frequently need to repair equipment at (or as near as possible to) the point where it was damaged. Equipment operators are responsible to properly diagnose the fault or damage. Such action ensures that the correct repair parts and maintenance personnel are sent forward to complete necessary repairs. In UO, particularly offensive and defensive operations, units may need to replace rather than repair equipment. Subsequently, sustainment personnel may need to plan for increased replacement of what might normally be repairable equipment, as well as plan for increased repair parts for man-portable items.

Add-On Protection Increases Wear

10-51. To increase protection against small arms, mines, rocket-propelled grenades, and lightweight antiarmor weapons, units may need to attach additional armor and improvised wire-mesh screens and cages to both wheeled and tracked vehicles operating in an urban environment. (Units can also sandbag vehicles to achieve a degree of increased protection.) These modifications, however, increase wear on brakes, springs, suspension, and tires, all of which are already vulnerable to the increased amount of debris present in the urban environment. Over time, this wear will begin to increase failure rates in major assemblies, such as engines, transmissions, and transfers. To offset the increase in wear and, subsequently repair parts requirements, commanders conducting UO should consider emphasizing these aspects during daily operator PMCS and increasing the frequency of scheduled services.

Increased Battle Damage Assessment and Repair

10-52. In UO, operators, crews, MSTs, and recovery teams execute BDAR far more than in other environments. BDAR quickly restores minimum essential combat capabilities for a specific mission (normally of short duration) or allows the equipment to self-recover by expediently fixing or bypassing components. Commanders may need to authorize supervised battlefield cannibalization and controlled exchange when units lack critical parts or cannot bring them forward. Sustainment commanders should consider the use of BDAR kits, specifically designed for the systems they support. The use of these kits, though providing only temporary repair, will extend the availability of combat systems until full repair capability is accessible.

Potential Urban Maintenance Resources

10-53. Although urban areas can complicate maintenance, they may contribute to this sustainment function. Analyzing the urban area in the commander's AO may reveal potential sources of parts, tools, equipment, and facilities, and the expertise necessary to fix equipment, repair components, and fabricate critical parts. Sites of potential value may include automotive repair shops, foundries, car dealerships, parts stores, junkyards, scrap-metal dealers, and machine shops. Urban areas may also serve as key sources for parts and facilities (and theater support contractor personnel) to repair automation and network communication equipment.

> **Host Nation Repair Operations: OPERATION IRAQI FREEDOM**
>
> During operations in Mosul, Iraq in the summer months of 2003, elements of the 2nd Brigade Combat Team (BCT), 101st Airborne Division (Air Assault) suffered from a decline in combat power that outpaced the ability of the distribution-based sustainment system to provide repair parts. As readiness across the BCT threatened to degrade operational capability, sustainment personnel from the brigade's forward support battalion (FSB)—drawing on lessons learned from the 1990-1991 Gulf War—scoured the local populace for repair parts, facilities, and expertise in an attempt to offset the losses suffered due to supply shortfalls.
>
> Initially, local retailers provided limited repair parts and facilities to the FSB, but the effect was significant and readiness rates began to stabilize. As time passed, however, the continued wear on equipment operating in the urban environment, coupled with the harsh summer weather, began to take a toll on reparable components and major assemblies. With limited availability in theater, and order-ship times as long as 60 days, requisitions would not be filled before the effect on combat fleet readiness reached a critical point.
>
> Recognizing a trend that threatened the success of the BCT's urban operations in Mosul, the FSB established a contract maintenance facility in the brigade support area, used its contingency contracting officer to hire a group of local mechanics certified through General Motors, and opened the first host nation repair facility in Iraq. The group repaired major assemblies and reparable components and systems, maintaining a failure rate consistent with major CONUS rebuild facilities. The effect was immediate: during the first four months of the program, a full two-thirds of the combat platforms repaired and returned to service operated with components or assemblies rebuilt in the host nation facility.
>
> Over time, the program was expanded to support other divisional units in northern Iraq. This effort provided maneuver commanders with responsive maintenance support and demonstrated the effect of sustainment flexibility and sustainability in urban operations. Furthermore, this action met the division commander's intent of infusing much-needed funds into the local economy while strengthening American-Iraqi relations in the region.

TRANSPORTATION SUPPORT

10-54. Transportation supports the concept of the urban operation by moving and transferring units, Soldiers, equipment, and supplies. Transportation incorporates military, commercial, host-nation, and urban area capabilities to establish a flexible system that expands to meet the needs of the force. Transportation includes movement control, terminal operations, and mode operations.

10-55. Urban areas are often critical to transportation operations. These areas may serve as a lodgment or support area for entry of Army forces and sustainment supplies. The existing transportation and distribution infrastructure may be essential to reception, staging, and onward movement. Theater support contracts, as well as host-nation support agreements, may greatly increase the ability of Army forces to use the urban area's facilities, which may include docks, airfields, warehouses, and cargo handling equipment. Urban support may also include skilled urban workers, such as longshoremen, truck drivers, and MHE operators.

Urban Terminals

10-56. In addition to serving as major seaports and aerial ports of debarkation, urban areas may provide additional terminals in the AO. Forces may use these terminals for further staging, loading, discharging, and transferring the handling between various inland transportation modes and carriers (motor, air, rail, and water). These urban terminals—with synchronized movement control—permit commanders to rapidly shift transportation modes and carriers. This increases flexibility and ensures the continued forward movement of equipment and supplies to support ongoing operations. Movement control, particularly in urban areas, relies heavily on support from military police in their maneuver and mobility support role. Without this support, urban LOCs may become congested, hinder movement and maneuver, and degrade force effectiveness (see FM 3-19.4). Urban commanders may need to establish multiple checkpoints, roadblocks, and traffic control points; restrict selected roads to military traffic; and reroute movement to unaffected road networks when civil support and refugee control operations compete for available routes. Again, military police operations are critical in this regard and will require continuous, close coordination with urban civilian police (if available).

Obstacles to Ground Transportation

10-57. Although urban areas can contribute to transportation operations, rubble and other damage can become obstacles to ground movement. Even in an undamaged urban environment, road and bridge weight restrictions, barriers and medians, steep embankments, large drainage systems, and other urban structures may limit transportation operations. Urban route maintenance, to include reinforcing bridges and constructing bypasses, may become a priority task for engineer units. Bypassed pockets of resistance and ambushes pose a constant threat along urban supply routes and routes in close proximity to urban areas. Urban convoy and resupply operations will often require increased security in the form of continuous route security operations, regular (daily, if necessary) mine and improvised explosive device (IED) clearance operations, numerous observation posts, attack helicopter support, a larger, more mobile tactical combat force, and adequate communications capabilities among convoy vehicles and with supporting forces and higher headquarters. While the level of security and force protection may vary, convoy operations are never considered an "administrative" operation (see FM 4-01.45). Security needs may increase manpower requirements for sustaining operations and potentially reduce resources from decisive operations. Moving critical supplies may require heavily armed convoys and armored vehicles. Drivers must be well trained in convoy operations, rehearsed, and alert. They must recognize and avoid potential mines and IEDs (stationary and vehicle-borne) and be able to react rapidly to ambushes. Additionally, commanders may need to institute deception measures such as randomly closing and opening supply routes to confuse threat pattern analysis. As before, aerial resupply can alleviate problems due to ground obstacles, but the air defense threat and proximity of threat forces may preclude their routine use.

Population Effects

10-58. The ability of Army forces to use vital urban transportation facilities depends largely on the civilians and the threat. The civilian population can affect the transportation system if they do not support the goals of Army operations. Urban transportation systems—such as ports, railroads, and rivers—require many specialists to operate. Without these specialists, the system's utility is degraded and may not function at all. In urban stability or civil support operations, Army forces will share the system with civilians and other agencies. Civilian authorities may refuse to allow Army forces to use any portions of an urban area's transportation system. Negotiating for access to that system under these circumstances then becomes a command priority.

Threat Effects

10-59. The threat can significantly affect urban transportation systems. Many are composed of smaller subsystems. Each subsystem is vulnerable to attack, which in turn often shuts down the whole system. A large canal system, for example, may have entrance and exit facilities, the canal itself, a means to pull the vessel along such as a locomotive engine, and the civilians that run each of these subsystems. Both an attacker and defender understand the components of the particular transportation system. If important to current or subsequent operations, defending forces then develop plans and allocate forces to protect these subsystems. Attacking forces, on the other hand, often avoid collateral damage to the system, while simultaneously preventing enemy destruction of the facilities.

FORCE HEALTH PROTECTION

10-60. Force health protection (FHP) consists of all services performed, provided, or arranged to promote, improve, conserve, or restore the mental or physical well-being of personnel in the Army and, as directed, for other services, agencies, and organizations. FHP conserves the force by—

- Preventing DNBIs and controlling combat stress.
- Clearing the urban area of casualties.
- Providing forward medical treatment and hospitalization and en route care during medical evacuation.
- Providing required veterinary, dental, and laboratory services.
- Ensuring that adequate Class VIII supplies, medical equipment, and blood are available.

10-61. FHP operations minimize the effects of wounds, injuries, disease, urban environmental hazards, and psychological stresses on unit effectiveness, readiness, and morale. Effective UO require acclimated Soldiers trained in specific urban tactics, techniques, and procedures. FHP helps maintain the health of urban forces, thereby conserving that trained manpower. Historically, urban combat operations have generated three- to six-times greater casualty rates than operations in any other type environment. (Greater enemy wounded and a need to provide care for them will also exacerbate this situation.) FHP operations that maintain the health and readiness of Army forces reduce the strain on the replacement and evacuation systems. Such care allows Soldiers to concentrate on the task at hand instead of the increased risks associated with UO.

10-62. As part of the overall LPT, commanders and medical planners must analyze and continuously assess the urban area. They must identify medical threats, required medical resources, and the quality and availability of medical facilities and resources (to include civilian medical personnel). This assessment prevents duplicated services and permits more effectively and efficiently organized medical resources.

10-63. FHP analysis also identifies available hospitals, clinics, medical treatment facilities, and medical supplies and equipment (including production facilities) in the urban area. It may also identify NGOs capable of providing medical services and supplies. (However, stringent federal regulations, standards of medical care, and a need for unavailable advanced technologies may limit their use by Army forces. But, this analysis will still be crucial in determining whether urban medical facilities can support civilian requirements.) FHP personnel keep abreast of the operational situation and its impact on FHP and the urban population. Peace operations, for example, may rapidly transition to high-intensity offensive and defensive operations requiring medical support able to handle potential mass casualty scenarios which may include civilians.

Care of Civilians

10-64. During combat operations, the military normally does not provide injured civilians with medical care, instead relying on civilian medical personnel to provide this service. However, based on METT-TC and requirements under Geneva Conventions, commanders may need to recover, evacuate, and treat numerous civilians (particularly in urban civil support or stability operations) until local civilian medical personnel and facilities can be reconstituted and supplied. In urban civil support operations involving weapons of mass destruction, the primary focus of Army support may be FHP.

10-65. If commanders provide civilian medical support, they adhere to the UO fundamental of transitioning control and transferring responsibilities for medical care of civilians to another agency or into civilian hands as soon as is practical. In UO, commanders consider and address the medical treatment of civilians (enemy and friendly) early in the planning process. Any initial assessment or survey teams therefore contain medical representatives. Of significant concern, commanders may need to ensure that adequate supplies and equipment are on hand to treat the urban population's children as combat health service support is normally focused on the treatment of adult Soldiers. Additionally, follow-on military care for civilians after treatment at a combat support hospital is often nonexistent or extremely limited to a case-by-case basis necessitating the quick restoration of civilian hospital capabilities. Finally, military health practitioners must also assess and consider cultural factors related to civilian medical treatment. (For example, male doctors may not be able to examine or treat indigenous women.) To alleviate confusion and preclude avoidable negative repercussions and unrealistic expectations, commanders may need to develop specific medical policies, directives, and standing operating procedures. These "medical rules of engagement" (ROE) ensure that subordinates know how much medical care they may provide to—

- The urban population.
- Other host-nation and third-country civilians.
- Coalition and host-nation forces.
- Contractor personnel.

10-66. Commanders should also consider that medical aid to the local urban population and other civilians can be a powerful influential tool. Properly planned and executed, its potential benefits for Army forces include enhanced security and intelligence collection created by increasing the willingness of civilians to interact with Soldiers. These effects cannot be left to happenstance. They should be a conscious part of the information operations campaign nested within the overall UO plan. Consideration must also be given, however, to mitigating the potential negative consequences involved in the inevitable reduction or curtailing of medical assistance by Army FHP personnel. As with any aspect of UO, providing care without considering the longer-term consequences could result in far greater disadvantages than benefits.

Disease and Nonbattle Injury

10-67. DNBI is a major medical threat during all operations and presents additional risk during UO. Urban pollution hazards and potential exposure to TIM may increase the risk to soldier health. Some urban areas, particularly those in developing countries, are already large sources of communicable diseases, such as tuberculosis, cholera, typhus, hepatitis, malaria, dengue, and acquired immune deficiency syndrome (AIDS). Other factors that increase the number of disease vectors and the potential for the transmission of infectious diseases might include—

- Physical damage or deterioration of urban infrastructure—such as electricity, water, and sewage services.
- Industries that use or produce hazardous materials.
- Increased habitats for rodents and other pests due to rubble, debris, and uncollected trash.
- Inability for the threat or local authorities to recover and dispose of human remains and animal carcasses.
- Greater threat of animal diseases transmittable to humans due to larger numbers of animals wounded or injured by the effects of UO.
- Greater susceptibility of urban civilians to infectious diseases—increasing the potential of person-to-person disease transmission—due to declining public health capabilities and nutrition levels.

10-68. The density of the environment extends these urban medical risks to Army forces. Commanders should establish a medical epidemiological surveillance system early to assess the health of the force and promptly identify unusual or local occurrences that may signal preventive medicine problems or the influence of biological or chemical agents. These potential hazards, particularly the release of TIM, will influence the type of medical supplies needed by medical personnel and will also necessitate critical planning and preparation for potential mass casualties (civilian as well as military).

10-69. Adhering to the UO fundamental of preserving critical infrastructure will likely prevent surges in DNBI. Preventive medicine personnel must identify the diseases and recommend control and preventive measures. In urban areas, they may also conduct civilian health screening, health education, and immunization programs. Medical screening of military personnel, particularly multinational forces, may be required to prevent introducing new diseases (especially drug-resistant strains) into an urban area. A new disease may tax the medical system and introduce a new medical problem into an area already in crisis. Lastly, field sanitation training (to include a general emphasis on hand washing; training in the use of barrier protection such as latex gloves when rendering care to any person; and animal, rodent, and pest control), equipment, and supplies are part of overall preventive medicine measures and considerations.

Combat Stress

10-70. Stress occurs in every operation and type of environment; while limited stress levels are beneficial, too much is harmful. Controlled combat stress can invoke positive, adaptive reactions such as loyalty, selflessness, and heroism. On the other hand, uncontrolled combat stress can result in negative, harmful behavior and combat stress casualties. Such behaviors and casualties—battle fatigue, misconduct stress behaviors, and post-traumatic stress disorder—can interfere with the unit's mission. Physical and mental factors leading to combat stress result from the environment and the deliberate enemy actions aimed at killing, wounding, or demoralizing Soldiers. However, many stressors are also generated from the Soldier's own leaders and mission demands.

Combat Stress: Chechnya – 1994 to 1996

Russia's 1994-1996 conflict with Chechnya, a republic in the southwestern part of the Russian Federation, produced an increased number of psychological trauma and combat stress casualties. One medical survey found 72 percent of the soldiers screened had some sort of psychological disorder symptoms. Of that, 46 percent exhibited asthenic depression (a weak, apathetic, or retarded motor state). The other 26 percent exhibited psychotic reactions such as high states of anxiety, excitement, or aggressiveness and a deterioration of moral values or interpersonal relations. The statistics showed more troops experienced combat stress disorders than during their 1980s war in Afghanistan. One primary difference was that in Chechnya, Russian forces conducted combat mostly in cities rather than in mountains, valleys, and other rural areas.

Combat always invokes fear in soldiers. However, poor training and planning, uncertainty in their cause, and urban populations that resented their presence exacerbated the psychological climate for the Russian forces in Chechnya. Acts of subversion and terrorism by Chechen guerrillas kept the Russians in a constant, high state of readiness and anxiety; the soldiers viewed every civilian—young or old, male or female—as a potential enemy. This psychological pressure was not simply a by-product, but an objective of information operations and a prime reason for taking the fight into the close confines of Chechnya's urban areas. Torture and mutilation of prisoners; immediate execution of captured pilots; imitative electromagnetic deception (Chechens mimicked Russian radio transmissions and directed Russian close air support against their own forces); and propaganda to convince civilians that Russia's actions had a religious bias against Muslims and Islam were conducted to exert intense, unremitting psychological pressure on Russian forces—with great success.

The characteristics of urban areas combined with Chechen insurgent activities and information operations, civilians that did not welcome foreigners, and an unpopular and poorly supported conflict with an open-ended mission reinforced the need for a national will to strengthen and support the urban fight. These characteristics also reinforce the need for clear objectives, proper leadership and training, and available medical assets able to properly diagnose and treat combat stress casualties.

10-71. During UO, compartmented urban combat leads to physical isolation while difficulties in transmitting radio signals lead to communications isolation; together, they create an overwhelming sense of loneliness. Snipers, mines, and IEDs combined with the closeness and high intensity of urban combat contribute to an unremitting fear of attack from any quarter that further increases stress casualties. Additionally, seeing and perhaps accidentally inflicting casualties on civilians (especially women and children) increase battle fatigue. If civilians are hostile or a threat uses the population as cover and concealment, then the potential for misconduct stress behaviors often increases. Urban areas may provide temptations for looting, alcohol and substance abuse, black marketeering, and harmful social interactions; these temptations may increase misconduct stress behaviors. Leaders can prevent or rapidly identify, successfully manage, and treat stress-related casualties (see FM 22-51 and FM 6-22.5) and prevent misconduct stress behaviors as well as potential violations of the law of war. To reduce the number of stress casualties, commanders must—

- Provide tough, realistic training in urban combat (and combat stress management).
- Plan for rotation of forces and adequate rest and relaxation opportunities.
- Establish effective ROE.
- Create unit cohesion.
- Ensure adequate mental health support is available.
- Provide strong, capable leadership. Leaders, forward on the battlefield, will determine whether the necessary discipline perseveres.

Evacuation

10-72. Transportation restrictions may preclude the medical evacuation (MEDEVAC) of urban casualties. MEDEVAC by air ambulance remains the preferred method; however, it may not be feasible. The threat from small arms fire, rocket-propelled grenades, and other shoulder-fired weapons may prohibit the employment of helicopters. Aeromedical evacuation can also be hampered by the lack of landing zones due to narrow streets, heavily damaged buildings, and barricaded or booby-trapped roofs. If the employment of air ambulances is not feasible, a ground evacuation operation will be necessary. In UO, ground ambulances should often be—

- Hardened to improve the safety of the crew and patients as well as to increase the survivability of the evacuation platform.
- Field-sited as close to the operation as is tactically and physically possible.
- Maneuverable in tight spaces and narrow streets.

10-73. Evacuation by ground ambulance may have their own set of obstacles to surmount such as rubble; debris; barricades; mines, booby-traps, and IEDs; damaged or destroyed vehicles, and flooded or snow-filled streets. If enough ground ambulances are unavailable due to increased casualties or the tactical situation does not allow, units will need to execute casualty evacuation (CASEVAC) operations.

10-74. In UO, CASEVAC may entail evacuation by litter and the expedient use of a vehicle of opportunity or other military vehicle not dedicated for medical evacuation. Based on the threat situation, CASEVAC may also be limited to armored vehicles, conducted only during limited visibility, or both. For evacuation by litter, units may require more litter bearers to move the injured to a point where they can be further evacuated by ground or air ambulance. Lengthy evacuation routes require more litter bearers, as multiple litter relay teams will be necessary to conserve energy and expedite evacuation. However, unless augmented or relieved of this responsibility by another unit, these litter bearers will come from the casualties' own unit thereby diminishing the unit's strength levels necessary to accomplish its primary mission. Instead, tactical commanders may need to wait until a lull in or a cessation of the battle occurs. Depending on projected casualty rates, higher-level commanders may augment units with additional personnel to perform evacuation or may assign maneuver units this mission. Evacuation personnel will require specific training in urban evacuation techniques (moving casualties from subsurface and supersurface levels to and along the surface level).

10-75. Overall, urban commanders at all echelons must develop detailed casualty evacuation plans that include the establishment of treatment facilities as far forward as the situation allows. Engineers are critical to clear routes for medical resupply and evacuation. Based on the threat situation, additional armored vehicles may need to accompany ambulances in a multi-vehicle convoy not only to and from the patient pickup point but also through to a higher level of care. Army forces may also need to plan for and acquire unique capabilities, equipment, and skills now typically seen in civilian urban search-and-rescue teams to clear debris and search for urban casualties. For example, specially trained dogs may play a vital role in locating victims among urban rubble.

Treatment

10-76. First aid training will have increased significance in UO. The compartmented nature of UO, transportation restrictions, communications difficulties, and the low density of trauma specialists (combat medics) may limit the urban casualty's initial treatment administered by nonmedical personnel or to self-treatment measures. To mitigate this risk, units identify and train combat lifesavers (CLS) to perform in the absence of trauma specialists. Since the likelihood of Army forces performing UO continues to increase, commanders strive to meet or exceed Army standards for the number of fully trained CLS required for their specific unit. CLS (and trauma specialist) training must be realistic, hands-on, and conducted at appropriate intervals to ensure CLS can perform lifesaving tasks to include starting intravenous lines. This increase in self, buddy, and CLS care, as well as longer delays in evacuation, may also increase requirements for additional first aid, CLS, and trauma specialist-carried supplies. (Ultimately, commanders may have to consider weighting the main effort with greater overall FHP support for units conducting UO than for units conducting operations in other environments.)

10-77. The increased potential for delayed evacuation during UO mandates that Army trauma specialists be skilled in prolonged casualty care. During the 3-4 October 1993 battle in Somalia, for example, seven trauma specialists managed 39 casualties for more than 14 hours before they could be evacuated. Evacuation delays significantly increase the potential for infection; therefore, trauma specialists should also be trained to administer antibiotics on the battlefield.

10-78. FHP personnel must be capable of recognizing and treating injuries due to IEDs and incendiary or thermobaric weapons (fuel-air explosives)—favored, urban-oriented threat weapons (see Chapter 3). Thermobaric weapons explode; create a cloud of volatile gases, liquids, or powders; and then ignite, creating an immense fireball consuming oxygen and creating enormous overpressure. When employed in an urban structure, the blast wave or overpressure of any explosive is greatly amplified. Injuries resulting from these weapons are massive burns, broken or crushed bones, concussions and traumatic brain injuries, missile injuries, and internal injuries. To counter this threat, medical personnel must be aware of the capabilities of these blast weapons, wounding patterns, and treatment methods.

10-79. The increased use of body armor during UO will help prevent penetrating chest and abdomen wounds. As such, FHP personnel should expect more groin, pelvis, head, and extremity injuries. Furthermore, when fighting soldiers that are known to use body armor, a threat (particularly snipers) can be expected to target the head and face more often than other anatomic areas resulting in more head injuries. Medical personnel must also anticipate and be prepared to treat an increase in hearing loss due to the firing of weapons (particularly recoilless) in enclosed spaces with little ventilation. Medical personnel (and engineers) should also be trained on how to properly remove rubble that has fallen on a Soldier or civilian so as not to increase the extent of his injuries. Lastly, FHP personnel will need to research and develop unique treatments for injuries due to an ever-increasing arsenal of nonlethal weapons.

EXPLOSIVE ORDNANCE DISPOSAL SUPPORT

10-80. Explosive ordnance disposal (EOD) support provides the capability to neutralize domestic or foreign conventional explosive hazards which include unexploded explosive ordnance (UXO), booby traps, IEDs, captured enemy ammunition (CEA), and bulk explosives. Such material and devices threaten military operations and military and civilian facilities, equipment, and personnel. Explosive hazards create a much greater risk during UO than operations in any other environment. Confined spaces, hard surfaces, and more personnel (both Soldiers and civilians) in the vicinity may magnify the detonating effects of explosives. Dense terrain makes all types of explosive hazards more difficult to locate. Moreover, terrorists select urban settings for their booby traps and IEDs to potentially kill and destroy more, thus gaining greater visibility for their message.

10-81. 10-81. EOD units detect, mark, identify, render safe, and dispose of explosive hazards. Importantly, EOD specialists will work closely with unit intelligence personnel to conduct explosive forensics—technical intelligence collection and exploitation—to help identify the makers of the devices as well as their ever-evolving methods of construction, placement, concealment, and detonation. Chemical reconnaissance with their array of chemical sensors will also work closely with EOD units to reduce the uncertainty of the chemical threat. This integrated forensic approach is a proactive method to take the advantage away from a terrorist or insurgent threat that seeks to employ these devices as a primary tactic.

10-82. Urban operations will increase EOD specialists' role as advisors and instructors on explosive hazards, protection measures, and disposal techniques. They will advise and train other Army forces, other services, multinational partners, and civilian authorities. EOD specialists will also work closely with public affairs and psychological operations personnel to increase public awareness of the horrific effects of these munitions and devices. This education process is designed not only to teach the urban population to identify and avoid explosive hazards, especially UXO, booby traps, and IEDs, but to gain civilian assistance in reporting their locations. (In this regard, rewards for information leading to the discovery of unexploded ordnance may be beneficial. However, the largest reward should normally be reserved for information leading to the arrest or capture of bomb makers.)

10-83. In an urban environment where there are large amounts of UXO, IEDs, and CEA, the sheer amount of explosives may overwhelm the capabilities of available EOD personnel. Because of their training in demolitions and explosives, commanders will often use combat engineers to augment the destruction and clearance effort. Engineers are normally given responsibility for UXO clearance and disposal while EOD teams use their technical skills for IEDs. Commanders may also consider embedding EOD expertise within engineer units to provide the knowledge necessary to determine the correct amount and type of explosive necessary to detonate UXO and to understand potential risks and effects of detonation of various munitions. For example, EOD specialists would better understand the second-order effects and risk reduction measures necessary to detonate a large missile containing numerous submunitions.

10-84. Unfortunately, there is no single device or technique that can eliminate every explosive hazard encountered. Clearance operations are dangerous and time-consuming. Based on METT-TC, commanders may consider the following options when dealing with these explosive hazards:

- Report, mark, and bypass the hazard.
- Employ tactical breaching procedures.
- Self-extract to alternate routes or positions.
- Restrict routes or positions from further use.
- Clear specific areas or positions, or conduct an extensive route clearance.
- Lastly, accept risk of casualties and continue with the assigned mission.

10-85. Effective tactics, techniques, and procedures (TTP) for addressing explosive hazards must be captured and shared with friendly forces and coalition partners. Commanders recognize, however, that adaptive threats will continually modify their own techniques to counter the unit's response. See FM 3-34.119, FM 20-32, FM 3-100.38, FM 4-30.51, and FM 4-30.16.

Human Resources Support

10-86. Human resources support (HRS) encompasses the following functions: manning the force, personnel support, and personnel services. These activities include personnel readiness; replacement operations; personnel accounting; personnel information management; casualty operations; essential personnel services; postal operations; morale, welfare, and recreation; and band support provided to Soldiers, their families, Department of Army civilians, and contractors.

10-87. Successful UO require HRS functions. Clearly, HRS focuses on caring for the needs of people: Soldiers and civilians who support them. Since a critical component of the urban environment is the population, these activities, when required, may also support them. For example, HRS elements, in conjunction with mortuary affairs, military police, and CA, may provide administrative support to assist in the accountability of displaced persons and civilian internees and the establishment of other populace and resources control measures. This support may include the provision of identification and registration cards, travel permits, and other documents, and administrative support to casualty operations. Human resources managers may also assist with community activities and coordination with the Red Cross. If necessary, a postal platoon may assist urban officials in training and reestablishing civilian postal operations. Overall, this type of assistance and support helps to care for the immediate needs of the urban population and, ultimately, speed transition back to legitimate civilian responsibility and control.

10-88. In offensive and defensive UO, HRS personnel will need to account for more casualties and more frequent reconstitution. In all UO, success relies on training individual replacements. Urban combat requires Soldiers skilled in specific urban combat TTP. They understand the societal aspects of the urban population and have training in crowd control to avoid escalating potentially volatile situations. Training these replacements prior to their actual deployment or while in the replacement system frees urban commanders from having to do so. It also helps reduce Soldier (and supporting civilian) isolation, anxiety, and fear associated with urban areas and operations.

Financial Management Support

10-89. Money and its management are often critical to the success of UO. As such, financial management support makes resources available when and where they are needed and assist the urban commander in maintaining fiscal responsibilities. Financial management operations are necessary to conduct the contracting necessary to restore and rebuild infrastructure and provide real-time information, accounting, and services. Financial management operations ensure that urban operational policies and procedures adhere to laws and regulations, develop command resource requirements, and leverage appropriate fund sources to meet them.

10-90. UO will likely include other U.S. military services, governmental agencies, and contractors; multinational forces; NGOs; and various elements of the host nation. So many actors complicate financial accounting, resource management, and cost capturing. Yet, these activities are vital to accomplishing UO and maintaining legal requirements. Army forces will receive support from and provide support to these participants. Financial managers provide crucial expertise to the urban commander on the obligation authority. They also furnish the documentation necessary to obtain and pay for local goods and services using contract or commercial vendor services. However, commanders (with financial management assistance) must assess the economic impact of UO on the economy. An appropriate analysis includes how well the urban economics and commerce infrastructure can support the deployed force and how an influx of U.S. currency may affect the overall economy either positively or negatively.

Labor, Reward, and Solatia Programs

10-91. Financial management elements can also provide joint pay support and non-U.S. pay support for host-nation employees and day laborers supporting UO, as well as for civilian internees. These units can support reward programs and solatia (financial compensation). Commanders can use reward programs to purchase weapons, radios, information, and other items from the urban inhabitants. However, they must use caution when paying for critical human intelligence. First, they must ensure that multiple collection agencies in the command avoid paying the same source for the same information and interpreting these repeated inputs as validation that information is accurate and reliable. Second, they must establish a price or pay scale so subordinate units (particularly in multinational UO) equally compensate their sources for information. Commanders may also need to make solatia to alleviate grief, suffering, and anxiety resulting from injuries and property or personal loss resulting from some aspect of UO. At other times, commanders make solatia to meet cultural expectations of the urban population. These payments are nominal in amount and made without an admission of liability by the Army. Significantly, they may help maintain the legitimacy of the operation and preserve the support of the urban populace.

Financial Support to Civil-Military Projects

> *...[I]n an endeavor like that in Iraq, money is ammunition. In fact, depending on the situation, money can be more important than real ammunition—and that has often been the case in Iraq since early April 2003 when Saddam's regime collapsed and the focus rapidly shifted to reconstruction, economic revival, and restoration of basic services. Once money is available, the challenge is to spend it effectively and quickly to rapidly achieve measurable results. This leads to a related observation that the money needs to*

be provided as soon as possible to the organizations that have the capability and capacity to spend it in such a manner.

Lieutenant General David H. Petraeus
Commander, 101st Airborne Division (Air Assault) and, later,
Commander, Multi-National Security Transition Command-Iraq
during OPERATION IRAQI FREEDOM

10-92. Other financial resources separate from operational funds may be appropriated to support civil-military operations (CMO) projects during stability operations. (During OPERATION IRAQI FREEDOM, these funds were referred to as commander's emergency response program [CERP] funds.) These funds are typically disbursed as bulk currency to tactical units, where the funds are allocated against lower-cost, high-impact projects that support the immediate needs of civil authorities and their citizens. As such, commanders must be prepared to receive, account, secure, and properly disburse these resources in accordance with established procedures. Commanders have an inherent responsibility for ensuring the proper use and disbursement of these funds and must execute appropriate judgment, fiscal responsibility, and supervisory oversight of any resources appropriated in support of CMO. While commanders establish appropriate guidelines to accomplish this fiscal responsibility, they must provide their subordinate commanders—who have the need, capacity, and capability to put these funds to use—reasonable flexibility in how they spend at least a portion of the money, so that they can address emerging needs within their AO.

Contracting Expertise

10-93. Due to the nature of UO, tactical-level units will be likely required to plan and initiate contracts not only for reconstruction projects but to supplement their own sustainment needs. Therefore, higher-level commands must be prepared to assist tactical-level units with contracting expertise. This expertise is required to overcome difficulties in delineating the scope of work, defining construction standards, and establishing viable delivery dates. For small projects and purchases, a standard contractual format may be sufficient. However, for large projects, commanders must provide their subordinates with knowledgeable contracting personnel, who can write contracts with clearly-understood specifications, create purchase orders that specify quality and delivery times, and help monitor work for quality control and assurance. Payment guidelines and standards among all services and members of a coalition are essential to eliminate competition and potential overpayment. Translators will be also essential to accurately translate requirements and standards in addition to trusted local agents to assist in selecting quality engineering and construction firms and workers. Overall, subordinate commanders need adequate contracting support and expertise to meet civilian needs as well as their expectations for timeliness and quality. Without it, commanders may risk losing critical opportunities to positively influence civilian perceptions. (See also Legal Support to Operations below.)

Religious Support

10-94. Religious support entails providing and performing religious support operations for commanders to protect the free exercise of religion for Soldiers, family members, and authorized civilians. This includes personal delivery of rites, sacraments, ordinances, pastoral and spiritual care, and religious education. Such support also consists of:

- Advising the commander on matters of religion, morals, and morale as affected by religion.
- The impact of local religions on threat courses of action, the urban society, and, subsequently, the mission.
- The ethical impact of command decisions.

Moral and Ethical Climate

10-95. Chaplains can help commanders sense the moral and ethical climate in Army units and understand potential moral dilemmas associated with planned UO. The urban environment affects Soldiers' mental health and can increase combat stress casualties, especially misconduct stress behaviors. Chaplains are an important part of combat stress support. The chaplain's presence and faith can sustain Soldiers throughout periods of great trauma. Religious support contributes to the total well being of Soldiers and aids in their return to combat readiness. Chaplains help bolster Soldiers' moral and ethical behaviors through spiritual fitness training so that Soldiers may better cope with ambiguous moral and ethical situations. Chaplains can help identify ethical concerns before they become critical command problems. To this end, their observations help develop and modify ROE, which—due to ambiguity or overly restraining rules—may be part of the problem.

Influence of Local Religions

10-96. Chaplains advise commanders on matters of religion as it affects the Soldiers within their units. They also explain the influences of local religions on threat courses of action and the urban populace and, subsequently, their potential effects on Army forces and UO. Religion is a crucial aspect of assessing the societal component of the urban environment. Understanding the major tenets and concepts of the religions and the impact of faith on civilians' lives may help commanders understand what motivates the threat and the urban populace. (Historically, when religion was a critical factor in armed conflict, the level of lethality has been significantly higher.) To understand threat strategy and tactics, commanders understand their ideology, which, in many cases, involves strong religious beliefs. This understanding also helps commanders to appreciate the inhabitants' attitudes toward other races, religions, and cultures and to identify unacceptable kinds of social interaction (particularly between Soldiers and civilians). Failure to recognize and respect religious beliefs can rapidly erode the perceived legitimacy of the mission and increase the length of the operation. In addition to its influence on threat courses of action, a thorough analysis of the urban environment also includes the degree of influence religion and religious leaders have on the area's government and economy.

Legal Support to Operations

10-97. Legal support provides operational law support in all legal disciplines (including military justice, international law, civil law—composed of contract, fiscal, and environmental law, claims, and legal assistance). This support assists in command and control, sustainment, and HRS of UO. Legal considerations are important in any operation; they take on added significance during UO. They form the foundation for establishing ROE and are critical in the targeting process. (For example, with the increased likelihood that the threat will use the media in an attempt to categorize Army attacks as unlawful, legal support will be critical in determining the protected status of targets.) Adequate legal instruction and training may help subordinate commanders and their Soldiers better understand some aspects ROE and, with this understanding, help them make the rapid decisions often required in a complex urban environment.

10-98. Legal considerations also affect how units acquire goods and services from urban areas and provide support to other agencies and organizations operating in an area. The environment's complex nature requires commanders and their staffs to review and closely consider applicable legal constraints when developing and executing friendly courses of action. Most urban areas have a highly developed legal system nested within county, provincial, state, and national systems. The SJA support to commanders addresses this urban system in conjunction with other legal systems and their potential to affect, positively or negatively, the success of UO.

10-99. International, host-nation, and U.S. law and other regulatory guidelines may vary in their applicability by time and place; actions permissible in one situation may be prohibited in another. These exceptions and complexities increase requirements for SJA, often working with CA personnel, to identify and resolve technical legal issues. Therefore, the SJA must actively advise and participate in all aspects of UO from predeployment training and initial planning through transition and redeployment. FM 27-100 contains detailed legal guidelines affecting UO.

International and Host-Nation Law

10-100. International law consists primarily of agreements, treaties, and customary law to include the law known as the law of war (see FM 27-10). The law of war consists of four general principles applicable when conducting any operation but requiring particular attention during UO. Figure 10-4 lists and describes the four principles: military necessity, discrimination (or distinction), unnecessary suffering (or humanity), and proportionality.

10-101. International law may affect urban operational issues, such as the right of entry, base operations, use of urban infrastructure, and overflight and landing rights. Status-of-forces agreements (SOFAs) exist or can be negotiated to resolve legal issues, such as the status of Soldiers (and, as necessary, contractors) operating in foreign areas to include criminal and civil jurisdictions, taxation concerns, and claims for damages and injuries. Unless a SOFA or other convention exists, Soldiers operating in foreign urban areas have the same legal status as tourists; they are subject to the laws and judicial process of the host nation. (During armed conflict, however, Soldiers always have the rights afforded to them by the Geneva Conventions). Commanders are responsible for understanding the international and host-nation agreements and laws that influence foreign UO. If local law hinders the operation, commanders may be able to inform the local U.S. diplomatic mission and request that it negotiate a solution.

10-102. Commanders may encounter civilian resistance groups whose actions may range from providing enemies with sustainment support to actively fighting Army and coalition forces. Members of such resistance groups must be dealt with in accordance with applicable provisions of the law of war. Commanders should seek legal guidance concerning detention and disposition of persons participating in various acts harmful to friendly forces and detrimental to the mission.

U.S. Law

10-103. Commanders conducting UO must also comply with U.S. law whether it is in the form of a statute, executive order, regulation, or other directive from a federal branch or agency. U.S. law influences UO by governing the acquisition of supplies and services for Army forces, regulating the assistance that can be rendered to foreign nations, and controlling intelligence activities. The Posse Comitatus Act, for example, makes it a crime for Army forces to enforce civil law. Similarly, portions of the Foreign Assistance Act prohibit Soldiers from performing law enforcement activities in foreign urban areas. However, circumstances—expressly authorized by the Constitution, acts of Congress, and other exceptions to these statutes—exist that allow the Army to support civilian law enforcement. Although not nearly all-inclusive, the above demonstrates how U.S. law further complicates urban operations, particularly stability operations, and increases the need for proactive SJA advice and counsel in all facets of UO.

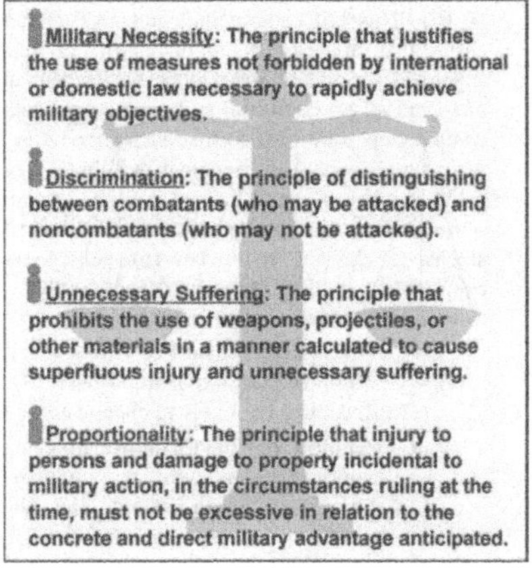

Figure 10-4. General principles of the law of war

Legal Aspects of Nonlethal Force

10-104. Nonlethal capabilities can augment the means of deadly force and extend urban firepower options. They enhance the commander's ability to apply force in proportion to the threat and to allow discrimination in its use. The range of nonlethal capabilities includes offensive information operations, smoke and obscurants, irritants (such as chemical riot control agents), nonpenetrating projectiles, high-pressure water devices, and directed-energy weapons. These continually expanding capabilities give commanders more options to confront situations that do not warrant deadly force but require Soldiers to employ overwhelming decisive power. However, nonlethal capabilities are subject to the same legal constraints as lethal force (in fact, some nonlethal capabilities can cause serious injury and death, particularly if not employed properly) and undergo the same legal review. Like lethal force, nonlethal capabilities show military necessity, distinguish between combatants and noncombatants, distinguish between military objectives and protected property, are used proportionally, and do not result in unnecessary suffering.

10-105. Of special note, commanders cannot employ chemical herbicides or riot control agents as a method of warfare without prior Presidential approval. However, the Secretary of Defense and regional combatant commanders have authority to use herbicides and riot control agents in limited circumstances during armed conflict and in peacetime operations. Considering the complexity of this issue, commanders of major operations must streamline the approval process so that subordinate commanders do not discount use of these agents simply because of the high-level approval authority. Subordinate commanders must, however, be able to anticipate and clearly justify their potential use in UO. Also, any approval does not mean that these agents must be employed.

Civilians Accompanying the Force

10-106. LOGCAP personnel and other civilian support contractors provide a variety of sustainment functions for the Army. These functions may range from providing unskilled labor, transportation support, and health care to technical support of sophisticated equipment and weapons systems. Commanders must ensure that civilians providing support in their AO are not placed in positions of jeopardy, but if they are, these civilians must understand the risks that they assume when they engage in activities that might be misconstrued as direct or active participation in hostilities.

10-107. This last caution indicates an important legal consideration for commanders. They must try to ensure that threat forces can distinguish civilian support contractors from combatants and thereby remove any justification for them to be targeted or attacked. (As Army forces seek to work closely with the members of governmental and nongovernmental agencies and organizations in urban stability operations, commanders may need to take steps necessary to mitigate the risk of attack caused by their involvement with Army forces.) In the event that threat forces capture civilian support contractors, this same concern helps to ensure that, as lawful civilians accompanying the force, they will be afforded prisoner of war status. As the role of contractor support to Army operations continues to expand, commanders must stay abreast of changes in the law (to include evolving interpretative distinctions) as it affects the use of contractors on the urban battlefield. Other important considerations for commanders with civilian contractors supporting their force are—

- **Minimize Exposure to Harm**. Civilian contractors should be assigned duties at locations that minimize their exposure to harm. Commanders will need to periodically review these locations since they are likely to change as the urban operation evolves and transitions.
- **Prohibit the Performance of Combat Functions**. Civilian support personnel should not be working in situations that involve combat operations where they might be conceived as combatants. As an example, they should not perform force protection for Army forces such as fortification construction and checkpoint security.
- **Prohibit the Use of Arms**. Civilians should not be armed, unless approved by the combatant or service component commander on a limited, by-exception basis for personal defense.
- **Provide Identification Cards**. Civilians should be provided, and carry on their person, a Geneva Convention identification card identifying them as civilians authorized to accompany military forces and entitling them to be treated, if captured, as prisoners of war.
- **Restrict the Wear of Military Uniforms**. Civilians should not normally wear distinctive U.S. military uniforms unless the combatant or service component commander authorizes them to do so. Regardless of their clothing, however, civilians will wear a symbol that establishes their civilian status.

General Engineering Support

10-108. General engineering support will be essential during UO. This support helps assess, construct, maintain, and restore essential LOCs and urban facilities to sustain Army forces, the urban population, or both. Using civilian resources and investing Army general engineering resources requires careful consideration by commanders and staff planners. Since all elements of the urban infrastructure interconnect, general engineering support touches each category to some degree. Figure 10-5 illustrates how urban-specific, general engineering tasks align primarily with the transportation and distribution, energy, and administration and human services components of the urban infrastructure. These engineering tasks are significant and readily apply to UO. The last two, providing fire fighting support and waste management, have not been previously addressed and require more specific consideration.

Construct, Maintain, or Restore

- Roads and Highways
- Over-the-Shore Facilities
- Ports
- Railroad Facilities
- Airports and Heliports
- Fixed Bridges
- Electric Power Facilities
- Petroleum Pipelines and Storage Facilities
- Water Facilities

and Provide

- Mobile Electric Power
- Construction Support
- Fire Fighting Support
- Waste Management

Figure 10-5. General engineering support tasks

The Use or Investment of Resources

10-109. During urban offensive and defensive operations, Army engineer units accomplish tasks to sustain or improve movement and mobility, protection, and sustainment of U.S. and allied forces. These units should maximize the existing urban facilities, host-nation support, civilian contractors, and joint engineer assets. However, commanders should consider how using urban facilities to support military forces may negatively affect the population. On the other hand, construction and repair may benefit both Army units and the urban inhabitants. Restoring the urban transportation network not only improves military LOCs, but may also allow needed commerce to resume. Repairing urban airfields or ports increases throughput capabilities for military supplies, facilitates medical evacuation operations to the support base, accelerates needed relief efforts, and allows international commerce to proceed. Commanders may first invest resources and conduct general engineering tasks to restore facilities for civilian use. Such actions stem future drains on operational resources or facilitate later transition of control back to civilian authorities. For example, repairing police stations, detention facilities, and marksmanship ranges may help urban governments reestablish law and order after completing urban offensive or defensive operations. During stability or civil support operations, the focus of general engineering will often be to support and assist the urban population rather than Army forces.

Fire Fighting Support

10-110. Fire protection and prevention, as well as fire fighting, takes on added significance during UO, particularly offensive and defensive operations. Most ordnance produces heat and flame. This, coupled with an abundance of combustible material (buildings, furniture, gasoline, oil, and propane), poses a serious risk to Soldiers, civilians, and the urban operation. Large shantytowns can exacerbate this problem. In highly combustible areas, commanders may even need to limit or preclude the use of small-arms tracer ammunition. Fire threats to urban areas can be categorized as—

- **Isolated Fires.** These are restricted to a single structure or a specific area within a structure.
- **Area Fires.** These fires consume two or more structures and may extend to encompass an entire block. Generally, streets will serve as firebreaks and help contain the fire within a single block.
- **Firestorms.** These are the most violent and dangerous fires capable of rapidly consuming large areas by creating windstorms and intense heat. Firestorms are often inextinguishable until they have consumed all available combustible materials.

- **Explosive Hazards**. These are present in areas containing fuels and chemicals as well as military explosive hazards.

10-111. When analyzing the administration and human services component of the infrastructure, commanders must determine the adequacy of existing civilian fire fighting support. A deteriorated or nonexistent infrastructure that cannot support the urban area will likely fail to handle the increased risk due to military operations. Commanders may need to provide fire fighting teams to support their own forces and civilians.

10-112. A military force task organized with multiple fire fighting teams (even with maximum use of available civilian fire-fighting assets) will only be able to fight some fires in the AO. Water distribution systems damaged during operations, chemical and other TIM, and hostile activities will further complicate and limit fire fighting capabilities. Commanders must develop priorities for equipment, facility, and infrastructure protection. All Soldiers need training in fire prevention and initial or immediate response fire fighting. Such training includes planning covered and concealed movement, withdrawal, and evacuation routes. Soldiers should be trained to identify and remove ignition and fuel sources and provided additional fire fighting material such as extinguishers, sand, and blankets (see FM 5-415 and the U.S. Department of Transportation's current version of the Emergency Response Guidebook).

Waste Management

10-113. Management of all forms of waste, particularly human, putrescible (such as food), and medical, may become a critical planning consideration for Army forces. This particularly applies if the urban waste management infrastructure was previously inadequate or damaged during UO or natural disasters, the Army force is operating in the urban area for an extended period of time, and a significant number of the urban population remains. Failure to adequately consider this aspect, possibly coupled with the decaying remains of humans and animals and an inadequate or tainted water supply (such as may be caused by flooding), may create unacceptable sanitary and hygiene conditions and subsequently increase DNBIs as well as civilian casualties.

Appendix A

Siege of Beirut: An Illustration of the Fundamentals of Urban Operations

The [Israeli Defense Forces] IDF had neither the strategy nor the experience nor the configuration of forces to fight and sustain a house-to-house campaign in Beirut.

Richard A. Gabriel
Operation Peace for Galilee: The Israeli-PLO War in Lebanon

OVERALL STRATEGIC SITUATION

A-1. In 1982, Israel launched OPERATION PEACE FOR GALILEE designed to destroy the Palestine Liberation Organization (PLO) presence in southern Lebanon. On 1 June, the IDF launched a massive assault across the border into southern Lebanon. The Israeli attack focused on the PLO, but the operations quickly involved major ground and air combat between Israel and Syrian forces.

A-2. In the first few weeks, Israeli forces quickly pushed back both the Syrians and the PLO. However, except for some PLO forces isolated in bypassed urban areas, such as Tyre and Sidon, most of the PLO fell back into Beirut (see Figure A-1). By 30 June, Israeli forces had reached the outskirts of southern Beirut, occupied East Beirut, isolated the city from Syria and the rest of Lebanon, and blockaded the sea approaches to the city. Even so, with most of the PLO intact inside and with significant military and political capability, the Israelis had yet to achieve the objective of OPERATION PEACE FOR GALILEE. The Israeli command had to make a decision. It had three choices: permit the PLO to operate in Beirut; execute a potentially costly assault of the PLO in the city; or lay siege to the city and use the siege to successfully achieve the objective. The Israelis opted for the latter.

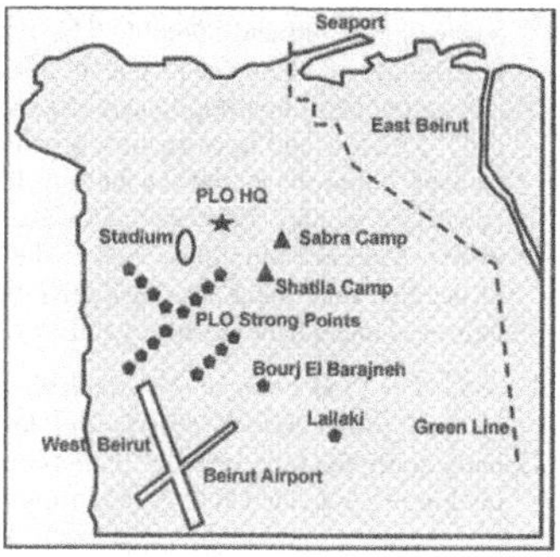

Figure A-1. The city of Beirut

Israeli Military Position

A-3. The Israelis had an excellent position around Beirut. They occupied high ground to the south and west, virtually dominating the entire city. Israeli naval forces controlled the seaward approaches to Beirut. The Israelis' position was also strong defensively, capable of defeating any attempt to break out of or into the city from northern Lebanon or Syria. The Israeli air force had total and complete air superiority. The Israelis controlled the water, fuel, and food sources of West Beirut. Although the PLO forces had stockpiles of food and supplies, the Israelis regulated the food, water, and generating power for the civil population.

A-4. Despite the superior positioning of Israeli forces, the IDF faced significant challenges to include the combat power of the PLO, Syria, and other threats in Beirut. Israeli doctrine and training did not emphasize urban operations. Additionally, Israel was constrained by its desire to limit collateral damage and friendly and noncombatant casualties. Organizationally, the Israeli army was not optimized to fight in urban terrain.

Armor and self-propelled artillery formations dominated the Israeli forces, and most Israeli infantry was mechanized. The Israeli forces had only a few elite formations of traditional dismounted infantry.

PLO MILITARY POSITION

A-5. Despite being surrounded and cut off from support, the PLO position in Beirut offered numerous advantages in addition to the characteristic advantages of urban defense. The PLO had long anticipated an Israeli invasion of southern Lebanon; it had had months to prepare bunkers, obstacles, and the defensive plan of Beirut and other urban areas. Approximately 14,000 Arab combatants in West Beirut readied to withstand the Israeli siege. This was done with the advice of Soviet, Syrian, and east European advisors. The preparation included stockpiling essential supplies in quantities sufficient to withstand a six-month siege. Also, the PLO fighters integrated into the civil populations of the urban areas. Often their families lived with them. The civil population itself was friendly and provided both information and concealment for PLO forces. PLO fighters were experienced in urban combat and knew the urban terrain intimately. PLO forces had been involved in urban fighting against Syrian conventional forces and Christian militias in Beirut several years prior to the Israeli invasion. Finally, the organization of the PLO—centered on small teams of fighters armed with machine-guns and antitank weapons, and trained in insurgent, hit-and-run tactics—was ideally suited to take maximum advantage of the urban environment.

ROLE OF CIVILIANS

A-6. Various ethnic and religious groups make up the civil population of southern Lebanon. However, West Beirut's population was heavily Palestinian and Lebanese. The civil population of West Beirut was between 350,000 and 500,000. The Palestinian population supported the PLO. The Lebanese population may be described as friendly neutral to the Israelis. Although unhappy under Palestinian dominance, this population was unwilling to actively support Israel. The civilian population was a logistic constraint on the PLO, which would have become significant had the siege lasted longer. The civilians in West Beirut were an even larger constraint on the Israelis. The presence of civilians significantly limited the ability of the Israelis to employ firepower. However, the Palestine refugee camps located in West Beirut were both civilian centers and military bases. The Israeli constraints on artillery and other systems against these parts of the city were much less restrictive than in other parts of West Beirut where the population was mostly Lebanese and where fewer key military targets existed.

A-7. The PLO knew of the Israeli aversion to causing civil casualties and purposely located key military centers, troop concentrations, and logistics and weapons systems in and amongst the population—particularly the refugee Palestinian population in the southern part of West Beirut. Tactically, they used the civilians to hide their forces and infiltrate Israeli positions.

A-8. The friendly Palestinian population provided intelligence to the PLO while the friendly Lebanese population provided intelligence for the IDF. Throughout the siege, the IDF maintained a policy of free passage out of Beirut for all civilians. This policy was strictly enforced and permitted no weapons to leave the city. Some estimates are that as many as 100,000 refugees took advantage of this policy.

INFORMATION OPERATIONS

A-9. The siege of Beirut involved using information operations (IO) to influence the media. PLO information operations were aimed at controlling the media and hence the international perception of the operation. This was done by carefully cultivating a select group of pro-PLO media years before hostilities even began. Once hostilities started, only these media sources were permitted to report from the besieged portions of the city, and they were only shown activities that portrayed the IDF negatively. The IDF did not vigorously counter the PLO plan. In fact, the IDF contributed to it by limiting media access to their activities. The PLO information operations had a successful impact. The international community was constantly pressuring the Israeli government to end hostilities. This put pressure on the IDF to conduct operations rapidly and to limit firepower and casualties.

CONDUCT OF THE URBAN OPERATIONS

A-10. The siege of Beirut began 1 July (see Figure A-2). By 4 July, Israeli forces occupied East Beirut, the Green Line separating East and West Beirut, and dominating positions south of the airport. IDF naval forces also controlled the sea west and north of Beirut. On 3 and 4 July, IDF artillery and naval fire began a regular campaign of firing on military targets throughout West Beirut. On 4 July, the IDF cut power and water to the city.

A-11. From 5 to 13 July, the Israeli fires continued to pound PLO targets in West Beirut. The PLO gave one significant response, firing on an Israeli position south of the city and causing several casualties. On 7 July, reacting to international pressure, the IDF returned power and water to West Beirut's civil population. On 11 July, the IDF launched its first attack, probing the southern portion of the airport with an armored task force (see Figure A-3). The PLO repulsed this attack and destroyed several IDF armored vehicles.

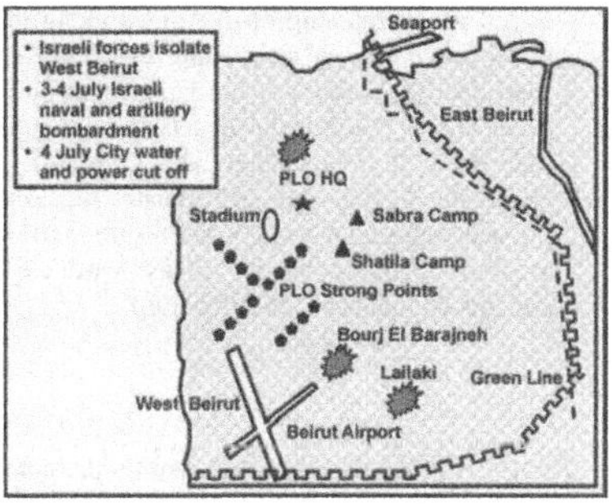

Figure A-2. Initial conduct of the urban operation

A-12. On 13 July, both sides entered into a cease-fire that lasted until 21 July. They began negotiations, mediated by international community, to end the siege. The PLO used this period to continue to fortify Beirut. The Israelis used the time to train their infantry and other arms in urban small unit tactics in Damour, a town the Israeli paratroopers had captured.

A-13. The cease-fire ended on 21 July as PLO forces launched three attacks on IDF rear areas. The Israelis responded with renewed and even more vigorous artillery, naval, and air bombardment of PLO positions in the city. The IDF attacks went on without respite until 30 July. On 28 July, the IDF renewed its ground attack in the south around the airport (see Figure A-4). This time IDF forces methodically advanced and captured a few hundred meters of ground establishing a toehold.

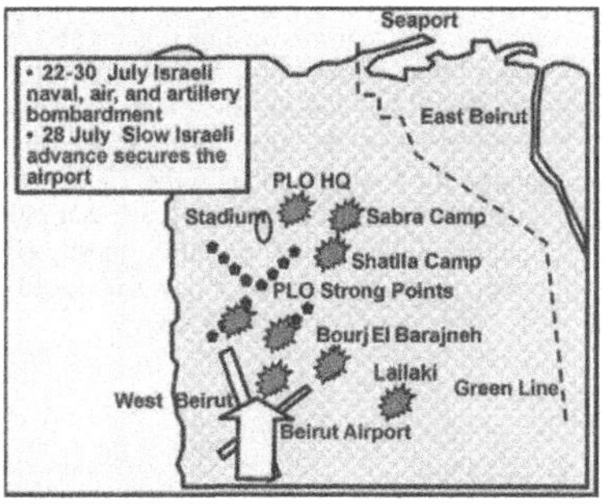

Figure A-3. Israeli probe of PLO defenses

A-14. The Israeli bombardment stopped on 31 July. However, on 1 August the IDF launched its first major ground attack, successfully seizing Beirut airport in the south (see Figure A-5). Israeli armored forces began massing on 2 August along the green line, simultaneously continuing the attack from the south to the outskirts of the Palestinian positions at Ouzai. On 3 August, the Israeli forces continued to reinforce both their southern attack forces and forces along the green line to prepare for continuing offensive operations. On 4 August, the IDF attacked at four different places. This was the much-anticipated major Israeli offensive.

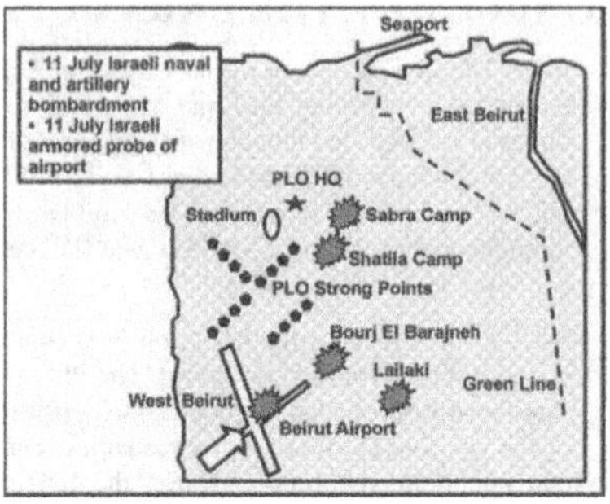

Figure A-4. Initial Israeli attack

A-15. The Israeli attack successfully disrupted the coherence of the PLO defense. The southern attack was the most successful: it pushed PLO forces back to their camps of Sabra and Shatila and threatened to overrun PLO headquarters. Along the green line the IDF attacked across three crossing points. All three attacks made modest gains against stiff resistance. For this day's offensive, the Israelis suffered 19 killed and 84 wounded, the highest single day total of the siege, bringing the total to 318 killed. Following the major attacks on 4 August, Israeli forces paused and, for four days, consolidated their gains and prepared to renew the offensive. Skirmishes and sniping continued, but without significant offensive action. On 9 August, the IDF renewed air and artillery attacks for four days. This activity culminated on 12 August with a massive aerial attack that killed over a hundred and wounded over 400—mostly civilians. A cease-fire started the next day and lasted until the PLO evacuated Beirut on 22 August.

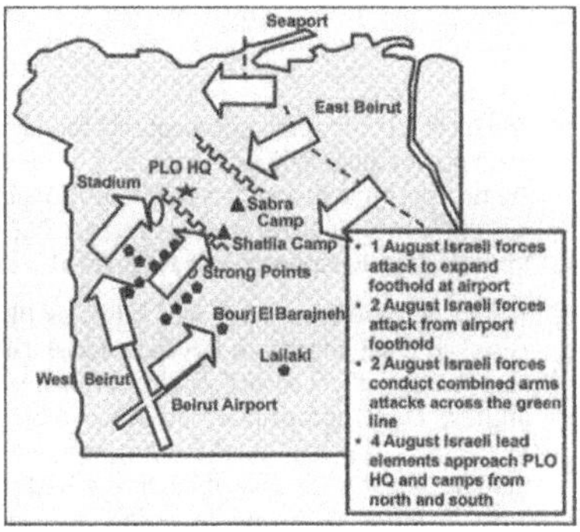

Figure A-5. Final Israeli attack

LESSONS

A-16. The Israeli siege of West Beirut was both a military and a political victory. However, the issue was in doubt until the last week of the siege. Military victory was never in question; the issue in doubt was whether the Israeli government could sustain military operations politically in the face of international and domestic opposition. On the other side, the PLO faced whether they could last militarily until a favorable political end could be negotiated. The answer was that the PLO's military situation became untenable before the Israeli political situation did.

A-17. This favorable military and political outcome stemmed from the careful balance of applying military force with political negotiation. The Israelis also balanced the type of tactics they employed against the domestic aversion to major friendly casualties and international concern with collateral damage.

Perform Aggressive Information Operations

A-18. The PLO devoted considerable resources and much planning on how to use IO to their best advantage. They chose to focus on media information sources as a means of influencing international and domestic opinion.

A-19. The PLO's carefully orchestrated misinformation and control of the media manipulated international sentiment. The major goal of this effort was to grossly exaggerate the claims of civilian casualties, damage, and number of refugees—and this was successfully accomplished. Actual casualties among the civilians were likely half of what the press reported during the battle. The failure of the IDF to present a believable and accurate account of operations to balance PLO efforts put tremendous pressure on the Israeli government to break off the siege. It was the PLO's primary hope for political victory.

A-20. In contrast to the overall weak performance in IO, the IDF excelled in psychological operations. IDF psychological operations attacked the morale of the PLO fighter and the Palestinian population. They were designed to wear down the will of the PLO to fight while convincing the PLO that the IDF would go to any extreme to win. Thus defeat was inevitable. The IDF used passive measures, such as leaflet drops and loudspeaker broadcasts. They used naval bombardment to emphasize the totality of the isolation of Beirut. To maintain high levels of stress, to deny sleep, and to emphasize their combat power, the IDF used constant naval, air, and artillery bombardment. They even employed sonic booms from low-flying aircraft to emphasize the IDF's dominance. These efforts helped to convince the PLO that the only alternative to negotiation on Israeli terms was complete destruction.

Maintain a Close Combat Capability

A-21. The ground combat during the siege of Beirut demonstrated that the lessons of tactical ground combat learned earlier in the twentieth century were still valid. Small combined arms teams built around infantry, but including armor and engineers, were the key to successful tactical combat. Artillery firing in direct fire support of infantry worked effectively as did the Vulcan air defense system. The Israeli tactical plan was sound. The Israelis attacked from multiple directions, segmented West Beirut into pieces, and then destroyed each individually. The plan's success strongly influenced the PLO willingness to negotiate. Tactical patience based on steady though slow progress toward decisive points limited both friendly and noncombatant casualties. In this case, the decisive points were PLO camps, strong points, and the PLO headquarters.

A-22. The Israeli willingness to execute close combat demonstrated throughout the siege, but especially in the attacks of 4 August, was decisive. Decisive ground combat was used sparingly, was successful and aimed at decisive points, and was timed carefully to impact on achieving the political objectives in negotiations. The PLO had hoped that their elaborate defensive preparations would have made Israeli assaults so costly as to convince the Israelis not to attack. That the Israelis could successfully attack the urban area convinced the PLO leadership that destruction of their forces was inevitable. For this reason they negotiated a cease-fire and a withdrawal on Israeli terms.

Avoid the Attrition Approach

A-23. The Israelis carefully focused their attacks on objects that were decisive and would have the greatest impact on the PLO: the known PLO headquarters and refugee centers. Other areas of West Beirut were essentially ignored. For example, the significant Syrian forces in West Beirut were not the focus of Israeli attention even though they had significant combat power. Selectively ignoring portions of the urban area allowed the Israelis to focus their combat power on the PLO and limit both friendly casualties and collateral damage.

Control the Essential and Preserve Critical Infrastructure

A-24. The Israeli siege assured Israeli control of the essential infrastructure of Beirut. The initial Israeli actions secured East Beirut and the city's water, power, and food supplies. The Israelis also dominated Beirut's international airport, closed all the sea access, and controlled all routes into and out of the city. They controlled and preserved all that was critical to operating the city and this put them in a commanding position when negotiating with the PLO.

Minimize Collateral Damage

A-25. The Israeli army took extraordinary steps to limit collateral damage, preserve critical infrastructure, and put in place stringent rules of engagement (ROE). They avoided randomly using grenades in house clearing, limited the use of massed artillery fires, and maximized the use of precision weapons. With this effort, the Israelis extensively used Maverick missiles because of their precise laser guidance and small warheads.

A-26. The strict ROE, however, conflicted with operational guidance that mandated that Israeli commanders minimize their own casualties and adhere to a rapid timetable. The nature of the environment made fighting slow. The concern for civilian casualties and damage to infrastructure declined as IDF casualties rose. They began to bring more field artillery to bear on Palestinian strong points and increasingly employed close air support. This tension underscores the delicate balance that Army commanders will face between minimizing collateral damage and protecting infrastructure while accomplishing the military objective with the least expenditure of resources—particularly soldiers. ROE is but one tool among many that a commander may employ to adhere to this UO fundamental.

Understand the Human Dimension

A-27. The Israelis had a noteworthy (although imperfect and at times flawed) ability to understand the human dimension during their operations against the PLO in Beirut. This was the result of two circumstances. First, the PLO was a threat with which the Israeli forces were familiar after literally decades of conflict. Second, through a close alliance and cooperation with Lebanese militia, the Israelis understood a great deal regarding the attitudes and disposition of the civil population both within and outside Beirut.

Separate Noncombatants from Combatants

A-28. Separating combatants from noncombatants was a difficult but important aspect of the Beirut operation. The Israelis made every effort to positively identify the military nature of all targets. They also operated a free passage system that permitted the passage of all civilians out of the city through Israeli lines. The need to impose cease-fires and open lanes for civilians to escape the fighting slowed IDF operations considerably. Additionally, Israeli assumptions that civilians in urban combat zones would abandon areas where fighting was taking place were incorrect. In many cases, civilians would try to stay in their homes, leaving only after the battle had begun. In contrast, the PLO tied their military operations closely to the civilian community to make targeting difficult. They also abstained from donning uniforms to make individual targeting difficult.

A-29. Earlier in OPERATION PEACE FOR GALILEE when the IDF attacked PLO forces located in Tyre, Israeli psychological operations convinced 30,000 Lebanese noncombatants to abandon their homes and move to beach locations outside the city. However, the IDF was subsequently unable to provide food, water, clothing, shelter, and sanitation for these displaced civilians. IDF commanders compounded the situation by interfering with the efforts by outside relief agencies to aid the displaced population (for fear that the PLO would somehow benefit). Predictably, many civilians tried to return to the city complicating IDF maneuver and targeting—that which the separation was designed to avoid. IDF commanders learned that, while separation is important, they must also adequately plan and prepare for the subsequent control, health, and welfare of the noncombatants they displace.

Restore Essential Services

A-30. Since essential services were under Israeli command, and had been since the beginning of the siege, the Israelis had the ability to easily restore these resources to West Beirut as soon as they adopted the cease-fire.

Transition Control

A-31. In the rear areas of the Israeli siege positions, the Israeli army immediately handed over civic and police responsibility to civil authorities. This policy of rapid transition to civil control within Israeli lines elevated the requirement for the Israeli army to act as an army of occupation. The Israeli army believed the efficient administration of local government and police and the resulting good will of the population more than compensated for the slightly increased force protection issues and the increased risk of PLO infiltration.

A-32. Upon the cease-fire agreement, Israeli forces withdrew to predetermined positions. International forces under United Nations (UN) control supervised the evacuation of the PLO and Syrian forces from Beirut. These actions were executed according to a meticulous plan developed by the Israeli negotiators and agreed to by the PLO. Israeli forces did not take over and occupy Beirut as a result of the 1982 siege (an occupation did occur later but as a result of changing situations).

SUMMARY

A-33. The Israeli siege of West Beirut demonstrates many of the most demanding challenges of urban combat. Apart from Israel's poor understanding of the strategic influence of the media on operations (a lesson that they learned and subsequently applied to future operations), the IDF's successful siege of Beirut emerged from their clear understanding of national strategic objectives and close coordination of diplomatic efforts with urban military operations. A key part of that synchronization of capabilities was the understanding that the efforts of the IDF would be enhanced if they left an escape option open to the PLO. This option was the PLO's supervised evacuation that occurred after the siege. Although the PLO was not physically destroyed, the evacuation without arms and to different host countries effectively shattered the PLO's military capability. Had Israel insisted on the physical destruction of the PLO in Beirut, the overall operation might have failed because that goal may not have been politically obtainable in view of the costs in casualties, collateral damage, and international opinion.

This page intentionally left blank.

Appendix B

An Urban Focus to the Intelligence Preparation of the Battlefield

Maneuvers that are possible and dispositions that are essential are indelibly written on the ground. Badly off, indeed, is the leader who is unable to read this writing. His lot must inevitably be one of blunder, defeat, and disaster.

Infantry in Battle

The complexity of the urban environment and increased number of variables (and their infinite combinations) increases the difficulty of providing timely, relevant, and effective intelligence support to urban operations (UO). Conducted effectively, however, the intelligence preparation of the battlefield (IPB) allows commanders to develop the situational understanding necessary to visualize, describe, and direct subordinates in successfully accomplishing the mission.

IPB is the systematic process of analyzing the threat and environment in a specific geographic area—the area of operations (AO) and its associated area of interest (see figure B-1). It provides the basis for intelligence support to current and future UO, drives the military decision-making process, and supports targeting and battle damage assessment. The procedure (as well as each of its four steps) is performed continuously throughout the planning, preparation, and execution of an urban operation.

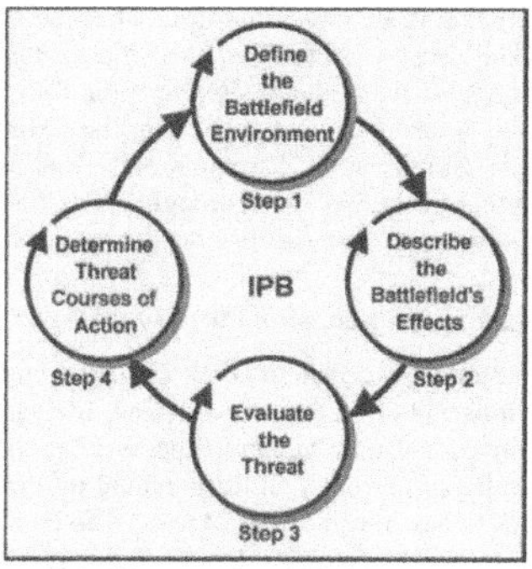

Figure B-1. The steps of IPB

UNAFFECTED PROCESS

B-1. The IPB process is useful at all echelons and remains constant regardless of the operation or environment. However, an urban focus to IPB stresses some aspects not normally emphasized for IPBs conducted for operations elsewhere. The complex urban mosaic is composed of the societal, cultural, or civil dimension of the urban environment; the overlapping and interdependent nature of the urban infrastructure; and the multidimensional terrain. This mosaic challenges the conduct of an urban-focused IPB. There is potential for full spectrum operations to be executed near-simultaneously as part of a single major operation occurring in one urban area with multiple transitions. Such multiplicity stresses the importance of a thorough, non-stop IPB cycle aggressively led by the commander and executed by the entire staff. Overall, the art of applying IPB to UO is in properly employing the steps to the specific operational environment. In UO, this translates to analyzing the significant characteristics of the environment, the role that its populace has in threat evaluation, and understanding how these affect the planning and execution of UO. FM 2-01.3 details how to conduct IPB; FM 34-3 has the processes and procedures for producing all-source intelligence. This appendix supplements the information found there; it does not replace it.

Increased Complexity

B-2. Uncovering intricate relationships takes time, careful analysis, and constant refinement to determine actual effects on friendly and threat courses of action (COAs). These relationships exist among—

- Urban population groups.
- The technical aspects of the infrastructure.
- The historical, cultural, political, or economic significance of the urban area in relation to surrounding urban and rural areas or the nation as a whole.
- The physical effects of the natural and man-made terrain.

B-3. A primary goal of any IPB is to accurately predict the threat's likely COA (step four—which may include political, social, religious, informational, economic, and military actions). Commanders can then develop their own COAs that maximize and apply combat power at decisive points. Understanding the decisive points in the urban operation allows commanders to select objectives that are clearly defined, decisive, and attainable.

Reducing Uncertainty and Its Effects

B-4. Commanders and their staffs may be unfamiliar with the intricacies of the urban environment and more adept at thinking and planning in other environments. Therefore, without detailed situational understanding, commanders may assign missions that their subordinate forces may not be able to achieve. As importantly, commanders and their staffs may miss critical opportunities because they appear overwhelming or impossible (and concede the initiative to the threat). They also may fail to anticipate potential threat COAs afforded by the distinctive urban environment. Commanders may fail to recognize that the least likely threat COA may be the one adopted precisely because it is least likely and, therefore, may be intended to maximize surprise. Misunderstanding the urban environment's effect on potential friendly and threat COAs may rapidly lead to mission failure and the unnecessary loss of Soldiers' lives and other resources. A thorough IPB of the urban environment can greatly reduce uncertainty and contribute to mission success.

Training, Experience, and Functional Area Expertise

B-5. Not all information about the urban environment is relevant to the situation and mission—hence the difficulty and the reason for IPB and intelligence analysis. Although it may appear daunting, institutional education, unit training, and experience at conducting intelligence support to UO will improve the ability to rapidly sort through all the potential information to separate the relevant from merely informative. (This applies to any new or difficult task.) The involvement and functional expertise of the entire staff will allow commanders to quickly identify the important elements of the environment affecting their operations. Fortunately, IPB as part of the entire intelligence process (plan, prepare, collect, process, and produce) is comprehensive enough to manage the seemingly overwhelming amounts of information coming from many sources. Accomplished properly, it allows commanders to recognize opportunities often without complete information.

B-6. As in any operational environment, tension exists between the desire to be methodical and the need to create the tempo necessary to seize, retain, and exploit the initiative necessary for decisive UO. Quickly defining the significant characteristics of the urban environment requiring in-depth evaluation (not only what we need to know but what is possible to know) allows rapid identification of intelligence gaps (what we know versus what we don't know). Such identification leads to information requirements and will drive the intelligence, surveillance, and reconnaissance (ISR) plan (how will we get the information we need). Commanders must carefully consider how to develop focused priority information requirements (PIR) to enable collectors to more easily weed relevant information from the plethora of information. Commanders can then make better decisions and implement them faster than urban threats can react.

Amplified Importance of Civil (Societal) Considerations

B-7. The Army focuses on warfighting. The experiences in urban operations gained at lower echelons often center on the tactics of urban offensive and defensive operations where the influences of terrain and enemy frequently dominate. At higher echelons, the terrain and enemy are still essential considerations, but the societal component of the urban environment is considered more closely and throughout the operational process. Moreover, the human or civil considerations gain importance in civil support or stability operations regardless of the echelon or level of command. In addition to the echelon and the type of operation, a similar relationship exists between the key elements of the urban environment and other situational factors. These factors can include where the operation lies within the range of operations or the level of war and the conventional or unconventional nature of the opposing threat. Figure B-2 graphically represents the varying significance of these elements to urban IPB. Overall, population effects are significant in how they impact the threat, Army forces, and overall accomplishment of strategic and operational goals.

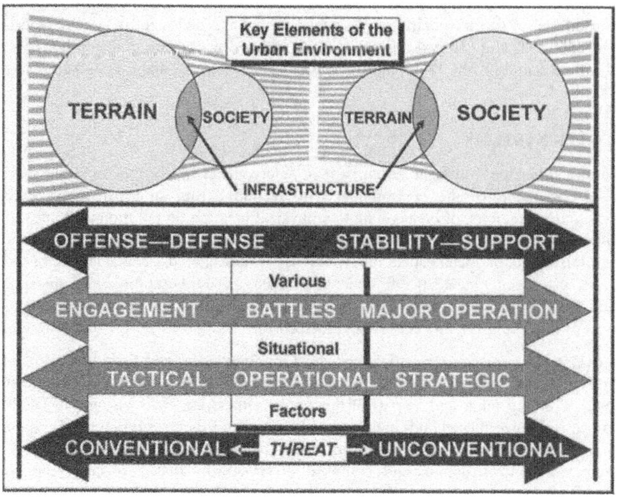

Figure B-2. Relativity of key urban environment elements

B-8. Describing the battlefield's effects—step two in IPB—ascribes meaning to the characteristics analyzed. It helps commanders understand how the environment enhances or degrades friendly and threat forces and capabilities. It also helps commanders understand how the environment supports the population. It also explains how changes in the "normal" urban environment (intentional or unintentional and because of threat or friendly activities) may affect the population. Included in this assessment are matters of perception. At each step of the IPB process, commanders must try to determine the urban society's perceptions of ongoing activities to ensure Army operations are viewed as intended. Throughout this process, commanders, staffs, and analysts cannot allow their biases—cultural, organizational, personal, or cognitive—to markedly influence or alter their assessment (see FM 34-3). This particularly applies when they analyze the societal aspect of the urban environment. With so many potential groups and varied interests in such a limited area, misperception is always a risk.

SIGNIFICANT CHARACTERISTICS

B-9. Urban intelligence analysis must include consideration of the urban environment's distinguishing attributes—man-made terrain, society, and infrastructure—as well as the underlying natural terrain (to include weather) and the threat. Because the urban environment is so complex, it is often useful to break it into categories. Then commanders can understand the intricacies of the environment that may affect their operations and assimilate this information into clear mental images. Commanders can then synthesize these images of the environment with the current status of friendly and threat forces and develop a desired end state. Then they can determine the most decisive sequence of activities that will move their forces from the current state to the end state. Identifying and understanding the environment's characteristics (from a friendly, threat, and noncombatant perspective) allows commanders to establish and maintain situational understanding. Then they can develop appropriate COAs and rules of engagement that will lead to decisive mission accomplishment.

B-10. Figures B-3, B-4, and B-5 are not intended to be all-encompassing lists of urban characteristics. Instead, they provide a starting point or outline useful for conducting an urban-focused IPB and analysis that can be modified to fit the specific operational environment and meet the commander's requirements. Commanders and staffs can compare the categories presented with those in the civil affairs area study and assessment format found in FM 41-10 and the IPB considerations for stability and civil support operations found in ST 2-91.1.

INTERCONNECTED SYSTEMS

B-11. Since the urban environment comprises an interconnected "system of systems," considerations among the key elements of the environment will overlap during an urban intelligence analysis. For example, boundaries, regions, or areas relate to a physical location on the ground. Hence, they have urban terrain implications. These boundaries, regions, or areas often stem from some historical, religious, political, administrative, or social aspect that could also be considered a characteristic of the urban society. Overlaps can also occur in a specific category, such as infrastructure. For instance, dams are a consideration for their potential effects on transportation and distribution (mobility), administration and human services (water supply), and energy (hydroelectric).

B-12. This overlap recognition is a critical concern for commanders and their staffs. In "taking apart" the urban environment and analyzing the pieces, commanders and staffs cannot lose perspective of how each piece interacts with any other and as part of the whole. Otherwise, their vision will be shortsighted, and they will fail to recognize the second- and third-order effects of their proposed COAs; the actual end state differing dramatically from the one envisioned by the commander. The increased density of combatants and noncombatants, infrastructure, and complex terrain means that a given action will likely have unintended consequences—positive or negative. Those consequences will be more widely felt and their impact will spread in less time than in other environments. These unintended results may have important strategic and operational consequences. The multiple ways these dynamic urban elements and characteristics combine make it necessary to approach each urban environment as a unique challenge for intelligence analysis.

URBAN TERRAIN AND WEATHER

Terrain

B-13. Earlier admonitions that civil considerations are more closely considered in UO do not necessarily mean that consideration for urban terrain is de-emphasized. In every urban operation, terrain and its effects on both threat and friendly forces must be assessed and understood. Then commanders can quickly choose and exploit the terrain (and weather conditions) that best supports their missions. Effective terrain analysis thoroughly assesses urban structures as well as the ground on which they stand (see figure B-3 and FM 5-33). An analysis of urban terrain first considers broader urban characteristics and effects and then progresses to a more detailed examination.

Natural Terrain

B-14. The natural terrain features beyond the urban area and beneath urban structures significantly influence unit operations. They dictate where buildings can be constructed, the slope and pattern of streets, and even the broad urban patterns that develop over longer periods of time, thereby influencing a unit's scheme of maneuver. The military aspects of terrain—observation and fields of fire, avenues of approach, key terrain, obstacles, and cover and concealment (OAKOC)—remain critical to the analysis of natural terrain in, under, and around urban areas. Fortunately, commanders and their staffs are normally accustomed to this type of analysis.

Forms and Functions

- Cores
- Industrial areas
  - Toxic industrial material production and storage facilities
  - Standard signs and markings for toxic chemicals
- Outlying high-rise areas
- Residential areas and shantytowns
- Commercial ribbon areas
- Forts and military bases

Broad Urban Patterns

- Types
  - Satellite
  - Network
  - Linear
  - Segment
- Dominant or central hub (if any)
- Area covered (square miles)

Street Patterns

- Basic Types
  - Radial
  - Grid
  - Irregular (planned and unplanned)
- Variations
  - Rayed
  - Radial-Ring
  - Contour-forming
  - Combined
- Widths

Construction and Placement

- Construction
  - Mass or framed
  - Light or heavy clad
  - Material (dirt, wood, stone, brick, cinder block, concrete, steel, and glass)
  - Density and thickness (roofs, floors, and interior and exterior walls)
  - Load bearing walls and columns
  - Height (floors)
  - Doors, windows, fire escapes, and other openings
  - Interior floor plan (including crawl spaces, elevators, and stairs)
- Placement
  - Random
  - Close-orderly block
  - Dispersed
- Ownership

Military Aspects of Terrain: OAKOC

- Observation and fields of fire
  - Smoke (fire), dust (explosions), and flying debris
  - Rubble
  - Engagement ranges (including minimum safe distances and backblast factors) and obliquity/angle (ricochets)
  - Elevation and depression considerations
  - Lasers and reflective concerns
- Avenues of approach (mobility corridors)
  - Airspace
  - Surface
  - Supersurface
  - Subsurface
- Key terrain
  - Landmarks
  - Buildings of significant cultural, social, political, historical, or economic significance
- Obstacles
  - Rubble and vehicles
  - Steep embankments
  - Medians
  - Inadequate bridges and overpasses (destroyed, weight-restricted, or narrow)
  - Tunnels and underpasses (destroyed or narrow)
  - Mines and roadside improvised explosive devices
  - Masking of fires
  - Burning buildings or other fire hazards
  - Rivers and lakes
- Cover and concealment
  - Building protection
  - Weapon penetration (single shot and multiple rounds) considerations
  - Rubble and vehicles

Figure B-3. Significant urban terrain characteristics

Analysis of an Urban Area's Underlying Terrain: Mitrovica, Kosovo – 1999

An urban area's underlying terrain provides many clues into its history, economy, society, and current situation. Mitrovica, Kosovo is an illustrative example. The Ibar River creates a natural line of communications through the middle of the city as well as an obstacle that bisects the urban area. This bisection naturally divides the two resident ethnic groups: Albanians and Serbs. The separation became significant at both the strategic and tactical levels during 1999 deployments to Kosovo. Army forces had to ensure that the Orthodox Church located south of the Ibar was accessible to Serbs residing in the north. North Atlantic Treaty Organization (NATO)

> peacekeepers built a footbridge across the river that allowed reliable, safe passage. The natural feature separating the two groups assisted NATO troops in maintaining stability in the region.

Man-Made Terrain

B-15. Building composition, frontages, placement, forms and functions, size, floor plans, and window and door locations affect maneuver, force positioning, and weapons deployment considerations. Angles, displacement, surface reflection, and antenna locations influence command and control. Structures also influence ISR operations. The increased density and volume created by man-made structures increases how much information commanders and their staffs collect and assess as well as the number of forces required. Building materials and construction will also influence force structures to include weapons and equipment required. Heavily constructed buildings combined with hot and cold extremes may affect target identification for thermal sights. Thick walls, for example, may make combat vehicle identification difficult by distorting hotspots. Additionally, the increased use of heaters and warming fires may clutter thermal sights with numerous hotspots. The ability to maneuver through the urban dimensions—airspace, supersurface, surface, and subsurface—and shoot through walls, ceilings, and floors also creates increased psychological stress. The physical characteristics of man-made terrain can also be analyzed using OAKOC.

Weather

B-16. Weather and its effects are often considered when examining the military aspects of terrain. Military aspects of weather include temperature (heat and cold), light conditions, precipitation (cloud cover, rain, snow, fog, and smog), and wind. Their military effects during UO are similar to any operational environment (see FM 34-81 and FM 34-81-1). Extremes of heat and cold affect weapon systems and the Soldiers that operate them. The extra luminescence provided by the ambient light of an urban area, unless controlled, may affect night vision capabilities and the ability of the Army to "own the night." Precipitation affects mobility and visibility. Smog inversion layers are common over cities. An inversion layer may trap smoke and chemicals in the air and increase ground and air temperatures to the detriment of Soldiers' health. (If the conditions are severe enough, it might require the use of protective masks.) Changes in temperature as a result of air inversions can also affect thermal sights during crossover periods of warm to cold and vice versa. Winds, which may increase as they are funneled through urban canyons, may—

- Increase other weather effects (for example, wind chill).
- Decrease visibility (blowing debris, sand, rain, and snow).
- Spread radiation, biological, and chemical hazards.
- Adversely affect low-altitude air mobility and airborne ISR platforms.

B-17. Commanders also analyze weather for its potential effect on civilians and civilian infrastructure as well as Soldiers and military equipment. Rain might create sewage overflow problems in urban areas with collapsed sewage infrastructure and in refugee camps, increasing disease and even creating panic. Rain and flooding may also make some subsurface areas impassable or extremely hazardous to civilian and military forces alike. Chemical agents and TIM may be washed into underground systems creating toxic hotspots. These effects become more pronounced as chemical agents or TIM are absorbed by brick and unsealed concrete sewers. Other weather effects on UO can include—

- Heavy snowfall in an urban area that may paralyze area transportation and distribution infrastructure, hindering the urban administration's ability to provide vital human services (police, fire fighting, medical, and rescue). Heavy rains and flooding may have similar effects especially on poorly designed and constructed roads or roads that have been damaged by tracked vehicles.
- Extreme hot and cold weather climates that increase the dependence (and military significance) of many elements of the infrastructure. For example, the energy infrastructure may be critical; without it, civilians may not be adequately cooled or heated or they may not be able to cook their food.

- In urban areas located in tropical regions, it can rain at the same time each day during the wet season. Threat forces may attack during these periods knowing aircraft will have difficulty responding. Bad weather also reduces the effectiveness of surveillance, direct and indirect fire, and logistic support.
- Inclement weather may preclude demonstrations or rallies by threats. Good weather may mean a maximum turnout of civilians for events such as festivals, sporting events, and other social, cultural, or religious activities.
- Severe weather may affect psychological, civil-military and humanitarian assistance operations. Heavy rains and severe dust storms may disrupt leaflet drops, construction projects, food and water distribution, and medical and veterinary assistance programs.

URBAN SOCIETY

B-18. This manual shows that societal considerations take on added importance in UO. Critical to operational success is knowing and understanding which groups live in an urban area, what relationships exist among them, and how each population group will respond to friendly and threat activities. Often determining any of this is very difficult. Cultural acuity is also essential in helping commanders and their staffs to view the urban area as the residents view it. The demographics presented depict what conditions exist, while the other categories help to explain the root causes or why conditions exist (see figure B-4 on page B-8). Other categories besides basic demographics that are important to gain this understanding include health, history, leadership, ethnicity and culture, religion, and government and politics.

From Cultural Awareness to Understanding and Competence

B-19. This societal understanding must go beyond a superficial awareness of gestures and taboos to developing a thorough understanding of the organizing principles that make the urban society hold together and function as a coherent entity—or not. In UO, particularly longer-term stability operations, commanders use this knowledge and insight to design and plan operations that work effectively within that society. For example in OPERATION IRAQI FREEDOM, Army units, as they shifted to long-term stability operations, were required to recruit, organize, equip, and train indigenous security forces in order to be able to transition security responsibility from U.S. forces to the burgeoning Iraqi government. Army commanders were successful—to a point. The first units that they built and trained attained a remarkable level of unity and tactical ability in a relatively short period of time. The newly formed units effectively fought alongside Army forces. However when prodded into conducting operations with minimal Army supervision, many Iraqi units lost unity and coherence. Army commanders initially failed to account for a basic Iraqi societal consideration: their society is patriarchal—each member or organization in Iraqi society always seeks a patron or sponsor to support them and provide for their needs. When Iraqi units trained and operated with Army forces, this relationship was satisfied. Operating on their own, however, Iraqi commanders and units lost this patronage without an equivalent Iraqi replacement. In hindsight, the establishment of an effective and more robust ministry of defense, either as a precursor or simultaneously with the establishment of tactical units, may have provided the Iraqi security forces the patronage that their culture required.

Resource the Societal Analysis

B-20. Aside from friendly and threat forces, the society is the only thinking component of the urban environment able to rapidly impact the urban operation. (Even people going about their daily routines can unwittingly hamper the mission.) Urban residents create conditions for restrictive rules of engagement, increase stress on Soldiers and logistic capabilities, and confuse threat identification (see Threat Considerations in this appendix). Demographic, health, safety, ethnic, and cultural concerns will be essential considerations in most UO. Other situational factors—the mission, enemy, and time available—dictate the balance between the level of detail and analysis to support the overall urban operation with the level of detail that commanders and their staffs can achieve. However, an analysis that fails to devote enough time and resources to understanding the urban society can find large elements of the population turned against the Army force. Detailed analysis of the urban society will help save time and ISR resources. While not necessarily a resource saver in the short term, a thorough understanding of the urban society and its culture ultimately contributes to faster achievement of strategic goals.

Population Demographics

- General population size
  - Village
  - Town
  - City
  - Metropolis
  - Megalopolis
- Group size based on race age, sex, political affiliation, economics, religion, tribe, clan, gang, criminal activities, or other significant grouping
  - Significant US or allied populations
  - Distribution, densities, and physical boundaries and overlaps
  - Majority, minority, and dominant groups
- Increasing or decreasing migration trends
  - Dislocated civilians
- Nongovernmental organizations (NGOs)
  - Local
  - National
  - International
- Languages (distribution, dialects, relationship to social structure)
- Educational levels and literacy rates
- Crime rates
- Birth and death rates
- Labor statistics and considerations
  - Skilled and unskilled
  - Imported and exported
  - Unemployment
  - Standard wages and per capita income
  - Workday and workweek norms

Health

- Diseases
- Nutritional deficiencies
- Local standards of care
- Pollution and environmental hazards (air, water, food, and soil)
- Health workers (types, numbers, and degree of skill)

History

- General and for a specific group
  - Internal or external
  - Recent conflicts
- Relationship with US, allies, and other participating multinational forces
- Applicable international treaties
- Status-of-forces agreements
- Antagonists/protagonists
- Heroes
- Events, facts, and dates considered important or celebrated
- Urban area's historical importance

Leadership and Prominent Personalities

- Identification, location, and prioritization of influential leaders (exploitation, evacuation, protection, etc.)
- Affiliation (ethnic, religion, military, government, industry, criminal, or entertainment)
- Education attained
- Organization and distribution of power
- Associations among different leaders and groups

Ethnicity and Culture

- Values, moral codes, taboos, and insults (verbal and non-verbal)
- Attitudes towards age, sex, and race (including same-sex interaction)
- Role of the clan, tribe, or family
- Biases between ethnic groups
- Privacy and individuality
- Recreation, entertainment, and humor
- Fatalism or self-determination
- Exchanges of gifts
- Displays of emotion
- Lines of authority
- Dating and marriage
- Greetings, leave-takings, and gestures
- Visiting practices
- Alcohol and drug use
- Important holidays, festivals, sporting, or entertainment events
- Eating and dietary practices
- Significance of animals and pets
- Urban-rural similarities and differences
- Driving habits
- Clothing

Religion

- Sects, divisions, and overlaps
- Religious biases and problems
- Relationship and influence on government, politics, economics, and education
- Impact on ethnic and cultural beliefs
- Key events or celebrations (daily, weekly, monthly, or annually)
- Funeral and burial practices

Government and Politics

- Present and past forms
- Organization and powers (executive, legislative, judicial, and administrative divisions)
- Scheduled elections and historical turnouts
- Degree of control over the population
  - Identification required
  - Border-crossing procedures
- Relationship with US or multinational governments, national governments, and criminal elements
- Political factions and boundaries
- Political traditions
- Grievances
- Censorship
- Nepotism and other clan, tribal, or social ties
- Civil defense and disaster preparedness (organization, plans, training, equipment, and resources)
  - Evacuation routes
- Legal system
  - System of laws
  - Applicable treaties
  - Courts and tribunals
  - Procedures
  - Records (birth and deeds)
- Property control
- Monetary system (formal and informal)
- Domestic and foreign trade
  - Taxation and tariffs
  - Customs requirements
  - Rationing and price controls
  - Economic performance and contribution to gross national product
- Economic aid
- Perception of relative deprivation
- Trade unions
- Competition with the black market and organized crime

Figure B-4. Significant urban societal characteristics

Transportation and Distribution

- Water
  - Shipyards and other port and harbor facilities
  - Inland waterways, canals, and locks
  - Offshore pipeline berths
  - Cargo storage and handling
  - Types and number of ships, boats, and ferries
  - Dams
- Streets and roads
  - Bridges and fords
  - Over- and underpasses
  - Raised embankments, tunnels, culverts, and other subterranean features (widths and clearances)
  - Parking areas (surface, subsurface, and supersurface)
  - Weight restrictions
  - Traffic light operations
  - Traffic patterns
  - Widths
  - Surface materials
- Rail
  - Lines
  - Terminals
  - Switchyards and junctions
  - Subways, bridges, elevated rail lines, and underpasses (clearances)
  - Track gauges
  - Types and number of rolling stock
  - Electrification
- Air
  - Airfields and runways (including capabilities)
  - Heliports and helipads (including rooftop)
  - Types and number of aircraft
  - Cargo storage and handling
- Trucking companies and delivery services
- Available material-handling equipment
- Rush hour and market time considerations
- Seasonal (weather) effects
- Rubble effects
- Impact of dislocated civilians and migration patterns
- Likely population congregation points
- Identifiable primary and alternate lines of communication

Economics and Commerce

- Industries
  - Types and locations
  - Important companies (including US or allied)
  - Military production facilities
- Sources of raw materials
- Use of toxic industrial materials and biological agents
  - Agriculture (insecticides, herbicides, and fertilizers)
  - Manufacturing
  - Cleaning
  - Research
- Food types, quantities, and sources
  - Requirements and availability
  - Storage and processing
  - Cleanliness standards
- Stores, shops, restaurants, hotels, and strip malls
- Recreation facilities
  - Outdoor and amusement parks
  - Stadiums and other sports facilities
- Machine shops
- Brick and lumber yards
- Banking and investment institutions

Administration and Human Services

- Police and fire protection
  - Headquarters, station, and key facilities locations
  - Organization and strengths
  - Equipment
  - Functions, authority, and jurisdictional boundaries
  - Contract guard services
- Welfare and public assistance
  - Monetary assistance
  - Orphanages
  - Elderly care facilities
- Water supply systems
  - Water sources and storage (lakes, reservoirs, cisterns, pools, and public baths)
  - Water treatment and quality
  - Pumping stations and other distribution methods (trucks, bottles)
  - Hydrant locations
- Snow removal capabilities
- Street light operations
- Health facilities
  - Hospitals
  - Emergency medical services
  - Mental institutions
  - Medical supplies and equipment
  - Research and pharmaceutical buildings
  - Blood banks
- Governmental buildings
  - Embassies
  - Capitol building
  - Legislative, judicial, and ministry buildings
  - Hall of records
- NGOs providing assistance
- Effects of military control measures on providing vital human services
- Waste and sanitation
  - Types (solid, sewage, and toxic)
  - Collection, processing, and disposal
  - Dumps or landfills
  - Drainage systems

Cultural

- Religious buildings and places of worship
- Shrines, monuments and other historical structures
- Schools
- Museums
- Theaters
- Libraries

Energy

- Types
  - Electric
  - Oil
  - Coal
  - Natural gas
  - Nuclear
  - Solar
  - Hydroelectric
  - Geothermal
- Facilities
  - Production and processing
  - Storage
- Distribution
  - Pipelines (above and below ground)
  - Power lines (overhead and underground)
  - Water, rail, and road
  - Potential hazards

Communication and Information

- Print media
  - Newspapers, periodicals, and pamphlets
  - Billboards and posters
  - Postal facilities
- Telephone facilities
  - Wire or wireless
  - Facsimile machines
- Telegraph facilities
- Radio facilities
- Police, fire, and rescue systems
- Security systems
- Television facilities
- Computers and the Internet
- Antennas, towers, relay stations, and lines (surface and subsurface)
- Integration of space-based capabilities
- Public forums and speech
- Low-technology media (car horns, drums, graffiti, and burning fires)
- Key media organizations and reporters
  - Local
  - International
  - US

Figure B-5. Significant urban infrastructure characteristics

URBAN INFRASTRUCTURE

B-21. Functional and analytical interdependence is readily apparent when examining urban infrastructures (see figure B-5 on page B-9). They are composed of physical structures or facilities and people. Hence, much of the analysis conducted for terrain and society can apply when assessing the urban infrastructure. For example, commanders, staffs, and analysts could not effectively assess the urban economic and commercial infrastructure without simultaneously considering labor. All aspects of the society relate and can be used to further analyze the urban work force since they are a sub-element of the urban society. Similarly, the OAKOC aspects used to evaluate terrain may also apply to the urban infrastructure, especially considerations of key terrain. Depending on the mission, elements of the urban infrastructure may become key terrain. While infrastructure analysis is not an engineer-specific task, Army engineers may provide the general and geospatial engineering expertise necessary to better understand the urban terrain and infrastructure.

THREAT CONSIDERATIONS

> *... [T]he adversaries of freedom ... send arms, agitators, aid, technicians and propaganda to every troubled area.... [S]ubversives and saboteurs and insurrectionists . ... [possess] the power to conscript talent and manpower for any purpose. and long experience in the techniques of violence and subversion....It is a contest of will and purpose as well as force and violence—a battle for minds and souls as well as lives and territory.*
>
> John F. Kennedy
> Message to Congress, 5 May 1961

B-22. Chapter 3 outlines the instability and uncertainty of the strategic environment. Commanders, staffs, and analysts identify and analyze the threat's composition, strength, disposition, leadership, training, morale, weapons and capabilities, vulnerabilities, internal logistics and external support, doctrine (if any), strategy or modus operandi, and tactics. The threat can take a variety of forms:

- Conventional military forces.
- Paramilitary forces.
- Guerrillas and insurgents.
- Terrorists.
- Militia or special police organizations.

B-23. A general study of guerrilla and insurgent tactics, techniques, and procedures may prove beneficial to many types of operations regardless of the actual composition or type of threat forces. Insurgent strategies and tactics may work especially well in this complex environment and will likely be a part of any threat COA. Particularly, commanders should understand how a threat might restrict itself by the laws of land warfare and similar conventions, or exploit the use of these conventions to its own gain. Commanders can refer to FM 31-20-3 and FMI 3-07.22 for more information. For many of the above threats, no doctrinal templates may exist. Intelligence staffs will need to identify and track common threat patterns and tactics, techniques, and procedures as they develop and change during the course of the operation. Understanding culture will be critical to determining threat motives that drive decisions. Comman415081ders, staffs, and analysts must evaluate, update (or create), and manage threat databases early (and continuously) during the operational process.

URBAN ENVIRONMENTAL THREATS

B-24. While threats vary, they share a common characteristic: the capability and intent to conduct violence against Army forces to negatively influence mission accomplishment. These threats are often the most recognizable for forces trained for warfighting—these are often the enemy. In order for Army units to be able to conduct full spectrum urban operations, commanders broaden their concept of the threat when analyzing the urban environment's terrain, societal, and infrastructure characteristics. This analysis includes many environmental dangers (potentially affecting both sides of a conflict as well as noncombatants) such as—

- Natural disasters (earthquakes, tornados, hurricanes, volcanic eruptions, fires, floods, tidal waves or tsunamis, and heavy snows).
- Food shortages.
- Water shortages (both quantity and quality).
- Rampant disease.
- Pollution and toxic industrial materials.

B-25. A critical difference between the latter forms of threat and the former is the lack of intent to do harm. The latter may stand alone as threats, or these conditions may be created, initiated, or used by the enemy or a hostile civilian group as a weapon or tool. Threat analysis includes identifying and describing how each relevant characteristic of the area of operations can hinder mission accomplishment. This analysis, particularly during stability or civil support operations, will require extensive coordination and cooperation with urban civil authorities, law enforcement, and numerous governmental and nongovernmental organizations.

CIVILIANS

B-26. In a major combat operation or campaign where offensive and defensive operations are conducted against a conventional enemy, threat identification is more readily accomplished. However, adaptation may be necessary to even further broaden the threat concept to include specific elements of the urban society and, in some instances, nongovernmental organizations (NGOs) and other civilian agencies working in the urban area. In many stability operations, this modification can account for opposing armed forces that are not an enemy but are a threat to the mission. As discussed in Chapter 9, Army forces in many stability or civil support operations must avoid classifying or thinking of these threats as the enemy. Further, commanders ensure that Soldiers do not shift the animosity that they feel toward urban threats onto the civilian population. In OPERATION IRAQI FREEDOM, for example, some Soldiers began to refer to Iraqi civilians as "hajjis" in reference to the annual pilgrimage—the Hajj—that forms one of the tenets or pillars of Islamic faith. Those who have completed the pilgrimage are entitled to add the phrase al-Hajj or hajji (pilgrim) to their name. However, the Soldiers were not using the term in its proper context but rather to denigrate. In response, commanders prohibited the inappropriate use of this term as one of many measures to stress and develop cultural understanding and respect for the people that Army forces were there to assist.

Need for More Accurate Categories

B-27. Urban commanders recognize that threat evaluation is not solely a straightforward assessment of the capabilities of a known, armed enemy. Instead, they develop a system of categories that mitigates potential situational uncertainty and better indicates the level of threat (or utility) that civilian groups pose for Army forces conducting UO. Categories can range from a simple construct—threat, neutral, or friendly—to a more refined one (see figure B-6 on page B-12). In most UO, a more intricate approach is necessary to detect and monitor shifts in key relevant relationships. While this analysis is important in any urban operations, it is especially critical when Army forces are opposed by urban insurgent and guerrilla forces.

B-28. In any system, commanders consider that the classification of a group is relative to the perspective from which it is viewed. This is an especially important consideration in multinational and interagency UO. During operations in Somalia, for example, U.S. Army forces may have viewed a particular clan as a hostile element. The United Nations' Italian contingent, with their colonial background in the region, may have considered the same clan as neutral or even an ally.

Shifting and Overlapping Groups

B-29. Commanders recognize that no system of categorization will precisely classify any given group; no system can reflect the overall nature and complexity of the urban society. A single group may fit in a particular category. It may also have components in two or more categories simultaneously. Often, it can shift among categories during an operation. A given group may have individuals in it who have interests identical to or different from that group and these individual interests may change over time.

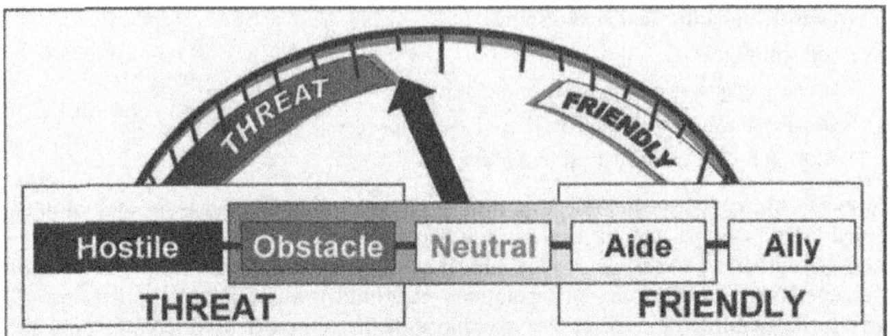

Hostile: An element (*individual, group, or organization*) that has the intent to negatively influence mission accomplishment by supporting enemy forces (or another hostile element) or exploiting friendly vulnerabilities but is not engaging in combat operations—otherwise, they lose any civilian status and become an enemy force.

Obstacle: An element whose presence or activities serve to unintentionally impede, disrupt, or constrain friendly operations by imposing additional losses in personnel, time, equipment, and resources. Its interests may or may not be compatible with friendly objectives.

Neutral: An element whose presence, activities, and overall interests do not conflict with friendly, enemy, or hostile elements.

Aide: An element whose *present* intentions are—either wittingly or unwittingly—compatible with friendly force objectives.

Ally: An element whose *fundamental* interests and intent align with friendly force objectives.

Figure B-6. A refined approach to civilian allegiance

B-30. A peace enforcement operation illustrates the varying nature of groups. An identified criminal group might be classified as an obstacle to the commander's mission because its illegal activities impede unit progress. Its compelling interest, however, is to make money rather than resist the actions of friendly forces. In the same operation, one of the armed belligerents may be intent on disrupting the peace process and would be, therefore, classified as a hostile. (Again, they are not an enemy unless the belligerents are engaged in planned combat operations against the peacekeeping force. Commanders should consider, however, that under international law, if any civilian takes a direct or active role in hostilities, that civilian can be subject to attack.) The belligerent force may finance the criminal organization to assist in further obstructing the peace mission. In this case, the criminal organization moves from being an obstacle to being a hostile.

B-31. This classification effort, therefore, is not a one-time undertaking; commanders must constantly review and update it. Groups or individuals can be influenced into assisting either the friendly or opposing force. People will also act opportunistically, shifting support and alliances as perceived advantages arise. Even seemingly passive and law abiding members of the urban society may conduct themselves in unexpected ways given the right conditions (mob violence, for example).

Shifting Civilian Interests and Intent

Among other applicable lessons (see also the vignette in Chapter 9), the Los Angeles riots of 1992 illustrate how urban population groups can shift their relative positions due to changing conditions in an urban AO. Several gangs exist in the Los Angeles area. Usually, these gangs are hostile to one another. During the riots, however, several rival gangs formed a "united front" against what was seen as a larger obstacle to their own interests: law enforcement. As a result, the hostile gangs became one another's aides during that time.

Similarities, Differences, Capabilities, and Vulnerabilities

B-32. Similarities and differences in interests and interdependencies between groups are often focal points for analysis (and the allocation of ISR assets). They may indicate how commanders may influence, coerce, or align civilian interests and intentions with mission objectives. Simultaneously, commanders should consider an analysis of the civilian element's (individual, group, or organization) capability to influence the accomplishment of friendly objectives. They should also consider civilians' vulnerabilities and dependencies. If a civilian group's fundamental interests align with friendly objectives and this group has the intent to assist friendly forces, it is clearly an ally. However, with limited or no capability, a specific group will not help move the commander any closer to achieving his desired end state and accomplishing the mission. Then the commander would normally limit the resources expended on this group to those necessary for maintaining their commitment to common goals and objectives.

B-33. In contrast, commanders may provide a group with resources to enhance or create the capability to assist in mission accomplishment. They may do this if they felt that the assistance gained (or reduction to threat support) exceeds the potential diminishment of their own force's capabilities from losing those same resources. Commanders would also consider the group's dependencies (such as food, infrastructure, and protection) and overall vulnerability to outside influence. If vulnerable to friendly influence or control (understanding urban societal considerations and matters of perception are critical in this regard), then forces are likely susceptible to enemy or hostile manipulation. Even if commanders can generate extra resources (and not significantly affect their own force's capabilities), they should still conduct this same cost-benefit analysis to determine which civilian group (if any) should receive resources.

Greatest Potential Nearer the Center

B-34. The most critical population sectors often are those nearest the middle of the spectrum shown in figure B-6, particularly if their capabilities (or potential) significantly degrade or enhance mission accomplishment. These are the obstacle, neutral, or aide categories. If their beliefs and interests can be adequately understood, commanders may have great chance to influence the population segment and significantly contribute to mission accomplishment. In the fight for information superiority, persuading a neutral population to become an aide may provide critical information required about the urban environment and threat forces operating within.

Political and Strategic Concerns

B-35. The aide category may be of significant political or strategic concern. An aide group may be invaluable for accomplishing intermediate objectives but become a vulnerability to accomplishing a larger stability operation. (Even an urban offensive or defensive operation is likely to transition to a stability operation.) For example, commanders may provide resources to a criminal organization to assist in defeating insurgent forces during urban combat operations. Once these forces have been subdued, the interests (monetary gain and defeat of the threat) linking friendly forces with this criminal element disappear. What may remain is a criminal organization with more power than a reconstituted or newly established law enforcement agency and a truly destabilizing force. This also illustrates the second- and third-order thinking that will be required of commanders and their staffs during UO.

URBAN INTELLIGENCE TOOLS AND PRODUCTS

B-36. Developing tools and products to assist with situational understanding is not the task of intelligence sections alone. Commanders and all elements of their staffs must develop and adapt products and tools suited to their particular requirements. Listed below are some of the analytic tools and products that may help meet those requirements in a complex environment (see figure B-7 on page B-14 and FMI 2-91.4). Standard tools and products include: modified combined obstacle overlays and doctrinal, situation, event, and decision support templates or matrices. In addition to these standards aids, staffs and analysts may develop or produce other innovative tools to assist commanders in their situational understanding of the complex urban environment. These tools may contain overlapping information as different types of information are compared to determine patterns among them. Staffs and analysts may also initiate requests for products (or information) from their higher headquarters or other agencies with the technical means or control over assets when the capability lies outside the Army force's means. The tools that may be developed or requested include—

- Imagery.
- Three-dimensional representations.
- Infrastructure blueprints.
- Hydrographic surveys.
- Psychological profiles.
- Matrices, diagrams, or charts.
- Various urban overlays.

B-37. There are many software applications available to the Army that can be used to conduct intelligence analysis and create many of the relevant products described above. These applications currently range from such programs as the Analyst Notebook and Crimelink which have link analysis, association matrix, and pattern analysis software tools to the Urban Tactical Planner developed by the Topographic Engineering Center as an operational planning tool. The focus of the following information, however, is the various types of tools that can be developed and used to understand the urban environment rather than the constantly evolving and changing software or hardware that may be used to create them.

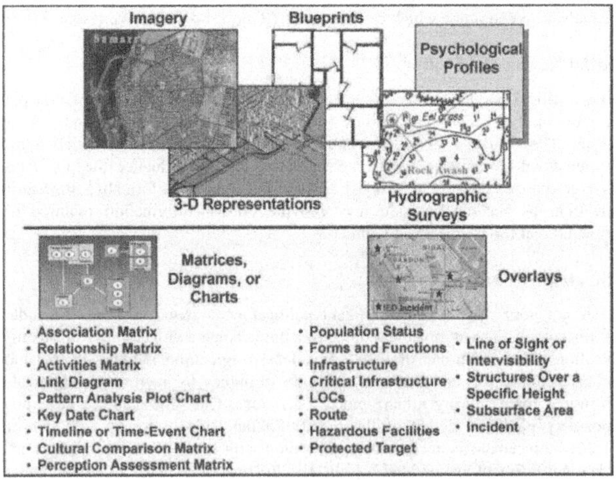

Figure B-7. Urban intelligence tools and products

Imagery

B-38. Recent satellite imagery or aerial photography will be required for most types of UO. Such images clarify vague and inaccurate maps and other graphic representations. Satellite assets provide responsive data input into the geographic information systems (GIS). (The National Geospatial Intelligence Agency [NGA] is one important source of imagery and map data.) GIS will often form the basis for creating the three-dimensional representations and the various overlays described below. Frequently updated (or continuous real-time) satellite or aerial imagery may be required for detailed pattern analysis and maintaining accurate situational understanding. For example, imagery taken during an area's rainy season may appear significantly altered during the summer months. Finally, digital, hand-held still and video cameras, particularly at the tactical level, will also be critical in developing and attaining situational understanding in a complex urban environment.

Three-Dimensional Representations

B-39. Often, physical or computer-generated (virtual) three-dimensional representations may be required to achieve situational understanding. These models or computer representations include specific sections of the urban area or specific buildings or structures. Such detail is particularly important for special operating forces and tactical-level units. These units require detail to achieve precision, increase the speed of the operation, and lessen friendly casualties and collateral damage.

Infrastructure Blueprints

B-40. Urban police, fire, health, public utilities, city engineer, realty, and tourist agencies, and other urban organizations often maintain current blueprints and detailed maps. Such documents may prove useful to update or supplement military maps or to clarify the intricacies of a specific infrastructure. They may prove critical in operations that require detailed information to achieve the speed and precision required for success. Without such detail, analysts determine interior configurations based on a building's outward appearance. That task generally becomes more difficult as the building size increases.

Hydrographic Surveys

B-41. Many urban areas are located along the world's littoral regions and major rivers. Therefore, commanders may need current hydrographic surveys to support amphibious, river crossing, riverine, and sustainment operations.

Psychological Profiles

B-42. Psychological profiles analyze how key groups, leaders, or decision makers think or act—their attitudes, opinions, and views. They include an analysis of doctrine and strategy, culture, and historical patterns of behavior. The degree to which the attitudes, beliefs, and backgrounds of the military either reflect or conflict with the urban populace's (or civilian leadership's) core values is extremely important in this analysis. Psychological profiles help to assess the relative probability of a threat (or noncombatant group) adopting various COAs as well as evaluating a threat's vulnerability to deception. These profiles are derived from open-source intelligence as well as signals and human intelligence.

Matrices, Diagrams, and Charts

B-43. Matrices, diagrams, and charts help to identify and understand key relationships among friendly and threat forces and other significant elements of the urban environment. While similar, each looks at information in a different way to uncover hidden patterns and connections useful in understanding the complex urban environment, particularly the threat and the urban society. These tools and products might include—

- **Association Matrix**. The association matrix helps identify the nature and relationship between individuals. Association matrices also help to identify those personalities and associations needing a more in-depth analysis necessary to establish the degree of relationship, contacts, or knowledge between individuals. Threat organizational structure and functions are uncovered as connections between personalities are made.

- **Relationship Matrix**. Relationship matrices are an extension of association matrices described above. They are used to examine the relationship between groups and organizations—threat forces, friendly forces, NGOs, the media, and the various elements of the urban population. Urban terrain and infrastructure may also be included as analytical factors to help expose relationships. These matrices graphically depict how human elements of the urban area interact with the physical elements to promote or degrade mission success. A significant relationship matrix may be a comparison of cultural perspectives—ideology, politics, religion, acceptable standards of living, and mores—between urban population groups and Army (and multinational) forces to help understand and accurately predict a civilian element's actions. The relationships discovered may suggest ways that commanders can shape the environment. For example, conflicting ideals or issues between two disparate threat forces may be emphasized to cause threats to focus their resources against each other rather than Army forces. Conversely, stressing common ideals identified between opposing noncombatant organizations may cause these groups to unite and work harmoniously toward a goal reflecting their commonalities.
- **Activities Matrix**. An activities matrix connects individuals to organizations, professions, events, activities, or addresses. Information from this matrix (combined with information from association and relationship matrices) also helps to link personalities, uncover the structure of an organization (threat or otherwise), and recognize differences, similarities, and dependencies for possible exploitation.
- **Link Diagram**. This tool graphically depicts many of the relationships and associations described above—people, groups, events, locations, or other factors deemed significant in any given situation. It helps commanders and analysts visualize how people and other factors are interrelated to determine key links.
- **Pattern Analysis Plot Chart**. This chart depicts the times and dates of a selected activity (such as ambushes, sniper and mortar attacks, bombings, and demonstrations) to search for patterns of activity for predictive purposes as well as to discern intent. This analysis can be conducted using a time-event wheel as well as by plotting events on maps using multiple historical overlays (analog or digital). (Compare with Incidents Overlay covered later in this chapter.) During pattern analysis, commanders and analysts consider not only what is occurring but also what is not occurring (or ceases to occur).
- **Key Event Chart**. In many urban operations, particularly stability operations, key holidays, historic events, and other significant cultural or political events can be extremely important for commanders to understand. More than just dates, these charts depict what can be expected to happen on each particular event. In Bosnia, for example, weddings are often held on Fridays and celebratory fire is a common occurrence. Understanding this cultural phenomenon could reduce collateral damage and accompanying civilian deaths.
- **Timeline or Time-Event Chart**. Timelines are a list of significant historical dates with relevant information and analysis that provide the commander with a record of past activities necessary to understand current operational conditions. A timeline may highlight a specific feature of the present situation such as population movements or political shifts or outline the general chronological record of the urban area, perhaps highlighting the activities of a certain population sector. Understanding past events also helps commanders predict future reactions to proposed COAs. Similarly to link diagrams, time-event charts display large amounts of information in a small space. These charts may help commanders discover larger-scale patterns of activity and relationships.
- **Cultural Comparison Matrix**. Commanders must avoid ethnocentricity—assuming that only one cultural perspective exists (often their own) and using that as the single lens with which to view the situation. To avoid this common obstacle to an accurate understanding of the urban society, commanders use this matrix to compare local ideology, attitudes, beliefs, and acceptable standards of living with associated U.S. and Army norms and values. This matrix not only highlights differences but also similarities.

- **Perception Assessment Matrix**. All of the above tools contribute to developing the perception assessment matrix; they allow commanders to see the urban environment from the perspective of its inhabitants. The perception assessment matrix is another tool that uses this refined perspective and the framework outlined in the matrix to predict the urban population's perception of a proposed COA. Although perceptions are not actions, they—more than reality—drive decision-making and resultant civilian activity. (As an example, legitimacy is a critical factor in developing many COAs that affect the civilian population, particularly during stability operations. This legitimacy must be viewed from the perspective of the inhabitants and not based on the Army commander's own culturally-shaped perceptions. Moreover, the legitimacy of an operation may need to be considered from other perspectives as well. These might include those of the American public, coalition partners and their publics, and those of neighboring states in the region.) This same matrix can later be used as a foundation to track and gauge the effectiveness of the chosen COA with favorable perception as one criterion of success. Commanders must be wary that perceptions are not fixed. They may change based on factors that the commander may not be able to control or influence. Therefore, commanders must continually monitor perceptions for deviation.

Various Urban Overlays

B-44. Staffs can produce various map overlays. These overlays depict physical locations of some aspect critical to the planning and conduct of the urban operation. Given adequate lead time, NGA can produce many overlays as an integrated map product (including satellite imagery). These overlays can include the—

- **Population Status Overlay**. This tool depicts the physical location of various groups identified by any significant social category such as religion, tribe, or language. During offensive and defensive operations, it may simply be where significant numbers of people are "huddled" or located throughout the battlefield. Population dispersal can vary significantly through the day, particularly at night, and must be considered as part of the overall analysis leading to the development of this tool.
- **Forms and Functions Overlay**. Based on the urban model, this overlay depicts the urban core or central business district, industrial areas, outlying high-rise areas, commercial ribbon areas, and residential areas, to include shantytowns.
- **Infrastructure Overlay**. This overlay is actually a series of overlays. It depicts identifiable subsystems in each form of urban infrastructure: communications and information, transportation and distribution, energy, economics and commerce, administration and human services, and cultural. Each subsystem can be broken down into more detail. Infrastructure data may be used to develop three other overlays—
  - **Critical Infrastructure Overlay**. This tool displays specific elements of the urban infrastructure that, if harmed, will adversely affect the living conditions of the urban society to the detriment of the mission. These elements may include power generation plants, water purification plants and pumping stations, and sewage treatment plants. This information could be coded as part of the overall infrastructure overlay.
  - **Lines of Communications (LOCs) Overlay**. The LOCs overlay highlights transportation systems and nodes, such as railways, road, trails, navigable waterways, airfields, and open areas for drop zones and landing zones. It also includes subsurface areas and routes such as sewage, drainage, and tunnels and considers movement between supersurface areas. The LOCs overlay and the route overlay (below) consider traffic conditions, times, and locations, to include potential points where significant portions of the urban population may congregate.
  - **Route Overlay**. This overlay emphasizes mobility information to assist commanders and planners in determining what forces and equipment can move along the urban area's mobility corridors. Pertinent data includes street names, patterns, and widths; bridge, underpass, and overpass locations; load capacities; potential sniper and ambush locations (which may be its own overlay); and key navigational landmarks such as major roads and highways, rivers and canals, cemeteries, bridges, stadiums, and churches. The structures over a specific height

overlay and subsurface overlay may assist in its development. As with the LOCs overlay, commanders, planners, and analysts think in all dimensions.

- **Line of Sight or Intervisibility Overlay**. This product creates a profile view (optical or electronic) of the terrain from the observer's location to other locations or targets. It can show trajectory or flight-line masking as well as obstructed or unobstructed signal pathways.
- **Structures over a Specific Height Overlay**. This level of detail may also be critical to communications, fires, and Army airspace command and control (air mobility corridors especially low-level flight profiles). Incorporated as part of this overlay, it may include floors or elevations above limitations for particular weapon systems at various distances from the structure.
- **Subsurface Area Overlay**. As an alternate to the building or structure height overlay, this product provides the locations of basements, underground parking garages, sewers, tunnels, subways, naturally-occurring subterranean formations, catacombs, and other subsurface areas. Similar to elevation "dead spaces," this overlay may show areas that exceed depression capabilities of weapon systems and potential threat ambush locations—again, affecting maneuver options.
- **Urban Logistic Resources Overlay**. This product identifies the locations of urban logistic resources that may contribute to mission accomplishment. It may contain specific warehouse sites, hospitals and medical supply locations, viable food stores, building material locations, fuel storage areas, car or truck lots, maintenance garages, and appliance warehouses. (NGO locations, taken from an NGO relationship matrix, may be an essential, overlapping element of this overlay.)
- **Hazardous Facilities Overlay**. This overlay identifies urban structures with known or suspected chemical, biological, or radiological features, such as nuclear power plants, fertilizer plants, oil refineries, pharmaceutical plants, and covert locations for producing weapons of mass destruction. These locations are critical to maneuver and fire planning.
- **Protected Target Overlay**. This overlay depicts terrain that should not be destroyed or attacked based on restrictions due to international, host-nation, or U.S. law and subsequent rules of engagement. These may include schools, hospitals, historical or other culturally significant monuments, and religious sites. This overlay may incorporate no-fire areas, such as special operations forces locations, critical infrastructure, logistic sources, and hazardous sites that must be protected as part of the commander's concept of the operation.
- **Incident Overlay**. Similar to the pattern analysis plot chart, this product depicts the location of different threat actions and types of tactics employed to uncover recurring routines, schemes, methods, tactics, or techniques and overall threat interests, objectives, or the desired end state.

B-45. The above IPB tools and products constitute a small sampling of what staffs and analysts can produce. They are limited only by their imaginations and mission needs (not all tools presented above may be relevant or necessary to every operation). Many products can be combined into a single product or each can generate further products of increasing level of detail. This is similar to transparent overlays positioned one atop another on a map. As discussed earlier, technology—software and hardware—will allow for more urban data to be combined, compared, analyzed, displayed, and shared—ultimately allowing scaleable products that can expand and contract to meet the intelligence requirements of any echelon of command down to the individual Soldier. The challenge will remain, however, to provide timely, accurate, complete, and relevant information in an understandable, usable, and share-capable form—share-capable among different levels of command as well as among multinational partners and various governmental and (as required) nongovernmental organizations—without overloading the commander.

Appendix C

Operations in Somalia: Applying the Urban Operational Framework to Stability Operations

It's impossible for an American mother to believe that a Somali mother would raise children to avenge the clan.

Major General Thomas M. Montgomery

GENERAL SITUATION

C-1. Following decades of political unrest and the fall of Somali dictator Siad Barre, a civil war broke out as 14 clans vied for power. The resulting nation composed of hostile social factions was held together by weak political alliances—none strong enough to unite and lead the country to national reconciliation. An ongoing drought led to famine and compounded the ethnic tensions and political instability. This volatile situation rapidly led to a phased U.S. involvement (see figure C-1). Army forces combined, sequenced, and proportionally emphasized the different types of operations to accomplish changing political objectives. Throughout all operations in Somalia, urban areas were critical to achieving mission success.

SOMALI OPERATIONS

INITIAL UN RESPONSE

C-2. The United Nations (UN) initially responded to requests for assistance from international relief organizations by sending supplies and other forms of humanitarian aid to Somalia. However, widespread looting, fighting between gangs, and other lawlessness prevented supplies from reaching the hungry and sick. Only 20 percent of the food entering the country reached the people who needed it. An estimated 25 percent of Somalia's 6 million people died of starvation or disease. In April 1992, the UN issued Security Council Resolution (UNSCR) 751 which authorized 50 unarmed observers, but the action had little effect. Under continuing pressure for additional measures to ensure the delivery of supplies and relief, the UN authorized 500 armed peacekeepers (furnished by Pakistan and transported by U.S. sea- and airlift) to protect humanitarian workers. The battalion's limited mission, designated UN Operations in Somalia I (UNOSOM I), encompassed safeguarding the unloading of ships and providing convoy security.

PROVIDE RELIEF (UNOSOM I)

C-3. In July 1992, the UN requested an increased airlift of supplies and the United States quickly responded. U.S. Central Command (USCENTCOM) activated joint task force (JTF) OPERATION PROVIDE RELIEF. Based on careful mission analysis, USCENTCOM limited the JTF's actions to—

- Deploying a humanitarian assistance survey team to assessing relief requirements.
- Providing an emergency airlift of supplies.
- Using Air Force cargo aircraft for daily relief sorties into Somalia.

C-4. USCENTCOM restricted the sorties to flying during daylight hours and to locations that would provide a permissive and safe environment. In mid-September 1992, the United States prudently expanded its role by stationing the amphibious ready group, USS Tarawa, offshore to provide support to the Pakistani security battalion and to provide security for U.S. airlift operations. The 11th Marine expeditionary unit (MEU) was on board the USS Tarawa to rapidly respond to any change in mission (see Appendix D for a description and the capabilities of a MEU).

Operation	Dates	UN Security Council Resolution	Relative Proportionality Between Types of Operations
PROVIDE RELIEF (UNOSOM I)	Aug 92 – Dec 92	UNSCR# 751 24 Apr 92	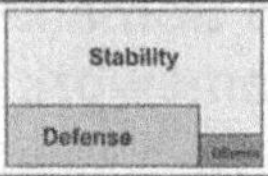
RESTORE HOPE (UNITAF)	Dec 92 – May 93	UNSCR# 794 3 Dec 92	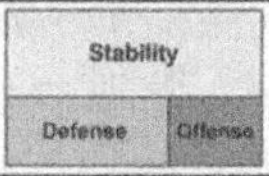
CONTINUED HOPE (UNOSOM II)	May 93 – Mar 94	UNSCR# 814 26 Mar 93	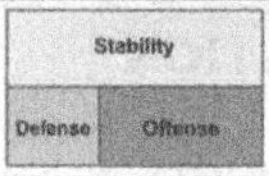

Figure C-1. Relative proportionality between types of operations

RESTORE HOPE (UNITAF)

C-5. By November 1992, the magnitude of the task, UN organizational deficiencies, and a continued lack of security precluded delivery of sufficient supplies to the needy. Notably, a ship laden with relief supplies was fired on in the harbor at Mogadishu, forcing its withdrawal before the supplies could be brought ashore, and a Pakistani peacekeeper was shot when his car was hijacked. Subsequently, the United States offered to provide forces and lead a UN-sponsored operation to reopen the flow of food to where it was needed most. In December 1992, the UN issued UNSCR 794, which authorized member states "to use all necessary means to establish a secure environment for humanitarian relief operations in Somalia" and demanded all "factions in Somalia immediately cease hostilities." To allay concerns of colonialism by a number of African countries, the UN Secretary-General was given oversight of the operation. The resolution also required soldiers to be withdrawn once order was restored; however, it provided no exit strategy. As clearly as possible, the USCENTCOM mission statement (below) for OPERATION RESTORE HOPE reflected the UN mandate:

> *When directed by the [President or the Secretary of Defense], [Commander, USCENTCOM] will conduct joint/combined military operations in Somalia to secure the major air and sea ports, key installations and food distribution points, to provide open and free passage of relief supplies, provide security for convoys and relief organization operations, and assist UN/[nongovernmental organizations] in providing humanitarian relief under UN auspices. Upon establishing a secure environment for uninterrupted relief operations, [Commander, USCENTCOM] terminates and transfers relief operations to UN peacekeeping forces.*

C-6. Mogadishu was the largest port in the country and the focal point of previous humanitarian relief activities of nongovernmental organizations (NGOs). It was also the headquarters of the coalition of 20 nations and over 30 active humanitarian relief organizations. As such, Mogadishu became the entry point for the operational buildup of the multinational force known as Unified Task Force (UNITAF) and the key logistic hub for all operations in Somalia. UNITAF immediately gained control over the flow of relief supplies into and through Mogadishu and stabilized the conflict among the clans. In less than a month, UNITAF forces expanded control over additional ports and interior airfields. They secured additional distribution sites in other key urban areas in the famine belt to include Baidoa, Baledogle, Gialalassi, Bardera, Belet Uen, Oddur, Marka, and the southern town of Kismayo (see figure C-2). With minimal force, the U.S.-led UNITAF established a secure environment that allowed relief to reach those in need, successfully fulfilling its limited—yet focused—mandate.

CONTINUED HOPE (UNOSOM II)

C-7. In March 1993, the UN issued UNSCR 814 establishing a permanent peacekeeping force, UNOSOM II. However, the orderly transition from UNITAF to UNOSOM II was repeatedly delayed until May 1993. (The UN Secretary-General urged the delay so that U.S. forces could effectively disarm bandits and rival clan factions in Somalia.) This resolution was significant in two critical aspects:

- It explicitly endorsed nation building with the specific objectives of rehabilitating the political institutions and economy of Somalia.
- It mandated the first ever UN-directed peace enforcement operation under the Chapter VII enforcement provisions of the Charter, including the requirement for UNOSOM II to disarm the Somali clans. The creation of a peaceful, secure environment included the northern region that had declared independence and had hereto been mostly ignored.

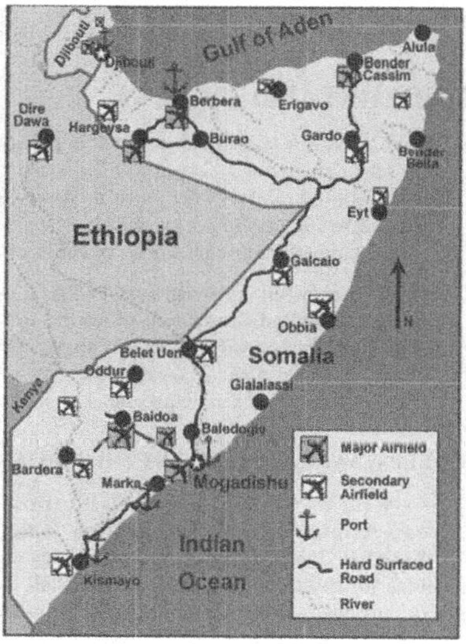

Figure C-2. Map of Somalia

C-8. These far-reaching objectives exceeded the limited mandate of UNITAF as well as those of any previous UN operation. Somali clan leaders rejected the shift from a peacekeeping operation to a peace enforcement operation. They perceived the UN as having lost its neutral position among rival factions. A more powerful clan leader, General Mohammed Farah Aideed (leader of the Habr Gidr clan), aggressively turned against the UN operation and began a radio and propaganda campaign. This campaign characterized UN soldiers as an occupation force trying to recolonize Somalia.

C-9. The mounting crisis erupted in June 1993. Aideed supporters killed 24 Pakistani soldiers and wounded 57 in an ambush while the soldiers were conducting a short-warning inspection of one of Aideed's weapons arsenals. UNSCR 837, passed the next day, called for immediately apprehending those responsible and quickly led to a manhunt for Aideed. The United States deployed 400 rangers and other special operations forces (SOF) personnel to aid in capturing Aideed, neutralizing his followers, and assisting the quick reaction force (QRF), composed of 10th Mountain Division units, in maintaining the peace around Mogadishu.

PHASED WITHDRAWAL

C-10. On 3 October 1993, elements of Task Force (TF) Ranger (a force of nearly 100 Rangers and SOF operators) executed a raid to capture some of Aideed's closest supporters. Although tactically successful, 2 helicopters were shot down, 75 soldiers were wounded, and 18 soldiers were killed accomplishing the mission. The U.S. deaths, as well as vivid scenes of mutilation to some of the soldiers, increased calls to Congress for withdrawing U.S. forces from Somalia. The President then ordered reinforcements to protect U.S. Forces, Somalia (USFORSOM) as they began a phased withdrawal with a 31 March deadline. The last contingent sailed from Mogadishu on 25 March, ending OPERATION CONTINUED HOPE and the overall U.S. mission in Somalia.

C-11. Although U.S. forces did not carry out the more ambitious UN goals of nation building, they executed their missions successfully, relieving untold suffering through humanitarian assistance with military skill and professionalism. Operations in Somalia occurred under unique circumstances, yet commanders may glean lessons applicable to future urban stability operations, as well as civil support operations. In any operations, commanders balance changing mission requirements and conditions.

UNDERSTAND

C-12. Although accomplished to varying degrees, U.S. forces failed to adequately assess and understand the urban environment, especially the society. Somali culture stresses the unity of the clan; alliances are made with other clans only when necessary to elicit some gain. Weapons, overt aggressiveness, and an unusual willingness to accept casualties are intrinsic parts of the Somali culture. Women and children are considered part of the clan's order of battle.

C-13. Early in the planning for OPERATION RESTORE HOPE, U.S. forces did recognize the limited transportation and distribution infrastructure in Mogadishu. The most notable was the limited or poor airport and harbor facilities and its impact on the ability of military forces and organizations to provide relief. Therefore, a naval construction battalion made major improvements in roads, warehouses, and other facilities that allowed more personnel, supplies, and equipment to join the relief effort faster.

UNDERSTANDING THE CLAN (THE HUMAN DIMENSION)

C-14. During OPERATION RESTORE HOPE, the UNITAF worked with the various clan leaders as the only recognized leadership remaining in the country. The UNITAF was under the leadership of LTG Robert B. Johnston and U.S. Ambassador to Somalia, Robert Oakley. In addition, UNITAF forces also tried to reestablish elements of the Somali National Police—one of the last respected institutions in the country that was not clan-based. This reinstated police force manned checkpoints throughout Mogadishu and provided crowd control at feeding centers. Largely because of this engagement strategy, the UNITAF succeeded in its missions of stabilizing the security situation and facilitating humanitarian relief. Before its termination, the UNITAF also worked with the 14 major Somali factions to agree to a plan for a transitional or transnational government.

C-15. The UN Special Representative of the Secretary-General, retired U.S. Navy Admiral Jonathon Howe, worked with the UNOSOM II commander, Turkish General Cevik Bir. During OPERATION CONTINUED HOPE, Howe and General Bir adopted a philosophy and operational strategy dissimilar to their UNITAF predecessors. Instead of engaging the clan leaders, Howe attempted to marginalize and isolate them. Howe initially attempted to ignore Aideed and other clan leaders in an attempt to decrease the warlord's power. Disregarding the long-established Somali cultural order, the UN felt that, in the interest of creating a representative, democratic Somali government, they would be better served by excluding the clan leadership. This decision ultimately set the stage for strategic failure.

THREAT STRATEGY AND TACTICS

C-16. During OPERATION RESTORE HOPE, U.S. forces also failed to properly analyze and understand their identified threat's intent and the impact that the urban environment would have on his strategy, operations, and tactics. The UN began to view eliminating Aideed's influence as a decisive point when creating an environment conducive to long-term conflict resolution. Aideed's objective, however, remained to consolidate control of the Somali nation under his leadership—his own brand of conflict resolution. He viewed the UN's operational center of gravity as the well-trained and technologically advanced American military forces, which he could not attack directly. He identified a potential American vulnerability—the inability to accept casualties for an operation not vital to national interests—since most Americans still viewed Somalia as a humanitarian effort. If he could convince the American public that the price for keeping troops in Somalia would be costly, or that their forces were hurting as many Somalis as they were helping, he believed they would withdraw their forces. If U.S. forces left, the powerless UN would leave soon after, allowing Aideed to consolidate Somalia under his leadership.

VULNERABILITY AND RISK ASSESSMENT

C-17. U.S. forces failed to assess, understand, and anticipate that Aideed would adopt this asymmetric approach and attack the American public's desire to remain involved in Somalia. By drawing U.S. forces into an urban fight on his home turf in Mogadishu, he could employ guerrilla insurgency tactics and use the urban area's noncombatants and its confining nature. Such tactics made it difficult for the U.S. forces to employ their technological superiority. If U.S. forces were unwilling to risk harming civilians, his forces could inflict heavy casualties on them, thereby degrading U.S. public support for operations in Somalia. If, on the other hand, the U.S. forces were willing to risk increased civilian casualties to protect themselves, those casualties would likely have the same effect.

C-18. However, an assessment and understanding of the Somali culture and society should have recognized the potential for Aideed's forces to use women and children as cover and concealment. Accordingly, the plan should have avoided entering the densely populated Bakara market district with such restrictive rules of engagement. As legitimacy is critical to stability operations, TF Ranger should have been prepared and authorized to employ nonlethal weapons, to include riot control gas, as an alternative to killing civilians or dying themselves.

C-19. U.S. forces also failed to assess and understand the critical vulnerability of their helicopters in an urban environment and the potential impact on their operations. TF Ranger underestimated the threat's ability to shoot down its helicopters even though they knew Somalis had attempted to use massed rocket-propelled grenade (RPG) fires during earlier raids. (Aideed brought in fundamentalist Islamic soldiers from Sudan, experienced in downing Russian helicopters in Afghanistan, to train his men in RPG firing techniques). In fact, the Somalis had succeeded in shooting down a UH-60 flying at rooftop level at night just one week prior to the battle. Instead, TF Ranger kept their most vulnerable helicopters, the MH-60 Blackhawks, loitering for forty minutes over the target area in an orbit that was well within Somali RPG range. The more maneuverable AH-6s and MH-6s could have provided the necessary fire support. Planning should have included a ready ground reaction force, properly task organized, for a downed helicopter and personnel recovery contingency.

C-20. Information operations considerations apply throughout the entire urban operational framework; however, operations security (OPSEC) is critical to both assessment and shaping. OPSEC requires continuous assessment throughout the urban operation particularly as it transitions among the spectrum of operations. As offensive operations grew during OPERATION CONTINUED HOPE, U.S. forces did little to protect essential elements of friendly information. Combined with the vulnerability of U.S. helicopters, Aideed's followers used U.S. forces' inattention to OPSEC measures to their advantage. The U.S. base in Mogadishu was open to public view and Somali contractors often moved about freely. Somalis had a clear view both day and night of the soldiers' billets. Whenever TF Ranger would prepare for a mission, the word rapidly spread through the city. On 3 October 1993, Aideed's followers immediately knew that aircraft had taken off and, based on their pattern analysis of TF Ranger's previous raids, RPG teams rushed to the rooftops along the flight paths of the task force's Blackhawks.

SHAPE

C-21. One of the most critical urban shaping operations is isolation. During OPERATION CONTINUED HOPE, U.S. forces largely discounted other essential elements of friendly information and did not establish significant public affairs and psychological operations (PSYOP) initiatives. In fact, Army forces lacked a public affairs organization altogether. Consequently, Aideed was not isolated from the support of the Somali people. This failure to shape the perceptions of the civilian populace coupled with the increased use of lethal force (discussed below) allowed Aideed to retain or create a sense of legitimacy and popular support. Ironically, the focus on Aideed helped create a popular figure among the warlords through which many noncompliant, hostile, and even neutral Somalis could rally behind.

C-22. During OPERATION RESTORE HOPE, Aideed conducted propaganda efforts through "Radio Aideed"—his own radio station. UNITAF countered these efforts with radio broadcasts. This technique proved so effective that Aideed called MG Anthony C. Zinni, UNITAF's director of operations, over to his house on several occasions to complain about UNITAF radio broadcasts. General Zinni responded, "If he didn't like what we said on the radio station, he ought to think about his radio station and we could mutually agree to lower the rhetoric." This approach worked.

ENGAGE

C-23. The complexity of urban operations requires unity of command to identify and effectively strike the center of gravity with overwhelming combat power or capabilities. Complex command and control relationships will only add to the complexity and inhibit a commander's ability to dominate and apply available combat power to accomplish assigned objectives. Stability operations as seen in Somalia required commanders to dominate within a supporting role and, throughout, required careful, measured restraint.

UNITY OF COMMAND (EFFORT)

C-24. During OPERATION RESTORE HOPE, UNITAF successfully met unity of command challenges through three innovations. First, they created a civil-military operations center (CMOC) to facilitate unity of effort between NGOs and military forces. Second, UNITAF divided the country into nine humanitarian relief sectors centered on critical urban areas that facilitated both relief distribution and military areas of responsibility. Third, to establish a reasonable span of control, nations that provided less than platoon-sized contingents were placed under the control of the Army, Marine Corps, and Air Force components.

C-25. On the other hand, during OPERATION CONTINUED HOPE, UNOSOM II command and control relationships made unity of command (effort) nearly impossible. The logistic components of USFORSOM were under UN operational control, while the QRF remained under USCENTCOM's combatant command—as was TF Ranger. However, the USCENTCOM commander was not in theater. He was not actively involved in planning TF Ranger's missions or in coordinating and integrating them with his other subordinate commands. It was left to TF Ranger to coordinate with the QRF as needed. Even in TF Ranger, there were dual chains of command between SOF operators and the Rangers. This underscores the need for close coordination and careful integration of SOF and conventional forces (see Chapter 4). It also emphasizes overall unity of command (or effort when command is not possible) among all forces operating in a single urban environment.

C-26. Following TF Ranger's 3 October mission, the command structure during OPERATION CONTINUED HOPE was further complicated with the new JTF-Somalia. This force was designed to protect U.S. forces during the withdrawal from Somalia. JTF-Somalia came under the operational control of USCENTCOM, but fell under the tactical control of USFORSOM. Neither the JTF nor USFORSOM controlled the naval forces that remained under USCENTCOM's operational control. However unity of effort (force protection and a rapid, orderly withdrawal) galvanized the command and fostered close coordination and cooperation among the semiautonomous units.

MEASURED RESTRAINT

C-27. During OPERATIONS PROVIDE RELIEF and RESTORE HOPE, U.S. forces dominated within their supporting roles. Their perseverance, adaptability, impartiality, and restraint allowed them to provide a stable, secure environment. Hence, relief organizations could provide the food and medical care necessary to reduce disease, malnourishment, and the overall mortality rate. However, during OPERATION CONTINUED HOPE, U.S. operations became increasingly aggressive under the UN mandate. However, peace enforcement also requires restraint and impartiality to successfully dominate and achieve political objectives. The increased use of force resulted in increased civilian casualties, which in turn reduced the Somalis' perception of U.S. legitimacy. As a result, most moderate Somalis began to side with Aideed and his supporters. Many Somalis, without effective information operations to shape their perceptions, felt that it was fine for foreign military forces to intervene in their country to feed the starving and even help establish a peaceful government, but not to purposefully target specific Somali leaders as criminals.

CONSOLIDATE AND TRANSITION

C-28. Across the spectrum of conflict, Army forces must be able to execute full spectrum operations not only sequentially but, as in the case of operations in Somalia, simultaneously. OPERATION PROVIDE RELIEF began primarily as a stability operation in the form of foreign humanitarian assistance. Later this operation progressed to include peacekeeping (another stability operation), defensive operations to protect

UN forces and relief supplies, and minimum offensive operations. As operations transitioned to OPERATION RESTORE HOPE, it became apparent that while foreign humanitarian assistance was still the principal operation, other operations were necessary. More forceful peacekeeping, show of force, and arms control (all forms of stability operations), and offensive and defensive operations grew increasingly necessary to establish a secure environment for uninterrupted relief operations. In the final phase of U.S. involvement during OPERATION CONTINUED HOPE, major changes to political objectives caused a major transition in the form of the stability operation to peace enforcement with an even greater increase in the use of force, offensively and defensively, to create a peaceful environment and conduct nation building. Finally, the TF Ranger raid demonstrated the need to maintain a robust, combined arms force capable of rapidly transitioning from stability operations (primarily peace operations) to combat operations.

SUMMARY

C-29. OPERATIONS PROVIDE RELIEF and RESTORE HOPE were unquestionably successes. Conversely, during OPERATION CONTINUED HOPE, the 3-4 October battle of Mogadishu (also known as the "Battle of the Black Sea") was a tactical success leading to an operational failure. TF Ranger succeeded in capturing 24 suspected Aideed supporters to include two of his key lieutenants. Arguably, given the appropriate response at the strategic level, it had the potential to be an operational success. After accompanying Ambassador Oakley to a meeting with Aideed soon after the battle, MG Zinni described Aideed as visibly shaken by the encounter. MG Zinni believed Aideed and his subordinate leadership were tired of the fighting and prepared to negotiate. Unfortunately, the U.S. strategic leadership failed to conduct the shaping actions necessary to inform and convince the American public (and its elected members of Congress) of the necessity of employing American forces to capture Aideed. The President was left with little recourse after the battle of Mogadishu but to avoid further military confrontation.

C-30. Despite this strategic failing, the operational commanders might have avoided the casualties, and any subsequent public and Congressional backlash, had they better communicated among themselves and worked with unity of effort. Recognizing the separate U.S. and UN chains of command, the UN Special Representative, along with the USCENTCOM, USFORSOM, and TF Ranger commanders, should have established the command and control architecture needed. This architecture would have integrated planning and execution for each urban operation conducted. These commanders failed to "operationalize" their plan. They did not properly link U.S. strategic objectives and concerns to the tactical plan. The TF Ranger mission was a direct operational attempt to obtain a strategic objective in a single tactical action. Yet, they failed to understand the lack of strategic groundwork, the threat's intent and capabilities, and the overall impact of the urban environment, to include the terrain and society, on the operation. Such an understanding may not have led to such a high-risk course of action but instead to one that de-emphasized military operations and emphasized a political solution that adequately considered the clans' influence.

Appendix D

Joint and Multinational Urban Operations

[Joint force commanders] synchronize the actions of air, land, sea, space, and special operations forces to achieve strategic and operational objectives through integrated, joint campaigns and major operations. The goal is to increase the total effectiveness of the joint force, not necessarily to involve all forces or to involve all forces equally.

JP 3-0

As pointed out earlier, Army forces, brigade size and larger, will likely be required to conduct operations in, around, and over large urban areas in support of a joint force commander (JFC). The complexity of many urban environments, particularly those accessible from the sea, requires unique leveraging and integration of all the capabilities of U.S. military forces to successfully conduct the operation. This appendix discusses many of these capabilities; JP 3-06 details joint urban operations.

PURPOSE

D-1. In some situations, a major urban operation is required in an inland area where only Army forces are operating. Army commanders determine if the unique requirements of the urban environment require forming a joint task force (JTF) or, if not, request support by joint capabilities from the higher joint headquarters. Sometimes the nature of the operation is straightforward enough or the urban operation is on a small enough scale that conventional intraservice support relationships are sufficient to meet the mission requirements.

D-2. Most major urban operations (UO), however, require the close cooperation and application of joint service capabilities. A JTF may be designated to closely synchronize the efforts of all services and functions in an urban area designated as a joint operations area (JOA). If a large urban area falls in the context of an even larger ground force area of operations, a JTF dedicated to the urban operation may not be appropriate. These situations still require joint capabilities. In such cases, the responsible JFC designates support relations between major land units and joint functional commands. The major land units can consist of Army forces, Marine Corps forces, or joint forces land component command. The joint functional commands can consist of the joint special operations task force, joint psychological operations task force, or joint civil-military operations task force.

D-3. This appendix describes the roles of other services and joint combatant commands in UO. It provides an understanding that enables Army commanders to recommend when to form a JTF or to request support from the JFC. It also provides information so commanders can better coordinate their efforts with those of the JFC and the commanders of other services or components conducting UO. Lastly, this appendix describes some considerations when conducting UO with multinational forces.

SISTER SERVICE URBAN CAPABILITIES

D-4. Army forces conducting UO rely on other services and functional joint commands for specialized support in the urban environment. These capabilities are requested from and provided through the commanding JFC. Army forces request the assets and capabilities described in this annex through their higher headquarters to the joint command. The JFC determines if the assets will be made available, the appropriate command relationship, and the duration of the support. Army forces prepare to coordinate planning and execution with other services and to exchange liaison officers. These capabilities can greatly increase the Army's ability to understand, shape, engage, consolidate, and transition within the context of UO.

AIR FORCE

D-5. Air Force support is an important aspect of the Army force concept for urban operations. Air Force elements have a role to play in UO across the spectrum of operations.

D-6. Air Force intelligence, surveillance, and reconnaissance (ISR) systems contribute significantly to understanding the urban environment. These ISR systems include the E-8 Joint Surveillance Target Attack Radar System (JSTARS) (see figure D-1), U2 Dragon Lady, RC-135 Rivet Joint, or RQ-4 Global Hawk and RQ/MQ-1 Predator unmanned aerial systems. Air Force ISR systems can provide vital data to help assess threat intentions, threat dispositions, and an understanding of the civilian population. These systems also can downlink raw information in real-time to Army intelligence processing and display systems, such as the common ground station or division tactical exploitation system.

Figure D-1. USAF E-8 JSTARS platform

D-7. Air interdiction (AI) can be a vital component of shaping the urban operational environment. Often, AI of the avenues of approach into the urban area isolates the threat by diverting, disrupting, delaying, or destroying portions of the threat force before they can be used effectively against Army forces. AI is especially effective in major combat operations where restrictions on airpower are limited and the threat is more likely to be a conventionally equipped enemy. In 1991 during OPERATION DESERT STORM, AI helped prevent the Iraqi 5th Mechanized Division from reaching Khafji.

D-8. Precision-guided munitions (PGMs) delivered from combat aircraft can positively and directly affect the conduct of Army close combat actions in the urban area. Special munitions designed to penetrate reinforced concrete can provide unique support to land forces executing UO. Although laser-guided bombs are extremely accurate with an effective laser lock-on, environmental factors such as smoke, haze, and weather can adversely affect successful laser designation. If the launching aircraft can achieve a successful laser designation and lock-on, these weapons have devastating effects, penetrating deep into reinforced concrete before exploding with great force. If launched without a lock-on, or if the laser spot is lost, these weapons are unpredictable and can travel long distances before they impact. However, continual advancements in tactics and weapons systems, such as joint direct attack munition (JDAM) weapons, help decrease the adverse effects of these environmental factors. (For example, in 2003 during OPERATION IRAQI FREEDOM, the Air Force supported the Army's attack on An Najaf with JDAM-equipped F 16s and B 1s during a blinding sandstorm. With near zero visibility, JDAM weapons provided significant fire support, allowing Army forces to complete their mission and continue the attack north.) When PGMs are unavailable, general-purpose bombs from 500 to 2,000 pounds can also prove effective. However, there may be risk factors associated with their use such as collateral damage due to decreased accuracy, threat exposure to the delivering aircraft due to lower delivery altitudes, or reduced weapons effectiveness due to shallower penetration angles. (See FM 3-06.1 for a more detailed description of various aircraft weapons and capabilities.)

D-9. In addition to shaping and engaging the UO through firepower, commanders can use Air Force capabilities to improve and augment the urban transportation and distribution infrastructure. Air Force units can repair or improve airfields, revitalize civil aviation maintenance facilities, manage air-delivered cargo, and control civil and military air traffic. These latter capabilities particularly enhance urban stability and civil support operations. These capabilities may even be decisive. Air-delivered cargo and air traffic management, for example, were the decisive factors in U.S. forces' successful resistance of the Soviet blockade of Berlin in 1948. In OPERATION RESTORE HOPE, from December 1992 to May 1993, Air Force operations in Mogadishu were critical to airlifting and staging supplies and forces. The Air Force determined the ultimate success of the humanitarian assistance operation (see Appendix C).

D-10. In unique situations, such as the Berlin Blockade and OPERATION RESTORE HOPE, exercising Air Force urban capabilities may be the decisive action of the operation. Most often, however, Air Force capabilities will play a shaping role, sustaining role, or both in joint urban operations because of the inherent requirement to occupy terrain and interface with the population.

Marine Corps

D-11. The Marine Corps can assault across water obstacles into a defended urban environment. This capability is an invaluable tactical and operational tool. The mere threat of this capability can divert many enemy forces from other avenues of approach and obscure the true nature of an attack. The impact of the threat of amphibious assault was vividly demonstrated during OPERATION DESERT STORM where embarked Marine Forces diverted several Iraqi divisions to defensive positions along the coast and near Kuwait City.

D-12. The presence of Marine amphibious equipment, apart from Marine infantry, provides Army forces unique capabilities. In UO, the amphibious operation is often not an assault from the sea, but rather an assault river crossing. In 1950, the 7th U.S. Infantry Division used amphibious tractor support from the 1st Marine Division to conduct an assault river crossing of the Han River into downtown Seoul. See FM 3-31.1.

D-13. The worldwide deployment of Marine air-ground task forces (MAGTFs) enables a short notice response into any urban areas accessible from the sea. Typically, a deployed MAGTF is a Marine expeditionary unit (special operations capable) (MEU[SOC]). The MEU(SOC) can perform forcible entry operations, seize lodgments, and may execute these tasks anticipating reinforcement by Army or joint forces. They are also well positioned and equipped to rapidly reinforce Army forces already deployed in theater. The special-operations-capable training that these units accomplish before deploying includes urban warfare training and contributes to their value in UO. The MEU(SOC) is relatively small (its core unit is a Marine infantry battalion), is forward deployed, and has a wide spectrum of organic capabilities. It is an important asset in crisis stability and civil support operations.

D-14. The Marine expeditionary brigade is the MAGTF between a MEU and a Marine expeditionary force (the largest MAGTF). All MAGTFs include an aviation combat element, which is particularly skilled at providing accurate and timely close air support. The responsiveness and accuracy of Marine close air support aptly suits it to UO where the lethality of combat and the close range of engagements demand accurate and responsive fires.

D-15. Marine forces that conduct UO work well in littoral urban areas because of their unique relationship with naval forces and thus their capability to closely integrate land and sea operations. A supporting arms liaison team may be attached to Army forces at battalion level to provide ship-to-shore communications and coordination for naval gunfire support.

Navy

D-16. Many major urban areas are accessible from the sea. Therefore, Army commanders should understand how sea power can influence and support UO. The Navy brings several major capabilities to UO. These include naval gunfire support, naval air support, and port and coastal security. In addition, the Navy can offer secure command post sites and meeting areas; provide limited emergency shelter and critical care for Soldiers and, as necessary, civilians; and store and transport critical supplies ashore providing flexibility in meeting support requirements. The ability to provide support from "over the horizon" not only enhances force protection but also the legitimacy of urban and host nation authorities during transition.

D-17. Naval surface fire support (which includes naval gunfire support) particularly applies to forcible entry operations in littoral urban areas. It provides an initial indirect fire support capability until Army forces land ashore. However, naval gunfire support lacks a precision munitions capability. In defensive operations, naval gunfire support can also add major fires, especially during retrograde operations through an urban area. Naval gunfire support of the 3rd Infantry Division was the primary fire support for the last two days of the X Corps retrograde operation out of the port of Hungnam, North Korea, in December 1950.

D-18. Destroyers and cruisers, which mount one and two 127mm MK45 lightweight gun systems respectively (see figure D-2), are capable of providing naval gunfire support. This gun system can provide a rate of fire of 20 rounds per minute per gun to a range of approximately 23 kilometers. Destroyers, cruisers, and attack submarines are capable of precision strikes with Tomahawk cruise missiles. Depending on the location of the urban area, naval air can provide responsive aviation support to UO alleviating the need for fixed-wing aviation bases.

Figure D-2. USN MK45 lightweight gun system

D-19. Due to its flat trajectory, terrain masking affects naval gunfire more than field artillery. Naval gunfire also results in large range probable errors (the dispersion pattern of the naval gun is roughly elliptical with the long axis in the direction of fire). Hence, coverage of targets such as roads and airfields is most effective when the gun-target line (GTL) coincides with the long axis of the target. Very close supporting fire can be delivered when the GTL is parallel to the front line of troops. Oppositely, a GTL perpendicular to the front trace can endanger friendly forces. Within the limits of hydrographic conditions, the ship can maneuver to achieve a better GTL, but ship movement also makes it difficult to adjust fire. Overall, naval and air threats, bad weather, and large range probable errors make naval gunfire difficult and can cause cancellation of supporting fires.

D-20. Water terminals located in urban areas are usually the debarkation points for the bulk of Army forces as well as a joint force theater logistics requirement. Army forces are responsible for water terminal operations while naval capabilities protect these strategic and operationally vital facilities.

D-21. Naval coastal warfare (NCW) is the responsibility of the JFC and is often exercised through the Navy component commander. He may assign a naval coastal warfare commander for an appropriate geographic area. NCW includes coastal sea control, port security, and harbor defense. While coastal sea control is conducted in the environment of the open seas, port security and harbor security include the urban environment. Port security is the safeguarding of vessels, harbors, ports, waterfront facilities, and cargo from internal threats. It includes destruction, loss, or injury from sabotage or other subversive acts; accidents; thefts; or other causes of similar nature. The Navy's role in protecting essential urban infrastructure is often key. This task is also important when executing stability operations if the threats against urban infrastructure will likely be unconventional.

D-22. Harbor defense protects harbor approaches, harbors, anchorages, and ports from external threats. Harbor defense focuses on the conventional defense of port infrastructure. It is a task appropriate in a major operation or campaign and often includes port security as a subtask. The JFC executes NCW (focused on harbor defense and port security) using combined Navy capabilities including surface warfare, aviation, and naval special operations. See also discussions in this appendix on transportation command and special operations command.

COAST GUARD

D-23. The Coast Guard, like the Navy, can significantly influence the conduct of UO when the urban area is accessible from the sea. The U.S. Coast Guard (USCG) is the federal authority for port security and harbor defense of domestic facilities. When directed by the President, the USCG can augment the Navy in operations overseas. Historically, the entire USCG was under Navy control during both World Wars I and II. USCG elements deployed overseas and operated under Navy control during the Vietnam War and during OPERATIONS DESERT SHIELD and DESERT STORM.

D-24. The USCG uses surface warfare systems and aerial reconnaissance systems to conduct its missions. Its air systems are unarmed, whereas its surface systems are armed for self-defense and law enforcement operations. The USCG is experienced and adept at supporting other agencies, local governments, and law enforcement. Its capabilities can best support Army UO in civil support operations or the civil support aspects of an operation. The USCG works effectively against an unconventional threat and threats with on-water capability.

URBAN FUNCTIONAL COMBATANT COMMAND CAPABILITIES

D-25. Three commands provide urban functional combatant command capabilities. Transportation Command manages the global transportation system. Strategic Command integrates plans for and employs space forces ad assets. Special Operations Command is responsible for providing special operations forces.

TRANSPORTATION COMMAND

D-26. U.S. Transportation Command (USTRANSCOM) provides strategic air, land, and sea transportation for the Department of Defense to deploy, employ, sustain, and redeploy U.S. military forces worldwide. USTRANSCOM provides global transportation management, using an integrated transportation system across the spectrum of operations through its transportation component commands (TCCs). The TCC consists of Air Mobility Command (AMC), Military Sealift Command (MSC), and Surface Deployment and Distribution Command (SDDC).

D-27. During urban operations, USTRANSCOM, through its TCC, can provide common-user terminal services in support of strategic transportation movements to a theater of operations. AMC provides common-user airlift, air refueling, and aeromedical evacuation services. It is the worldwide aerial port manager and, where designated, the operator of common-user aerial ports of embarkation and aerial ports of debarkation. MSC provides common-user sealift services between seaports of embarkation (SPOEs) and seaports of debarkation (SPODs). SDDC provides common-user ocean terminal service and, where designated, serves as the single port manager at SPOEs and SPODs. The urban transport system, in the form of railheads, ports, and airfields, is integral to many urban operations across the spectrum of Army operations and often the objective of Army UO. Army forces planning, preparing, and executing urban operations engage the supported geographic commander of a combatant command or his component commands in all aspects of conducting UO when the objective is transport related.

D-28. In offensive operations, commanders use transportation expertise to identify the urban transportation infrastructure (both the terrain and social [human] aspects) that is secured or that can affect current and future operations. This analysis also includes second- and third-order support systems. USTRANSCOM is consulted regarding the degree of acceptable damage that the system can sustain and still meet mission requirements. During execution, USTRANSCOM units may integrate into the operation so they can begin operating the transportation systems as early as possible. During the Inchon landing of September 1950, Army forces had begun rail operations on D+1 and port operations under way by D+3 of the forcible entry.

D-29. In defensive and stability or civil support operations, USTRANSCOM units safeguard and prevent disruption of the transport system by the conduct of defensive operations or stability tasks.

USTRANSCOM advises Army commanders of the impact of defensive tactics, techniques, and procedures (TTP) on USTRANSCOM operations as well as security requirements for USTRANSCOM facilities. The JFC provides guidance to deconflict any issues that may arise from these potentially divergent missions and tasks.

STRATEGIC COMMAND

D-30. The United States Strategic Command (USSTRATCOM) is the combatant command that integrates and synchronizes space operations, information operations, integrated missile defense, global command and control (C2); ISR, global strike, and strategic deterrence to provide. A unified source for greater understanding of specific threats around the world and the means to rapidly respond to those threats including weapons of mass destruction. USSTRATCOM operates assigned forces through its service component commands: U.S. Army Space Command/Army Forces Strategic Command; Air Force Space Command, and Naval Net Warfare Command.

D-31. USSTRATCOM provides critical space systems that offer global coverage and potential for real-time and near real-time support to urban operations. Space systems are unconstrained by political boundaries. Commanders can use space systems during peacetime or times of crisis to monitor an urban area before inserting friendly forces. Space systems enhance operations and assist commanders in overcoming some of the physical challenges of the urban environment. These enhancement operations include ISR; positioning and navigation; environmental monitoring; and communications. Space-based imagery and sensors are important ISR capabilities that contribute to situational understanding throughout the depth, breadth, and height of the urban area. ISR systems can provide route and target information for mission planning, locate presurveyed missile launch sites, detect camouflage, assess threat operations and movements, and warn of hostile acts and reconnaissance. Positioning and navigation systems also assist situational understanding through links to digital information systems (INFOSYS) while assisting tactical navigation in ambiguous terrain found in some urban areas. These enhancements can include precise location and position information for urban fires, ingress and egress routes, and rendezvous coordination. Environmental monitoring systems can provide weather and ionospheric information needed to assess weapon selection, air routes, ground and water trafficability, and communications. Communications systems provide secure, survivable links between elements of Army and joint forces to disseminate plans, orders, and warnings. These systems may form a critical link in the INFOSYS that transmit data to assessment centers and intelligence to key decision makers. However, some of the same environmental influences and degradation as ground-based systems may affect space-based systems. Commanders should refer to FM 3-14 for more information on space support to Army operations.

SPECIAL OPERATIONS COMMAND

D-32. U.S. Special Operations Command (USSOCOM) exercises combatant command of all active and reserve special operations forces stationed in the United States. USSOCOM also provides trained and combat-ready special operations forces (SOF) to the geographical combatant commanders and, when directed by the President or the Secretary of Defense, command-designated special operations.

D-33. SOF provide commanders with capabilities critical to success in the urban environment. The density of this environment in both space and time requires the careful integration of SOF and conventional forces. Army forces conducting UO have a clear and unambiguous C2 relationship with the SOF in the urban area to ensure coordination, massing of effects, and unity of effort.

D-34. Each service has unique special operations (SO) capabilities. For example, SOF can identify and seize or destroy key terrain or infrastructure in denied areas; secure or capture key personnel; identify supportive civilians (aides and allies) who can facilitate future infiltrations and the cache of critical supplies; counter urban insurgencies; and conduct unconventional warfare in enemy-held urban areas. SOF can also emplace covert sensors and surveillance devices, provide clandestine intelligence collection, and provide target acquisition information or emplace beacons (for weapons guidance or navigation) in the highly restrictive terrain of the urban environment. In multinational UO, they can provide coalition support teams with trained, culturally aware, language proficient, military liaison personnel with organic communications connectivity. Army leaders understand the SO capabilities available to ensure that they request the right support for Army UO and to ensure unity of effort within the urban JOA.

Army SOF

D-35. The Army provides five types of SOF units to USSOCOM. Two of these, psychological operations (PSYOP) and civil affairs, are discussed in Chapter 4.

Ranger Forces

D-36. The U.S. Army 75th Ranger Regiment can rapidly deploy light infantry forces from company through regimental size. Rangers specialize in direct action (DA) SO missions and focus on airfield seizure and raids. Typically, Ranger units turn over their objectives to conventional units upon mission completion. The Ranger DA capability especially applies to UO because of the many critical infrastructures that often prompt Army forces to engage in UO. Ranger capabilities are ideal for seizing critical facilities to preempt their defense or destruction. The size and combat power of Ranger units permit execution of offensive and defensive operations against enemy conventional units for periods of limited duration; austere combat service support capabilities limit the Ranger regiment's ability to sustain combat action without extensive augmentation.

Special Forces

D-37. U.S. Army Special Forces Command (Airborne) trains and prepares Army Special Forces (SF) to deploy and execute operational requirements for geographical combatant commanders. SF units are small and capable of extended operations in remote and hostile locations. SF units execute seven basic missions: foreign internal defense (FID), unconventional warfare (UW), counterproliferation of weapons of mass destruction, special reconnaissance (SR), DA, counterterrorism, and information operations. Important collateral activities consist of coalition support, combat search and rescue, counterdrug activities, humanitarian demining activities, humanitarian assistance, security assistance, and special activities.

D-38. SR, DA, and coalition support are particularly important in UO. SF units may physically penetrate an urban area to conduct SR to determine threat strengths, dispositions, and intentions. In some situations, SF units, due to their language and cultural training, can gather accurate information regarding the disposition and attitudes of the population. SF special reconnaissance can also determine or verify the functional status of urban infrastructure as well as conduct target acquisition, area assessment, and poststrike reconnaissance.

D-39. SF teams can execute terminal guidance and with qualified joint tactical air controllers can control operations for Army and joint precision fires. Although these teams possess limited organic combat power for their short-duration, DA missions, they can bring significant effects to bear against high-payoff targets. Yet, SF units have no capability to conduct a sustained defense of such targets.

D-40. SF units can advise, train, and assist urban indigenous movements already in existence to conduct unconventional warfare and possibly accompany these groups into combat. The upsurge in urban insurgency and terrorism has caused worldwide concern since it is not confined to developing countries. Present day dissident groups are well aware of this situation. These groups realize that to be successful, they must center the insurgent activities on the major cities or political center of their countries.

Special Operations Aviation Forces

D-41. Army special operations aviation provided by the 160th Special Operations Aviation Regiment operates primarily to support SO missions. They execute insertion, extraction, and resupply missions to support SO. Similar to conventional aviation, this capability may be more vulnerable to concealed air defenses when operating over hostile or unsecured urban terrain than in many other environments.

Navy SOF

Sea-Air-Land Teams

D-42. Navy sea-air-land teams (SEALs) specialize in water approaches to targets. They operate in small, squad-size teams and have many of the same capabilities as Army Special Forces (see above). Navy SEALs do not typically have the cultural and language training of Army Special Forces. Their capability to insert from sea gives them a unique ability to penetrate into urban areas that are accessible from the sea.

Special Boat Units

D-43. Special boat units (SBUs) employ, operate, and maintain various surface combatant craft (see figure D-3) to conduct and support naval and joint special operations, riverine warfare, and coastal patrol and interdiction. The SBU can infiltrate and exfiltrate forces; provide small-caliber gunfire support; and conduct coastal patrol, surveillance, harassment, and interdiction of maritime lines of communications. These units are ideal in UO that include ports and rivers. They can assist in port security, conduct river patrols, and participate in harbor defense. They are well suited for preventing sea infiltration by unconventional threats.

Figure D-3. USN MK V special

Operations CraftAir Force SOF

D-44. The Air Force has three primary SOF elements: special operations air units, battlefield Airmen, and combat aviation advisory units.

Special Operations Wings and Groups

D-45. Air Force special operations aviation elements operate both fixed-wing and rotary-wing aircraft (which include variants of both C-130s and MH-53s). Air Force SOF conduct core missions which include specialized air mobility, precision aerospace fires, aerospace surface interface, personnel recovery operations, specialized refueling, combat aviation advisory, information operations, and PSYOP. As a result, Air Force SOF conduct infiltrations, exfiltrations, resupplies, close air support of SOF ground elements, electronic warfare, and aerial C2 support.

D-46. The AC-130 U and AC-130 H model aircraft (see figure D-4) are designed specifically to provide close air support to ground SOF. However, they will be an invaluable asset to any ground force operating in an urban environment. Hence, Army commanders seek opportunities to train with AC-130 gunships prior to the conduct of UO. AC-130s are armed with one 40mm autocannon and one 105mm howitzer. The AC-130 U is also armed with one 25mm autocannon. This aircraft has night capability and is extremely accurate. Its fires are responsive and can be decisive in close urban combat. Its cannon and howitzer are accurate enough to concentrate fire onto a single spot to create a rooftop breach that allows fire to be directed deeper into the building. The aircraft can provide excellent covert illumination with its infrared spotlight. The AC-130 is very vulnerable to air defense systems, which friendly forces must suppress or destroy to effectively use this system. (During OPERATION DESERT STORM in 1991, an AC-130 was shot down over Kuwait.)

Battlefield Airmen

D-47. Battlefield Airmen consist of special tactics teams (STTs) and special operations weather teams (SOWTs). Combat controllers and pararescue personnel constitute STTs. STTs are specially tailored to meet mission criteria and may vary from a small three-man team to a larger twenty-five-man element. STTs support the UO Army commanders by—

- Performing air-land-sea personnel recovery operations.
- Providing terminal attack control or guidance.
- Establishing and operating navigational aids and beacons.
- Providing liaison to ground commanders.
- Providing visual flight rules and limited instrument flight rules air traffic control.
- Providing positive control of the terminal objective area aviation environment during SOF operations.

Note: SOWTs are normally attached to Army SOF to provide weather observation and limited tactical forecasting.

Combat Aviation Advisory (CAA) Operations

D-48. CAA operations are part of an interagency process aimed toward freeing and protecting the host nation's population from subversion, lawlessness, and insurgency. Specifically, CAA operations are tailored to assess, train, advise, and assist foreign aviation forces in air operations employment and sustainability. CAA operations support geographical combatant commanders throughout the spectrum of conflict primarily by facilitating the availability, reliability, safety and interoperability, and integration of friendly and allied aviation forces supporting joint, interagency, and multinational forces and organizations. CAA operations provide assistance in the interrelated areas of FID, coalition support, UW, and humanitarian and disaster relief. CAA operations also include a liaison role in coalition support.

MULTINATIONAL CONSIDERATIONS

D-49. Army UO in foreign urban areas will often be joint and likely have a multinational component. When properly executed, integrating multinational forces into UO greatly enhances the operation's military (as well as political) effectiveness. Properly integrating multinational forces into UO requires a thorough understanding of both the urban environment and the nature of individual national forces. This understanding includes the political, cultural, and historical characteristics of the other national forces. Such understanding also includes the national force's doctrine and military capabilities, strengths, and weaknesses. In UO, these considerations are critical because these factors will alter the urban population's attitude toward multinational forces and the behavior of such forces as they interact with the urban population. Combining this understanding with effective C2 and an equal understanding of the urban environment results in effective multinational UO (see FM 100-8).

CULTURAL COMPATIBILITY

D-50. When evaluating and assigning UO tasks, Army commanders also consider the degree of interaction with the civilian population. A national force from a Muslim-majority country may work better with a civilian population that is also Muslim than a force having a different religion. In such a situation, the national force with the same religion as the urban population may be assigned tasks that require close relations with civilians. Army forces, in this situation, may be assigned tasks that are more remote from the population. National forces that have a national history of animosity to the civil population (or that sympathize with antagonists within the civilian population) are not used in tasks requiring diplomacy and close cooperation with the civilians or government.

Military Capabilities and Restrictions

D-51. Army forces are responsible to understand the military capabilities of national forces with which they work. Some national forces, as part of their normal capabilities, are adept at police functions that enable them to operate with little training in a law enforcement role. Other national forces specialize in small-unit, light infantry patrolling. These forces may be ideal in a stability operation. In contrast, a national force composed of conscripts and trained primarily in conventional warfare techniques may best work as a reinforcing force or may require extensive training before mission execution in an urban environment or a stability situation. Army commanders also consider the type of weapon systems with which participating multinational forces are equipped; they may be more or less effective in an urban environment. For example, some countries may still possess the Vulcan antiaircraft gun system (or similar weapon) that can be very useful in urban offensive and defensive operations.

D-52. Commanders in multinational UO must also understand nation-specific rules of engagement (ROE). Some nations will severely limit their participating military's use of lethal and nonlethal fires. On the other hand, other nations may not be as constrained as Army units. Ideally, commanders attempt to create ROE that are the same across the coalition but recognize that this may not be achievable. Hence, in an environment where minimization of collateral damage will be paramount, Army commanders will need to carefully consider national limitations on the employment of forces when planning and executing UO. Relatedly, commanders take into account each nation's ability to achieve precision effects with their available forces and systems.

Intelligence Sharing

D-53. Due to classification issues, sharing intelligence with coalition partners may be challenging; the United States may have close intelligence ties with some countries within the coalition and few or none with others. In many cases intelligence personnel from other countries have unique skills (and cultural perspectives) that can significantly contribute to the friendly intelligence effort. Reports from some coalition members may fill intelligence gaps for Army forces and the coalition as a whole. On the other hand, few countries have the sophisticated intelligence collection assets available to Army forces and information that Army units may provide could be critical to a coalition partner's force protection and overall mission success.

Information Management

D-54. National capabilities to collect, process, store, display, and disseminate information and establish a common operational picture vary considerably between coalition partners. As the capabilities of Army information systems increase, those of our future partners may not keep pace. Even when capabilities are similar, equipment may not be compatible. The complex nature of the urban environment and coalition language differences (and linguist shortages) will only serve to exacerbate incompatibilities and limitations. Therefore commanders must understand capabilities and limitations to ensure that orders, control measures, and information can be appropriately disseminated, displayed, and understood. Significantly, they must understand the coalition's ability to accurately locate and track friendly forces as well as identify friend from foe. Commanders must take the appropriate steps to ensure that essential intelligence and relevant information is provided to coalition partners in a form (and language) that they can accept and use. This will be critical to establishing the common operational picture necessary to facilitate rapid decision making and effective command and control across the coalition.

Logistic Requirements

D-55. Logistically, agreement among the multinational forces should include support consolidation whenever possible. Creating a multinational logistics office or coordination center works best when accomplished early. This office can coordinate local contracts as well as already agreed upon host-nation support. This coordination among participating nations will reduce the competition for local assets that could otherwise have detrimental effects on one or more participating nations. Additionally, many nations providing forces for a coalition effort may not be able to support them logistically and will rely almost entirely on the United States or another coalition army for logistic support. Finally, pooled resources such as transportation assets and maintenance personnel could potentially reduce deployment requirements for one or more coalition partners. As the coalition matures, the role of this office or center may be expanded to include command activities.

This page intentionally left blank.

Source Notes

These are sources quoted or paraphrased in this publication. They are listed by page number. Quotations are identified by the first few word of the quote. Where a quote is embedded within a paragraph, the paragraph number is listed. Boldface indicates the title of historical vignettes.

Chapter 1–The Urban Outlook

1-1 "Today's security environment demands more...": FM 1, *The Army*, 14 June 2005: 1-19.

1-4 **Rome: A Microcosm of Urban Warfare**: R. Ernest Dupuy and Trevor N. Dupuy, *The Encyclopedia of Military History: From 3500 B.C. to the Present* (San Francisco: Harper & Row, 1977).

1-6 **The Three Block War**: as quoted in Russell W. Glenn, Steven Hartman, and Scott Gerwehr, *Urban Combat Service Support Operations: The Shoulders of Atlas* (Santa Monica, CA: RAND, 2003): 26.

1-8 **Winning the Peace in Iraq: The Requirement for Full Spectrum Operations**: Peter W. Chiarelli and Patrick R. Michaelis, "Winning the Peace: The Requirement for Full-Spectrum Operations," *Military Review* (July/August 2005): 4.

Chapter 2–Understanding the Urban Environment

2-1 "From a planning perspective, commanders view cities....": FM 3-0, *Operations*, 14 June 2001: 6-19.

2-16 "Me and Somalia Against the World": cited in Andrew S. Natsios, "Food Through Force: Humanitarian Intervention and U.S. Policy," *The Washington Quarterly* (winter 1994): 136.

2-18 **Understanding the Effects of Unit and Soldier Actions in Iraq**: *Initial Impressions Report: Operation Iraqi Freedom—Stability Operations-Support Operations,* Center for Army Lessons Learned (CALL) Newsletter No. 04-13 (Fort Leavenworth, KS: Center for Army Lessons Learned, U.S. Training and Doctrine Command [TRADOC], 2004): ii.

2-21 "Whoever coined the phrase... ": General Sir Rupert Smith, "Wars in Our Time–A Survey of Recent and Continuing Conflicts," *World Defence Systems*, volume 3:2 (London: Royal United Services Institute for Defence Studies, 2001): 121.

2-25 **Understanding the Urban Environment: Paris – 1944**: Russell W. Glenn, "Urban Combat is Complex," *Proceedings* (February 2002): 65. Russell W. Glenn, Steven Hartman, and Scott Gerwehr, *Urban Combat Service Support Operations: The Shoulders of Atlas* (Santa Monica, CA: RAND, 2003): 10-11.

Chapter 3–Understanding the Urban Threat

3-1 "...[T]he United States could be forced to intervene...": *Quadrennial Defense Review Report*, 30 September 2001: 6.

3-3 "...Iraq made no direct effort...": Gregory Fontenot, E. J. Degen, and David Tohn, *On Point: US Army in Operation IRAQI FREEDOM* (Fort Leavenworth, KS: Combat Studies Institute Press, 2004): 388.

3-4 **Tempo**: 26th Infantry Regimental Association, *Aachen. Military Operations in Urban Terrain* (Lititz, PA: 26th Infantry Regimental Association, 1999); Irving Werstein, *The Battle of Aachen* (New York: Thomas Y. Crowell Company, 1962); Charles Whiting, *Bloody Aachen* (New York: Stein and Day, 1976).

3-6 "Chechen fighters sometimes disguised...": Olga Oliker, *Russia's Chechen Wars 1994–2000: Lessons from Urban Combat* (Santa Monica, CA: RAND, 2001): 21.

3-7 **Identifying Threats from Noncombatants**: Olga Oliker, *Russia's Chechen Wars 1994–2000: Lessons from Urban Combat* (Santa Monica, CA: RAND, 2001).

3-9 **Information and the Media**: Eric Hammel, *Fire in the Streets: The Battle for Hue, Tet 1968* (New York: Dell Publishing, 1991).

3-12 **Cultural and Religious Instability**: Sid Heal, "Crowds, Mobs and Nonlethal Weapons," *Military Review* (March/April 2000): 45–50.

3-13 **Food and Water Shortages**: Leif R. Rosenberger, "The Strategic Importance of the World Food Supply," *Parameters* (spring 1997): 84–105.

3-15 **Insurgencies and the Urban Society**: Roger Trinquier, *Modern Warfare: A French View of Counterinsurgency* (New York: Praeger Publisher, 1964): 16-17.

3-16 **Crime and Criminal Organizations**: Eugene Linden, "The Exploding Cities of the Developing World," *Foreign Affairs* (January/February 1996): 52–65; Andrew S. Natsios, "Commander's Guidance: A Challenge of Complex Humanitarian Emergencies," *Parameters* (summer 1996): 50–66.

Chapter 4–Understanding the Urban Environment's Effects on Warfighting Functions and Tactics

4-1 "War is, above all things, an art...": Francis V. Greene, "The Important Improvements in the Art of War During the Past Twenty Years and Their Probable Effect on Future Military Operations," *Journal of the Military Service Institution of the United States* 4, no. 13 (1883): 41.

4-13 "Fighting in a city is much...": Vasili I. Chuikov, *The Battle for Stalingrad* (New York: Holt, Rinehart, and Winston, 1964): as cited by JP 3-06, *Doctrine for Urban Operations*, 16 September 2002: 1-8.

4-14 "I heard small-arms fire and...": Mark A. B. Hollis, "Platoon Under Fire," *Infantry* (January/April 1998): 29–30.

4-17 **Example of Simple Communications Innovation: Israel's Six-Day War – 1967**: Trevor N. Dupuy and Paul Martell, *Flawed Victory: The Arab-Israeli Conflict and the 1982 War in Lebanon* (Fairfax, VA: Hero Books, 1986); Richard A. Gabriel, *Operation Peace for Galilee: The Israeli-PLO War in Lebanon* (New York: Hill and Wang, 1984).

Chapter 5–Contemplating Urban Operations

5-1 "We based all our further calculations...": S. M. Shtemenko, *The Soviet General Staff at War, 1941-1945: Book One* (Moscow: Progress Publishers, 1981): 317–318.

5-19 "Four hostile newspapers...": Napoleon Bonaparte, Justin Wintle (ed.), *The Dictionary of War Quotations* (New York: The Free Press, 1989): 73.

5-24 **Conventional and Special Forces Integration**: Gregory Fontenot, E. J. Degen, and David Tohn, *On Point: US Army in Operation IRAQI FREEDOM* (Fort Leavenworth, KS: Combat Studies Institute Press, 2004): 403.

Chapter 6–Foundations for Urban Operations

6-1 "Utilities such as electricity…": Richard Connaughton, John Pimlott, and Duncan Anderson, *The Battle for Manila: The Most Devastating Untold Story of World War II* (Novata, CA: Presidio Press, 1995): 103.

6-7 **Applying the Urban Operational Framework: Panama – 1989**: Ronald H. Cole, *Operation Just Cause* (Washington DC: Joint History Office, 1995); Thomas Donnelly, Margaret Roth, and Caleb Baker, *Operation Just Cause: The Storming of Panama* (New York: Lexington Books, 1991); John T. Fishel, *The Fog of Peace: Planning and Executing the Restoration of Panama* (Carlisle, PA: Strategic Studies Institute, 1992); Malcolm McConnell, *Just Cause: The Real Story of America's High-Tech Invasion of Panama* (New York: St. Martin's Press, 1991); John Embry Parkerson, Jr., "United States Compliance with Humanitarian Law Respecting Civilians During Operation Just Cause," *Military Law Review* (1991): 31–140; Jennifer Morrison Taw, *Operation Just Cause: Lessons for Operations Other Than War* (Santa Monica, CA: RAND, 1996).

6-12 "I'm talking about attacking...": William S. Wallace quoted in Michael R. Gordon, "Baghdad Targets Picked if Hussein Holes Up There," *New York Times* (7 March 2003): A11.

6-13 "Do not try to..." T. E. Lawrence, "Twenty-Seven Articles," *The Arab Bulletin* (20 August 1917): as cited in David H. Petraeus, "Learning Counterinsurgency: Observations from Soldiering in Iraq," *Military Review* (January/February 2006): 4–5.

Chapter 7–Urban Offensive Operations

7-1 "…Capture Suez City...": Abraham Adan, *On the Banks of the Suez* (Novato, CA: Presidio Press, 1980): 409.

7-2 **Operational Context of Urban Operations: Brittany Ports – 1944**: Kent H. Butts, "The Strategic Importance of Water," *Parameters* (spring 1997): 65–83; Paul Carell, *Stalingrad: The Defeat of the German 6th Army* (Atglen, PA: Schieffer Publishing, 1993); R. Ernest Dupuy and Trevor N. Dupuy, *The Encyclopedia of Military History: From 3500 B.C. to the Present* (San Francisco: Harper & Row, 1977); Ralph Peters, "Our Soldiers, Their Cities," *Parameters* (spring 1996): 43–50; John J. Peterson, *Into the Cauldron* (Clinton, MD: Clavier House, 1973).

7-10 **Forms of Attack in the Urban Offense: Metz – 1944**: John Colby, *War From the Ground Up: The 90th Division in WWII* (Austin, TX: Nortex Press, 1991); Anthony Kemp, *The Unknown Battle: Metz, 1944* (New York: Stein and Day, 1981).

7-17 **Isolating an Urban Area: Hue, Vietnam – 1968**: James R. Arnold, *Tet Offensive 1968: Turning Point in Vietnam* (London: Osprey, 1990); Eric Hammel, *Fire in the Streets: The Battle for Hue, Tet 1968* (New York: Dell Publishing, 1991).

7-21 "Often what seems to be...": Russell W. Glenn, *An Attack on Duffer's Downtown* (Santa Monica, CA: RAND, 2001): 28.

7-22 **Creative Task Organization: Using Artillery in the Direct Fire Role**: 26th Infantry Regimental Association, *Aachen. Military Operations in Urban Terrain* (Lititz, PA: 26th Infantry Regimental Association, 1999); Ronald H. Cole, *Operation Just Cause* (Washington DC: Joint History Office, 1995); Thomas Donnelly, Margaret Roth, and Caleb Baker, *Operation Just Cause: The Storming of Panama* (New York: Lexington Books, 1991); David Eshel, *Mid-East Wars: The Lebanon War 1982* (Hod Hasharon, Israel: Eshel-Dramit, 1983); Richard A. Gabriel, *Operation Peace for Galilee: The Israeli-PLO War in Lebanon* (New York: Hill and Wang, 1984); Malcolm McConnell, *Just Cause: The Real Story of America's High-Tech Invasion of Panama* (New York: St. Martin's Press, 1991); Irving Werstein, *The Battle of Aachen* (New York: Thomas Y. Crowell Company, 1962); Charles Whiting, *Bloody Aachen* (New York: Stein and Day, 1976); Bruce Allen Watson, *Sieges: A Comparative Study* (Westport, CT: Praeger, 1993).

7-26 **Bold Operational Maneuver to Seize an Urban Area: Inchon and Seoul, Korea – 1950**: Robert Debs Heinl, Jr., *Victory at High Tide: The Inchon-Seoul Campaign* (Annapolis, MD: Nautical & Aviation Publishing Company of America, 1979); Douglas MacArthur, *Reminiscences* (New York: McGraw-Hill, 1964); Shelby L. Stanton, *Ten Corps in Korea, 1950* (Novato, CA: Presidio Press, 1989).

Chapter 8–Urban Defensive Operations

8-1 "Generally, a modern city magnifies...": Christopher R. Gabel, "Military Operations on Urbanized Terrain: The 2d Battalion, 26th Infantry, at Aachen, October 1944," *Urban Combat Operations: tactics, Techniques and Procedure.* Center for Army Lessons Learned (CALL) Newsletter No. 99-16 (Fort Leavenworth, KS: Center for Army Lessons Learned, U.S. Training and Doctrine Command [TRADOC], November 1999): 1-9.

8-5 **Urban Defense in a Major Operation: Stalingrad –1942 to 1943**: Anthony Beevor, *Stalingrad* (New York: Penguin Books, 1998); Vasili I. Chuikov, *The Battle for Stalingrad* (New York: Holt, Rinehart, and Winston, 1964); William Craig, *Enemy at the Gates: The Battle for Stalingrad* (New York: Reader's Digest Press, 1973).

8-8 "What is the position...": Michael Dewar, *War in the Streets: The Story of Urban Combat from Calais to Khafji* (New York: Sterling Publishers, 1992): 8.

8-10 **Defensive Combat Power: Suez – 1973**: Abraham Adan, *On the Banks of the Suez* (Novato, CA: Presidio Press, 1980).

Chapter 9–Urban Stability and Civil Support Operations

9-1 "The Rangers were bound...": Mark Bowden, *Black Hawk Down: A Story of Modern War* (New York: Atlantic Monthly Press, 1999): 18.

9-5 "In wars of intervention...": Antoine-Henri Jomini, "The Art of War," in *Roots of Strategy: Book 2* (Harrisburg, PA: Stackpole Books, 1987): 441.

9-8 **Assessment of Security and Force Protection: Belfast, Northern Ireland**: Michael Dewar, *British Army in Northern Ireland* (New York: Sterling Publishers, 1985); Desmond Hamill, *Pig in the Middle: The Army in Northern Ireland, 1969-1984* (London: Methuen, 1985).

9-11 "If there is any lesson...": Frederick Funston, "How the Army Worked to Save San Francisco," *Cosmopolitan Magazine* (July 1906). Available at http: //www.sfmuseum.org/1906/cosmo.html.

9-12 **Civil Support and Coordination with Civilian Authorities: Los Angeles – 1992**: James D. Delk, *Fires & Furies: The LA Riots* (Palm Springs, CA: Etc. Publications, 1995); William W. Mendel, *Combat in Cities: The LA Riots and Operation Rio* (Fort Leavenworth, KS: Foreign Military Studies Office, 1996).

9-14 **Restraint: An Najaf, Iraq – 2003**: Gregory Fontenot, E. J. Degen, and David Tohn, *On Point: US Army in Operation IRAQI FREEDOM* (Fort Leavenworth, KS: Combat Studies Institute Press, 2004).

9-16 "I met immediately with the...": Jeffrey B. Van Sickle, "Stability Operations in Northern Iraq: Task Force Altun Kupri," *Infantry* (January/February 2005): 26.

Chapter 10–Urban Sustainment

10-1 "Even supply is different…" Ralph Peters, "Our Soldiers, Their Cities," *Parameters* (spring 1996): 48.

10-2 **Base Security: Tan Son Nhut, Vietnam – 1968**: James R. Arnold, *Tet Offensive 1968: Turning Point in Vietnam* (London: Osprey, 1990); Keith William Nolan, *The Battle for Saigon: Tet 1968* (New York: Pocket Books, 1996).

10-4 "[Sustainment] planning accounts for...": FM 3-0, *Operations*, 14 June 2001: 6-19.

10-8 "Greater friendly force density would appear...": Russell W. Glenn, Steven Hartman, and Scott Gerwehr, *Urban Combat Service Support Operations: The Shoulders of Atlas* (Santa Monica, CA: RAND, 2003): 4.

10-13 **Host Nation Repair Operations: OPERATION IRAQI FREEDOM**: Steven Leonard, personal experience written specifically for FM 3-06.

10-18 **Combat Stress: Chechnya – 1994 to 1996**: Timothy L. Thomas and Charles P. O'Hara, "Combat Stress in Chechnya: 'The Equal Opportunity Disorder'," *Army Medical Department Journal* (January/March 2000): 46–53.

10-22 "…[I]n an endeavor like that in Iraq…": David H. Petraeus, "Learning Counterinsurgency: Observations from Soldiering in Iraq," *Military Review* (January/February 2006): 4–5.

Appendix A–Siege of Beirut: An Illustration of the Fundamentals of Urban Operations

A-1 **Siege of Beirut: An Illustration of the Fundamentals of Urban Operations**: Trevor N. Dupuy and Paul Martell, *Flawed Victory: The Arab-Israeli Conflict and the 1982 War in Lebanon* (Fairfax, VA: Hero Books, 1986); David Eshel, *Mid-East Wars: The Lebanon War 1982* (Hod Hasharon, Israel: Eshel-Dramit, 1983); Richard A. Gabriel, *Operation Peace for Galilee: The Israeli-PLO War in Lebanon* (New York: Hill and Wang, 1984).

A-1 "The [Israeli Defense Forces] IDF had...": Richard A. Gabriel, *Operation Peace for Galilee: The Israeli-PLO War in Lebanon* (New York: Hill and Wang, 1984): 128.

Appendix B–An Urban Focus to the Intelligence Preparation of the Battlefield

B-1 "Maneuvers that are possible...": *Infantry in Battle, 2d ed.* (Richmond, VA: Garret & Massie, 1939; Reprint, Fort Leavenworth, KS: U.S. Army Command and General Staff College, 1981): 69.

B-5 **Analysis of an Urban Area's Underlying Terrain: Mitrovica, Kosovo – 1999**: Jamison Jo Medby and Russell W. Glenn, *Street Smart: Intelligence Preparation of the Battlefield for Urban Operations* (Santa Monica, CA: RAND, 2002).

B-10 "…[T]he adversaries of freedom…": John F. Kennedy, Message to Congress, 25 May 1961. Available at <http://www.jfklibrary.org/Historical+Resources/Archives>.

B-12 **Shifting Civilian Interests and Intent**: James D. Delk, *Fires & Furies: The LA Riots* (Palm Springs, CA: Etc. Publications, 1995); Jamison Jo Medby and Russell W. Glenn, *Street Smart: Intelligence Preparation of the Battlefield for Urban Operations* (Santa Monica, CA: RAND, 2002).

Appendix C–Operations in Somalia: Applying the Urban Operational Framework to Stability Operations

C-1 **Operations in Somalia: Applying the Urban Operational Framework to Stability and Reconstruction Operations**: Kenneth Allard, *Somalia Operations: Lessons Learned* (Washington, DC: National Defense University Press, 1995).

C-1 "It's impossible for an American mother...": Thomas M. Montgomery, interview, "Ambush in Mogasdishu," *Frontline*, WGBH Boston, 29 September 1998, <http://www.pbs.org/wgbh/pages/frontline/programs/1998.html>.

C-2 C-5 …" to use all necessary…": *United Nations Security Council Resolution 794* (1992) (New York: United Nations, 1992), <http://daccessdds.un.org/doc/UNDOC/GEN/N92/772/11/PDF>.

C-2 C-5 "When directed by the…": CENTCOM mission statement as cited in Kenneth Allard, *Somalia Operations: Lessons Learned* (Washington, DC: National Defense University Press, 1995): 13-14.

Appendix D–Joint and Multinational Urban Operations

D-1 "[Joint force commanders] synchronize the actions…" JP 3-0, *Doctrine for Joint Operations*, 10 September 2001: II-4.

This page intentionally left blank.

Glossary

SECTION I – ACRONYMS AND ABBREVIATIONS

AI	air interdiction
AIDS	acquired immune deficiency syndrome
AMC	Air Mobility Command
AO	area of operations
ARVN	Army of the Republic of Viet Nam
ASCOPE	areas, structures, capabilities, organizations, people, and events
BCT	brigade combat team
BDAR	battle damage assessment and repair
BENELUX	Belgium, Netherlands, and Luxembourg
C2	command and control
CA	civil affairs
CAA	combat aviation advisory
CASEVAC	casualty evacuation
CBRN	chemical, biological, radiological, nuclear
CBRNE	chemical, biological, radiological, nuclear, high-yield explosive
CCIR	commander's critical information requirement
CI	counterintelligence
CEA	captured enemy ammunition
CERP	commander's emergency response program
CLS	combat lifesaver
CMO	civil-military operations
CMOC	civil-military operations center
CNA	computer network attack
CND	computer network defense
CNE	computer network exploitation
CNO	computer network operations
COA	course of action
COG	center of gravity
CONUS	continental United States
COP	common operational picture
DA	direct action
DNBI	disease and nonbattle injury
DOTMLPF	doctrine, organization, training, materiel, leadership and education, personnel, and facilities
EAC	echelons above corps
EEFI	essential elements of friendly information

EOD	explosive ordnance disposal
EW	electronic warfare
FHP	force health protection
FID	foreign internal defense
FM	field manual; frequency modulation
FMI	field manual, interim
FSB	forward support battalion
G-9	assistant chief of staff, civil affairs
GIS	geographic information system
GTL	gun-target line
HRS	human resources support
HUMINT	human intelligence
IDF	Israeli Defense Forces
IED	improvised explosive device
IMINT	imagery intelligence
INFOSYS	information system
IO	information operations
IPB	intelligence preparation of the battlefield
ISR	intelligence, surveillance, and reconnaissance
JDAM	Joint Direct Attack Munition
JFC	joint force commander
JOA	joint operations area
JP	joint publication
JTF	joint task force
JUO	joint urban operations
LAN	local area network
LNO	liaison officer
LOC	line of communications
LOGCAP	logistics civilian augmentation program
LOS	line of sight
LPT	logistics preparation of the theater
MAGTF	Marine air-ground task force
MANPADS	man-portable air defense system
MEDEVAC	medical evacuation
METT-TC	mission, enemy, terrain and weather, troops and support available, time available, civil considerations
MEU	Marine expeditionary unit
MEU(SOC)	Marine expeditionary unit (special operations capable)
MHE	materials handling equipment
MLRS	Multiple Launch Rocket System
MP	military police

MSC	Military Sealift Command
MST	maintenance support team
NATO	North Atlantic Treaty Organization
NCW	naval coastal warfare
NGA	National Geospatial-Intelligence Agency
NGO	nongovernmental organization
NKPA	North Korean People's Army
NVA	North Vietnamese Army
OAKOC	observation and fields of fire, avenues of approach, key terrain, obstacles, and cover and concealment
ODA	operational detachment alpha
OIF	Operation Iraqi Freedom
OPSEC	operations security
PA	public affairs
PDF	Panamanian Defense Force
PGM	precision-guided munitions
PLO	Palestinian Liberation Organization
PMCS	preventive maintenance checks and services
POW	prisoner of war
PSYOP	psychological operations
QRF	quick reaction force
ROE	rules of engagement
RPG	rocket-propelled grenade
SBU	special boat unit
SDDC	Surface Deployment and Distribution Command
SEAL	sea-air-land (team)
SF	special forces
SIGINT	signals intelligence
SJA	Staff Judge Advocate
SO	special operations
SOF	special operations forces
SOFA	status-of-forces agreement
SOP	standing operating procedure
SOWT	special operations weather team
SPOD	seaport of debarkation
SPE	seaport of embarkation
SR	special reconnaissance
STT	special tactics team
SWET	sewer, water, electricity, trash
SWEAT	sewer, water, electricity, academics trash
SWEAT-MS	sewer, water, electricity, academics trash-medical, security

TACON	tactical control
TC	training circular
TCC	transportation component command
TF	task force
TIM	toxic industrial material
TTP	tactics, techniques, and procedures
UAS	unmanned aircraft system
UN	United Nations
UNITAF	Unified Task Force
UNOSOM	United Nations Operations in Somalia
UNSCR	UN Security Council resolution
UO	urban operations
USACE	United States Army Corps of Engineers
USAID	United States Agency for International Development
USCENTCOM	United States Central Command
USFORSOM	United States Forces, Somalia
USCG	United States Coast Guard
USTRANSCOM	United States Transportation Command
UXO	unexploded ordnance
UW	unconventional warfare
VC	Viet Cong
WMD	weapons of mass destruction

SECTION II – Terms and Definitions

agility – The ability to move and adjust quickly and easily. (FM 3-0)

air interdiction – Air operations conducted to destroy, neutralize, or delay the enemy's military potential before it can be brought to bear effectively against friendly forces at such distance from friendly forces that detailed integration of each air mission with the fire and movement of friendly forces is not required. (JP 1-02)

area defense – A type of defensive operation that concentrates on denying enemy forces access to designated terrain for a specific time rather than destroying the enemy outright. (FM 3-0)

area of interest – That area of concern to the commander, including the area of influence, areas adjacent thereto, and extending into enemy territory to the objectives of current or planned operations. This area also includes areas occupied by enemy forces who could jeopardize the accomplishment of the mission. (JP 1-02)

area of operations – An operational area defined by the joint force commander for land and naval forces. Areas of operations do not typically encompass the entire operational area of the joint force commander, but should be large enough for component commanders to accomplish their missions and protect their forces. (JP 1-02)

ARFOR – The senior Army headquarters and all Army forces assigned or attached to a combatant command, subordinate joint force command, joint functional command, or multinational command. (FM 3-0)

assessment – (Army) The continuous monitoring of the current situation and progress of an operation.(FMI 5-0.1)

asymmetry – Dissimilarities in organization, equipment, doctrine, and values between other armed forces (formally organized or not) and US forces. Engagements are symmetric if forces, technologies, and weapons are similar; they are asymmetric if forces, technologies, and weapons are different, or if a resort to terrorism and rejection of more conventional rules of engagement are the norm. (FM 3-0)

attack – An offensive operation that destroys or defeats enemy forces, seizes and secures terrain, or both. (FM 3-0)

attainability - One of the eight characteristics of combat service support: generating the minimum essential supplies and services necessary to begin operations. (FM 4-0)

attrition - The reduction of the effectiveness of a force caused by loss of personnel and materiel. (JP 1-02)

avenue of approach – An air or ground route of an attacking force of a given size leading to its objective or to key terrain in its path. (JP 1-02)

axis of advance – (Army) The general area through which the bulk of a unit's combat power must move. (FM 3-90)

battle – A set of related engagements that lasts longer and involves larger forces than an engagement. (FM 3-0)

battlefield organization – The allocation of forces in the area of operations by purpose. It consists of three all-encompassing categories of operations: decisive, shaping, and sustaining. (FM 3-0)

battle rhythm – The sequencing of command and control activities within a headquarters and throughout the force to facilitate effective command and control. (FMI 5-0.1)

branch – (Army) A contingency plan or course of action (an option built into the basic plan or course of action) for changing the mission, disposition, orientation, or direction of movement of the force to aid success of the current operation, based on anticipated events, opportunities, or disruptions caused by enemy actions. Army forces prepare branches to exploit success and opportunities, or to counter disruptions caused by enemy actions. (FM 3-0)

breakout – An operation conducted by an encircled force to regain freedom of movement or contact with friendly units. It differs from other attacks only in that a simultaneous defense in other areas of the perimeter must be maintained. (FM 3-90)

buffer zone – A defined area controlled by a peace operations force from which disputing or belligerent forces have been excluded. A buffer zone is formed to create an area of separation between disputing or belligerent forces and reduce the risk of renewed conflict. Also called **area of separation** in some United Nations operations. (JP 1-02)

bypass – A tactical mission task in which the commander directs his unit to maneuver around an obstacle, position, or enemy force to maintain the momentum of the operation while deliberately avoiding combat with an enemy force. (FM 3-90)

campaign – A series of related military operations aimed at accomplishing a strategic or operational objective within a given time and space. (JP 1-02)

canalize – (Army) A tactical mission task in which the commander restricts enemy movement to a narrow zone by exploiting terrain coupled with the use of obstacles, fires, or friendly maneuver. (FM 3-90)

capability - The ability to execute a specified course of action (a capability may or may not be accompanied by an intention) (JP 1-02)

casualty evacuation – (Army) A term used by nonmedical units to refer to the movement of casualties aboard nonmedical vehicles or aircraft. (FM 8-10-6)

centers of gravity – Those characteristics, capabilities, or sources of power from which a military force derives its freedom of action, physical strength, or will to fight. (JP 1-02)

civil affairs – Designated Active and Reserve component forces and units organized, trained, and equipped specifically to conduct civil affairs activities and to support civil-military operations. (JP 1-02)

civil considerations – The influence of manmade infrastructure, civilian institutions, and attitudes and activities of the civilian leaders, populations, and organizations within an area of operations on the conduct of military operations. (FM 6-0)

civil disturbance – Group acts of violence and disorder prejudicial to public law and order. (JP 1-02)

civil-military operations – The activities of a commander that establish, maintain, influence, or exploit relations between military forces, governmental and nongovernmental civilian organizations and authorities, and the civilian populace in a friendly, neutral, or hostile operational area in order to facilitate military operations, to consolidate and achieve US objectives. Civil-military operations may include performance by military forces of activities and functions normally the responsibility of the local, regional, or national government. These activities may occur prior to, during, or subsequent to other military actions. They may also occur, if directed, in the absence of other military operations. Civil-military operations may be performed by designated civil affairs, by other military forces, or by a combination of civil affairs and other forces. (JP 1-02)

civil-military operations center – An ad hoc organization, normally established by the geographic combatant commander or subordinate joint force commander, to assist in the coordination of activities of engaged military forces, and other United States Government agencies, nongovernmental organizations, and regional and international organizations. There is no established structure, and its size and composition are situation dependent. (JP 1-02)

civil support operations – Domestic operations that address the consequences of man-made or natural accidents and incidents beyond the capabilities of civilian authorities. (FM 1)

clear – (Army) 1. A tactical mission task that requires the commander to remove all enemy forces and eliminate organized resistance within an assigned area. (FM 3-90) 2. The total elimination or neutralization of an obstacle that is usually performed by follow-on engineers and is not done under fire. (FM 3-34.2)

close air support – Air action by fixed- and rotary-wing aircraft against hostile targets that are in close proximity to friendly forces and that require detailed integration of each air mission with the fire and movement of those forces. (JP 1-02)

close combat – Combat carried out with direct fire weapons, supported by indirect fire, air-delivered fires, and nonlethal engagement means. Close combat defeats or destroys enemy forces or seizes and retains ground. (FM 3-0)

coalition – An ad hoc arrangement between two or more nations for common action. (JP 1-02)

collateral damage – Unintentional or incidental injury or damage to persons or objects that would not be lawful military targets in the circumstances ruling at the time. Such damage is not unlawful so long as it is not excessive in light of the overall military advantage anticipated from the attack. (JP 1-02)

combatant command – A unified or specified command with a broad continuing mission under a single commander established and so designated by the President, through the Secretary of Defense and with the advice and assistance of the Chairman of the Joint Chiefs of Staff. Combatant commands typically have geographic or functional responsibilities. (JP 1-02)

combat power – The total means of destructive and/or disruptive force which a military unit/formation can apply against the opponent at a given time. (JP 1-02)

combined arms – The synchronized or simultaneous application of several arms—such as infantry, armor, field artillery, engineers, air defense, and aviation—to achieve an effect on the enemy that is greater than if each arm were used against the enemy separately or in sequence. (FM 3-0)

combined arms team – (Army) Two or more arms mutually supporting one another, usually consisting of infantry, armor, cavalry, aviation, field artillery, air defense artillery, and engineers. (FM 3-90).

command and control system – (Army) The arrangement of personnel, information management, procedures, and equipment and facilities essential for the commander to conduct operations. (FM 6-0)

commander's critical information requirements – (Army) – Elements of information required by commanders that directly affect decisionmaking and dictate the successful execution of military operations. (FM 3-0)

commander's intent – (Army) A clear, concise statement of what the force must do and the conditions the force must meet to succeed with respect to the enemy, terrain, and civil considerations that represent the operation's desired end state. (FMI 5-0.1)

commander's visualization – The mental process of developing situational understanding, determining a desired end state, and envisioning how the force will achieve that end state. (FMI 5-0.1)

common operational picture – (Army) An operational picture tailored to the user's requirements, based on common data and information shared by more than one command. (FM 3-0)

computer network attack – Actions taken through the use of computer networks to disrupt, deny, degrade, or destroy information resident in computers or computer networks, or the computers and networks themselves. (JP 1-02)

computer network defense – Actions taken through computer networks to protect, monitor, analyze, detect and respond to unauthorized activity within Department of Defense information systems and computer networks. (JP 1-02)

concept of operations – (Army) How commanders see the actions of subordinate units fitting together to accomplish the mission. As a minimum, the description includes the scheme of maneuver and concept of fires. The concept of operations expands the commander's selected course of action and expresses how each element of the force will cooperate to accomplish the mission. (FM 3-0)

constraint – A restriction placed on the command by a higher command. A constraint dictates an action or inaction, thus restricting the freedom of action a subordinate commander has for planning. (FM 5-0)

control – (Army) 1. The regulation of forces and warfighting functions to accomplish the mission in accordance with the commander's intent. (FMI 5-0.1) 2. A tactical mission task that requires the commander to maintain physical influence over a specified area to prevent its use by an enemy. (FM 3-90)

controlled exchange – The removal of serviceable parts, components, or assemblies from unserviceable, economically reparable equipment and their immediate reuse in restoring a like item of equipment to a combat operable or serviceable condition. (FM 4-30.3)

control measure – A means of regulating forces or warfighting functions. (FMI 5-0.1)

counterattack – (Army) A form of attack by part or all of a defending force against an enemy attacking force with the general objective of denying the enemy his goal in attacking. See FM 3-0.

counterdeception – Efforts to negate, neutralize, diminish the effects of, or gain advantage from a foreign deception operation. Counterdeception does not include the intelligence function of identifying foreign deception operations. (JP 1-02)

counterdrug – Those active measures taken to detect, monitor, and counter the production, trafficking, and use of illegal drugs. (JP 1-02)

counterinsurgency – Those military, paramilitary, political, economic, psychological, and civic actions taken by a government to defeat insurgency. (JP 1-02)

counterintelligence – Information gathered and activities conducted to protect against espionage, other intelligence activities, sabotage, or assassinations conducted by or on behalf of foreign governments or elements thereof, foreign organizations, or foreign persons, or international terrorist activities. (JP 1-02)

countermobility operations – The construction of obstacles and emplacement of minefields to delay, disrupt, and destroy the enemy by reinforcement of the terrain. The primary purpose of countermobility operations is to slow or divert the enemy, to increase time for target acquisition, and to increase weapon effectiveness. (JP 1-02)

counterpropaganda – Programs of products and actions designed to nullify propaganda or mitigate its effects. (FM 3-05.30)

counterterrorism – Operations that include the offensive measures taken to prevent, deter, preempt, and respond to terrorism. (JP 1-02)

course of action – 1. Any sequence of activities that an individual or a unit may follow. 2. A possible plan open to an individual or a commander that would accomplish or is related to the accomplishment of a mission. 3. The scheme adopted to accomplish a job or mission. 4. A line of conduct in an engagement. 5. A product of the Joint Operation Planning and Execution System concept development phase. (JP 1-02)

cover – (Army) 1. Protection from the effects of fires. (FM 6-0) 2. A form of security operation whose primary task is to protect the main body by fighting to gain time while also observing and reporting information and preventing enemy ground observation of and direct fire against the main body. Unlike a screening or guard force, the covering force is a self-contained force capable of operating independently of the main body. (FM 3-90)

dead space – 1. An area within the maximum range of a weapon, radar, or observer, which cannot be covered by fire or observation from a particular position because of intervening obstacles, the nature of the ground, the characteristics of the trajectory, or the limitations of the pointing capabilities of the weapon. 2. An area or zone which is within range of a radio transmitter, but in which a signal is not received. 3. The volume of space above and around a gun or guided missile system into which it cannot fire because of mechanical or electronic limitations. (JP 1-02)

debarkation – The unloading of troops, equipment, or supplies from a ship or aircraft. (JP 1-02)

decision point – (Army) An event, area, or point in the battlespace where and when the friendly commander will make a critical decision. (FM 5-0)

decision support matrix – An aid used by the commander and staff to make battlefield decisions. This matrix is a staff product of the wargaming process that lists the decision point, location of the decision point, the criteria to be evaluated at the point of decision, the action or operations to occur at the decision point, and the unit or element that is to act and has responsibility to observe and report the information affecting the criteria for the decision. (FM 5-0)

decision support template – (Army) A staff product initially used in the wargaming process that graphically represents the decision points and projected situations, and indicates when, where, and under what conditions a decision is most likely to be required to initiate a specific activity or event. (FM 5-0)

decisive engagement – In land and naval warfare, an engagement in which a unit is considered fully committed and cannot maneuver or extricate itself. In the absence of outside assistance, the action must be fought to a conclusion and either won or lost with the forces at hand. (JP 1-02)

decisive operation – The operation that directly accomplishes the task assigned by the higher headquarters. Decisive operations conclusively determine the outcome of major operations, battles, and engagements. (FM 3-0)

decisive point – A geographic place, specific key event, critical system or function that allows commanders to gain a marked advantage over an enemy and greatly influence the outcome of an attack. (JP 1-02)

defeat – A tactical mission task that occurs when an enemy force has temporarily or permanently lost the physical means or the will to fight. The defeated force's commander is unwilling or unable to pursue his adopted course of action, thereby yielding to the friendly commander's will, and can no longer interfere to a significant degree with the actions of friendly forces. Defeat can result from the use of force or the threat of its use. (FM 3-90)

defeat in detail – Concentrating overwhelming combat power against separate parts of a force in sequence rather than defeating the entire force at once. (FM 3-90)

defense in depth – The siting of mutually supporting defense positions designed to absorb and progressively weaken attack, prevent initial observations of the whole position by the enemy, and to allow the commander to maneuver the reserve. (JP 1-02)

defensive information operations – The integration and coordination of policies and procedures, operations, personnel, and technology to protect and defend information and information systems. Defensive information operations ensure timely, accurate, and relevant information access while denying adversaries the opportunity to exploit friendly information and information systems for their own purposes. (FM 3-0)

defensive operations – Operations that defeat an enemy attack, buy time, economize forces, or develop conditions favorable for offensive operations. Defensive operations alone normally cannot achieve a decision. Their purpose is to create conditions for a counteroffensive that allows Army forces to regain the initiative. (FM 3-0)

delay – A form of retrograde in which a force under pressure trades space for time by slowing the enemy's momentum and inflicting maximum damage on the enemy without, in principle, becoming decisively engaged. (FM 3-90)

depth – The extension of operations in time, space, and resources. (FM 3-0)

destroy – 1. A tactical mission task that physically renders an enemy force combat-ineffective until it is reconstituted. 2. To damage a combat system so badly that it cannot perform any function or be restored to a usable condition without being entirely rebuilt. (FM 3-90)

direct action – Short-duration strikes and other small-scale offensive actions conducted as a special operation in hostile, denied, or other politically sensitive environments and which employ specialized military capabilities to seize, destroy, capture, exploit, recover, or damage designated targets. Direct action differs from conventional offensive actions in the level of physical and political risk, operational techniques, and the degree of discriminate and precise use of force to achieve specific objectives. (JP 1-02)

direct approach – The application of combat power directly against the enemy center of gravity or the enemy's principal strength. (FM 3-90)

direct fire – Fire delivered on a target using the target itself as a point of aim for either the weapon or the director. (JP 1-02)

disinformation – Information disseminated primarily by intelligence organizations or other covert agencies designed to distort information or deceive or influence United States decisionmakers, United States forces, coalition allies, key actors, or individuals via indirect or unconventional means. (FM 3-13)

displaced person – (Army) A civilian who is involuntarily outside the national boundaries of his or her country or as an internally displaced person is a civilian involuntarily outside his area or region within his country. (FM 2-0)

display – (Army) An information management activity: to represent relevant information in a usable, easily understood audio or visual form tailored to the needs of the user that conveys the common operational picture for decisionmaking and exercising command and control functions (FM 6-0)

disposition – 1. Distribution of the elements of a command within an area, usually the exact location of each unit headquarters and the deployment of forces subordinate to it. (JP 1-02)

disrupt – 1. A tactical mission task in which a commander integrates direct and indirect fires, terrain, and obstacles to upset an enemy's formation or tempo, interrupt his timetable, or cause his forces to commit prematurely or attack in piecemeal fashion. (FM 3-90) 2. An engineer obstacle effect that focuses fire planning and obstacle effort to cause the enemy to break up his formation and tempo, interrupt his timetable, commit breaching assets prematurely, and attack in a piecemeal effort. (FM 90-7)

distribution system – That complex of facilities, installations, methods, and procedures designed to receive, store, maintain, distribute, and control the flow of military materiel between the point of receipt into the military system and the point of issue to using activities and units. (JP 1-02)

doctrinal template – A model based on known or postulated adversary doctrine. Doctrinal templates illustrate the disposition and activity of adversary forces and assets conducting a particular operation unconstrained by the effects of the battlespace. They represent the application of adversary doctrine under ideal conditions. Ideally, doctrinal templates depict the threat's normal organization for combat, frontages, depths, boundaries and other control measures, assets available from other commands, objective depths, engagement areas, battle positions, and so forth. Doctrinal templates are usually scaled to allow ready use with geospatial products. (JP 1-02)

doctrine – Fundamental principles by which the military forces or elements thereof guide their actions in support of national objectives. It is authoritative but requires judgment in application. (JP 1-02)

economy – One of the eight characteristics of combat service support: providing the most efficient support at the least cost to accomplish the mission. (FM 4-0)

economy of force – One of the nine principles of war: allocate minimum essential combat power to secondary efforts. (FM 3-0)

effect – (Army) A result, outcome, or consequence of an action. (FMI 5-0.1)

electromagnetic deception – The deliberate radiation, re-radiation, alteration, suppression, absorption, denial, enhancement, or reflection of electromagnetic energy in a manner intended to convey misleading information to an enemy or enemy electromagnetic-dependent weapons, thereby degrading or neutralizing the enemy's combat capability. Among the types of electromagnetic deception are: a. **manipulative electromagnetic deception**—Actions to eliminate revealing, or convey misleading, electromagnetic telltale indicators that may be used by hostile forces; b. **simulative electromagnetic deception**—Actions to simulate friendly, notional, or actual capabilities to mislead hostile forces; c. **imitative electromagnetic deception**—The introduction of electromagnetic energy into enemy systems that imitates enemy emissions. (JP 1-02)

electromagnetic spectrum – The range of frequencies of electromagnetic radiation from zero to infinity. It is divided into 26 alphabetically designated bands. (JP 1-02)

electronic attack—That division of electronic warfare involving the use of electromagnetic energy, directed energy, or antiradiation weapons to attack personnel, facilities, or equipment with the intent of degrading, neutralizing, or destroying enemy combat capability and is considered a form of fires. Electronic attack includes: 1. actions taken to prevent or reduce an enemy's effective use of the electromagnetic spectrum, such as jamming and electromagnetic deception, and 2. employment of weapons that use either electromagnetic or directed energy as their primary destructive mechanism (lasers, radio frequency weapons, particle beams), or antiradiation weapons. (JP 1-02)

electronic warfare – Any military action involving the use of electromagnetic and directed energy to control the electromagnetic spectrum or to attack the enemy. The three major subdivisions within electronic warfare are: electronic attack, electronic protection, and electronic warfare support. (JP 1-02)

electronic warfare support—That division of electronic warfare involving actions tasked by, or under direct control of, an operational commander to search for, intercept, identify, and locate or localize sources of intentional and unintentional radiated electromagnetic energy for the purpose of immediate threat recognition, targeting, planning, and conduct of future operations. Thus, electronic warfare support provides information required for immediate decisions involving electronic warfare operations and other tactical actions such as threat avoidance, targeting, and homing. Electronic warfare support data can be used to produce signals intelligence, provide targeting for electronic or destructive attack, and produce measurement and signature intelligence. (JP 1-02)

embarkation – The process of putting personnel and/or vehicles and their associated stores and equipment into ships and/or aircraft. (JP 1-02)

encirclement – An operation where one force loses its freedom of maneuver because an opposing force is able to isolate it by controlling all ground lines of communications. (FM 3-0)

end state – (Army) At the operational and tactical levels, the conditions that, when achieved, accomplish the mission. At the operational level, these conditions attain the aims set for the campaign or major operation. (FM 3-0)

engagement – A tactical conflict, usually between opposing lower echelon maneuver forces. (JP 1-02)

envelopment – (Army) – A form of maneuver in which an attacking force seeks to avoid the principal enemy defenses by seizing objectives to the enemy rear to destroy the enemy in his current positions. At the tactical level, envelopments focus on seizing terrain, destroying specific enemy forces, and interdicting enemy withdrawal routes. (FM 3-0).

essential elements of friendly information – (Army) The critical aspects of a friendly operation that, if known by the enemy, would subsequently compromise, lead to failure, or limit success of the operation, and, therefore, must be protected from enemy detection. (FM 3-13)

essential task – A specified or implied task that must be executed to accomplish the mission. Essential tasks are always included in the unit's mission statement. (FM 5-0)

exfiltration – The removal of personnel or units from areas under enemy control by stealth, deception, surprise, or clandestine means. (JP 1-02)

exploitation – 1. Taking full advantage of success in military operations, following up initial gains, and making permanent the temporary effects already achieved. 2. An offensive operation that usually follows a successful attack and is designed to disorganize the enemy in depth. (JP 1-02)

explosive ordnance disposal – The detection, identification, on-site evaluation, rendering safe, recovery, and final disposal of unexploded explosive ordnance. It may also include explosive ordnance which has become hazardous by damage or deterioration. (JP 1-02)

firepower – 1. The amount of fire which may be delivered by a position, unit, or weapon system. 2. Ability to deliver fire. (JP 1-02)

fires – The effects of lethal or nonlethal weapons. (JP 1-02)

fire support coordinating measure – A measure employed by land or amphibious commanders to facilitate the rapid engagement of targets and simultaneously provide safeguards for friendly forces. (JP 1-02)

fix – (Army) 1. A tactical mission task where a commander prevents the enemy from moving any part of his force from a specific location for a specific period of time. 2. An engineer obstacle effect that focuses fire planning and obstacle effort to slow an attacker's movement within a specified area, normally an engagement area. (FM 3-90)

flexibility – One of the eight characteristics of combat service support: being able to adapt combat service support structures and procedures to changing situations, missions, and concepts of operations. (FM 4-0)

force projection – The ability to project the military element of national power from the continental United States (CONUS) or another theater in response to requirements for military operations. Force projection operations extend from mobilization and deployment of forces to redeployment to CONUS or home theater. (JP 1-02)

force protection – Actions taken to prevent or mitigate hostile actions against Department of Defense personnel (to include family members), resources, facilities, and critical information. These actions conserve the force's fighting potential so it can be applied at a decisive time and place and incorporates the coordinated and synchronized offensive and defensive measures to enable the effective employment of the joint force while degrading opportunities for the enemy. Force protection does not include actions to defeat the enemy or protect against accidents, weather, or disease. (JP 1-02)

force tailoring – The process of determining the right mix and sequence of units for a mission. (FM 3-0)

foreign humanitarian assistance – Programs conducted to relieve or reduce the results of natural or manmade disasters or other endemic conditions such as human pain, disease, hunger, or privation that might present a serious threat to life or that can result in great damage to or loss of property. Foreign humanitarian assistance (FHA) provided by US forces is limited in scope and duration. The foreign assistance provided is designed to supplement or complement the efforts of the host nation civil authorities or agencies that may have primary responsibility for providing FHA. FHA operations are those conducted outside the United States, its territories, and possessions. (JP 1-02)

foreign internal defense – Participation by civilian and military agencies of a government in any of the action programs taken by another government to free and protect its society from subversion, lawlessness, and insurgency. (JP 1-02)

fratricide – The unintentional killing or wounding of friendly personnel by friendly firepower. (FM 3-0)

frontal attack – (Army) A form of maneuver in which the attacking force seeks to destroy a weaker enemy force or fix a larger enemy force in place over a broad front. (FM 3-0)

full spectrum operations – The range of operations Army forces conduct in war and military operations other than war. (FM 3-0)

graphic control measure – A symbol used on maps and displays to regulate forces and warfighting functions. (FMI 5-0.1)

gun-target line – An imaginary straight line from gun to target. (JP 1-02)

hazardous material – Any substance which has a human health hazard associated with it. Special storage, use, handling, and shipment safety procedures and protocols must be followed to help protect against accidental exposure. Hazardous materials are specifically identified under federal law. (FM 3-100.4)

high angle fire – (Army) Fire delivered to clear an obstacle (such as a hill) that low-angle fire cannot, or fire delivered to attack targets on the reverse side of an obstacle (such as a hill) that cannot be attacked with low-angle or direct fire. (FM 6-30).

host nation – A nation that receives the forces and/or supplies of allied nations, coalition partners, and/or NATO organizations to be located on, to operate in, or to transit through its territory. (JP 1-02)

host-nation support – Civil and/or military assistance rendered by a nation to foreign forces within its territory during peacetime, crises or emergencies, or war based on agreements mutually concluded between nations. (JP 1-02)

human intelligence – A category of intelligence derived from information collected and provided by human sources. (JP 1-02)

imagery intelligence – Intelligence derived from the exploitation of collection by visual photography, infrared sensors, lasers, electro-optics, and radar sensors such as synthetic aperture radar wherein images of objects are reproduced optically or electronically on film, electronic display devices, or other media. (JP 1-02)

implied task – A task that must be performed to accomplish a specified task or mission, but is not stated in the higher headquarters order. (FM 5-0)

indirect fire – Fire delivered on a target that is not itself used as a point of aim for the weapons or the director. (JP 1-02)

infiltration – (Army) 1. A form of maneuver in which an attacking force conducts undetected movement through or into an area occupied by enemy forces to occupy a position of advantage in the enemy rear while exposing only small elements to enemy defensive fires. (FM 3-90)

information – (Army) 1. In the general sense, the meaning humans assign to data. 2. In the context of the cognitive hierarchy, data that have been processed to provide further meaning. (FM 6-0)

information fratricide – The results of employing information operations elements in a way that causes effects in the information environment that impede the conduct of friendly operations or adversely affect friendly forces. (FM 3-13)

information management – The provision of relevant information to the right person at the right time in a usable form to facilitate situational understanding and decisionmaking. It uses procedures and information systems to collect, process, store, display, and disseminate information. (FM 3-0)

information operations – (Army) The employment of the core capabilities of electronic warfare, computer network operations, psychological operations, military deception, and operations security, in concert with specified supporting and related capabilities, to affect and defend information and information systems and to influence decisionmaking. (FM 3-13)

information requirements – (Army) All information elements the commander and staff require to successfully conduct operations, that is, all elements necessary to address the factors of METT-TC. (FM 6-0)

information superiority – The operational advantage derived from the ability to collect, process, and disseminate an uninterrupted flow of information while exploiting or denying an adversary's ability to do the same. (JP 1-02)

information system – (Army) The equipment and facilities that collect, process, store, display, and disseminate information. This includes computers—hardware and software—and communications, as well as policies and procedures for their use. (FM 3-0)

insurgency – An organized movement aimed at the overthrow of a constituted government through the use of subversion and armed conflict. (JP 1-02)

integration – One of the eight characteristics of combat service support: the total inclusion of Army combat service support into the operations process (plan, prepare, execute, assess), as well as into other logistics comnponents of the unified force. (FM 4-0).

intelligence – 1. The product resulting from the collection, processing, integration, analysis, evaluation, and interpretation of available information concerning foreign countries or areas. 2. Information and knowledge about an adversary obtained through observation, investigation, analysis, or understanding. (JP 1-02)

intelligence preparation of the battlefield – The systematic, continuous process of analyzing the threat and environment in a specific geographic area. Intelligence preparation of the battlefield (IPB) is designed to support the staff estimate and military decisionmaking process. Most intelligence requirements are generated as a result of the IPB process and its interrelation with the decisionmaking process. (FM 34-130)

intelligence, surveillance, and reconnaissance – An activity that synchronizes and integrates the planning and operation of sensors, assets, and processing, exploitation, and dissemination systems in direct support of current and future operations. This is an integrated intelligence and operations function. (JP 1-02)

intelligence, surveillance, and reconnaissance plan – An integrated plan for collection of information from all available sources and analysis of that information to produce intelligence to meet requirements. Specifically, a logical plan for transforming priority intelligence requirements into orders or requests to reconnaissance and surveillance assets to collect pertinent information within a required time limit. (FM 34-3)

interagency – Activities or operations conducted by or through coordination with two or more agencies or an agency and one or more Services of the same nation. (FM 3-07)

interdict – A tactical mission task where the commander prevents, disrupts, or delays the enemy's use of an area or route. (FM 3-90)

interior lines – A force operates on interior lines when its operations diverge from a central point. (FM 3-0)

isolate – A tactical mission task that requires a unit to seal off—both physically and psychologically—an enemy from his sources of support, deny an enemy freedom of movement, and prevent an enemy unit from having contact with other enemy forces. (FM 3-90)

joint force – A general term applied to a force composed of significant elements, assigned or attached, of two or more Military Departments, operating under a single joint force commander. (JP 1-02)

joint force commander – A general term applied to a combatant commander, subunified commander, or joint task force commander authorized to exercise combatant command (command authority) or operational control over a joint force. (JP 1-02)

joint operations – A general term to describe military actions conducted by joint forces, or by Service forces in relationships (e.g., support, coordinating authority), which, of themselves, do not create joint forces. (JP 1-02)

joint task force – A joint force that is constituted and so designated by the Secretary of Defense, a combatant commander, a subunified commander, or an existing joint task force commander. (JP 1-02)

key terrain – Any locality, or area, the seizure or retention of which affords a marked advantage to either combatant. (JP 1-02)

landing zone – Any specified zone used for the landing of aircraft. JP 1-02)

law of war – That part of international law that regulates the conduct of armed hostilities. (JP 1-02)

liaison – That contact or intercommunication maintained between elements of military forces or other agencies to ensure mutual understanding and unity of purpose and action. (JP 1-02)

line of communications – A route, either land, water, and/or air, that connects an operating military force with a base of operations and along which supplies and military forces move. (JP 1-02)

lines of operations – (Army) A line that defines the orientation of the force in time and space, or purpose, in relation to an enemy or objective. (FMI 5-0.1).

logistics – The science of planning and carrying out the movement and maintenance of forces. In its most comprehensive sense, those aspects of military operations which deal with: a. design and development, acquisition, storage, movement, distribution, maintenance, evacuation, and disposition of materiel; b. movement, evacuation, and hospitalization of personnel; c. acquisition or construction, maintenance, operation, and disposition of facilities; and d. acquisition and furnishing of services. (JP 1-02)

logistics-over-the-shore operations – The loading and unloading of ships without the benefit of deep draft-capable, fixed port facilities; or as a means of moving forces close to tactical assembly areas dependent on threat force capabilities. (JP 1-02)

logistics preparation of the theater – Actions taken by combat service support personnel to optimize means—force structure, resources, and strategic lift—of supporting the joint force commander's plan. (FM 4-0)

major operation – A series of tactical actions (battles, engagements, strikes) conducted by various combat forces of a single or several Services, coordinated in time and place, to accomplish operational and, sometimes, strategic objectives in an operational area. These actions are conducted simultaneously or sequentially in accordance with a common plan and are controlled by a single commander. (JP 1-02)

maneuver – One of the nine principles of war: place the enemy in a disadvantageous position through the flexible application of combat power. (FM 3-0)

Marine air-ground task force – The Marine Corps principal organization for all missions across the range of military operations, composed of forces task-organized under a single commander capable of responding rapidly to a contingency anywhere in the world. The types of forces in the Marine air-ground task force (MAGTF) are functionally grouped into four core elements: a command element, an aviation combat element, a ground combat element, and a combat service support element. The four core elements are categories of forces, not formal commands. The basic structure of the MAGTF never varies, though the number, size, and type of Marine Corps units comprising each of its four elements will always be mission dependent. The flexibility of the organizational structure allows for one or more subordinate MAGTFs to be assigned. In a joint or multinational environment, other Service or multinational forces may be assigned or attached. (FM 1-02)

Marine expeditionary force – The largest Marine air-ground task force (MAGTF) and the Marine Corps principal warfighting organization, particularly for larger crises or contingencies. It is task-organized around a permanent command element and normally consists of one or more Marine divisions, Marine aircraft wings, and Marine force service support groups. The Marine expeditionary force is capable of missions across the range of military operations, including amphibious assault and sustained operations ashore in any environment. It can operate from a sea base, a land base, or both. In a joint or multinational environment, it may contain other Service or multinational forces assigned or attached to the MAGTF. (FM 1-02)

Marine expeditionary unit – A Marine air-ground task force (MAGTF) that is constructed around an infantry battalion reinforced, a helicopter squadron reinforced, and a task-organized combat service support element. It normally fulfills Marine Corps' forward sea-based deployment requirements. The Marine expeditionary unit provides an immediate reaction capability for crisis response and is capable of limited combat operations. In a joint or multinational environment, it may contain other Service or multinational forces assigned or attached to the MAGTF. (FM 1-02)

mass – One of the nine principles of war: the effects of combat power at the decisive place and time. (FM 3-0)

mass casualty – Any large number of casualties produced in a relatively short period of time, usually as the result of a single incident such as a military aircraft accident, hurricane, flood, earthquake, or armed attack that exceeds logistical support capabilities. (JP 1-02)

measure of effectiveness – (Army) A criterion used to assess changes in system behavior, capability, or operational environment that is tied to measuring the attainment of an end state, achievement of an objective, or creation of an effect. (FMI 5-0.1)

medical evacuation – The timely and efficient movement of the wounded, injured, or ill while providing en route medical care to and between medical treatment facilities. (FM 4-02)

meeting engagement – A combat action that occurs when a moving force engages an enemy at an unexpected time and place. (FM 3-0)

METT-TC – A memory aid used in two contexts: (1) In the context of information management, the major subject categories into which relevant information is grouped for military operations: mission, enemy, terrain and weather, troops and support available, time available, civil considerations. (2) In the context of tactics, the major factors considered during mission analysis. (FM 6-0)

military deception – Actions executed to deliberately mislead adversary military decisionmakers as to friendly military capabilities, intentions, and operations, thereby causing the adversary to take specific actions (or inactions) that will contribute to the accomplishment of the friendly mission. (JP 1-02)

military decisionmaking process – A process that integrates the activities of the commander, staff and subordinate commanders in developing and operation plan or order. It establishes procedures for analyzing a mission; developing, analyzing, and comparing courses of action; selecting the best course of action; and producing an operation plan or order (FMI 5-0.1).

misinformation – Incorrect information from any source that is released for unknown reasons or to solicit a response or interest from a nonpolitical or nonmilitary target. (FM 3-13)

mission – 1. The task, together with the purpose, that clearly indicates the action to be taken and the reason therefor. 2. In common usage, especially when applied to lower military units, a duty assigned to an individual or unit; a task. (JP 1-02)

mission command – The conduct of military operations through decentralized execution based upon mission orders for effective mission accomplishment. Successful mission command results from subordinate leaders at all echelons exercising disciplined initiative within the commander's intent to accomplish missions. It requires an environment of trust and mutual understanding. (FM 6-0)

mission orders – A technique for completing combat orders that allows subordinates maximum freedom of planning and action in accomplishing missions and leaves the "how" of mission accomplishment to the subordinate. (FM 6-0)

mobile defense – (Army) A type of defensive operation that concentrates on the destruction or defeat of the enemy through a decisive attack by a striking force. (FM 3-0)

mobility corridor – Areas where a force will be canalized due to terrain restrictions. They allow military forces to capitalize on the principles of mass and speed and are therefore relatively free of obstacles. (JP 1-02)

mobility operations – Obstacle reduction by maneuver and engineer units to reduce or negate the effects of existing or reinforcing obstacles. The objective is to maintain freedom of movement for maneuver units, weapon systems, and critical supplies. (FM 3-34)

modified combined obstacle overlay – A joint intelligence preparation of the battlespace product used to portray the effects of each battlespace dimension on military operations. It normally depicts militarily significant aspects of the battlespace environment, such as obstacles restricting military movement, key geography, and military objectives. (JP 1-02)

monitoring – (Army) Continuous observation of the current situation to identify opportunities for the force, threats to the force, gaps in information, and progress according to the plan or order. (FMI 5-0.1)

movement to contact – A form of the offensive designed to develop the situation and to establish or regain contact. (JP 1-02)

naval coastal warfare – Coastal sea control, harbor defense, and port security, executed both in coastal areas outside the United States in support of national policy and in the United States as part of this Nation's defense. (JP 1-02)

naval gunfire support – Fire provided by Navy surface gun systems in support of a unit or units tasked with achieving the commander's objectives. A subset of naval surface fire support. (JP 1-02)

naval surface fire support – Fire provided by Navy surface gun and missile systems in support of a unit or units. (JP 1-02)

neutral – (Army) 1. An individual, group of individuals, organization, or nation that is not hostile to or in any way supportive of only one belligerent force in a hostile environment. (FM 3-07)

no-fire area – A land area, designated by the appropriate commander, into which fires or their effects are prohibited. (JP 1-02)

noncombatant – 1. An individual in an area of combat operations who is not armed and is not participating in any activity in support of any of the factions or forces involved in combat. 2. An individual, such as chaplain or medical personnel, whose duties do not involve combat. (FM 3-07)

noncombatant evacuation operations – Operations directed by the Department of State, the Department of Defense, or other appropriate authority whereby noncombatants are evacuated from foreign countries when their lives are endangered by war, civil unrest, or natural disaster to safe havens or to the United States. (JP 1-02)

nongovernmental organization – A private, self-governing, not-for-profit organization dedicated to alleviating human suffering; and/or promoting education, health care, economic development, environmental protection, human rights, and conflict resolution; and/or encouraging the establishment of democratic institutions and civil society. (JP 1-02)

nonlethal fires – Any fires that do not directly seek the physical destruction of the intended target and are designed to impair, disrupt, or delay the performance of enemy operational forces, functions, and facilities. Psychological operations, electronic warfare (jamming), and other command and control countermeasures are all nonlethal fire options. (FM 6-20)

nonlethal weapons – Weapons that are explicitly designed and primarily employed so as to incapacitate personnel or materiel, while minimizing fatalities, permanent injury to personnel and undesired damage to property and the environment. a. Unlike conventional lethal weapons that destroy their targets through blast, penetration, and fragmentation, nonlethal weapons employ means other than gross physical destruction to prevent the target from functioning. b. Nonlethal weapons are intended to have one, or both, of the following characteristics: (1) They have relatively reversible effects on personnel or materiel. (2) They affect objects differently within their area of influence. (JP 1-02)

objective – (Army) 1. One of the nine principles of war: direct every military operation toward a clearly defined, decisive and attainable objective. (FM 3-0) 2. A location on the ground used to orient operations, phase operations, facilitate changes of direction, and provide for unity of effort. (FM 3-90)

offensive – One of the nine principles of war: seize, retain, and exploit the initiative. (FM 3-0)

offensive information operations – The integrated use of assigned and supporting capabilities and activities, mutually supported by intelligence, to affect enemy decisionmakers or to influence others to achieve or promote specific objectives. (FM 3-0)

offensive operations – Operations which aim at destroying or defeating an enemy. Their purpose is to impose US will on the enemy and achieve decisive victory. (FM 3-0)

operation – 1. A military action or the carrying out of a strategic, operational, tactical, service, training, or administrative military mission. 2. The process of carrying on combat, including movement, supply, attack, defense, and maneuvers needed to gain the objectives of any battle or campaign. (JP 1-02)

operational control – Command authority that may be exercised by commanders at any echelon at or below the level of combatant command. Operational control is inherent in combatant command (command authority) and may be delegated within the command. When forces are transferred between combatant commands, the command relationship the gaining commander will exercise (and the losing commander will relinquish) over these forces must be specified by the Secretary of Defense. Operational control is the authority to perform those functions of command over subordinate forces involving organizing and employing commands and forces, assigning tasks, designating objectives, and giving authoritative direction necessary to accomplish the mission. Operational control includes authoritative direction over all aspects of military operations and joint training necessary to accomplish the missions assigned to the command. Operational control should be exercised through the commanders of subordinate organizations. Normally this authority is exercised through subordinate joint force commanders and Service and/or functional component commanders. Operational control normally provides full authority to organize commands and forces and to employ those forces as the commander in operational control considers necessary to accomplish assigned missions; it does not, in and of itself, include authoritative direction for logistics or matters of administration, discipline, internal organization, or unit training. (JP 1-02)

operational framework – The arrangement of friendly forces and resources in time, space, and purpose with respect to each other and the enemy or situation. It consists of the area of operations, battlespace, and battlefield organization. (FM 3-0)

operations process – The major command and control activities performed during operations: planning, preparation, execution with continuous assessment. These activities occur continuously throughout an operation, overlapping and recurring as required. (FMI 5-0.1)

operations security – A process of identifying critical information and subsequently analyzing friendly actions attendant to military operations and other activities to: a. identify those actions that can be observed by adversary intelligence systems; b. determine indicators hostile intelligence systems might obtain that could be interpreted or pieced together to derive critical information in time to be useful to adversaries; and c. select and execute measures that eliminate or reduce to an acceptable level the vulnerabilities of friendly actions to adversary exploitation. (JP 1-02)

paramilitary forces – Forces or groups distinct from the regular armed forces of any country, but resembling them in organization, equipment, training, or mission. (JP 1-02)

patrol – A detachment of ground, sea, or air forces sent out for the purpose of gathering information or carrying out a destructive, harassing, mopping-up, or security mission. (JP 1-02)

peace enforcement – Application of military force, or the threat of its use, normally pursuant to international authorization, to compel compliance with resolutions or sanctions designed to maintain or restore peace and order. (JP 1-02)

peacekeeping – Military operations undertaken with the consent of all major parties to a dispute, designed to monitor and facilitate implementation of an agreement (ceasefire, truce, or other such agreement) and support diplomatic efforts to reach a long-term political settlement. (JP 1-02)

peace operations – A broad term that encompasses peacekeeping operations and peace enforcement operations conducted in support of diplomatic efforts to establish and maintain peace. (JP 1-02)

penetration – (Army) A form of maneuver in which an attacking force seeks to rupture enemy defenses on a narrow front to disrupt the defensive system. (FM 3-90).

perceptions – Mental images the commander wants the deception target to believe are real. (FM 3-13)

phase line – A line utilized for control and coordination of military operations, usually an easily identified feature in the operational area. (JP 1-02)

physical destruction – The application of combat power to destroy or degrade adversary forces, sources of information, command and control systems, and installations. It includes direct and indirect forces from ground, sea, and air forces. Also included are direct actions by special operations forces. (FM 3-13)

physical security – 1. That part of security concerned with physical measures designed to safeguard personnel; to prevent unauthorized access to equipment, installations, material, and documents; and to safeguard them against espionage, sabotage, damage, and theft. 2. In communications security, the component that results from all physical measures necessary to safeguard classified equipment, material, and documents from access thereto or observation thereof by unauthorized persons. (JP 1-02)

pickup zone – A geographic area used to pick up troops or equipment by helicopter. (FM 90-4)

plan – A design for a future or anticipated operation. (FM 5-0)

planning – The process by which commanders (and staffs, if available) translate the commander's visualization into a specific course of action for preparation and execution, focusing on the expected results. (FMI 5-0.1)

port of debarkation – (DOD) The geographic point at which cargo or personnel are discharged. This may be a seaport or aerial port of debarkation; for unit requirements, it may or may not coincide with the destination. (JP 1-02)

port of embarkation – (DOD) The geographic point in a routing scheme from which cargo and personnel depart. This may be a seaport or aerial port from which personnel and equipment flow to a port of debarkation; for unit and nonunit requirements, it may or may not coincide with the origin. (JP 1-02)

principles of war – Principles that provide general guidance for conducting war and military operations other than war at the strategic, operational, and tactical levels. The nine principles of war are: objective, offensive, mass, economy of force, maneuver, unity of command, security, surprise, and simplicity (FM 3-0)

propaganda – Any form of communication in support of national objectives designed to influence the opinions, emotions, attitudes, or behavior of any group in order to benefit the sponsor, either directly or indirectly. (JP 1-02)

psychological operations – Planned operations to convey selected information and indicators to foreign audiences to influence their emotions, motives, objective reasoning, and ultimately the behavior of foreign governments, organizations, groups, and individuals. The purpose of psychological operations is to induce or reinforce foreign attitudes and behavior favorable to the originator's objectives. (JP 1-02)

public affairs – Those public information, command information, and community relations activities directed toward both the external and internal publics with interest in the Department of Defense. (JP 1-02)

pursuit – An offensive operation designed to catch or cut off a hostile force attempting to escape, with the aim of destroying it. (JP 1-02)

push – In logistics, the delivery of a predetermined amount of supplies to a user on a scheduled basis without the user requesting them. (FM 4-0)

railhead – A point on a railway where loads are transferred between trains and other means of transport. (JP 1-02)

reachback – The process of obtaining products, services, and applications, or forces, or equipment, or material from organizations that are not forward deployed. (JP 1-02)

rear area – (Army) For any particular command, the area extending forward from its rear boundary to the rear of the area assigned to the next lower level of command. This area is provided primarily for the performance of support functions, and is where the majority of the echelon's sustaining functions occur. (FM 3-0)

refugee – A person who, by reason of real or imagined danger, has left his home country or country of nationality and is unwilling or unable to return. (JP 1-02)

relevant information – All information of importance to commanders and staffs in the exercise of command and control. (FM 3-0)

relief in place – (Army) An tactical enabling operation in which, by direction of higher authority, all or part of a unit is replaced in an area by the incoming unit. (FM 3-90)

reorganization – Action taken to shift internal resources within a degraded unit to increase its level of combat effectiveness. (FM 100-9)

reserve – Portion of a body of troops which is kept to the rear or withheld from action at the beginning of an engagement, in order to be available for a decisive movement. (JP 1-02)

responsiveness – One of the eight principles of combat service support: providing the right support at the right place at the right time. (FM 4-0)

retrograde – A type of defensive operation that involves organized movement away from the enemy. (FM 3-0)

riot control agent – (Army) A substance that produces temporary irritating or disabling physical effects that disappear within minutes of removal from exposure. There is no significant risk of permanent injury, and medical treatment is rarely required. (FM 3-11.11)

risk – (DOD) Probability and severity of loss linked to hazards. (JP 1-02)

risk management – The process of identifying, assessing, and controlling risk arising from operational factors, and making decisions that balance risk cost with mission benefits. (JP 1-02)

rules of engagement – Directives issued by competent military authority that delineate the circumstances and limitations under which United States forces will initiate and/or continue combat engagement with other forces encountered. (JP 1-02)

running estimate – A staff section's continuous assessment of current and future operations to determine if the current operation is proceeding according to the commander's intent and if future operations are supportable. (FMI 5-0.1)

search and attack – A technique of conducting a movement to contact that shares many of the characteristics of an area security mission. (FM 3-0)

secure – (Army) 1. A tactical mission task that involves preventing a unit, facility, or geographical location from being damaged or destroyed as a result of enemy action. (FM 3-90)

security – One of the nine principles of war: never permit the enemy to acquire an unexpected advantage (FM 3-0).

seize – (Army) A tactical mission task that involves taking possession of a designated area using overwhelming force. (FM 3-90)

sequel – (Army) An operation that follows the current operation. It is a future operation that anticipates possible outcomes – success, failure, or stalemate – of the current operation. (FM 3-0).

shaping operation – An operation at any echelon that creates and preserves conditions for the success of decisive operations. (FM 3-0)

show of force – An operation designed to demonstrate US resolve that involves increased visibility of US deployed forces in an attempt to defuse a specific situation that, if allowed to continue, may be detrimental to US interests or national objectives. (JP 1-02)

signals intelligence – 1. A category of intelligence comprising either individually or in combination all communications intelligence, electronic intelligence, and foreign instrumentation signals intelligence, however transmitted. 2. Intelligence derived from communications, electronics, and foreign instrumentation signals. (JP 1-02)

simplicity – One of the nine principles of war: prepare clear, uncomplicated plans, and clear, concise orders to ensure thorough understanding. (FM 3-0) [Note: "simplicity" is also one of the eight characteristics of combat service support.]

situational understanding – (Army) The product of applying analysis and judgment to the common operational picture to determine the relationship among the factors of METT-TC. (FM 3-0)

space operations – The employment of space system capabilities that provide the means to enhance command and control, facilitate the maneuver of forces, reduce the commander's uncertainty, and improve fire support, air defense, intelligence collection, and combat service support operations which will support strategic, operational, and tactical missions across the operational continuum in the near, mid, and far term. (FM 3-14)

special operations – Operations conducted in hostile, denied, or politically sensitive environments to achieve military, political, economic, and/or informational objectives employing military capabilities for which there is no broad conventional force requirement. These operations often require covert, clandestine, or low visibility capabilities. Special operations are applicable across the rnage of military operations. They can be conducted independently or in conjunction with operations of conventional forces or other government agencies and may include operations through, with, or by indigenous or surrogate forces. Special operations differ from conventional operations in degree of physical and political risk, operational techniques, mode of employment, independence from friendly support, and dependence on detailed operational intelligence and indigenous assets. (JP 1-02)

special operations forces – Those Active and Reserve Component forces of the Military Services designated by the Secretary of Defense and specifically organized, trained, and equipped to conduct and support special operations. (JP 1-02)

special reconnaissance – Reconnaissance and surveillance actions conducted by special operations forces in hostile, denied, or politically sensitive environments to collect or verify information of strategic or operational significance, employing military capabilities not normally found in conventional forces. These forces provide an additive capability for commanders and supplement other conventional reconnaissance and surveillance actions. (JP 1-02)

stability operations – Operations that promote and protect US national interests by influencing the threat, political, and information dimensions of the operational environment through a combination of peacetime developmental, cooperative activities and coercive actions in response to crisis. (FM 3-0)

status-of-forces agreement – An agreement that defines the legal position of a visiting military force deployed in the territory of a friendly state. Agreements delineating the status of visiting military forces may be bilateral or multilateral. Provisions pertaining to the status of visiting forces may be set forth in a separate agreement, or they may form a part of a more comprehensive agreement. These provisions describe how the authorities of a visiting force may control members of that force and the amenability of the force or its members to the local law or to the authority of local officials. To the extent that agreements delineate matters affecting the relations between a military force and civilian authorities and population, they may be considered as civil affairs agreements. (JP 1-02)

strategy – The art and science of developing and employing instruments of national power in a synchronized and integrated fashion to achieve theater, national and/or multinational objectives. (JP 1-02)

striking force – A committed force organized to conduct the decisive attack in a mobile defense. It normally comprises the maximum combat power available to the commander at the time of the attack. (FM 3-0)

strong point – (Army) A heavily fortified battle position tied to a natural or reinforcing obstacle to create an anchor for the defense or to deny the enemy decisive or key terrain. (FM 3-90)

surprise – One of the nine principles of war: strike the enemy at a time or place or in a manner for which he is unprepared. (FM 3-0)

survivability – (Army) 1. Concept which includes all aspects of protecting personnel, weapons, and supplies while simultaneously deceiving the enemy. Survivability tactics include building a good defense; employing frequent movement; using concealment, deception, and camouflage; and constructing fighting and protective positions for both individuals and equipment. Encompasses planning and locating position sites, designing adequate overhead cover, analyzing terrain conditions and construction materials, selecting excavation methods, and countering the effects of direct and indirect fire weapons. (FM 5-103) 2. One of the eight characteristics of combat service support: being able to shield support functions from destruction or degradation. (FM 4-0)

sustainability - One of the eight characteristics of combat service support: the ability to maintain continuous support, rear area and base security, movement control, terrain management, and infrastructure development. (FM 4-0)

sustaining operation – An operation at any echelon that enables shaping and decisive operations by providing combat service support, rear area and base security, movement control, terrain management, and infrastructure development. (FM 3-0)

synchronization – 1. The arrangement of military actions in time, space, and purpose to produce maximum relative combat power at a decisive place and time. See FM 3-0. 2. In the intelligence context, application of intelligence sources and methods in concert with the operation plan. (JP 1-02)

tactical combat force – A combat unit, with appropriate combat support and combat service support assets, that is assigned the mission of defeating Level III threats. (JP 1-02)

tactical control – Command authority over assigned or attached forces or commands, or military capability or forces made available for tasking, that is limited to the detailed direction and control of movements or maneuvers within the operational area necessary to accomplish missions or tasks assigned. Tactical control is inherent in operational control. Tactical control may be delegated to, and exercised at any level at or below the level of combatant command. When forces are transferred between combatant commands, the command relationship the gaining commander will exercise (and the losing commander will relinquish) over these forces must be specified by the Secretary of Defense. Tactical control provides sufficient authority for controlling and directing the application of force or tactical use of combat support assets within the assigned mission or task. (JP 1-02)

tactical level of war –The level of war at which battles and engagements are planned and executed to accomplish military objectives assigned to tactical units or task forces. Activities at this level focus on the ordered arrangement and maneuver of combat elements in relation to each other and to the enemy to achieve combat objectives. (JP 1-02)

target acquisition – The detection, identification, and location of a target in sufficient detail to permit the effective employment of weapons. (JP 1-02)

targeting – The process of selecting and prioritizing targets and matching the appropriate response to them, taking account of operational requirements and capabilities. (JP 1-02)

target reference point – An easily recognizable point on the ground (either natural or man-made) used to initiate, distribute, and control fires. Target reference points (TRPs) can also designate the center of an area where the commander plans to distribute or converge the fires of all his weapons rapidly. They are used by task force and below, and can further delineate sectors of fire within an engagement area. TRPs are designated using the standard target symbol and numbers issued by the fire support officer. Once designated, TRPs also constitute indirect fire targets. (FM 3-90)

task organization – A temporary grouping of forces designed to accomplish a particular mission. (FM 3-0)

task-organizing – (Army) The process of allocating available assets to subordinate commanders and establishing their command and support relationships. (FM 3-0)

tempo – The rate of military action. (FM 3-0)

terminal control – 1. The authority to direct the maneuver of aircraft which are delivering ordnance, pasengers, or cargo to a specific location or target. Terminal control is a type of air control. 2. Any electronic, mechanical, or visual control given to aircraft to facilitate target acquisition and resolution. (JP 1-02)

terrorism – The calculated use of unlawful violence or threat of unlawful violence to inculcate fear; intended to coerce or to intimidate governments or societies in the pursuit of goals that are generally political, religious, or ideological. (JP 1-02)

terrorist – An individual who uses violence, terror, and intimidation to achieve a result. (JP 1-02)

throughput distribution – The bypassing of one or more intermediate supply echelons in the supply system to avoid multiple handling. (FM 4-0)

time-phased force and deployment data – The Joint Operation Planning and Execution System database portion of an operation plan; it contains time-phased force data, non-unit-related cargo and personnel data, and movement data for the operation plan, including the following: a. In-place units; b. Units to be deployed to support the operation plan with a priority indicating the desired sequence for their arrival at the port of debarkation; c. Routing of forces to be deployed; d. Movement data associated with deploying forces; e. Estimates of non-unit related cargo and personnel movements to be conducted concurrently with the deployment of forces; and f. Estimate of transportation requirements that must be fulfilled by common-user lift resources as well as those requirements that can be fulfilled by assigned or attached transportation resources. (JP 1-02)

toxic industrial material – A generic term for toxic radioactive compounds in solid, liquid, aerosolized or gaseous form. These may be used, or stored for use, for industrial, commercial, medical, military, or domestic purposes. Toxic industrial materials may be chemical, biological, or radioactive and described as toxic industrial chemical, toxic industrial biological, or toxic industrial radiological. (FM 3-11.4)

track – 1. To display or record the successive positions of a moving object. 2. To keep a gun properly aimed, or to point continuously a target-locating instrument at a moving target. (JP 1-02)

troop leading procedures – A dynamic process used by small unit leaders to analyze a mission, develop a plan, and prepare for an operation. (FM 5-0)

turning movement – (Army) A form of maneuver in which the attacking force seeks to avoid the enemy's principal defensive positions by seizing objectives to the enemy rear and causing the enemy to move out of his current positions or divert major forces to meet the threat. (FM 3-0)

unconventional warfare – A broad spectrum of military and paramilitary operations, normally of long duration, predominantly conducted through, with, or by indigenous or surrogate forces who are organized, trained, equipped, supported, and directed in varying degrees by an external source. It includes, but is not limited to, guerrilla warfare, subversion, sabotage, intelligence activities, and unconventional assisted recovery. (JP 1-02)

unexploded explosive ordnance – Explosive ordnance which has been primed, fused, armed, or otherwise prepared for action, and which has been fired, dropped, launched, projected, or placed in such a manner as to constitute a hazard to operations, installations, personnel, or material, and remains unexploded either by malfunction or design or for any other cause. (JP 1-02)

unity of command – One of the nine principles of war: for every objective, ensure unity of effort under one responsible commander.

unity of effort – Coordination and cooperation among all military forces and other organizations toward a commonly recognized objective, even if the forces and nonmilitary organizations are not necessarily part of the same command structure. (FM 6-0)

unmanned aircraft system – A powered aerial vehicle that does not carry a human operator, uses aerodynamic forces to provide lift, can fly automomously or is remotely operated, can be expendable or recoverable, and can carry a lethal or nonlethal payload. Ballistic or semi-ballistic vehicles, cruise missiles, and artillery projectiles are not considered unmanned aircraft systems.

urban area – A topographical complex where manmade construction or high population density is the dominant feature. See also **urban environment**; **urban operations**. (FM 3-06)

urban environment – The physical urban area, as well as the complex and dynamic interaction among its key components—the terrain (natural and manmade), the population, and the supporting infrastructure—as an overlapping and interdependent system of systems. (FM 3-06)

urban operations – Offense, defense, stability, and support operations conducted in a topographical complex and adjacent natural terrain where manmade construction and high population density are the dominant features. (FM 3-0)

versatility – The ability of Army forces to meet the global, diverse mission requirements of full spectrum operations. (FM 3-0)

visualize – To create and think in mental images. (FM 6-0)

vulnerability – 1. The susceptibility of a nation or military force to any action by any means through which its war potential or combat effectiveness may be reduced or its will to fight diminished. 2. The characteristics of a system that cause it to suffer a definite degradation (incapability to perform the designated mission) as a result of having been subjected to a certain level of effects in an unnatural (manmade) hostile environment. 3. In information operations, a weakness in information system security design, procedures, implementation, or internal controls that could be exploited to gain unauthorized access to information or an information system. (JP 1-02)

warfighting function – A group of tasks and systems (people, organizations, information, and processes), united by a common purpose, that commanders use to accomplish missions and training objectives. The six warfighting functions are: a. **command and control** – the related tasks and systems that support the commander in exercising authority and direction; b. **fire support** – the related tasks and systems that provide collective and coordinated use of Army indirect fires, joint fires, and offensive information operations; c. **intelligence** – the related tasks and systems that facilitate understanding of the enemy, terrain, weather, and civil considerations; d. **movement and maneuver** – the related tasks and systems that move forces to achieve a position of advantage in relation to the enemy; e. **protection** – the related tasks and systems that preserve the force so the commander can apply maximum combat power; f. **sustainment** – the related tasks and systems that provide support and services to ensure freedom of action, extend operational reach, and prolon endurance. (FMI 5-0.1)

weapons of mass destruction – Weapons that are capable of a high order of destruction and/or of being used in such a manner as to destroy large numbers of people. Weapons of mass destruction can be high explosives or nuclear, biological, chemical, and radiological weapons, but exclude the means of transporting or propelling the weapon where such means is a separable and divisible part of the weapon. (JP 1-02)

This page intentionally left blank.

References

Military publications are listed by title. When a field manual has been published under a new number for the first time, the old number is provided in parenthesis after the new number.

REQUIRED PUBLICATIONS

These documents must be available to intended users of this publication.

JOINT PUBLICATIONS

Joint publications are available at <http://www.dtic.mil/doctrine/jel/>.

JP 1-02. *Department of Defense Dictionary of Military and Associated Terms*, 12 April 2001.

JP 3-06. *Doctrine for Joint Urban Operations*. 16 September 2002.

ARMY PUBLICATIONS

Army doctrinal publications are available at <http://www.apd.army.mil/>.

FM 1-02 (101-5-1). *Operational Terms and Graphics*. 21 September 2004.

FM 34-130. *Intelligence Preparation of the Battlefield*. 8 July 1994. FM 34-130 will be republished as FM 2-01.3

FM 3-0 (100-5). *Operations*. 14 June 2001.

FM 3-06.11 (90-10-1). *Combined Arms Operations in Urban Terrain*. 28 February 2002.

FM 3-07 (100-20). *Stability Operations and Support Operations*. 20 February 2003. When updated, FM 3-07 will be retitled *Stability Operations*.

FM 3-90. *Tactics*. 04 July 2001.

FM 5-19 (100-14). *Composite Risk Management*. 21 August 2006.

FM 7-0. *Training the Force*. 22 October 2002.

FM 7-1. *Battle Focused Training*. 15 September 2003.

FM 5-0 (101-5). *Army Planning and Orders Production*. 20 January 2005.

RELATED PUBLICATIONS

These sources contain relevant supplemental information.

JOINT PUBLICATIONS

Joint publications are available at <http://www.dtic.mil/doctrine/jel/>.

JP 3-0. *Doctrine for Joint Operations*. 10 September 2001.

JP 3-05. *Doctrine for Joint Special Operations*. 17 December 2003.

JP 3-08. *Interagency Coordination During Joint Operations* (2 Volumes). 9 October 1996.

JP 3-16. *Joint Doctrine for Multinational Operations*. 05 April 2000.

JP 4-0. *Doctrine for Logistic Support of Joint Operations*. 06 April 2000.

JP 5-00.2. *Joint Task Force Planning Guidance and Procedures*. 13 January 1999.

Joint Military Operations Historical Collection. 15 July 1997.

ARMY PUBLICATIONS

Army doctrinal publications are available at <http://www.apd.army.mil/>.

Army Regulations

FM 1. *The Army*. 14 June 2005.

FM 2-0 (34-1). *Intelligence*. 17 May 2004.

FM 2-22.3 (34-52). *Human Intelligence Collector Operations*. 6 September 2006.

FM 3-05 (100-25). *Army Special Operations Forces*. 20 September 2006.

FM 3-05.20 (31-20). *Special Forces Operations*. 1 April 2004.

FM 3-05.30 (33-1). *Psychological Operations*. 15 April 2005.

FM 3-05.105. *Multiservice Tactics, Techniques, and Procedures for Special Operations Forces in Nuclear, Biological, and Chemical Environments*. 28 September 2001.

FM 3-05.212 (TC 31-25). *Special Forces Waterborne Operations*. 31 August 2004.

FM 3-05.301 (33-1-1). *Psychological Operations Tactics, Techniques, and Procedures*. 31 December 2003.

FM 3-05.401 (41-10-1). *Civil Affairs Tactics, Techniques, and Procedures*. 23 September 2003.

FM 3-06 (90-10). *Urban Operations*. 01 June 2003.

FM 3-06.1. *Aviation Urban Operations: Multi-service Tactics, Techniques, and Procedures for Aviation Urban Operations*. 09 July 2005.

FM 3-07.2. *Antiterrorism*. (Future Publication).

FM 3-11.3 ( ). *Multiservice Tactics, Techniques, and Procedures for Chemical, Biological, Radiological, and Nuclear Contamination Avoidance*. 2 February 2006.

FM 3-11.9 (3-9). *Potential Military Chemical/Biological Agents and Compounds*. 10 January 2005.

FM 3-11.14 (3-14). *Multiservice Tactics, Techniques, and Procedures for Nuclear, Biological, and Chemical Vulnerability Assessment*. 28 November 2004.

FM 3-11.19 (3-19). *Multiservice Tactics, Techniques, and Procedures for Nuclear, Biological, and Chemical Reconnaissance*. 30 July 2004.

FM 3-11.21 (3-21). *Multiservice Tactics, Techniques, and Procedures for Nuclear, Biological, and Chemical Aspects of Consequence Management*. 12 December 2001.

FM 3-11.22. *Weapons of Mass Destruction Civil Support Team Tactics, Techniques, and Procedures*. 06 June 2003.

FM 3-13 (100-6). *Information Operations: Doctrine, Tactics, Techniques, and Procedures*. 28 November 2003.

FM 3-14 (100-18). *Space Support to Army Operations*. 18 May 2005.

FM 3-19.4 (19-4). *Military Police Leaders' Handbook*. 04 March 2002.

FM 3-19.15 (19-15). *Civil Disturbance Operations*. 18 April 2005.

FM 3-19.17. *Military Working Dogs*. 22 September 2005.

FM 3-19.40 (19-40). *Military Police Internment/Resettlement Operations*. 01 August 2001.

FM 3-19.50. *Police Intelligence Operations*. 21 July 2006.

FM 3-22.40 (90-40). *NLW: Tactical Employment of Nonlethal Weapons*. 15 January 2003.

FM 3-28. *Civil Support Operations*. (Future Publication).

FM 3-31. *Joint Force Land Component Commander Handbook (JFLCC)*. 13 December 2001.

FM 3-31.1 (90-31). *AMCI: Army and Marine Corps Integration in Joint Operations*. 21 November 2001.

FM 3-34 (5-100, 5-114). *Engineer Operations*. 02 January 2004.

FM 3-34.2. *Combined Arms Breaching Operations*. 31 August 2000. FM 3-34.2 will be republished as FM 3-90.11

FM 3-34.230 (5-105). *Topographic Operations*. 03 August 2000.

FM 3-61.1. *Public Affairs Tactics, Techniques and Procedures*. 01 October 2000.

FM 3-90.12 (90-13). *Combined Arms Gap Crossing Operations*. TBP.

FM 3-97.6 (90-6). *Mountain Operations*. 28 November 2000.

FM 3-100.12. *Risk Management: Multiservice Tactics, Techniques, and Procedures for Risk Management*. 15 February 2001.

FM 3-100.38. *UXO: Multi-service Tactics, Techniques and Procedures for Unexploded Ordnance Operations*. 18 August 2005.

FM 4-0 (100-10). *Combat Service Support*. 29 August 2003.

FM 4-01.45. *Tactical Convoy Operations: Multiservice Tactics, Techniques, and Procedures for Tactical Convoy Operations*. 24 March 2005.

FM 4-02.51 (8-51). *Combat and Operational Stress Control*. 6 June 2006.

FM 4-30.16. *EOD: Multiservice Procedures for Explosive Ordnance Disposal in a Joint Environment*. 27 October 2005.

FM 4-30.51. *Unexploded Ordnance (UXO) Procedures*. 13 July 2006.

FM 5-01.12 (90-41). *JTF LNO Integration: Multiservice Tactics, Techniques, and Procedures for Joint Task Force (JTF) Liaison Officer Integration*. 27 January 2003.

FM 5-33. *Terrain Analysis*. 8 September 1992. FM 5-33 will be republished as FM 3 34.330.

FM 5-103. *Survivability*. 10 June 1985. FM 5 103 will be republished as FM 5-34.112.

FM 5-104. *General Engineering*. 12 November 1986. FM 5-104 will be republished as FM 3-34.250.

FM 5-170. *Engineer Reconnaissance*. 05 May 1988. FM 5-170 will be republished as FM 3 34.212.

FM 5-415. *Fire-Fighting Operations*. 09 February 1999. FM 5-415 will be republished as FM 3-34.281.

FM 6-0. *Mission Command: Command and Control of Army Forces*. 11 August 2003.

FM 6-22.5 (22-9). *Combat Stress*. 23 June 2000.

FM 7-15. *The Army Universal Task List (Incl. changes 2)*. 6 July 2006.

FM 20-32. *Mine/Countermine Operations*. 1 April 2005.

FM 22-51. *Leaders' Manual for Combat Stress Control*. 29 September 1994. FM 22-51 will be republished as FM 4-02.22.

FM 22-100. *Army Leadership: Be, Know, Do*. 31 August 1999. FM 22-100 will be republished as FM 6-22.

FM 23-10. *Sniper Training*. 17 August 1994. FM 23-10 will be republished as FM 3-22.10.

FM 27-10. *The Law of Land Warfare*. 18 July 1956. FM 27-10 will be republished as FM 1-04.10.

FM 27-100. *Legal Support to Operations*. 01 March 2000. FM 27-100 will be republished as FM 1-04.

FM 31-20-3. *Foreign Internal Defense Tactics, Techniques, and Procedures for Special Forces*. 20 September 1994. FM 31-20-3 will be republished as FM 3-05.202.

FM 31-50. *Combat in Fortified and Built-Up Areas*. March 1964.

FM 34-3. *Intelligence Analysis*. 15 March 1990. FM 34-3 will be republished as FM 2-33.4.

FM 34-60. *Counterintelligence*. 03 October 1995. FM 34-60 will be republished as FM 2-01.2.

FM 34-81. *Weather Support for Army Tactical Operations*. 31 August 1989. FM 34-81 will be republished as FM 2-33.2.

FM 34-81-1. *Battlefield Weather Effects*. 23 December 1992. FM 34-81-1 will be republished as FM 2-33.201.

FM 41-10. *Civil Affairs Operations*. 14 February 2000. FM 41-10 will be republished as FM 3-05.40.

FM 44-8. *Combined Arms for Air Defense*. 01 June 1999. FM 44-8 will be republished as FM 3-01.8.

FM 46-1. *Public Affairs Operations*. 30 May 1997. FM 46-1 will be republished as FM 3-61.

FM 90-8. *Counterguerrilla Operations*. 29 August 1986. FM 90-8 will be republished as FM 3-07.11.

FM 90-10. *Military Operations on Urbanized Terrain (MOUT) (How to Fight)*. 15 August 1979.

FM 100-5. *Operations*. 1944.

FM 100-8. *The Army in Multinational Operations*. 24 November 1997. FM 100-8 will be republished as FM 3-16.

FM 100-11. *Force Integration*. 15 January 1998. FM 100-11 will be republished as FM 3-100.11.

FMI 2-91.4. *Intelligence Support to Operations in the Urban Environment*. 30 June 2005. FMI 2-91.4 will be republished as FM 2-91.4.

FMI 3-07.22. *Counterinsurgency Operations*. 01 October 2004. FMI 3-07.22 will be republished as FM 3-07.22.

FMI 3-34.119. *Improvised Explosive Device Defeat*. 21 September 2005.

FMI 4-30.50. *Modular Explosive Ordnance Disposal Operations*. 31 July 2006.

FMI 5-0.1. *The Operations Process*. 31 March 2006.

TC 3-15. *Nuclear Accident and Incident Response and Assistance (NAIRA)*. 27 December 1988.

TC 5-230. *Army Geospatial Guide for Commanders and Planners*. 28 November 2003.

TC 7-98-1. *Stability and Support Operations Training Support Package*. 05 June 1997.

TC 9-21-01. *Soldiers Improvised Explosive Device (IED) Awareness Guide: Iraq and Afghanistan Theaters of Operation*. 28 May 2004.

TC 90-1. *Training for Urban Operations*. 01 April 2002.

Department of Defense Publications

Defense Intelligence Agency Manual 58-11. *Defense Human Resources Intelligence Collection Management*. 20 August 1968.

Quadrennial Defense Review Report. 30 September 2001.

U.S. Army Training and Doctrine Command (TRADOC), Deputy Chief of Staff for Intelligence (DCSINT). *TRADOC DCSINT Handbook No. 1: A Military Guide to Terrorism in the Twenty-First Century* (Version2.0). Fort Leavenworth, KS: U.S. Government Printing Office, 2004.

Nonmilitary Publications

26th Infantry Regimental Association. *Aachen: Military Operations in Urban Terrain*. Lititz, PA: 26th Infantry Regimental Association, 1999.

Adan, Abraham. *On the Banks of the Suez*. Novato, CA: Presidio Press, 1980.

Akers, Frank H., Jr. and George B. Singleton. "Task Force Ranger: A Case Study Examining the Application of Advanced Technologies in Modern Urban Warfare." Oak Ridge, TN: National Security Program Office, November 2000.

Allard, Kenneth. *Somalia Operations: Lessons Learned*. Washington, DC: National Defense University Press, 1995.

Arnold, James R. *Tet Offensive 1968: Turning Point in Vietnam*. London: Osprey, 1990.

Bass, Bernard M. "Leading in the Army after Next." *Military Review* (March/April 1998): 46–58.

Beevor, Anthony. *Stalingrad*. New York: Penguin Books, 1998.

Bowden, Mark. *Black Hawk Down: A Story of Modern War*. New York: Atlantic Monthly Press, 1999.

Butts, Kent H. "The Strategic Importance of Water." *Parameters* (spring 1997): 65–83.

Carell, Paul. *Stalingrad: The Defeat of the German 6th Army*. Atglen, PA: Schieffer Publishing, 1993.

Celeski, Joseph D. "Joint Urban Operations: Special Forces in Urban Campaign Planning." *Special Warfare* (fall 2001): 2–13.

Chiarelli, Peter W. and Patrick R. Michaelis. "Winning the Peace: The Requirement for Full-Spectrum Operations." *Military Review* (July/August 2005): 4–17.

Chiarelli, Peter W., Patrick R. Michaelis, and Geoffrey A. Norman. "Armor in Urban Terrain: The Critical Enabler." *Armor* (March/April 2005): 7–12.

Chuikov, Vasili I. *The Battle for Stalingrad.* New York: Holt, Rinehart, and Winston, 1964.

Colby, John. *War From the Ground Up: The 90th Division in WWII.* Austin, TX: Nortex Press, 1991.

Cole, Ronald H. *Operation Just Cause.* Washington, DC: Joint History Office, 1995.

Connaughton, Richard, John Pimlott, and Duncan Anderson. *The Battle for Manila: The Most Devastating Untold Story of World War II.* Novata, CA: Presidio, 1995.

Craig, William. *Enemy at the Gates: The Battle for Stalingrad.* New York: Reader's Digest Press, 1973.

Delk, James D. *Fires & Furies: The LA Riots.* Palm Springs, CA: Etc. Publications, 1995.

Dewar, Michael. *British Army in Northern Ireland.* New York: Sterling Publishers, 1985.

______. *War in the Streets: The Story of Urban Combat from Calais to Khafji.* New York: Sterling Publishers, 1992.

Donnelly, Thomas, Margaret Roth, and Caleb Baker. *Operation Just Cause: The Storming of Panama.* New York: Lexington Books, 1991.

Dupuy, R. Ernest and Trevor N. Dupuy. *The Encyclopedia of Military History: From 3500 B.C. to the Present.* San Francisco: Harper & Row, 1977.

Dupuy, Trevor N. and Paul Martell. *Flawed Victory: The Arab-Israeli Conflict and the 1982 War in Lebanon.* Fairfax, VA: Hero Books, 1986.

Edwards, Sean J. A. *Freeing Mercury's Wings: Improving Tactical Communications in Cities.* Santa Monica, CA: RAND, 2001.

Eshel, David. *Mid-East Wars: The Lebanon War 1982.* Hod Hasharon, Israel: Eshel-Dramit, 1983.

Fishel, John T. *The Fog of Peace: Planning and Executing the Restoration of Panama.* Carlisle, PA: Strategic Studies Institute, 1992.

Fontenot, Gregory, E. J. Degen, and David Tohn. *On Point: US Army in Operation IRAQI FREEDOM.* Fort Leavenworth, KS: Combat Studies Institute Press, 2004.

Funston, Frederick. "How the Army Worked to Save San Francisco." *Cosmopolitan Magazine* (July 1906): <http://www.sfmuseum.org/1906/cosmo.html>.

Gabel, Christopher R. "Military Operations on Urbanized Terrain: The 2d Battalion, 26th Infantry, at Aachen, October 1944." *Urban Combat Operations: tactics, Techniques and Procedure.* Center for Army Lessons Learned (CALL) Newsletter No. 99-16. Fort Leavenworth, KS: Center for Army Lessons Learned, U.S. Training and Doctrine Command [TRADOC], November 1999: 1-9–1-13.

Gabriel, Richard A. *Operation Peace for Galilee: The Israeli-PLO War in Lebanon.* New York: Hill and Wang, 1984.

Glenn, Russell W. *An Attack on Duffer's Downtown.* Santa Monica, CA: RAND, 2001.

______. *Heavy Matter: Urban Operations' Density of Challenges.* Santa Monica, CA: RAND, 2000.

______. "Urban Combat is Complex." *Proceedings* (February 2002): 62–65.

______. "…We Band of Brothers": *The Call for Joint Urban Operations Doctrine.* Santa Monica, CA: RAND, 1999.

Glenn, Russell W., Steven Hartman, and Scott Gerwehr. *Urban Combat Service Support Operations: The Shoulders of Atlas.* Santa Monica, CA: RAND, 2003.

Gordon, Michael R. "Baghdad Targets Picked if Hussein Holes Up There." *New York Times* (7 March 2003): 1A2.

Grau, Lester W. "Something Old, Something New: Guerrillas, Terrorists, and Intelligence Analysis." *Military Review* (July/August 2004): 42–49.

Grau, Lester W. and Timothy Smith. *A 'Crushing' Victory: Fuel-Air Explosives and Grozny 2000.* Fort Leavenworth, KS: Foreign Military Studies Office, 2000.

Hamill, Desmond. *Pig in the Middle: The Army in Northern Ireland, 1969–1984.* London: Methuen, 1985.

Hammel, Eric. *Fire in the Streets: The Battle for Hue, Tet 1968*. New York: Dell Publishing, 1991.

Heal, Sid. "Crowds, Mobs and Nonlethal Weapons." *Military Review* (March/April 2000): 45–50.

Heinl, Robert Debs, Jr. *Victory at High Tide: The Inchon-Seoul Campaign*. Annapolis, MD: Nautical & Aviation Publishing Company of America, 1979.

Herzog, Cham. *The War of Atonement, October, 1973*. Boston: Little, Brown and Company, 1975.

Higham, Robin. *Bayonets in the Streets: The Use of Troops in Civil Disturbances*. Lawrence, KS: University Press of Kansas, 1969.

Hoar, Joseph P. "A CINC's Perspective." *JFQ Forum* (autumn 1993): 56–63.

Hollis, Mark A. B. "Platoon Under Fire." *Infantry* (January/April 1998): 27–34.

Infantry in Battle. 2d ed. Garret & Massie: Richmond, VA, 1939. Reprint, Fort Leavenworth, KS: U.S. Army Command and General Staff College, 1981.

Initial Impressions Report: Operation Iraqi Freedom—Stability Operations-Support Operations Center for Army Lessons Learned (CALL) Newsletter No. 04-13 (Fort Leavenworth, KS: Center for Army Lessons Learned, U.S. Army Training and Doctrine Command [TRADOC], 2004).

Initial Impressions Report: Operation Iraqi Freedom—Stability Operations-Support Operations. Center for Army Lessons Learned (CALL) Newsletter No. 03 (Fort Leavenworth, KS: Center for Army Lessons Learned, U.S. Army Training and Doctrine Command [TRADOC], 2003).

Jomini, Antoine-Henri. "The Art of War" in *Roots of Strategy: Book 2*. Harrisburg, PA: Stackpole Books, 1987.

Kemp, Anthony. *The Unknown Battle: Metz, 1944*. New York: Stein and Day, 1981.

Kennedy, John F. Message to Congress, 25 May 1961. Available at <http://www.jfklibrary.org/Historical+Resources/Archives>.

Linden, Eugene. "The Exploding Cities of the Developing World." *Foreign Affairs* (January/February 1996): 52–65.

Mabry, Robert L. et al. "United States Army Rangers in Somalia: An Analysis of Combat Casualties on an Urban Battlefield." *Journal of Special Operations Medicine* (fall 2001): 24–40.

MacArthur, Douglas. *Reminiscences*. New York: McGraw-Hill, 1964.

Manwaring, Max G. *Street Gangs: The New Urban Insurgency*. Carlisle, PA: Strategic Studies Institute, 2005.

Maxwell, Mark D. "The Law of War and Civilians on the Battlefield: Are We Undermining Civilian Protections?" *Military Review* (September/October 2004): 17–25.

McConnell, Malcolm. *Just Cause: The Real Story of America's High-Tech Invasion of Panama*. New York: St. Martin's Press, 1991.

Medby, Jamison Jo and Russell W. Glenn. *Street Smart: Intelligence Preparation of the Battlefield for Urban Operations*. Santa Monica, CA: RAND, 2002.

Mendel, William W. *Combat in Cities: The LA Riots and Operation Rio*. Fort Leavenworth, KS: Foreign Military Studies Office, 1996.

Monmonier, Mark. *How to Lie with Maps*. Chicago: University of Chicago Press, 1996.

Montgomery, Thomas M. Interview. "Ambush in Mogasdishu." *Frontline*, WGBH Boston. 29 September 1998. Available at <http://www.pbs.org/wgbh/pages/frontline/programs/1998.html>.

Moore, John C. "Sadr City: The Armor Pure Assault in Urban Terrain." *Armor* (November/December 2004): 31–37.

Natsios, Andrew S. "Commander's Guidance: A Challenge of Complex Humanitarian Emergencies." *Parameters* (summer 1996): 50–66.

______. "Food Through Force: Humanitarian Intervention and U.S. Policy." *The Washington Quarterly* (winter 1994): 136.

Nolan, Keith William. *The Battle for Saigon: Tet 1968*. New York: Pocket Books, 1996.

Oakley, Robert B. "An Envoy's Perspective." *JFQ Forum* (autumn 1993): 44–55.

Oliker, Olga. *Russia's Chechen Wars 1994–2000: Lessons from Urban Combat*. Santa Monica, CA: RAND, 2001.

Operation Restore Hope Lessons Learned Report: 3 December 1992 – 4 May 1993, Operations Other than War. Fort Leavenworth, KS: Center for Army Lessons Learned (CALL), U.S. Army Combined Arms Command (CAC), 1993. Reprint, Washington, DC: U.S. Government Printing Office, 1996.

Parkerson, John Embry, Jr. "United States Compliance with Humanitarian Law Respecting Civilians During Operation Just Cause." *Military Law Review* (1991): 31–140.

Peters, Ralph. "Our Soldiers, Their Cities." *Parameters* (spring 1996): 43–50.

Peterson, John J. *Into the Cauldron*. Clinton, MD: Clavier House, 1973.

Petraeus, David H. "Learning Counterinsurgency: Observations from Soldiering in Iraq." *Military Review* (January/February 2006): 2–12.

Ritchie, Elspeth C. and Robert Mott. "Caring for Civilians During Peace Keeping Missions: Priorities and Decisions." *Military Medicine* (Vol. 167, Supplement 3, 2002): 14–16.

Rosenberger, Leif R. "The Strategic Importance of the World Food Supply." *Parameters* (spring 1997): 84–105.

Schadlow, Nadia. "War and the Art of Governance." *Parameters* (autumn 2003): 85–94.

Shelton, H. Hugh and Timothy D. Vane. "Winning the Information War in Haiti." *Military Review* (November/December 1995): 3–9.

Shtemenko, S. M. *The Soviet General Staff at War, 1941-1945: Book One*. Moscow: Progress Publishers, 1981.

Smith, Rupert. "Wars in Our Time: A Survey of Recent and Continuing Conflicts." *World Defence Systems,* volume 3:2. London: Royal United Services Institute for Defence Studies, 2001.

Stanton, Shelby L. *Ten Corps in Korea, 1950*. Novato, CA: Presidio Press, 1989.

Taw, Jennifer Morrison. *Operation Just Cause: Lessons for Operations Other Than* War. Santa Monica, CA: RAND, 1996.

Thomas, Timothy L. "The Battle for Grozny: Deadly Classroom for Urban Combat." *Parameters* (summer 1999): 87–102.

Thomas, Timothy L. and Charles P. O'Hara. "Combat Stress in Chechnya: 'The Equal Opportunity Disorder'." *Army Medical Department Journal* (January/March 2000): 46–53.

Trinquier, Roger. *Modern Warfare: A French View of Counterinsurgency*. New York: Praeger Publisher, 1964.

United Nations Security Council Resolution 794 (1992). New York: United Nations, 1992. Available at <http://daccessdds.un.org/doc/UNDOC/GEN/N92/772/11/PDF>.

U.S. Agency for International Development, Bureau for Humanitarian Response, Office of Foreign Disaster Assistance. *Field Operations Guide for Disaster Assessment and Response, Version 3.0*. Washington, DC: U.S. Government Printing Office, 2003. Available at <http://www.usaid.gov/our_work/humanitarian_assistance/>.

U.S. Department of Transportation, Research and Special Programs Administration, Office of Hazardous Materials Initiative and Training. *2004 Emergency Response Guidebook*. Washington, DC: U.S. Government Printing Office, 2004. Available at <http://hazmat.dot.gov/pubs/erg/erg2004.pdf>.

U.S. Environmental Protection Agency, Office of Emergency and Remedial Response. *HAZMAT Team Planning Guidance*. Washington, DC: U.S. Government Printing Office, 1990.

Van Sickle, Jeffrey B. "Stability Operations in Northern Iraq: Task Force Altun Kupri." *Infantry* (January/February 2005): 25–29.

Watson, Bruce Allen. *Sieges: A Comparative Study*. Westport, CT: Praeger, 1993.

Werstein, Irving. *The Battle of Aachen*. New York: Thomas Y. Crowell Company, 1962.

Whiting, Charles. *Bloody Aachen*. New York: Stein and Day, 1976.

Wintle, Justin, ed. *The Dictionary of War Quotations*. New York: The Free Press, 1989.

Yates, Lawrence A. "Military Stability and Support Operations: Analogies, Patterns and Recurring Themes." *Military Review* (July/August 1997): 51–61.

RECOMMENDED READINGS

The following sources contain relevant supplemental information.

Alexander, John B. *Future War: Non-Lethal Weapons in Twenty-First-Century* Warfare. New York: St. Martin's Press, 1999.

Ashworth, G. J. *The City and War*. London: Routledge, 1991.

Barker, A. J. *Dunkirk: The Great Escape*. New York: David McKay, 1977.

Baumann, Robert F., Lawrence A. Yates, and Versalle F. Washington. "My Clan Against the World": *U.S. and Coalition Forces in Somalia, 1992-1994*. Fort Leavenworth, KS: Combat Studies Institute Press, 2004.

Birtle, Andrew J. *U.S. Army Counterinsurgency and Contingency Operations Doctrine: 1860-1941*. Washington, DC: Center of Military History, 2001.

Blaufarb, Douglas S. *The Counterinsurgency Era: U.S. Doctrine and Performance 1950 to Present*. New York: The Free Press, 1977.

Breuer, William B. *Hitler's Fortress Cherbourg: The Conquest of a Bastion*. New York: Stein and Day, 1984.

Brown, Stephen D. "PSYOP in Operation Uphold Democracy." *Military Review* (September/October 1996): 57–76.

Bulloch, Gavin. "Military Doctrine and Counterinsurgency: A British Perspective." *Parameters* (summer 1996): 4–16.

Urban Combat Operations: Tactics, Techniques and Procedures. Center for Army Lessons Learned (CALL) Newsletter No. 99-16 (Fort Leavenworth, KS: Center for Army Lessons Learned, U.S. Army Training and Doctrine Command [TRADOC], November 1999).

Cordesman, Anthony H. and Abraham R. Wagner. *The Lessons of Modern War—Volume II: The Iran-Iraq War*. Boulder, CO: Westview Press, 1990.

Corum, James S. and Wray R. Johnson. *Airpower in Small Wars: Fighting Insurgents and Terrorists*. Lawrence, KS: University Press of Kansas, 2003.

Desch, Michael C., ed. *Soldiers in Cities: Military Operations on Urban Terrain*. Carlisle Barracks, PA: Strategic Studies Institute, 2001.

Duffy, Christopher. *Fire & Stone: The Science of Fortress Warfare, 1660–1860*. Mechanicsburg, PA: Stackpole Books, 1996.

______. *The Fortress in the Age of Vauban and Frederick the Great, 1660–1789: Siege Warfare Volume I*. London: Routledge & Kegan Paul, 1985.

______. *Siege Warfare: The Fortress in the Early Modern World, 1494–1660*. London: Routledge & Kegan Paul, 1979.

Edwards, Sean J. A. *Mars Unmasked: The Changing Face of Urban Operations*. Santa Monica, CA: RAND, 2000.

Eikenberry, Karl W. "Take No Casualties." *Parameters* (summer 1996): 109–118.

Erickson, John. *The Road to Berlin: Continuing the History of Stalin's War with Germany*. Boulder, CO: Westview Press, 1983.

Gerwehr, Scott and Russell W. Glenn. *The Art of Darkness: Deception and Urban Operations*. Santa Monica, CA: RAND, 2000.

Glenn, Russell W. *Combat in Hell: A Consideration of Constrained Urban Warfare*. Santa Monica, CA: RAND, 1996.

______. *Marching Under Darkening Skies: The American Military and Impending Urban Operations Threat*. Santa Monica, CA: RAND, 1998.

Glenn, Russell W., ed. *Capital Preservation: Preparing for Urban Operations in the Twenty-First Century: Proceedings of the RAND Arroyo-TRADOC-MCWL-OSD Urban Operations Conference, March 22-23, 2000*. Santa Monica, CA: RAND, 2001.

Glenn, Russell W. et al. *Denying the Widow-Maker, Summary of Proceedings, RAND-DDBL Conference on Military Operations on Urbanized Terrain*. Santa Monica, CA: RAND, 1999.

______. *Getting the Musicians of Mars on the Same Sheet of Music: Army, Joint, Multinational, and Interagency C4ISR Interoperability*. Santa Monica, CA: RAND, 2000.

Glenn, Russell W. et al., eds. *The City's Many Faces: Proceedings of the RAND Arroyo-MCWL-J8 UWG Urban Operations Conference, April 13-14, 1999*. Santa Monica, CA: RAND, 2000.

Gouré, Leon. *The Siege of Leningrad*. Stanford, CA: Stanford University Press, 1962.

Grau, Lester W. and Jacob W. Kipp. "Urban Combat: Confronting the Specter." *Military Review* (July/August 1999): 9–17.

Groves, John R. "Operations in Urban Environments." *Military Review* (July/August 1998): 30–39.

Habeck, Mary R. Lecture. "Jihadist Strategies in the War on Terrorism." Washington, DC: The Heritage Foundation, 8 November 2004. Available at http://www.heritage.org/research/nationalsecurity/hl855.cfm.

Hahn, Robert F., II and Bonnie Jezior. "Urban Warfare and the Urban Warfighter of 2025." *Parameters* (summer 1999): 74–86.

Harman, Nicholas. *Dunkirk: The Patriotic Myth*. New York: Simon and Schuster, 1980.

Heckstall-Smith, Anthony. *Tobruk: The Story of a Siege*. New York: W. W. Norton, 1959.

Hibbert, Christopher. *The Battle of Arnhem*. London: Batsford, 1962.

House, John M. "The Enemy After Next." *Military Review* (March/April 1998): 22–28.

Jackson, W. G. F. *The Battle for Italy*. New York: Harper & Row, 1967.

Karslake, Basil. *1940 – The Last Act: The Story of the British Forces in France After Dunkirk*. Hamden, CT: Archon Books, 1979.

Kostof, Spiro. *The City Shaped: Urban Patterns and Meanings Through History*. London: Thames and Hudson, 1999.

______. *The City Assembled: The Elements of Urban Form Throughout History*. Boston: Little Brown and Company, 1999.

Kuehl, Daniel T. "Airpower vs. Electricity: Electric Power as a Target for Strategic Air Operations." *Journal of Strategic Studies* (March 1995): 237–266.

Larson, Eric. *Casualties and Consensus: The Historical Role of Casualties in Domestic Support of U.S. Military Operations*. Santa Monica, CA: RAND, 1996.

Lieven, Anatol. *Chechnya: Tombstone of Russian Power*. New Haven: Yale University Press, 1998.

______. "The World Turned Upside Down: Military Lessons of the Chechen War." *Armed Forces Journal International* (August 1998): 40–43.

McKee, Alexander. *Caen: Anvil of Victory*. New York: St. Martin's Press, 1964.

McLaurin, R. D. "Military Forces in Urban Antiterrorism." Aberdeen Proving Ground, MD: U.S. Army Human Engineering Laboratory, 1989.

McLaurin, R. D. and R. Miller. "Urban Counterinsurgency: Case Studies and Implications for U.S. Military Forces." Aberdeen Proving Ground, MD: U.S. Army Human Engineering Laboratory, 1989.

Mercier, Charles E. "Terrorists, WMD, and the U.S. Army Reserve." *Parameters* (autumn 1997): 98–118.

Milton, T. R. "Urban Operations: Future War." *Military Review* (February 1994): 37–46.

Moskin, J. Robert. *Among Lions: The Battle for Jerusalem, June 5–7, 1967*. New York: Ballantine Books, 1982.

Northam, Ray M. *Urban Geography*. New York: Wiley, 1975.

Peters, Ralph. "The Culture of Future Conflict." *Parameters* (winter 1995 96): 18–27.

______. "The Future of Armed Warfare." *Parameters* (autumn 1997): 50–59.

______. "The Human Terrain of Urban Operations." *Parameters* (spring 2000): 4–12.

Robertson, William G., ed. *Block by Block: The Challenges of Urban Operations*. Fort Leavenworth, KS: U.S. Army Command and Staff College Press, 2003.

Rosenau, William G. "'Every Room is a New Battle': The Lessons of Modern Urban Warfare." *Studies in Conflict and Terrorism* (October–December 1997): 371–394.

Scalard, Douglas. "People of Whom We Know Nothing: When Doctrine Isn't Enough." *Military Review* (July/August 1997): 4–11.

Scales, Robert H. "The Indirect Approach: How U.S. Military Forces Can Avoid the Pitfalls of Future Urban Warfare." *Armed Forces Journal International* (October 1998): 68–74.

Siegel, Adam B. *Eastern Exit: The Noncombatant Evacuation Operation (NEO) from Mogadishu, Somalia, in January 1991*. Alexandria, VA: Center for Naval Analyses, 1992.

Spiller, Roger J. *Sharp Corners: Urban Operations at Century's End*. Fort Leavenworth, KS: Combat Studies Institute, 2001.

Stech, Frank J. "Winning CNN Wars." *Parameters* (autumn 1994): 37–56.

Stevenson, Jonathan. *Losing Mogadishu: Testing U.S. Policy in Somalia*. Annapolis, MD: Naval Institute Press, 1995.

Taw, Jennifer Morrison and Bruce Hoffman. *The Urbanization of Insurgency: The Potential Challenge to U.S. Army Operations*. Santa Monica, CA: RAND, 1994.

Tully, Andrew. *Berlin: Story of a Battle*. Westport, CT: Greenwood Press, 1977.

United Nations Department of Economic and Social Affairs, Population Division. *World Urbanization Prospects: The 2003 Revision*. New York: United Nations, 2003.

Vick, Alan et al. *Aerospace Operations in Urban Environments: Exploring New Concepts*. Santa Monica, CA: RAND, 2000.

Wykes, Alan. *The Siege of Leningrad*. New York: Ballentine Books Inc., 1968.

Index

FM 3-06
26 October 2006

By Order of the Secretary of the Army:

PETER J. SCHOOMAKER
General, United States Army
Chief of Staff

Official:

JOYCE E. MORROW
Administrative Assistant to the
Secretary of the Army
0627803

DISTRIBUTION:

Active Army, Army National Guard, and U. S. Army Reserve: To be distributed in accordance with initial distribution number 111233, requirements for FM 3-06.

Field Manual
No. 3-19.30

Headquarters
Department of the Army
Washington, DC, 8 January 2001

PHYSICAL SECURITY

Contents

Distribution Restriction: Approved for public release; distribution is unlimited.

*This publication supersedes FM 19-30, 1 March 1979.

Preface

This field manual (FM) sets forth guidance for all personnel responsible for physical security. It is the basic reference for training security personnel. It is intended to be a "one-stop" physical-security source for the Department of Defense (DOD), the Department of the Army (DA), and other proponents and agencies of physical security.

Prevention and protection are the two primary concerns of physical security. Both serve the security interests of people, equipment, and property. These interests must be supported at all staff and command levels; and this support must be unified in joint, multinational, and interagency operations.

Support to joint, multinational, and interagency operations relies on the fact that the Army will not conduct operations alone. Additionally, force-projection operations conducted by the military will involve the integration of war-fighting capabilities with stability and support operations. This manual's primary focus is the articulation of a balanced understanding of physical security for joint, multinational, and interagency operations throughout the environments of peacetime, conflict, and war (whether in the continental United States [CONUS] or outside the continental United States [OCONUS]).

Physical security must integrate the various capabilities of joint, multinational, and interagency operations in pursuit of a seamless connection between the strategic, operational, and tactical levels of war. Physical security must also address an expanded range of threats that embraces not only traditional threat components of war, but also nontraditional threats generated by guerrillas, terrorists, criminals, and natural or man-made disasters. In addition, physical security must address the concept of Homeland Defense due to the aforementioned threats.

Homeland Defense is the military's role in the United States (US) government's principal task of protecting its territory and citizens. This is accomplished by joint, interagency, and multijurisdictional organizations. Homeland Defense includes—

- Supporting domestic authorities for crisis and consequence management with regard to weapons of mass destruction (WMD).
- Protecting national-security assets (such as installations) and deploying forces and ensuring the availability, integrity, and adequacy of other critical assets.
- Deterring and defending against strategic attacks while maintaining freedom of action through antiterrorism and force-protection operations.

With this in mind, it is essential to address the five pillars of force protection—combating terrorism, physical security, personal security, law enforcement, and operations security (OPSEC). Physical security is a central component of force protection and provides an integrated venue to express support for operations. Physical security is a primary-leader task and an inherent part of all operations to protect soldiers, family members, civilians, and resources. This function directly supports the Army's universal task list.

While the effects of these changes (when viewed individually) appear revolutionary, the basic activities remain relatively unchanged, though executed under different conditions and standards. Another component that remains unchanged is our reliance upon quality soldiers and leaders well versed in physical-security fundamentals. Leaders will be challenged to ensure that they are functionally proficient; possess an understanding of physical-security operations; are educated in joint, multinational, and interagency operations; and have the ability to perform physical-security functions in support of full-dimension operations.

Appendix A contains an English-to-metric measurement conversion chart. Appendix B is a sample installation crime-prevention handbook. This handbook is designed to assist commanders in developing crime-prevention programs for their installation and units.

The proponent of this publication is HQ TRADOC. Send comments and recommendations on DA Form 2028 directly to Commandant, US Army Military Police School (USAMPS), ATTN: ATSJ-MP-TD, Directorate of Training, 401 Engineer Loop, Suite 2060, Fort Leonard Wood, Missouri 65473-8926.

Unless this publication states otherwise, masculine nouns and pronouns do not refer exclusively to men.

Chapter 1

Physical-Security Challenges

Physical security is defined as that part of security concerned with physical measures designed to safeguard personnel; to prevent unauthorized access to equipment, installations, material, and documents; and to safeguard against espionage, sabotage, damage, and theft. As such, all military operations face new and complex physical-security challenges across the full spectrum of operations. Challenges relative to physical security include the control of populations, information dominance, multinational and interagency connectivity, antiterrorism, and the use of physical-security assets as a versatile force multiplier.

OVERVIEW

1-1. Reductions in manpower and funding are critical challenges to physical security. Manpower for supporting physical-security activities is reduced through deployments and cutbacks. The rapid evolution of physical-security-equipment technology also lends to physical-security challenges, which are exponentially multiplied by the introduction of the information age.

1-2. Physical-security challenges must be understood, and measures must be taken to minimize them to enhance force protection. Leaders must create order when coming upon a situation; and when they depart, some semblance of that order must remain. They must be aware of the human-dimension factors and ensure that their soldiers do not become complacent. It was human error rather than modern technology that took lives in the bombings of the African embassy. Warning was given, but not heeded. Complacency became a physical-security challenge.

AUTOMATED INFORMATION SYSTEMS

1-3. Success on past battlefields has resulted not so much from technological advances, but from innovative ways of considering and combining available and new technologies as they apply to war fighting. Some of these technologies dealt with disseminating and processing information. For example, the telegraph, the telephone, the radio, and now the computer have redefined the fire-support paradigm.

1-4. As the armed forces move into the technological age, a greater need for physical-security measures is required. The risks associated with automated information systems (AISs) are widespread because computers are used for everything. Army Regulation (AR) 380-19 outlines the requirements that commanders and managers need for processing unclassified and classified information and for securing media, software, hardware, and different systems.

1-5. The threat to AISs and information systems security (ISS) involves deliberate, overt, and covert acts. This includes the physical threat to tangible property, such as the theft or destruction of computer hardware. Also included is the threat of electronic, electromagnetic-pulse, radio-frequency (RF), or computer-based attacks on the information or communications components that control or make up critical Army command and control (C^2) infrastructures. In most cases, the threat's target is the information itself rather than the system that transmits it. The threat comes from a range of sources, including the following:

- Unauthorized users (such as hackers) are the main source of today's attacks, primarily against computer-based systems. The threat they pose to AIS networks and mainframe computers is growing.
- Insiders are those individuals with legitimate access to an AIS. They pose the most difficult threat to defend against. Whether recruited or self-motivated, the AIS insider has access to systems normally protected by ISS against an attack.
- Terrorists once had to operate in the immediate vicinity of a target to gain access to or collect intelligence on that target. The proximity to the target risked exposure and detection. Today, a terrorist can accomplish most target selection, intelligence collection, and preoperational planning by gaining access through a computer network. He can increase his probability of success by using computer systems to reduce his "time on target." Terrorist access to an AIS also increases the threat of critical-data destruction or manipulation. Although his presence would be virtual, the potential for damage to Army C^2 systems could be equal to or greater than that achieved by physical intrusion, especially when used as a force multiplier in conjunction with a traditional terrorist attack. Therefore, while traditional preventive measures are still needed to protect unwanted access to information, the information age has added additional concerns for the commander and new opportunities for those with hostile intent.
- Non-state- and state-sponsored groups provide additional challenges. In many cases, it is difficult to confirm state sponsorship of threat activity against an AIS, no matter how apparent the affiliation might seem. Activists of all persuasions are increasingly taking advantage of information-age technology. Neither AISs nor ISS are immune from an adversary's interest in exploiting US military information systems or disrupting communication infrastructures. The availability of low-cost technology and the proliferation of an AIS increase the risk to the Army by potential adversaries.
- Foreign-intelligence services (FIS), both civil and military, are continually active and are another source of contention concerning information systems. In peacetime, they are increasingly targeted against US commercial and scientific interests, rather than military information. With little effort, this peacetime intrusiveness could easily be refocused on AISs and ISS using a wide range of information operations tactics.

- Political and religious groups are other potential adversaries to AISs and ISS. The world's political climate is diverse and complicated. It embraces traditional mainstream political values, as well as radical religious fundamentalism and political extremism. When political or religious viewpoints also incorporate anti-US sentiment, US information infrastructures (including AISs) are increasingly at risk of penetration or exploitation by these potential adversaries.

1-6. When considering an AIS, physical security is more than just safeguarding the equipment. It includes the following elements:

- Software is marked for each system and secured when not in use.
- Initial logon is password-protected (at a minimum).
- Passwords are a minimum of eight characters, using a mixture of letters and numerals.
- Access to an AIS is allowed only to authorized and cleared personnel (per AR 380-19).

1-7. Classified material is entered and transmitted only on approved devices with the following considerations:

- Approved classified devices are operated in a secured environment.
- Classified devices are secured in appropriate containers when not in use.
- Secure telephone unit–III (STU-III) keys are secured in an appropriate safe when not in use (as outlined in AR 380-19).

1-8. Additional information regarding AISs can be found in ARs 380-5 and 380-19. Required training of personnel working with an AIS is located in AR 380-19.

OPSEC AND THE THREAT

1-9. OPSEC is a process of identifying critical information and subsequently analyzing friendly actions attendant to military operations and other activities. The threat is identified using the factors of mission, enemy, terrain, troops, time available, and civilian considerations (METT-TC). The threat defines the physical-security challenges. Implementing physical-security measures supports OPSEC. Providing soundproof rooms for conducting briefings is a simple but invaluable measure.

1-10. Another issue to consider when evaluating physical-security challenges is what actions to take in case of political implications interfering with physical-security measures. In the devastating event at Khobar Towers, a warning was given but not everyone received it. It took too long to evacuate the building after the warning was issued because a cohesive plan was not in place.

1-11. Commanders can minimize the challenges to physical security through proactive measures. They should periodically change the physical-security posture of their area of responsibility to throw off perpetrators.

Chapter 2

The Systems Approach

Commanders must ensure that appropriate physical-security measures are taken to minimize the loss of personnel, supplies, equipment, and material through both human and natural threats. Commanders commonly exercise those protective responsibilities through the provost marshal (PM) and/or physical-security officer and the force-protection officer. The force-protection officer must coordinate with several different agencies to complete his mission. For example, the Army's Intelligence and Counterintelligence Program (see Appendix C) provides information that will be used to complete the unit's crisis-management plan (see Appendix D).

PROTECTIVE SYSTEMS

2-1. The approach to developing protective measures for assets should be based on a systematic process resulting in an integrated protective system. The protective system focuses on protecting specific assets against well-defined threats to acceptable levels of protection. The system is organized in-depth and contains mutually supporting elements coordinated to prevent gaps or overlaps in responsibilities and performance.

2-2. Effective protective systems integrate the following mutually supporting elements:

- Physical protective measures, including barriers, lighting, and electronic security systems (ESSs).
- Procedural security measures, including procedures in place before an incident and those employed in response to an incident. (These include procedures employed by asset owners and those applied by and governing the actions of guards.)
- Terrorism counteraction measures that protect assets against terrorist attacks.

2-3. The following determinations are made when considering system-development procedures:

- The resources available.
- The assets to be protected.
- The threat to those assets.
- The risk levels applicable to those assets.
- The applicable regulatory requirements for protecting the assets.
- The applicable level of protection for those assets against the threat.
- Additional vulnerabilities to the assets (based on the threat).

SYSTEMS DEVELOPMENT

2-4. AR 190-51, DA Pamphlet (Pam) 190-51, and Technical Manual (TM) 5-853-1 are useful tools for developing protective systems using the systems approach. The key to applying these tools successfully is to use a team approach. A team may include physical-security, intelligence, and operations personnel; the installation engineers; and the user of the assets. It may also include representatives from the multinational, host-nation (HN), and local police as well as the regional security office from the embassy.

ASSETS

2-5. Protective systems should always be developed for specific assets. The goal of security is to protect facilities and buildings and the assets contained inside. The risk-analysis procedure in DA Pam 190-51 is used to identify assets. This procedure is applied to all mission-essential or vulnerable areas (MEVAs) according to AR 190-13. It represents the majority of assets with which DOD is commonly concerned. These assets include—

- Aircraft and components at aviation facilities.
- Vehicle and carriage-mounted or -towed weapons systems and components at motor pools.
- Petroleum, oil, and lubricants (POL).
- Controlled medical substances and other medically sensitive items.
- Communication and electronics equipment; test, measurement, and diagnostic equipment (TMDE); night-vision devices (NVDs); and other high-value precision equipment and tool kits.
- Organizational clothing and individual equipment stored at central-issue facilities.
- Subsistence items at commissaries, commissary warehouses, and troop-issue facilities.
- Repair parts at installation-level supply activities and direct-support (DS) units with authorized stockage lists.
- Facilities-engineering supplies and construction materials.
- Audiovisual equipment, training devices, and subcaliber devices.
- Miscellaneous pilferable assets (not included above) and money.
- Mission-critical or high-risk personnel.
- General military and civilian populations.
- Industrial and utility equipment.
- Controlled cryptographic items.
- Sensitive information (included in TM 5-853-1, but not included in DA Pam 190-51).
- Arms, ammunition, and explosives (AA&E).
- Installation banks and finance offices.

RISK LEVELS

2-6. DA Pam 190-51 provides a procedure for determining risk levels—assessing the value of the assets to their users and the likelihood of compromise. These factors are assessed by answering a series of questions leading to value and likelihood ratings.

2-7. Asset value is determined by considering the following three elements:

- The criticality of the asset for its user and the Army as a whole.
- How easily the asset can be replaced.
- Some measure of the asset's relative value.

2-8. The relative value differs for each asset. For some assets, the relative value is measured in terms of monetary cost.

2-9. The likelihood of the threat is assessed for each applicable aggressor category by considering the asset's value to the aggressor, the history of or potential for aggressors attempting to compromise the asset, and the vulnerability of the asset based on existing or planned protective measures.

REGULATORY REQUIREMENTS

2-10. The risk level is the basis for determining the required protective measures for assets covered in AR 190-51. For each asset type, there may be physical protective measures, procedural security measures, and terrorism counteraction measures. These measures are specified by risk level. The measures identified in AR 190-51 are the minimum regulatory measures that must be applied for the identified threat level. The minimum regulatory measures for AA&E are based on the risk category established in AR 190-11.

ANTITERRORISM/FORCE-PROTECTION CONSTRUCTION STANDARDS

2-11. In accordance with DOD Instruction 2000.16, the commanders in chief (CINCs) have developed standards for new construction and existing facilities to counter terrorism threat capabilities within the area of responsibility. These construction standards have specific requirements for such measures as standoff distance, perimeter barriers, building construction, and parking. The DOD construction standard provides for minimum standards that must be incorporated into all inhabited DOD structures regardless of the identified threat. These standards provide a degree of protection that will not preclude the direct effects of blast but will minimize collateral damage for buildings and people and will limit the progressive collapse of structures. These standards add relatively little cost, may facilitate future upgrades, and may deter acts of aggression. (All services have adopted common criteria and minimum standards to counter antiterrorism/force-protection [AT/FP] vulnerabilities and terrorism threats.) Protection to identified threat levels is described in the following paragraphs. Physical-security personnel must be familiar with the CINC and DOD AT/FP construction standards because these standards may affect elements of physical-security plans and how individual facilities are secured.

THREAT IDENTIFICATION

2-12. The threat must be described in specific terms to help determine the assets' vulnerabilities or to establish protective measures. This description should include the tactics that aggressors will use to compromise the asset (weapons, tools, and explosives are likely to be used in an attempt). For example, the threat might be described as a moving vehicle bomb consisting of a 4,000-pound vehicle containing a 500-pound explosive. Another example would be a forced-entry threat using specific hand, power, or thermal tools. These types of threat descriptions (called the design-basis threat) can be used to design detailed protective systems to mitigate the attacks. TM 5-853-1 and DA Pam 190-51 contain procedures for establishing design-basis threat descriptions in the format described above. These procedures can be used together or separately. Threats listed in the TM will be summarized later in this chapter. When using the TM as a lone source or in conjunction with DA Pam 190-51, the following actions occur:

- When the TM process is used alone, the user goes through an identical process to that in DA Pam 190-51 up to the point where the risk level would be determined. In TM 5-853-1, the value and likelihood ratings are used differently than in DA Pam 190-51. The likelihood rating is used to determine the weapons, tools, and explosives that will be used by a particular aggressor in carrying out a specific tactic. In this procedure, higher likelihood ratings result in more severe mixes of weapons, tools, and explosives. The assumption is that the more likely the attack, the more resources the aggressor is likely to use in carrying out the attack.
- When the procedure in TM 5-853-1 is used in conjunction with the results of the DA Pam 190-51 risk analysis, the likelihood rating is taken directly from the risk analysis and applied as described above.

LEVEL OF PROTECTION

2-13. The level of protection applies to the design of a protective system against a specified threat (for example, a bomb, breaking and entering, pilfering, and so forth). The level of protection is based on the asset's value rating from either DA Pam 190-51 or TM 5-853-1. The level increases as the asset's value rating increases. There are separate levels of protection for each tactic. TM 5-853-1 provides detailed guidance on how to achieve the levels of protection, and Chapter 3 of this manual provides a summary of the levels of protection as they apply to various tactics.

VULNERABILITIES

2-14. Vulnerabilities are gaps in the assets' protection. They are identified by considering the tactics associated with the threat and the levels of protection that are associated with those tactics. Some vulnerabilities can be identified by considering the general design strategies for each tactic described in TM 5-853-1 and as summarized in Chapter 3 of this manual. The general design strategies identify the basic approach to protecting assets against specific tactics. For example, the general design strategy for forced entry is to provide a way to detect attempted intrusion and to provide barriers to delay the aggressors until a response force arrives. Vulnerabilities may involve inadequacies in intrusion-detection systems (IDSs) and barriers. Similarly, the general design strategy for a moving vehicle bomb is to keep the vehicle as far from the facility as possible and to harden the facility to resist the explosive at that distance. Vulnerabilities may involve limited standoff distances, inadequate barriers, and building construction that cannot resist explosive effects at the applicable standoff distance.

PROTECTIVE MEASURES

2-15. Where vulnerabilities have been identified, protective measures must be identified to mitigate them. AR 190-13, AR 190-51, DA Pam 190-51, and TM 5-853-1 are effective tools for developing protective measures. The key to effective development of protective systems is a partnership between physical-security personnel and the installation engineers. Appendix E of this manual discusses information for office security, which should be listed in the physical-security plan (see Appendix F). Appendix G discusses personal-protection measures.

THE INTEGRATED PROTECTIVE SYSTEM

2-16. Protective systems integrate physical protective measures and security procedures to protect assets against a design-basis threat. The characteristics of integrated systems include deterrence, detection, defense, and defeat.

DETERRENCE

2-17. A potential aggressor who perceives a risk of being caught may be deterred from attacking an asset. The effectiveness of deterrence varies with the aggressor's sophistication, the asset's attractiveness, and the aggressor's objective. Although deterrence is not considered a direct design objective, it may be a result of the design.

DETECTION

2-18. A detection measure senses an act of aggression, assesses the validity of the detection, and communicates the appropriate information to a response force. A detection system must provide all three of these capabilities to be effective.

2-19. Detection measures may detect an aggressor's movement via an IDS, or they may detect weapons and tools via X-ray machines or metal and explosive detectors. Detection measures may also include access-control elements that assess the validity of identification (ID) credentials. These control elements may provide a programmed response (admission or denial), or they may relay information to a response force. Guards serve as detection elements, detecting intrusions and controlling access.

2-20. Nuclear, biological, and chemical (NBC) detection systems must be used to measure and validate acts of aggression involving WMD. NBC detection systems should also be used to communicate a warning.

DEFENSE

2-21. Defensive measures protect an asset from aggression by delaying or preventing an aggressor's movement toward the asset or by shielding the asset from weapons and explosives. Defensive measures—

- Delay aggressors from gaining access by using tools in a forced entry. These measures include barriers along with a response force.
- Prevent an aggressor's movement toward an asset. These measures provide barriers to movement and obscure lines of sight (LOSs) to assets.
- Protect the asset from the effects of tools, weapons, and explosives.

2-22. Defensive measures may be active or passive. Active defensive measures are manually or automatically activated in response to acts of aggression. Passive defensive measures do not depend on detection or a response. They include such measures as blast-resistant building components and fences. Guards may also be considered as a defensive measure.

DEFEAT

2-23. Most protective systems depend on response personnel to defeat an aggressor. Although defeat is not a design objective, defensive and detection systems must be designed to accommodate (or at least not interfere with) response-force activities.

SECURITY THREATS

2-24. Security threats are acts or conditions that may result in the compromise of sensitive information; loss of life; damage, loss, or destruction of property; or disruption of mission. Physical-security personnel and design teams must understand the threat to the assets they are to protect in order to develop effective security programs or design security systems. Historical patterns and trends in aggressor activity indicate general categories of aggressors and the common tactics they use against military assets. Aggressor tactics and their associated tools, weapons, and explosives are the basis for the threat to assets.

THREAT SOURCES

2-25. There are many potential sources of threat information. Threat assessment is normally a military-intelligence (MI) responsibility. MI personnel commonly focus on such security threats as terrorists and military forces. Within the US and its territories, the Federal Bureau of Investigation (FBI) has primary responsibility for both foreign and domestic terrorists. The FBI, the US Army Criminal Investigation Command (USACIDC [CID]), and local law-enforcement agencies are good sources for physical-security personnel to obtain criminal threat information. Coordinating with these elements on a regular basis is essential to maintaining an effective security program.

THREAT CATEGORIES

2-26. Security threats are classified as either human or natural. Human threats are carried out by a wide range of aggressors who may have one or more objectives toward assets such as equipment, personnel, and operations. Aggressors can be categorized and their objectives can be generalized as described below. (See DA Pam 190-51 and TM 5-853-1 for more information.)

Aggressor Objectives

2-27. Four major objectives describe an aggressor's behavior. Any one of the first three objectives can be used to realize the fourth. These objectives include—

- Inflicting injury or death on people.
- Destroying or damaging facilities, property, equipment, or resources.
- Stealing equipment, materiel, or information.
- Creating adverse publicity.

Aggressor Categories

2-28. Aggressors are grouped into five broad categories—criminals, vandals and activists, extremists, protest groups, and terrorists. Hostile acts performed by these aggressors range from crimes (such as burglary) to low-intensity conflict threats (such as unconventional warfare). Each of these categories describes predictable aggressors who pose threats to military assets and who share common objectives and tactics.

- Criminals can be characterized based on their degree of sophistication. They are classified as unsophisticated criminals, sophisticated criminals, and organized criminal groups. Their common objective is the theft of assets; however, the assets they target, the quantities they seek, their relative efficiency, and the sophistication of their actions vary significantly. Vandals and activists may also be included under this category.
- Vandals and activists are groups of protesters who are politically or issue oriented. They act out of frustration, discontent, or anger against the actions of other social or political groups. Their primary objectives commonly include destruction and publicity. Their selection of targets will vary based on the risk associated with attacking them. The degree of damage they seek to cause will vary with their sophistication.
- Extremists are radical in their political beliefs and may take extreme, violent actions to gain support for their beliefs or cause.
- Protesters are considered a threat only if they are violent. Lawful protesters have to be considered, but significant protective measures and procedures are not normally needed to control their actions. The presence of extremists or vandals/activists at a peaceful protest increases the chance of the protest becoming violent.
- Terrorists are ideologically, politically, or issue oriented. They commonly work in small, well-organized groups or cells. They are sophisticated, are skilled with tools and weapons, and possess an efficient planning capability. There are three types of terrorists—CONUS, OCONUS, and paramilitary OCONUS.
  - CONUS terrorists are typically right- or left-wing extremists operating in distinct areas of the US.
  - OCONUS terrorists generally are more organized than CONUS terrorists. They usually include ethnically or religiously oriented groups.

- Paramilitary OCONUS terrorist groups show some military capability with a broad range of military and improvised weapons. Attacks by OCONUS terrorists are typically more severe.

2-29. Natural threats are usually the consequence of natural phenomena. They are not preventable by physical-security measures, but they are likely to have significant effects on security systems and operations. They may require an increase in protective measures either to address new situations or to compensate for the loss of existing security measures. They may reduce the effectiveness of existing security measures by such occurrences as collapsed perimeter fences and barriers, inoperable protective lighting, damaged patrol vehicles, and poor visibility. Natural threats and their effects relative to security include the following:

- Floods may result in property damage, destruction of perimeter fences, and damage to IDSs. Heavy rains or snowfalls may have similar effects even if they do not result in flooding.
- Storms, tornadoes, high winds, or rain may cause nuisance alarms to activate and cause damage to IDSs. They may limit the visibility of security personnel and may affect closed-circuit television (CCTV) systems. Winds may also disrupt power or communication lines and cause safety hazards from flying debris.
- Earthquakes may cause nuisance alarms to activate or may disrupt IDSs. They may also cause broken water or gas mains, fallen electrical or communication lines, and weakened or collapsed buildings.
- Snow and ice can make travel on patrol roads difficult, may delay responses to alarms, may impede the performance of IDSs, and may freeze locks and alarm mechanisms. Heavy ice may also damage power and communication lines.
- Fires may damage or destroy perimeter barriers and buildings, possibly leaving assets susceptible to damage or theft.
- Fog can reduce the visibility of security forces, thereby requiring additional security personnel. It may also increase the response time to alarms and reduce the effectiveness of security equipment such as CCTV systems.

Aggressor Tactics

2-30. Aggressors have historically used a wide range of offensive strategies reflecting their capabilities and objectives. These offensive strategies are categorized into 15 tactics that are specific methods of achieving aggressor goals (see TM 5-853-1). Separating these tactics into categories allows facility planners and physical-security personnel to define threats in standardized terms usable as a basis for facility and security-system design. Common aggressor tactics include—

- **Moving vehicle bomb.** An aggressor drives an explosive-laden car or truck into a facility and detonates the explosives. His goal is to damage or destroy the facility or to kill people. This is a suicide attack.
- **Stationary vehicle bomb.** An aggressor covertly parks an explosive-laden car or truck near a facility. He then detonates the explosives either by time delay or remote control. His goal in this tactic is the

same as for the moving vehicle bomb with the additional goal of destroying assets within the blast area. This is commonly not a suicide attack. It is the most frequent application of vehicle bombings.

- **Exterior attack.** An aggressor attacks a facility's exterior or an exposed asset at close range. He uses weapons such as rocks, clubs, improvised incendiary or explosive devices, and hand grenades. Weapons (such as small arms) are not included in this tactic, but are considered in subsequent tactics. His goal is to damage the facility, to injure or kill its occupants, or to damage or destroy assets.
- **Standoff weapons.** An aggressor fires military weapons or improvised versions of military weapons at a facility from a significant distance. These weapons include direct (such as antitank [AT] weapons) and indirect LOS weapons (such as mortars). His goal is to damage the facility, to injure or kill its occupants, or to damage or destroy assets.
- **Ballistics.** The aggressor fires various small arms (such as pistols, submachine guns, shotguns, and rifles) from a distance. His goal is to injure or kill facility occupants or to damage or destroy assets.
- **Forced entry.** The aggressor forcibly enters a facility using forced-entry tools (such as hand, power, and thermal tools) and explosives. He uses the tools to create a man-passable opening or to operate a device in the facility's walls, doors, roof, windows, or utility openings. He may also use small arms to overpower guards. His goal is to steal or destroy assets, compromise information, injure or kill facility occupants, or disrupt operations.
- **Covert entry.** The aggressor attempts to enter a facility or a portion of a facility by using false credentials or stealth. He may try to carry weapons or explosives into the facility. His goals include those listed for forced entry.
- **Insider compromise.** A person authorized access to a facility (an insider) attempts to compromise assets by taking advantage of that accessibility. The aggressor may also try to carry weapons or explosives into the facility in this tactic. His goals are the same as those listed for forced entry.
- **Visual surveillance.** The aggressor uses ocular and photographic devices (such as binoculars and cameras with telephoto lenses) to monitor facility or installation operations or to see assets. His goal is to compromise information. As a precursor, he uses this tactic to determine information about the asset of interest.
- **Acoustic eavesdropping.** The aggressor uses listening devices to monitor voice communications or other audibly transmitted information. His goal is to compromise information.
- **Electronic-emanations eavesdropping.** The aggressor uses electronic-emanation surveillance equipment from outside a facility or its restricted area to monitor electronic emanations from computers, communications, and related equipment. His goal is to compromise information.

- **Mail-bomb delivery.** The aggressor delivers bombs or incendiary devices to the target in letters or packages. The bomb sizes involved are relatively small. His goal is to kill or injure people.
- **Supplies-bomb delivery.** The aggressor conceals bombs in various containers and delivers them to supply- and material-handling points such as loading docks. The bomb sizes in this tactic can be significantly larger that those in mail bombs. His goal is to damage the facility, kill or injure its occupants, or damage or destroy assets. Appendix H addresses the actions to take when a bomb is suspected.
- **Airborne contamination.** An aggressor contaminates a facility's air supply by introducing chemical or biological agents into it. His goal is to kill or injure people.
- **Waterborne contamination.** An aggressor contaminates a facility's water supply by introducing chemical, biological, or radiological agents into it. These agents can be introduced into the system at any location with varying effects, depending on the quantity of water and the contaminant involved. His goal is to kill or injure people.

2-31. The aforementioned tactics are typical threats to fixed facilities for which designers and physical-security personnel can provide protective measures. However, some common terrorist acts are beyond the protection that facility designers can provide. They cannot control kidnappings, hijackings, and assassinations that take place away from facilities or during travel between facilities. Protection against these threats is provided through operational security and personal measures (see Appendices G and I), which are covered in doctrine relative to those activities and are under the general responsibility of the CID.

TACTICAL ENVIRONMENT CONSIDERATIONS

2-32. When determining the assets and threats, the same considerations should be given to the systems approach in the tactical environment as when in the cantonment area. The same process of determining the assets, their risk level, and any regulatory guidance apply. Identifying potential threats and the level of protection required for the assets are necessary. Commanders and leaders must also identify additional vulnerabilities and other required protective measures. Commanders are not expected to have the same physical protective measures due to the impact of resources, budget, location, and situations.

2-33. Commanders must consider the various tactics used by aggressors and use their soldiers' abilities to counteract these tactics. Considerations for specific assets (such as military-working-dog [MWD] and explosive-ordnance-disposal [EOD] teams and their abilities to detect and disassemble a bomb) must be identified. Units must have the ability to improvise in a tactical environment. Their training and resourcefulness will compensate for shortcomings in the field.

2-34. The systems approach to security provides focus and integration of resources. Protective systems are mutually supporting and systematically developed to negate the threat. Commanders conduct an intelligence preparation of the battlefield (IPB) and vulnerability assessments (VAs) to determine risks. Security resources and measures are applied to mitigate risks and to deter, detect, defend, and defeat the threat.

Chapter 3

Design Approach

Developing protective systems to protect assets depends on an effective partnership between engineers and physical-security personnel. Physical-security personnel need to understand the basic approaches the engineers will take in laying out protective systems. Engineers must understand the issues involved with ensuring that anything they lay out is compatible with security operations and the operations of the asset users. The best way to ensure a viable design is through teamwork. This chapter provides a summary of the basic approaches to protecting assets against threats (the design strategies). Understanding these strategies is critical to being an effective team member in developing protective systems.

DESIGN STRATEGIES

3-1. There are separate design strategies for protecting assets from each tactic described in Chapter 2. There are two types of strategies associated with each tactic—the general-design and specific-design strategies. The general-design strategy is the general approach to protecting assets against tactics. The specific-design strategy refines the general-design strategy to focus the performance of the protective system on a particular level of protection. (See TM 5-853-1 for more information.)

PROTECTIVE MEASURES

3-2. Protective measures are developed as a result of the general- and specific-design strategies. These protective measures commonly take the form of site-work, building, detection, and procedural elements.

- Site-work elements include the area surrounding a facility or an asset. Technically, they are associated with everything beyond 5 feet from a building. They can include perimeter barriers, landforms, and standoff distances.
- Building elements are protective measures directly associated with buildings. These elements include walls, doors, windows, and roofs.
- Detection elements detect such things as intruders, weapons, or explosives. They include IDSs, CCTV systems used to assess intrusion alarms, and weapon and explosive detectors. These elements can also include the guards used to support this equipment or to perform similar functions.
- Procedural elements are the protective measures required by regulations, TMs, and standing operating procedures (SOPs). These elements provide the foundation for developing the other three elements.

VEHICLE BOMBS

3-3. Vehicle-bomb tactics include both moving and stationary vehicle bombs. In the case of a moving vehicle bomb, the aggressor drives the vehicle into the target. This is commonly known as a suicide attack. In a stationary vehicle bomb, he parks the vehicle and detonates the bomb remotely or on a timed delay.

GENERAL-DESIGN STRATEGY

3-4. Blast pressures near an exploding vehicle bomb are very high, but they decrease rapidly with distance from the explosion. The design strategy for these tactics is to maintain as much standoff distance as possible between the vehicle bomb and the facility and then, if necessary, to harden the facility for the resulting blast pressures. Barriers on the perimeter of the resulting standoff zone maintain the required standoff distance. The difference between moving and stationary vehicle-bomb tactics is that the aggressor using the moving vehicle bomb will attempt to crash through the vehicle barriers; the aggressor using the stationary vehicle bomb will not. Therefore, vehicle barriers for the moving vehicle bomb must be capable of stopping a moving vehicle at the perimeter of the standoff zone. For a stationary vehicle bomb, vehicle barriers must mark the perimeter of the standoff zone, but they are not required to stop the moving vehicle. They only need to make it obvious if an aggressor attempts to breach the perimeter.

LEVELS OF PROTECTION

3-5. There are three levels of protection for vehicle bombs—low, medium, and high. The primary differences between the levels are the degree of damage allowed to the facility protecting the assets and the resulting degree of damage or injury to the assets.

- **Low.** The facility or the protected space will sustain a high degree of damage but will not collapse. It may not be economically repairable. Although collapse is prevented, injuries may occur and assets may be damaged.
- **Medium.** The facility or the protected space will sustain a significant degree of damage, but the structure will be reusable. Occupants and other assets may sustain minor injuries or damage.
- **High.** The facility or the protected space will sustain only superficial damage. Occupants and other assets will also incur only superficial injury or damage.

SITE-WORK ELEMENTS

3-6. The two primary types of site-work elements for vehicle bombs are the standoff distance and vehicle barriers. The vehicle's speed must also be taken into consideration.

Standoff Distance

3-7. The standoff distance is the maintained distance between where a vehicle bomb is allowed and the target. The initial goal should be to make that distance as far from the target facility as practical. Figure 3-1 shows the distances required to limit building damage to particular levels (including the levels of protection described above) for a range of bomb weights. All bomb weights are given in terms of equivalent pounds of trinitrotoluene (TNT), which is a standard way of identifying all explosives regardless of their composition. The example in Figure 3-1 is a building of conventional construction (common, unhardened construction). Buildings built without any special construction at these standoff distances will probably withstand the explosive effects. Conventionally constructed buildings at standoff distances of less than those shown in Figure 3-1 will not adequately withstand blast effects. (Refer to TM 5-853-1 for information on hardening buildings to resist a blast.) Do not allow vehicles to park within the established standoff distances. Recognize that this restriction can result in significant operational and land-use problems.

3-8. **Exclusive Standoff Zone.** When an exclusive standoff zone is established, do not allow vehicles within the perimeter unless they have been searched or cleared for access. The zone's perimeter is established at the distance necessary to protect the facility against the highest threat explosive. All vehicles should be parked outside the exclusive standoff zone; only

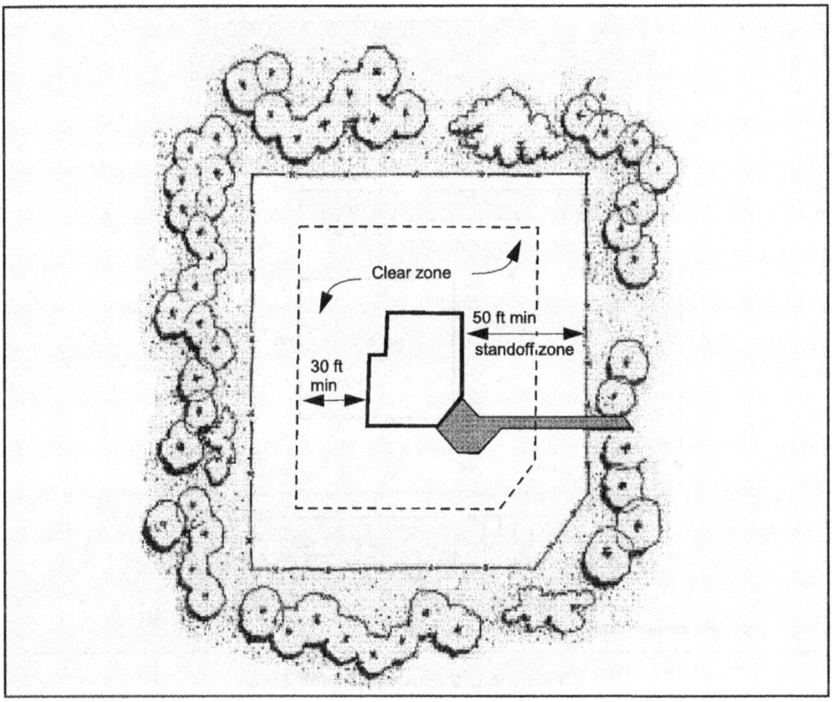

Figure 3-1. Standoff Distance

maintenance, emergency, and delivery vehicles should be allowed within the zone after being searched. Figure 3-2 shows an exclusive standoff zone.

3-9. **Nonexclusive Standoff Zone.** A nonexclusive standoff zone is established in a location having a mixture of cars and trucks (with relatively few trucks). A nonexclusive standoff zone takes advantage of aggressors being able to conceal a smaller quantity of explosives in a car than they can in a truck. Therefore, a nonexclusive standoff zone includes inner and outer perimeters. The inner perimeter is set at a distance corresponding to the weight of explosives that can be concealed in cars. The outer perimeter is set at a distance associated with the weight that can be placed in trucks (refer to TM 5-853-1). With these two perimeters, cars can enter the outer perimeter without being searched but they cannot enter the inner perimeter. Trucks cannot enter the outer perimeter, since it is established based on what they can carry. Figure 3-3 shows a nonexclusive standoff zone. The nonexclusive standoff zone provides the advantages of allowing better use of the parking areas and limiting the number of vehicles that need to be searched at the outer perimeter.

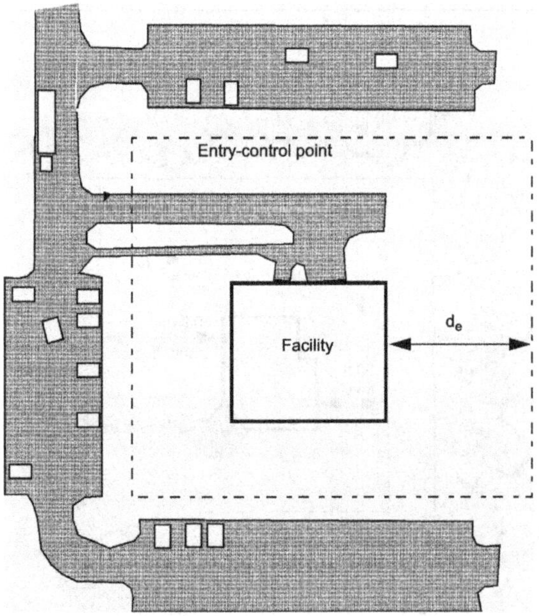

d_e = exclusive standoff-zone distance

Figure 3-2. Exclusive Standoff Zone

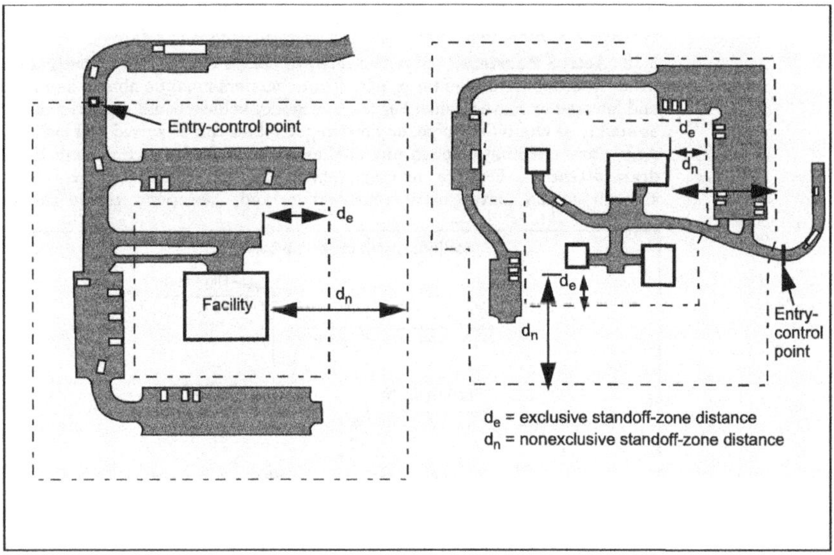

Figure 3-3. Nonexclusive Standoff Zone

Vehicle Barriers

3-10. Two types of vehicle barriers are used for vehicle bombs—perimeter and active barriers. The type of barrier used for a moving vehicle bomb differs from the barrier used for a stationary vehicle bomb. The barrier used for a stationary vehicle bomb does not have to stop a vehicle's motion. The goal for that barrier is to make anybody driving through the barrier noticeable. The assumption is that the aggressor's goal in the stationary vehicle bomb is to park the vehicle and sneak away without being noticed. Crashing through a barrier would be noticeable. Barriers for the moving vehicle bomb need to stop the vehicle's motion; they must be much more substantial.

3-11. **Perimeter Barriers.** Perimeter barriers are fixed barriers placed around the entire perimeter of a standoff zone. Anything that presents a fixed obstacle will work for the stationary vehicle bomb. Common applications include chain-link fences, hedges made of low bushes, and high (over 8 inches) curbs. Aggressors driving through such barriers are likely to be noticed. Barriers capable of stopping moving vehicles include chain-link fences reinforced with cable, reinforced concrete "Jersey barriers", pipe bollards, planters, ditches, and berms. When barriers such as the Jersey barriers and planters are used to stop moving vehicles, they must be anchored into the ground to be effective. The cables in the reinforced fence also have to be anchored into the ground or partially buried. Spaces between barriers should be no greater than 4 feet. Figure 3-4 shows common perimeter barriers for stationary or moving vehicle bombs. Refer also to TM 5-853-1.

3-12. **Active Barriers.** Active barriers are placed at openings in perimeters where vehicles need to enter or exit. These barriers must be able to be raised and lowered or moved aside. For the stationary vehicle bomb, barriers can be as simple as chain-link, pipe, or wooden gates that can be raised and lowered. Aggressors crashing through any of these or similar obstructions will likely draw attention. For the moving vehicle bomb, the barriers are heavy structures and have many construction and operations considerations

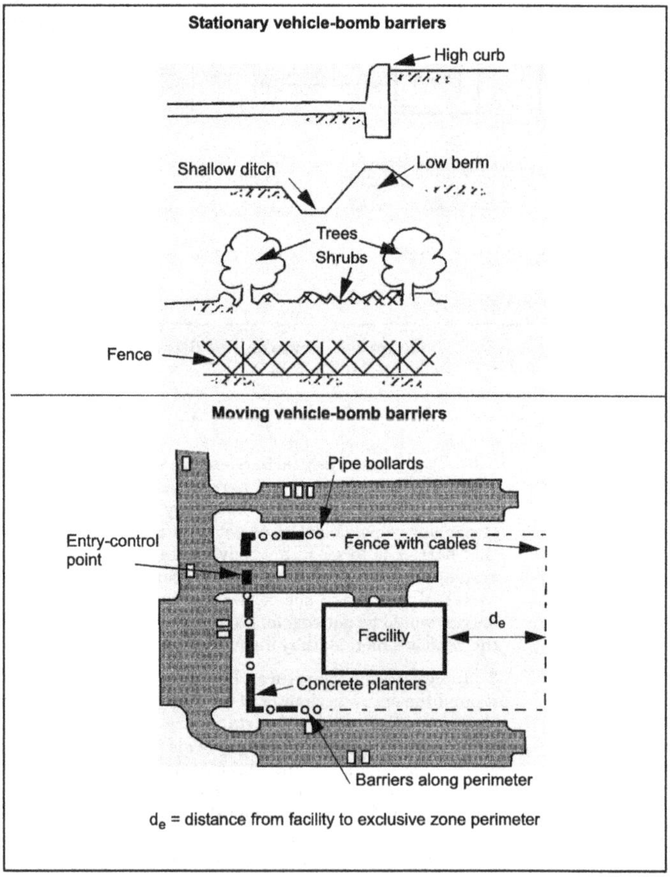

Figure 3-4. Perimeter-Barrier Application

associated with them. These barriers may stop vehicles weighing up to 15,000 pounds and travelling 50 miles per hour. They commonly cost tens of thousands of dollars (refer to TM 5-853-1). Some common active vehicle barriers are shown in Figure 3-5. For temporary or deployed conditions, park a vehicle across an opening and move it aside to grant access.

Speed Control

3-13. It is important to control the speed of a vehicle approaching a barrier used for a moving vehicle bomb. The energy from a vehicle that a barrier must stop increases as its speed increases. The energy also increases with more weight, but the effect of speed is much greater. Therefore, decreasing the vehicle's speed results in smaller and less costly barriers. The best way to limit a vehicle's approach speed to perimeter barriers is to place or retain obstacles in potential approach paths. The vehicles are forced to reduce speed when going around these obstacles. The same principle applies for road approaches. Placing obstacles in a serpentine pattern on the road forces a vehicle to reduce its speed (see Figure 3-6, page 3-8). If the vehicle hits the obstacles instead of going around them, they are still slowed down. Other means to slow vehicles include forcing them to make sharp turns and installing traffic circles.

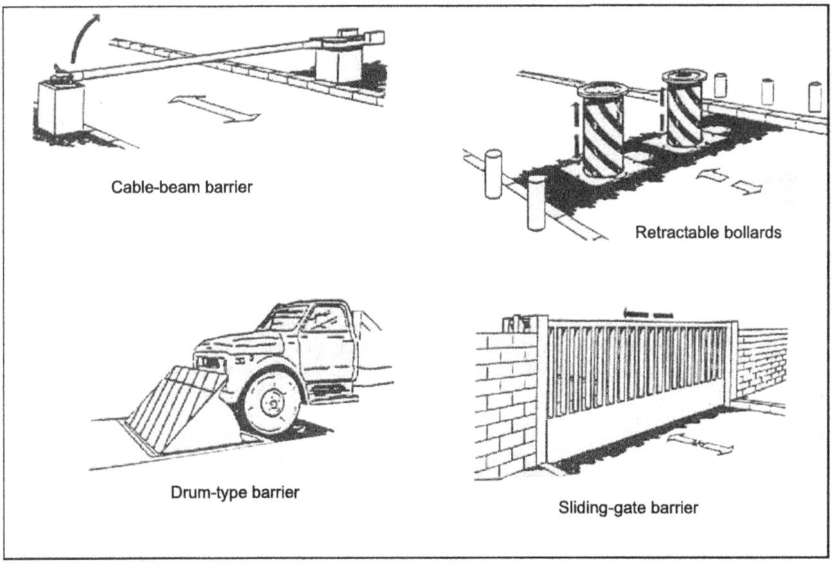

Figure 3-5. Active Vehicle Barriers

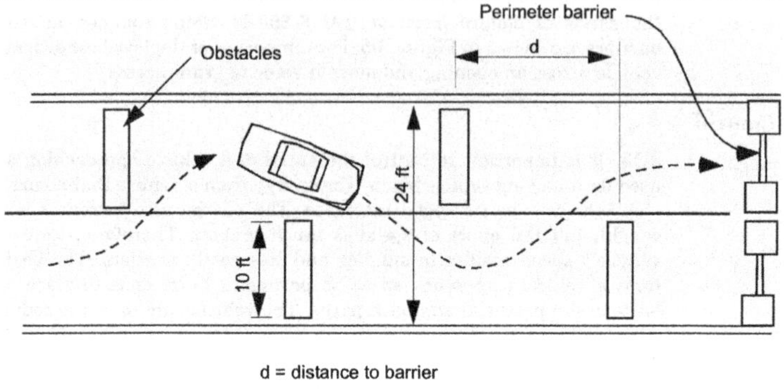

Figure 3-6. Serpentine Pattern

BUILDING ELEMENTS

3-14. Once the standoff distance is established and the site has been laid out, the designers can select the building components necessary to protect the assets against the threat explosives at the standoff distance. The building components include the walls, roofs, doors, and windows. Detailed design issues related to these building elements are covered in TM 5-853-1.

Walls and Roofs

3-15. If the distances shown for the desired damage levels in Figure 3-1, page 3-3, cannot be enforced, the building's walls and roofs will need to be strengthened. This can be achieved in new construction by using reinforced masonry or reinforced concrete in the walls and reinforced concrete in the roof. When the standoff distance is not available for existing construction, a more detailed analysis may be required to determine what the explosion's impact will be on the structure. When the construction is inadequate, more standoff distance should be investigated or the engineers should apply specialized techniques for retrofitting the construction to increase its strength.

Windows

3-16. Historically, glass fragments have caused about 85 percent of injuries and deaths in bomb blasts. There are two basic approaches to mitigating the effects of bomb blasts on glass—retrofitting the windows with film or curtains and using blast-resistant glazing.

3-17. **Retrofitting Windows**. One of the most common means of decreasing the hazards from broken glass is to install fragment-retention film on the glass. The film is a plastic (polyester) sheet that adheres to the window glass with a special adhesive. The film does not strengthen the glass; but when the glass breaks, it keeps the fragments from spreading throughout the room. The glass fragments stick to the film, and the film either stays in the window frame or falls into the room in one or more large, relatively nonhazardous pieces instead of many small, lethal pieces. Another retrofit approach is to install a blast curtain or a heavy drape behind the window. The curtain or drape catches the glass fragments. The curtains are generally used with fragment-retention film. Another retrofit technique is to use fragment-retention film with a metal bar placed across the window. This "catcher bar" catches the window. The designs for this and other types of retrofit devices are complicated and require specialized engineering-analysis tools. The retrofit techniques are generally thought of as providing a lower level of protection than the glazing replacement techniques. For deployed locations, removing the windows and covering them with plywood minimizes the danger.

3-18. **Blast-Resistant Glazing.** To achieve higher levels of protection, the window glass must be replaced and the window frame should be reinforced. Because of its expense, this procedure is generally limited to new construction and major renovations. Special blast-resistant glazing and frames are available that use either tempered glass or a plastic glazing (such as polycarbonate). Another promising type of blast-resistant glazing is laminated glass, in which several layers of common glass are adhered together with a special interlayer. The resulting laminated construction is usually stronger than common glass while retaining the same thickness. The interlayer acts similarly to fragment-retention film. For deployed locations, a means of minimizing the danger of windows is to remove them and replace them with plywood.

Doors

3-19. Doors are another building component particularly vulnerable to an explosive blast. Common metal and wood doors provide little resistance to a blast. The two ways to address the problem of doors is to install them in foyers or to replace them. Glass doors or doors containing windows should be avoided.

Foyers

3-20. Door hazards can be reduced by installing doors in foyers during construction or by adding foyers to existing buildings. When a door is located in a foyer and the outer door fails, the outer door flies into a wall instead of the building's interior (see Figure 3-7, page 3-10). The inner door then has a greater chance of remaining intact. This option generally provides a low level of protection.

3-21. Another option is to replace the doors with specially constructed blast-resistant doors and frames. These doors are commercially available and can provide a high level of protection, but they are very expensive and heavy. The doorframe must be made of the same type of material and provide the same level of protection as the door.

DETECTION ELEMENTS

3-22. Detection elements for vehicle bombs are limited to the use of guards to control access into standoff zones. The guards search vehicles seeking entry

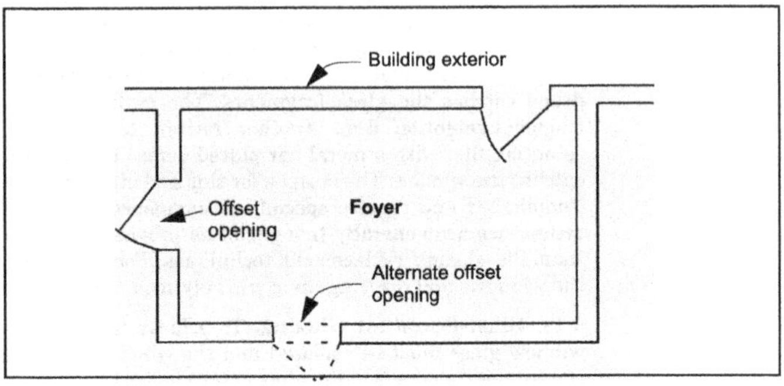

Figure 3-7. Door in Foyer

into the perimeter through an entry-control point. The recommended levels of searches depend on the required level of protection (see TM 5-853-1). Guards can be stationed at entry-control points continuously, or they can be summoned to an entry-control point when access is needed. The latter is commonly the case for the inner perimeter of exclusive standoff zones where only delivery and maintenance vehicles need access.

EXTERIOR ATTACK

3-23. An exterior attack is a physical attack using weapons such as rocks, clubs, improvised incendiary devices (IIDs) such as Molotov cocktails, explosives such as improvised explosive devices (IEDs), and hand grenades. The explosives can be thrown at or placed near a facility's exterior. Examples of IEDs for this tactic range from pipe bombs and hand grenades to briefcase-sized explosives.

GENERAL-DESIGN STRATEGY

3-24. Because the exterior attack is directed at a facility's exterior surfaces, the general-design strategy is to keep aggressors away from the facility (at a standoff distance) and, if necessary, to harden the facility's exterior components to resist the effects of weapons and explosives. A standoff distance from the facility reduces the degree of hardening required to resist weapons effects. When briefcase-sized bombs are a threat, an obstacle-free zone should be established around the facility and the explosives placed within should be detected and disarmed.

LEVELS OF PROTECTION

3-25. The levels of protection for exterior attacks are similar to those for vehicle bombs. Levels of protection vary based on the level of building damage and asset injury or damage allowed. However, due to the limited sizes of explosives involved in this tactic, the damage to the building will be much more localized and injuries or damage to assets will be confined to smaller areas.

SITE-WORK ELEMENTS

3-26. Site-work elements for exterior attacks are relatively limited because the explosive weights are more limited. Large standoff distances are not a consideration. The common approach to site-work elements is to lay out a standoff zone of about 50 feet and to provide a fence or perimeter barrier about 7 feet high. The purpose of the standoff is to make it harder for aggressors to throw pipe bombs and hand grenades at targets inside the perimeter. Trees can be left around the perimeter to make it harder for aggressors to throw explosives over the fence. The remaining component of site-work elements is a clear zone around the facility. A clear zone is applied so that anything placed in that area can be detected visually. This limits the aggressor's ability to place explosives near the target facility.

BUILDING ELEMENTS

3-27. Building elements for exterior attacks are similar to those for vehicle bombs. For small IEDs and IIDs, the building-element requirements do not increase the cost of the building significantly. For larger, briefcase-sized bombs, the measures are more significant than for incendiary devices but less than for vehicle bombs.

Walls and Roofs

3-28. Walls and roofs are not a problem with small explosives. Conventional construction normally provides adequate protection. Walls with 6-inch reinforced concrete or 8-inch, grout-filled, reinforced masonry will withstand the effects of typical pipe bombs or hand grenades. The corresponding roof construction is 6-inch reinforced concrete. In the case of briefcase-sized bombs, considerations similar to those discussed for vehicle bombs need to be employed.

Windows

3-29. A significant goal when constructing windows is to make them difficult to throw an explosive or incendiary device through, especially when considering smaller explosives. This is accomplished by constructing smaller windows or making narrow windows (see Figure 3-8, page 3-12). For existing windows, parts of the windows can be covered to achieve a narrow effect. These windows still may be susceptible to breakage due to explosive effects, even from the smaller explosives. This problem is solved by installing 3/4-inch-thick plastic (polycarbonate) glazing or by raising the windows over 6 feet high to develop a small standoff distance (as shown in Figure 3-9, page 3-12). A 3/4-inch glazing will also stop grenade fragments. Fragment-retention film, a blast curtain, or a heavy drape as described in vehicle-bomb tactics are also good applications for small bombs.

Conventionally Constructed Doors

3-30. Doors are not a significant problem with small bombs and incendiary devices. Generally, metal doors are adequate for incendiary devices, and doors placed in foyers (as shown in Figure 3-7) are adequate for pipe bombs and hand grenades. A similar application for briefcase-sized bombs would provide

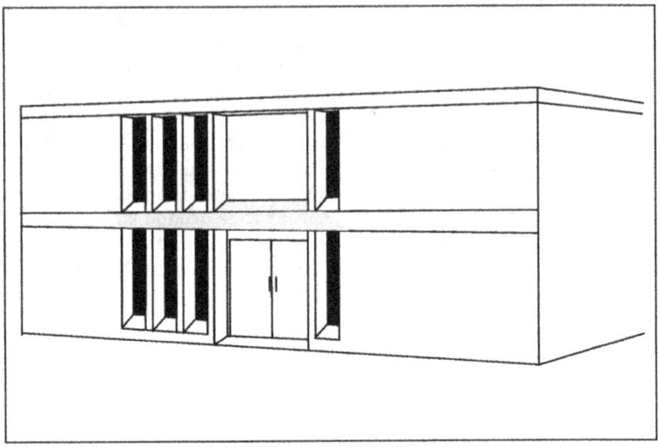

Figure 3-8. Narrow Recessed Windows

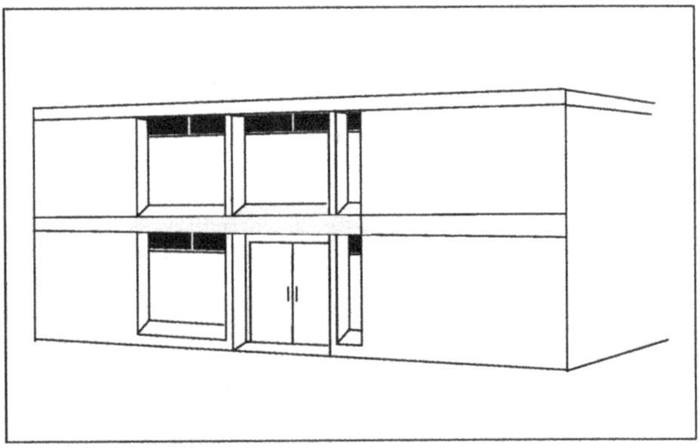

Figure 3-9. Narrow Raised Windows

only a low level of protection. To achieve higher levels of protection for briefcase-sized bombs, blast-resistant doors must be installed.

3-31. The requirements to meet the levels of protection for larger explosives are similar to those described for vehicle bombs, but they will not stop grenade fragments. Fragment-retention film and drapes or curtains can provide a low level of protection, but blast-resistant glazing is required to achieve a higher level of protection.

DETECTION ELEMENTS

3-32. Other than awareness of aggressor activity on or outside the site, detection is only a specific design goal where briefcase-sized bombs are anticipated. When that is the case, the clear zone around the building must be visually monitored so that any objects placed in it are detected. At higher levels of protection, visual surveillance is augmented by IDSs.

STANDOFF WEAPONS

3-33. The standoff-weapons tactic includes the use of AT weapons and mortars. In both of these tactics, the aggressor fires weapons at assets located in the protected facility from a distance. An AT-weapon attack requires a clear LOS to the target, while mortars can fire over obstacles and only need a clear line of flight.

GENERAL-DESIGN STRATEGY

3-34. Standoff-weapons attacks cannot be detected reliably before they occur. Protective design to resist these tactics relies on blocking LOSs to protected areas of a facility or hardening the facility to resist the particular weapon's effects. The approaches to protection against mortars and AT weapons differ from each other and will be discussed separately. Detection measures are not applicable for these tactics.

LEVELS OF PROTECTION

3-35. There are two levels of protection against both mortars and AT weapons. For AT weapons, the low level of protection depends on detonating the AT round before it hits the target facility. The high level of protection avoids the risk associated with that and hardens the building to resist the direct impact of the AT round.

3-36. For mortars, the low level of protection involves allowing some areas of the facility to be sacrificed. Those spaces provide a buffer to the assets to be protected. The assets within the sacrificial areas and the areas themselves may be destroyed. At the high level of protection, the building's exterior fully resists the mortar rounds and there are no sacrificial areas.

SITE-WORK ELEMENTS

3-37. The primary site-work element for standoff weapons is to obstruct LOSs from vantage points outside of the site. With AT weapons, the aggressor cannot hit what he cannot see. This is not true with mortars, but blocking LOSs from mortar firing points helps to make targeting more difficult. The LOSs are blocked by using trees, other buildings, vehicle parking areas, or fences. Another site-work element, a predetonation screen, applies only to an AT weapon. When using a predetonation screen, the AT round is detonated on the screen and its effects are dissipated in the distance between the screen and the target (see Figure 3-10, page 3-14). Any screen material (such as a

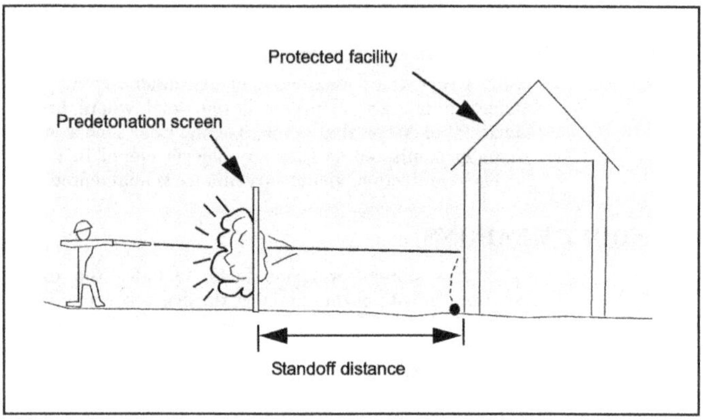

Figure 3-10. Predetonation Screen

wooden fence) will detonate the round unless it has spaces in it. The screen distances vary from less than 10 feet to almost 40 feet, depending on the building construction (see TM 5-853-1). This measure only applies to the low level of protection.

BUILDING ELEMENTS

3-38. Building elements for AT weapons and mortars involve the building's layout. This includes the materials used in the construction.

Layout

3-39. A building's interior layout is only an issue for the low level of protection against a mortar round. The layout issue involves designating sacrificial areas in which unimportant assets are located. The assets to be protected are located in a hardened interior layer. Figure 3-11 includes a plan view (from above). The sacrificial area has to be both around and above the protected area in case a mortar round comes from above. If such a layout is not feasible, other options include going to a higher level of protection and either hardening the entire building or building the facility underground (which are both very expensive).

Walls and Roofs

3-40. Walls and roofs must offer protection against both AT weapons and mortar rounds. The design of walls that protect against AT weapons varies with the level of protection. For the low level of protection where the round is predetonated, the walls can be of conventional construction, varying with the standoff distance from the predetonation screen to the wall. For higher levels of protection, the walls must resist the full effect of the round, requiring the walls to be 24-inch-thick reinforced concrete. Roofs are not an issue in

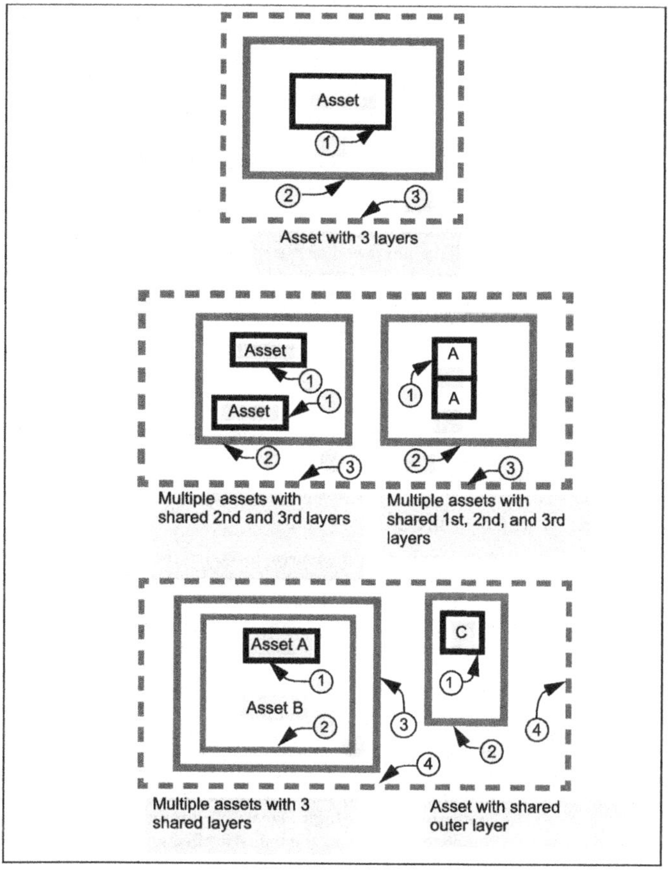

Figure 3-11. Assets Protected by Hardened Interior Layer

protecting against AT weapons because it is difficult to get direct LOSs to roofs. If such LOSs are possible, the roof should be designed like the walls.

3-41. To provide protection against mortar rounds, walls and roofs should be designed to resist the explosive effects in the rounds at the standoff distance that the sacrificial space provides. In the case of sacrificial areas, the walls can be of common construction. The interior protected-area walls are then designed of reinforced concrete or reinforced masonry for the standoff distance those sacrificial walls provide. When the walls must resist the full effect of the rounds (as in the higher level of protection), they are likely to be very thick (up to 30 inches of reinforced concrete for some improvised mortars). Similar considerations should be made for roofs. Roofs are designed to take the direct effects of the round or to take the round at the standoff distance provided by the sacrificial area.

Doors and Windows

3-42. It is impractical to provide doors and windows that are resistant to mortar rounds and AT weapons. Windows should only be used in sacrificial areas where there is a mortar threat. When there is an AT weapon threat, windows can only be used where the round is predetonated. The windows should be narrowed or raised to present a smaller target (see Figures 3-8 and 3-9, page 3-12). Doors should be placed in foyers (see Figure 3-7, page 3-10) for protection against AT rounds and to achieve a low level of protection against mortars. Blast-resistant doors are necessary to achieve a high level of protection against mortar rounds.

BALLISTICS

3-43. In a ballistics tactic, aggressors fire small arms at assets from vantage points outside of the target facility's control. Ballistic attacks cannot be detected reliably before they occur.

GENERAL-DESIGN STRATEGY

3-44. Protective measures to resist these tactics rely on blocking LOSs to protected areas of a facility or by hardening the facility to resist the ballistic effects. This strategy focuses on assets within buildings. Protecting people or property in the open is difficult and can only be addressed through operational measures. Detection measures are not applicable for this tactic.

LEVELS OF PROTECTION

3-45. There are only two levels of protection for this tactic. The low level of protection depends on blocking LOSs to assets. This strategy assumes that the aggressor cannot hit what he cannot see. The risk of an aggressor firing into a building randomly and hitting something is what makes this the low level of protection. The high level of protection involves hardening building components to resist the ballistic effects. These strategies can be thought of as either hardening or hiding.

SITE-WORK ELEMENTS

3-46. Site-work elements are of limited use for the ballistics tactic. When they are applied, they are used to obstruct LOSs from vantage points outside of the site, which is consistent with the low level of protection. The LOSs can be blocked using trees, other buildings, motor pools, or fences.

BUILDING ELEMENTS

3-47. Building elements are the principal means of protecting assets against a ballistics attack. They can be applied to achieve either the low or high level of protection.

Walls and Roofs

3-48. Walls and roofs are inherently opaque, so it is easy to achieve the low level of protection (hiding) with them. Achieving the high level of protection (hardening) for walls and roofs can be done within conventional construction using reinforced concrete, concrete-masonry units (CMUs), or clay brick. The material's required thickness is shown in Table 3-1. The thicknesses of CMUs and clay brick are nominal, meaning they do not represent the actual thickness of the material; they represent the thicknesses at which those materials are commercially available. Steel plates (mild steel and armor steel) and bullet-resistant fiberglass can be used to retrofit existing building components that would not provide the needed bullet resistance.

Table 3-1. Required Thicknesses, in Inches

Ballistics Type	Reinforced Concrete	Grouted CMU*	Clay Brick*	Steel Plate		Bullet-Resistant Fiberglass
				Mild	Armor	
.38 special	2	4	4	1/4	3/16	5/16
9 mm	2 1/2	4	4	5/16	1/4	7/16
7.62 and 5.56 mm	4	6	6	9/16	7/16	1 1/8
7.62-mm AP	6 1/2	8	8	13/16	11/16	N/A
*Nominal thicknesses						

Windows

3-49. Windows can include openings in walls and skylights, although skylights are only an issue where there are LOSs to them. When skylights require protection, treat them like windows. Achieving the low level of protection (hiding) for windows requires making it difficult to see through them, such as installing reflective film on the glass. An aggressor cannot see through the windows during daylight while it is lighter outside than inside, but he may see through them at night when the opposite might be true. Drapes or blinds that can be closed at night address that vulnerability. To achieve the high level of protection requires bullet-resistant window assemblies. These are commercially available for a wide range of ballistics types. They are purchased as manufactured-and-tested assemblies (including glazing and frames, both of which are equally bullet-resistant). The glazing materials and thicknesses and the framing details are proprietary to their manufacturers. The manufacturers make them according to industry test standards to ensure an effective product.

Doors

3-50. Doors without glass easily meet the requirements for the low level of protection. Meeting the high level of protection requires the installation of bullet-resistant door assemblies. Doors can be installed in foyers so that there is no direct LOS into assets within the building (see Figure 3-7, page 3-10).

FORCED ENTRY

3-51. In the forced-entry tactic, an aggressor tries to forcibly gain access to assets. He may use tools or explosives to breach building components or other barriers.

GENERAL-DESIGN STRATEGY

3-52. The general-design strategy for forced entry is to detect the aggressor early in the forced-entry attempt and delay him long enough for a response force to intercept him. The combination of detection and defensive measures must provide sufficient time for a response force to intercept the aggressor before he reaches the asset or before he escapes with it, depending on the protective goals for the asset. The first goal would apply where the asset is likely to be destroyed or where access to it is not acceptable. The second goal would be applied when the idea is to prevent it from being stolen.

LEVELS OF PROTECTION

3-53. Several levels of protection apply to forced entry. These levels vary in terms of system design, delay time, and response-force arrival time.

SITE-WORK ELEMENTS

3-54. Site-work elements do not normally play a major role in protecting against a forced entry. However, the site should be laid out and maintained so that an aggressor does not have a hiding place nearby that will conceal his attempts to break into the building. Another site-work element is the application of perimeter barriers, most commonly fences. Fences are effective at delineating a boundary and at keeping honest people honest, but they are ineffective for preventing a forced entry. The design strategy for forced entry is based on delaying the aggressor, and any serious aggressor could climb a fence in less than 4 seconds or can cut through a fence in less than 10 seconds. Therefore, fences are not used as delay elements, but they are used to establish boundaries and as platforms on which to hang sensors. The final site-work consideration is securing utility-access ports such as manholes. If there are utility tunnels through which aggressors can enter a building, those accesses should be locked using padlocks or locking bolts.

BUILDING ELEMENTS

3-55. Building elements are the principal construction elements of a system for protecting against a forced entry. The building elements are used to provide delay. The process for designing to resist forced entry involves laying out concentric "rings" of delay (called defensive layers). These defensive layers can include the facility's exterior, interior rooms within that layer, and containers within the interior rooms. The individual building components for each of the layers (walls, doors, windows, floors, ceilings, and roofs) provide the delay time (see TM 5-853-1).

DETECTION ELEMENTS

3-56. For a protective system to be effective against a forced entry, the aggressors must be detected at a point of adequate delay. Detection at that point can be achieved by using an IDS. Once a sensor detects an aggressor, the alarm annunciator communicates that event to security personnel, who then dispatch a response force. The alarm can be assessed through a guard response or via CCTV. Chapter 6 and TM 5-853-4 provide detailed discussion of IDSs, CCTV systems, and other elements of ESSs.

COVERT ENTRY AND INSIDER COMPROMISE

3-57. In the covert-entry tactic, an aggressor who is not authorized to be in the facility attempts to enter using false credentials. In the insider-compromise tactic, personnel with legitimate access to a facility try to compromise an asset. The insider may or may not have legitimate access to the asset itself. The purpose of the entry in either case can be to steal or otherwise compromise the asset or to destroy it. In the latter case, the aggressor may bring IEDs or IIDs.

GENERAL-DESIGN STRATEGY

3-58. The general-design strategy for both the insider-compromise and covert-entry tactics is to keep people from entering areas they are not authorized to enter. For covert entry, aggressors are denied access to controlled areas. For insider compromise, aggressors are denied access to assets within controlled areas based on their need to have access to them. The general-design strategy also includes detecting aggressors removing assets from protected areas and detecting aggressors carrying tools, weapons, and explosives into protected areas.

LEVELS OF PROTECTION

3-59. The levels of protection for these tactics address different issues, depending on whether the aggressor's goal is to steal or otherwise compromise an asset or to destroy it. When the goal is to steal or compromise an asset, the levels of protection vary with the number and sophistication of the access controls required to verify personnel access into a controlled area. When the goal is to destroy the assets, the levels of protection vary with the amount of damage the building (and the assets inside) are allowed to sustain and the sophistication of detecting weapons or explosives at entry points.

BUILDING ELEMENTS

3-60. Building elements vary with an aggressor's goal. To protect against theft or compromise of assets, building elements are used to establish and maintain controlled areas into which only authorized personnel can enter. For insider compromise, there may be an additional requirement that access be further limited among personnel otherwise authorized access to the controlled area. That access is based on the need to have access to a specific asset. The result is that the controlled area may be compartmentalized, and each compartmentalized area may have separate access requirements. There are no special construction requirements for these tactics if the goal is theft of compromise. The only requirement is that the building elements of controlled areas should provide enough resistance to require aggressors to force their way through them to gain entry and to provide evidence of the forced entry if it is attempted. Forcing entry would be contrary to the aggressor's assumed goal to be covert. In addition, a common design goal would be to limit the number of entrances into controlled areas because there will need to be access control at each entry.

3-61. To protect against the destruction of assets, building elements are used to shield assets from the effects of explosives going off at access-control points.

The basic approach is to lay out areas at access points in which guards can search for carried-in weapons, explosives, or incendiary devices. The construction of that area is designed to limit damage to the rest of the building if an explosive is detonated in that area. Those levels of damage are similar to those discussed in relation to vehicle bombs. The walls and doors between the access point and the protected area will be hardened, and the walls and doors to the outside will be of lightweight construction so that they may fail and vent the blast pressure away from the building. At the higher level of protection, the access-control area is located in a separate facility and the target building is hardened to resist an explosion in that separate facility.

DETECTION ELEMENTS

3-62. Detection elements for these tactics also vary based on the aggressor's goal. For theft, the detection elements are mainly related to access control. For destruction, the detection elements are used to detect weapons, explosives, or incendiary devices.

3-63. The main detection elements for theft or compromise are access-control devices. These can include procedural systems (such as guards checking ID), mechanical systems (such as keyed or combination locks), or electronic entry-control elements (such as electronic card readers, keypads, and biometric devices). Chapter 6 provides detailed discussion of electronic devices. The sophistication of these elements and the number used varies with the level of protection. For example, achieving the higher levels of protection requires the application of multiple forms of access-control elements such as a card reader and an electronic keypad for electronic-entry control or a badge check and badge exchange for a procedural system.

3-64. When destruction of the assets is the goal, detection is oriented toward detecting weapons, explosives, or incendiary devices. At the lower levels of protection, it is sufficient for guards to search for carried-in items. Achieving higher levels of protection requires the application of such equipment as metal detectors, X-ray machines, and explosive detectors.

SURVEILLANCE AND EAVESDROPPING

3-65. Surveillance and eavesdropping tactics include visual surveillance, acoustic eavesdropping, and electronic-emanations eavesdropping. In these tactics, aggressors remain outside of controlled areas and try to gather information from within those areas. The tools used for these tactics include ocular devices for the visual-surveillance tactic and listening devices and electronic-emanations-eavesdropping equipment for the eavesdropping tactic.

GENERAL-DESIGN STRATEGY

3-66. The general-design strategy for these tactics is to deny aggressors access to information assets. The kind of information (objects, operations, or files; secure conversations; or electronically processed data) and how it can be compromised differs for each tactic as do the specific protective strategies. Therefore, each tactic is addressed separately.

LEVELS OF PROTECTION

3-67. Each of these tactics has only one level of protection. Either one protects or fails to protect against these tactics.

SITE-WORK ELEMENTS

3-68. Site-work elements play a minor role in protecting assets from all surveillance or eavesdropping tactics. The main issue is to eliminate or control vantage points from which aggressors can surveil or eavesdrop on assets or operations. In addition, for the visual-surveillance tactic, a design goal can be to block LOSs from vantage points. Items used to block LOSs include trees, bushes, fences, and other buildings (see Figure 3-12).

BUILDING ELEMENTS

3-69. Building elements are the principal components of the protective strategies for surveillance and eavesdropping tactics. For visual surveillance,

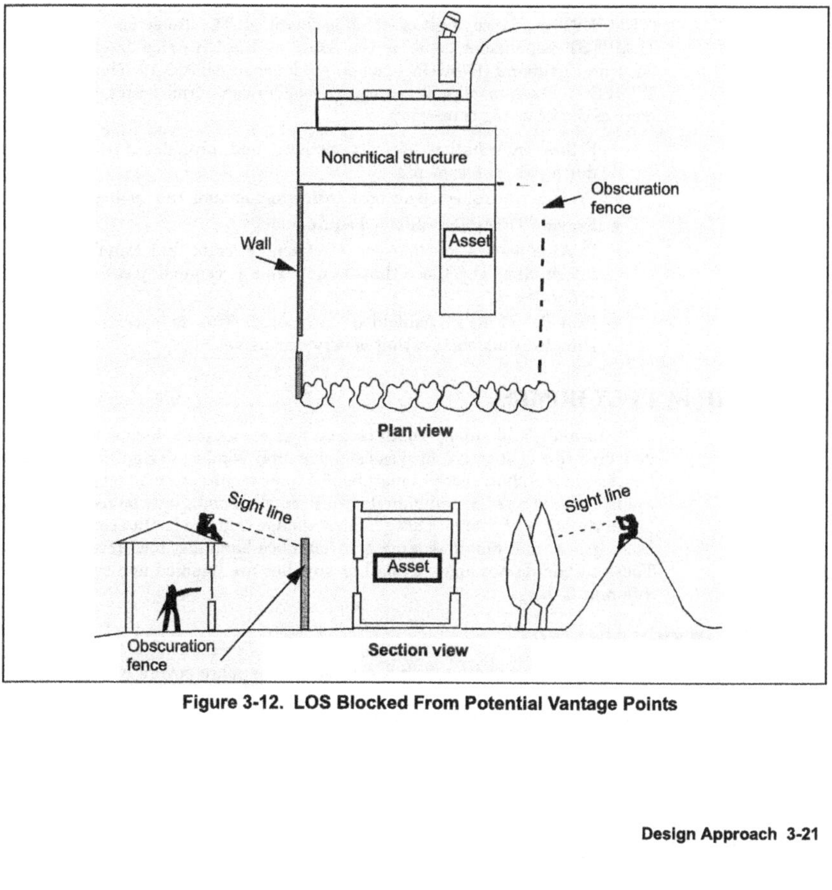

Figure 3-12. LOS Blocked From Potential Vantage Points

the building elements must block LOSs from outside the building. Walls and roofs perform this function effectively. Doors are only a problem when they have windows in them or are made of transparent materials. When this is the case, they can be treated like windows or they can be placed in foyers so that there are no direct LOSs through them. Windows can be treated with reflective film and drapes or blinds as described in the ballistics tactics. When there are LOSs through skylights, they should be treated like windows.

3-70. Building elements for acoustic eavesdropping relate to the construction of areas (preferably separated from the building exterior) that minimize the sound that can be transmitted through them. This requires specialized construction that has a sound-transmission-coefficient (STC) rating. Walls, floors, and ceilings can be constructed to achieve specific STC ratings using conventional construction materials as described in TM 5-853-1. Doors and windows that are STC rated are commonly manufactured and tested as assemblies. This type of design and construction can be expensive.

3-71. Protection against electronic-emanations eavesdropping involves the application of Terminal Electromagnetic-Pulse Emanation Standard (TEMPEST) guidance, most of which is classified. The protection is based on a TEMPEST assessment done for the Army by the US Army Intelligence and Security Command (INSCOM) and on guidance in AR 380-19. The results of a TEMPEST assessment will commonly lead to countermeasures from one or more of the following categories:

- Follow information security policies and procedures recommended during the assessments.
- Provide controlled space both inside and outside the facility.
- Provide TEMPEST-shielded equipment.
- Provide separation between electronic circuits that handle classified information and those that do not. This is commonly called red/black separation.
- Provide TEMPEST-shielded enclosures. This is specialized, metal-shielded construction that is very expensive.

MAIL AND SUPPLY BOMBS

3-72. In mail- and supply-bomb tactics, aggressors place bombs in materials delivered to a facility. Explosives used in supply bombs are significantly larger (briefcase size) than those in mail bombs (pipe bombs or smaller). Mail bombs are usually directed at individuals, while supply bombs may be used to target larger numbers of people. These tactics assume that the facility containing the asset has a mail-handling area or a supplies-handling and -receiving area. These tactics do not apply if mail or supplies are handled and screened in a different facility.

GENERAL-DESIGN STRATEGY

3-73. A bomb exploding within a building has more severe effects than the same size bomb exploding outside of the facility because the blast pressures cannot dissipate inside. Also, there is no standoff distance between the explosive and the facility to mitigate blast effects. The general-design strategy for mail and supply bombs is to detect delivered bombs before they explode and to harden the area where the explosion takes place. This minimizes the damage to the remainder of the facility. Occupants and contents within the mail room or supplies-handling area are likely to be killed or destroyed if an undetected bomb explodes.

LEVELS OF PROTECTION

3-74. The levels of protection for mail and supply bombs are based on the amount of damage allowed to the building and, therefore, the occupants of the building. They also vary based on the sophistication of the detection measures used.

BUILDING ELEMENTS

3-75. The purpose of building elements in relation to these bomb tactics is to shield assets from the effects of explosives going off at supply areas, receiving points, or mail rooms. The basic approach is to lay out either a mail room or a supplies-receiving area in which people can search suspicious packages for explosives or incendiary devices. Constructing this type of area will limit the damage to the rest of the building if an explosive is detonated there. Those levels of damage are similar to those discussed in relation to vehicle bombs.

Mail Rooms

3-76. Mail rooms should be located on the facility's exterior, away from any critical assets. The walls and ceiling between the mail room and the remainder of the building are hardened to keep the blast effects out of the facility. The exterior walls and doors should be of lightweight construction so that they may fail and vent the blast pressure away from the building. There may be an explosives container in the mail room where suspicious packages can be placed. If the package explodes, the container will keep its effects from causing damage or injury. The hardened construction will protect assets outside of the mail room if the explosion occurs outside of the container. Check with EOD personnel to determine the local policy for using explosive containers. At higher levels of protection, the mail room is constructed to completely contain the effects of an explosion either through hardened construction or by using a specialized construction called vented suppressive shielding. Mail rooms should not have windows into protected areas. Doors between the mail room and the rest of the building should be avoided, placed in foyers, or replaced with blast-resistant doors, depending on the desired level of protection.

Supplies-Handling Areas

3-77. Supplies-handling areas should also be on the building's exterior, away from critical areas of the facility. Walls and doors between the handling area and the protected area should be hardened, and the exterior walls and doors should be of lightweight construction so that they may fail and vent the blast pressure away from the building. There should be no windows between the handling area and the protected area. At the higher level of protection, the handling area is located in a separate facility and the target building is hardened to resist an explosion in that separate facility.

DETECTION ELEMENTS

3-78. Detection for these assets varies with the level of protection. At the lower levels of protection, bombs are detected by inspection. As the level of protection goes up, the sophistication of the detection increases. At the higher levels of protection, equipment such as X-ray examining devices, metal detectors, and explosives detectors can be used. Explosive-detection dogs are an alternative to explosive detectors.

CHEMICAL AND BIOLOGICAL CONTAMINATION

3-79. When using chemical- and biological-contamination tactics, aggressors introduce contaminants into the air or water supply to a facility or a group of facilities. Both airborne and waterborne contaminants include chemical, biological, and radiological agents. Aggressors may also forcibly enter a facility to contaminate water or air using the forced-entry tactic.

GENERAL-DESIGN STRATEGY

3-80. Both chemical and biological agents are difficult to detect in water and air supplies. Radiological agents are relatively easy to detect in water, but they are not commonly included in water-quality examinations. It is unlikely that all agents will be detected, so the general-design strategy for these tactics is to filter out suspected airborne contaminants or to shut off suspected waterborne contaminants. Also, because contaminants can easily be entered into the environment from inside a facility, the strategy includes limiting access to the facility (especially mechanical rooms, water intakes, and so forth).

LEVELS OF PROTECTION

3-81. The levels of protection for each of these tactics differ only in the frequency with which some protective measures are exercised. For the low level of protection, they are exercised only in response to a known threat. In the high level of protection, they are exercised continuously.

SITE-WORK ELEMENTS

3-82. Site-work elements are only significant for waterborne contamination. They include protecting water-treatment plants and water-storage structures. This protection may include constructing perimeter barriers (such as chain-link fences) and controlling access to the plant site. These measures are used because most contaminants require quantities on the order of truckloads to contaminate a water supply, so the focus of security is to keep such large vehicles under control. The perimeter barriers do not need to stop the vehicles because the assumption is that the aggressor wants to be covert. An overt act would alert people to avoid the water supply.

BUILDING ELEMENTS

3-83. Building elements for both tactics include controlling access so that aggressors cannot sneak in and plant devices in the building. Protection against airborne contamination at a facility involves making elements of the air-handling system (including air intakes) inaccessible and laying out toxin-free areas for people to be protected. A toxin-free area is an area in which the internal air pressure is higher than the external air pressure. Therefore, if a chemical, biological, or radiological device is set off outside, its contaminant will not be able to penetrate the protected area. Achieving that "net positive pressure" requires a significant air-handling system with air filters to filter contaminants out of the air. It also requires an air-lock entrance into the area so contaminants cannot enter through the door. At the low level of protection, the filters and the air-handling system are only used in response to a credible threat. At the high level of protection, that risk is not acceptable and the filters are run continuously.

3-84. The building-element issues for waterborne contamination are limited to providing protection against forced and covert entries into water-treatment plants and water-storage areas. These methods have been previously described. The only additional issue is the provision for alternative water sources. If it is suspected or detected that the water is contaminated, a backup water source should be in place (such as bottled water). For the high level of protection, bottled water should always be used for drinking.

Chapter 4

Protective Barriers

Protective barriers are used to define the physical limits of an installation, activity, or area. Barriers restrict, channel, or impede access and are fully integrated to form a continuous obstacle around the installation. They are designed to deter the worst-case threat. The barriers should be focused on providing assets with an acceptable level of protection against a threat.

OVERVIEW

4-1. Protective barriers form the perimeter of controlled, limited, and exclusion areas. Utility areas (such as water sources, transformer banks, commercial power and fuel connections, heating and power plants, or air-conditioning units) may require these barriers for safety standards. Protective barriers consist of two major categories—natural and structural.

- Natural protective barriers are mountains and deserts, cliffs and ditches, water obstacles, or other terrain features that are difficult to traverse.
- Structural protective barriers are man-made devices (such as fences, walls, floors, roofs, grills, bars, roadblocks, signs, or other construction) used to restrict, channel, or impede access.

4-2. Barriers offer important benefits to a physical-security posture. They create a psychological deterrent for anyone thinking of unauthorized entry. They may delay or even prevent passage through them. This is especially true of barriers against forced entry and vehicles. Barriers have a direct impact on the number of security posts needed and on the frequency of use for each post.

4-3. Barriers cannot be designed for all situations. Considerations for protective structural barriers include the following:

- Weighing the cost of completely enclosing large tracts of land with significant structural barriers against the threat and the cost of alternate security precautions (such as patrols, MWD teams, ground sensors, electronic surveillance, and airborne sensors).
- Sizing a restricted area based on the degree of compartmentalization required and the area's complexity. As a rule, size should be kept to a minimum consistent with operational efficiency. A restricted area's size may be driven by the likelihood of an aggressor's use of certain tactics. For example, protecting assets from a vehicle bomb often calls for a substantial explosives standoff distance. In these cases, mitigating the vehicle bomb would often be more important than minimizing the restricted area to the extent necessary for operational efficiency. Protective barriers should be established for—
  - Controlling vehicular and pedestrian traffic flow.

- Providing entry-control points where ID can be checked.
- Defining a buffer zone for more highly classified areas.
- Precluding visual compromise by unauthorized individuals.
- Delaying forced entry.
- Protecting individual assets.

4-4. If a secured area requires a limited or exclusion area on a temporary or infrequent basis, it may not be possible to use physical structural barriers. A temporary limited or exclusion area may be established where the lack of proper physical barriers is compensated for by additional security posts, patrols, and other security measures during the period of restriction. Temporary barriers (including temporary fences, coiled concertina wire, and vehicles) may be used. Barriers are not the only restrictive element, and they may not always be necessary. They may not be ideal when working with limited or exclusion areas or when integrated with other controls.

4-5. Because barriers can be compromised through breaching (cutting a hole through a fence) or by nature (berms eroded by the wind and rain), they should be inspected and maintained at least weekly. Guard-force personnel should look for deliberate breaches, holes in and under barriers, sand dunes building up against barriers, and the proper functioning of locks.

FENCING

4-6. Three types of fencing are authorized for use in protecting restricted areas—chain link, barbed wire, and barbed tape or concertina. The type used for construction depends primarily on the threat and the degree of permanence. It may also depend on the availability of materials and the time available for construction. Fencing may be erected for other uses besides impeding personnel access. It can impede observation, can serve as a means to defeat standoff-weapon systems (such as rocket-propelled grenades [RPGs]), and can serve as a barrier to hand-thrown weapons (such as grenades and firebombs).

4-7. Generally, chain-link fencing will be used for protecting permanent limited and exclusion areas. All three types of fencing may be used to augment or increase the security of existing fences that protect restricted areas. Examples would be to create an additional barrier line, to increase existing fence height, or to provide other methods that effectively add to physical security. It is important to recognize that fencing provides very little delay when it comes to motivated aggressors, but it can act as a psychological deterrent.

CHAIN LINK

4-8. Chain-link fence (including gates) must be constructed of 6-foot material, excluding the top guard. Fence heights for conventional arms and ammunition security must be 6 feet for standard chain-link, wire-mesh fencing. Chain-link fences must be constructed with 9-gauge or heavier wire. They must be galvanized with mesh openings not larger than 2 inches per side and have twisted and barbed selvages at the top and the bottom. The wire must be taut and securely fastened to rigid metal or reinforced-concrete posts set in concrete. It must reach within 2 inches of hard ground or pavement. On soft ground, it must reach below the surface deep enough to compensate for shifting soil or sand. Materials and construction must meet with the US Army Corps of Engineers (USACE) guide specifications shown in the USACE Standard (STD) 872-90 series. Weaknesses in the chain-link fence occur as a result of weather (rusting) or failure to keep it fastened to the post that affects the desired tightness. Damage to the fence and fence fabric may be the result of allowing vegetation and trees to grow on or near the fence. The interaction between the fence and the overgrowth often leads to fence damage and reduces the integrity and continuity of the fence as a perimeter boundary and barrier. The perimeter fence is the most obvious protective measure. A well-maintained fence indicates that the asset owner is dedicated to physical security.

BARBED WIRE

4-9. Standard barbed wire is twisted, double-strand, 13.5-gauge wire, with four-point barbs spaced an equal distance apart. Barbed-wire fencing (including gates) intended to prevent human trespassing should not be less than 6 feet high and must be affixed firmly to posts not more than 6 feet apart. The distance between strands should not exceed 6 inches, and at least one wire should be interlaced vertically and midway between posts. The ends must be staggered or fastened together, and the base wire must be picketed to the ground.

BARBED TAPE OR CONCERTINA

4-10. A barbed-taped obstacle (BTO) is fabricated from 0.025-inch stainless steel and is available in 24-, 30-, 40-, and 60-inch-diameter coils. The barbs shall have a minimum length of 1.2 inches, and the barb cluster's width shall be 1.21 inches. A BTO deploys tangle-free for fast installation. It may be recovered and used again. Fifty feet (plus or minus 2 inches) can be covered by 101 coil loops. Handling barbed tape requires the use of heavy barbed-tape gauntlets instead of standard barbed-wire gauntlets.

Barbed-Tape Concertina

4-11. Barbed-tape concertina (standard concertina barbed tape) is a commercially manufactured wire coil of high-strength-steel barbed wire that is clipped together at intervals to form a cylinder. When opened, it is 50 feet long and 3 feet in diameter. When used as the perimeter barrier for a restricted area, the concertina must be laid between poles with one roll on top of another or in a pyramid arrangement (with a minimum of three rolls).

4-12. Reinforced barbed-tape concertina consists of a single strand of spring-steel wire and a single strand of barbed tape. The sections between barbs of the barbed tape are securely clinched around the wire. Each coil is about 37 1/2 inches in diameter and consists of 55 spiral turns connected by steel clips to form a cylindrical diamond pattern when extended to a coil length of 50 feet. One end turn is fitted with four bundling wires for securing the coil when closed and each end turn is fitted with two steel carrying loops. The concertina extends to 50 feet without permanent distortion. When released, it can be retracted into a closed coil.

4-13. When possible, a top guard should be constructed on all perimeter fences and may be added on interior enclosures for additional protection. A top guard is an overhang of barbed wire or tape along the top of a fence, facing outward and upward at about a 45-degree angle. Placing barbed wire or tape above it can further enhance the top guard. Top-guard supporting arms will be permanently affixed to the top of fence posts to increase the overall height of the fence by at least 1 foot. (Due to liability issues in some locations, the top guards will not be allowed to face outward where the fence is adjacent to public areas.) Three strands of barbed wire spaced 6 inches apart must be installed on the supporting arms. The number of strands of wire or tape may be increased when required. The top guard of fencing adjoining gates may range from a vertical height of 18 inches to the normal 45-degree outward protection but only for sufficient distance along the fence to open the gates adequately. Bottom and top tension wires should be used in lieu of fence rails. A concrete sill may be cast at the bottom of the fence to protect against soil erosion. A bottom rail is used on high-security fences to prevent intruders from lifting the fence.

Gates and Entrances

4-14. The number of gates and perimeter entrances must be the minimum required for safe and efficient operation of the facility. Active perimeter entrances must be designed so that the guard force maintains full control. Semiactive entrances, such as infrequently used vehicular gates, must be locked on the inside when not in use. When closed, gates and entrances must provide a barrier structurally comparable to their associated barriers. Care must be afforded against the ability to crawl under gates. Top guards, which may be vertical, are required for all gates.

Triple-Standard Concertina (TSC) Wire

4-15. This type of fence uses three rolls of stacked concertina. One roll will be stacked on top of two rolls that run parallel to each other while resting on the ground, forming a pyramid. In many situations, this fence has been used effectively in place of a chain-link fence. (If perimeter fencing consists of TSC, a top guard is not feasible.)

Tangle-Foot Wire

4-16. Barbed wire or tape may be used in appropriate situations to construct a tangle-foot obstruction either outside a single perimeter fence or in the area between double fences to provide an additional deterrent to intruders. The wire or tape should be supported on short metal or wooden pickets spaced at irregular intervals of 3 to 10 feet and at heights between 6 and 12 inches. The wire or tape should be crisscrossed to provide a more effective obstacle. The space and materials available govern the depth of the field.

AIRCRAFT CABLE

4-17. Although not used very often, aircraft cable can be used as a temporary barrier. Refer to FM 5-34 for information required for determining the barrier's strength. The barrier is created using wire rope. Clips are spaced six times the diameter of the wire rope. Aircraft cable (deployed as described above or attached to a chain-link fence) can also be made to act as a barrier to moving vehicles. To do so, the cable must be anchored into the ground at both ends at about 200-foot intervals (see TM 5-853-1).

UTILITY OPENINGS

4-18. Sewers, air and water intakes and exhausts, and other utility openings of 10 inches or more in diameter that pass through perimeter barriers must have security measures equivalent to that of the perimeter (see TM 5-820-4). Specific requirements of various openings are discussed below:

- Manhole covers 10 inches or more in diameter must be secured to prevent unauthorized opening. They may be secured with locks and hasps, by welding them shut, or by bolting them to their frame. Ensure that hasps, locks, and bolts are made of materials that resist corrosion. Keyed bolts (which make removal by unauthorized personnel more difficult) are also available.
- Drainage ditches, culverts, vents, ducts, and other openings that pass through a perimeter and that have a cross-sectional area greater than 96 square inches and whose smallest dimension is greater than 6 inches will be protected by securely fastened welded bar grilles (refer to TM 5-853-3, Figure 8-1). As an alternative, drainage structures may be constructed of multiple pipes, with each pipe having a diameter of 10 inches or less. Multiple pipes of this diameter may also be placed and secured in the inflow end of a drainage culvert to prevent intrusion into the area. Ensure that any addition of grilles or pipes to culverts or other drainage structures is coordinated with the engineers so that they can compensate for the diminished flow capacity and additional maintenance that will result from the installation.

OTHER PERIMETER BARRIERS

4-19. Buildings less than two stories high that form part of a perimeter must have a top guard along the outside edge to deny access to the roof. When using masonry walls as part of a perimeter barrier, they must be at least 7 feet high and have a barbed-wire top guard. The top guard should be sloped outward at a 45-degree angle and carry at least three strands of barbed wire. This will increase the vertical height of the barrier by at least 1 foot.

4-20. Protect windows, active doors, and other designated openings by securely fastening bars, grilles, or chain-link screens. Fasten window barriers from the inside. If hinged, the hinges and locks must be on the inside. Building elements that provide delay against forced entry have stringent requirements. These elements should be designed according to TM 5-853-1.

SECURITY TOWERS

4-21. It is not acceptable to observe a perimeter from towers only. However, all towers should be located to provide maximum observation and should be constructed for protection from small-arms fire.

4-22. Mobile towers are useful in some temporary situations such as a large, open storage area where receiving and storing activities take place. All facilities using towers must have a support force available for emergencies. Tower personnel should be rotated at frequent intervals.

4-23. The height of a tower increases the range of observation during daylight hours and at night with artificial illumination. However, during inclement weather and during a blackout, towers lose this advantage and must be supplemented by on-ground observation.

4-24. The following considerations should be made when planning for the use of towers:

- Hardening the tower against small-arms effects by using sandbags, salvaged armor, or commercially fabricated bullet-resistant construction. This may require strengthening the tower supports, which should be performed only under the supervision of an engineer. The level of protection required must equate to the threat level identified during the IPB or the military decision-making process (MDMP). The best approach is to design for the worst identified threat rather than to try and modify the tower at a later date on short notice.
- Installing communications and alarm systems, both audible and visual (primary and alternate).
- Using appropriate surveillance, target-acquisition, and night-observation (STANO) equipment with the tower and perimeter barriers being surveilled. Infrared (IR) items may be especially valuable. Considerations for the selection and use of STANO equipment must be made while evaluating the effects of perimeter protective lighting.
- Providing security lighting for route protection to the tower. Security lighting also allows for support of the guard force entering or exiting the perimeter.
- Ensuring that the tower's height is determined according to the area of observation.
- Ensuring that towers have overlapping, mutually supporting fields of observation and fire.
- Providing towers with a backup fortified defensive fighting position, as appropriate.

INSTALLATION ENTRANCES

4-25. The number of installation or activity gates and perimeter entrances in active use should be limited to the minimum number required for safe and efficient operations. When necessary, install vehicle barriers in front of vehicle gates. Security lighting should be considered at entry points (see Chapter 5). Refer to TM 5-853-1 for the application and selection of these barriers.

4-26. Plans to use guards for controlling entry to an installation or activity must be predetermined based on the threat conditions (THREATCON). The construction of the guard post must be included in the security plan.

PERIMETER ENTRANCES

4-27. Active perimeter entrances should be designated so that security forces maintain full control without an unnecessary delay in traffic. This is accomplished by having sufficient entrances to accommodate the peak flow of pedestrian and vehicular traffic and having adequate lighting for rapid and efficient inspection. When gates are not operational during nonduty hours, they should be securely locked, illuminated during hours of darkness, and inspected periodically by a roving patrol. Additionally, warning signs should be used to warn drivers when gates are closed. Doors and windows on buildings that form a part of the perimeter should be locked, lighted, and inspected.

ENTRY-CONTROL STATIONS

4-28. Entry-control stations should be provided at main perimeter entrances where security personnel are present. Considerations for construction and use should be based on the information outlined in USACE STD 872-50-01.

4-29. Entry-control stations should be located as close as practical to the perimeter entrance to permit personnel inside the station to maintain constant surveillance over the entrance and its approaches. Additional considerations at entry-control stations include—

- Establishing a holding area for unauthorized vehicles or those to be inspected further. A turnaround area should be provided to keep from impeding other traffic.
- Establishing control measures such as displaying a decal on the window or having a specially marked vehicle.

4-30. Entry-control stations that are manned 24 hours each day should have interior and exterior lighting, interior heating (where appropriate), and a sufficient glassed area to afford adequate observation for personnel inside. Where appropriate, entry-control stations should be designed for optimum personnel ID and movement control. Each station should also include a telephone, a radio, and badge racks (if required).

4-31. Signs should be erected to assist in controlling authorized entry, to deter unauthorized entry, and to preclude accidental entry. Signs should be plainly displayed and be legible from any approach to the perimeter from a reasonable distance. The size and coloring of a sign, its letters, and the interval of posting must be appropriate to each situation.

4-32. Entry-control stations should be hardened against attacks according to the type of threat. The methods of hardening may include—

- Reinforced concrete or masonry.
- Steel plating.
- Bullet-resistant glass.
- Sandbags, two layers in depth.
- Commercially fabricated, bullet-resistant building components or assemblies.

WARNING SIGNS

4-33. A significant amount of warning signs should be erected to ensure that possible intruders are aware of entry into restricted areas. Warning signs augment control signs. They warn intruders that the area is restricted and that trespassing may result in the use of deadly force.

4-34. Warning signs should be installed along the limited area's physical barriers and at each entry point where they can be seen readily and understood by anyone approaching the perimeter. In areas where English is one of two or more languages commonly spoken, warning signs must contain the local language in addition to English. The wording on the signs will denote warning of a restricted area. The signs should be posted at intervals of no more than 100 feet. They must not be mounted on fences equipped with intrusion-detection equipment. Additionally, the warning signs prescribed in AR 190-13 should be posted at all entrances to limited, controlled, and exclusion areas. See Chapter 7 for more details.

OTHER SIGNS

4-35. Signs setting forth the conditions of entry to an installation or area should be plainly posted at all principal entrances. The signs should be legible under normal conditions at a distance not less than 50 feet from the point of entry. Such signs should inform the entrant of the provisions (search of the person, the vehicle, packages, and so forth) or prohibitions (such as against cameras, matches, and lighters and entry for reasons other than official business) that may be prescribed by the installation commander.

4-36. Signs or notices legibly setting forth the designation of restricted areas and provisions of entry should be plainly posted at all entrances and at other points along the perimeter line as necessary. The wording of these signs or notices is prescribed in AR 190-13.

INSTALLATION PERIMETER ROADS AND CLEAR ZONES

4-37. When the perimeter barrier encloses a large area, an interior all-weather perimeter road should be provided for security-patrol vehicles. Clear zones should be maintained on both sides of the perimeter barrier to provide an unobstructed view of the barrier and the ground adjacent to it. Roads within the clear zone should be as close to the perimeter barrier as possible without interfering with it. The roads should be constructed to allow effective road barriers to deter motor movement of unauthorized personnel during mobilization periods.

4-38. Clear zones should be kept clear of weeds, rubbish, or other material capable of offering concealment or assistance to an intruder attempting to breach the barrier. A clear zone of 20 feet or more should exist between the perimeter barrier and exterior structures, parking areas, and natural or man-made features. When possible, a clear zone of 50 feet or more should exist between the perimeter barrier and structures within the protected area, except when a building's wall constitutes part of the perimeter barrier. Ammunition supply points (ASPs) will have clear zones 12 feet outside of the ASP and 30 feet inside, and the vegetation will not exceed 8 inches (4 inches for high-threat and highly controlled areas). Refer to AR 190-11 and DOD 0-2000.12-H, Appendix EE, for further information.

4-39. When it is impossible to have adequate clear zones because of property lines or natural or man-made features, it may be necessary to increase the height of the perimeter barrier, increase security-patrol coverage, add more security lighting, or install an intrusion-detection device along that portion of the perimeter.

4-40. When considering the construction of a new site or perimeter, ensure that the plans include a fence located well inside the property line, thus permitting control of enough space outside the fence to maintain at least a minimal clear zone. The following considerations apply:

- On a large installation (such as a proving ground), it is unreasonable to construct an expensive perimeter fence and keep it under constant observation. Such an installation is usually established in a sparsely inhabited area. Its comparative isolation and the depth of the installation give reasonable perimeter protection. Under these circumstances, it is usually sufficient to post warning signs or notices, reduce access roads to a minimum, and periodically patrol the area between the outer perimeter and the conventionally protected vital area of the installation.
- An alternative to erecting new or replacing old chain-link fence involving an entire installation perimeter is to relocate or isolate the sensitive area or item by—
  - Relocating the item within a safe perimeter.
  - Consolidating the item with other items.
  - Erecting a chain-link fence (regulations permitting) around individual assets rather than the installation's perimeter.

ARMS-FACILITY STRUCTURAL STANDARDS

4-41. It is next to impossible to build a protective barrier that cannot be penetrated by a human or heavy armor. Therefore, as opposed to protecting a facility using only one barrier, enhance security by using a combination of barriers to increase delay. Multiple barriers also cause aggressors to expend more energy trying to breach all of the barriers. They also provide the appearance of additional security and may further deter some aggressors.

4-42. The interest of security must be kept in mind when constructing walls, ceilings, floors, and roofs. Facilities that house arms and ammunition are constructed as security barriers in the interest of deterring and delaying penetration. Construction guidelines for arms facilities are outlined in AR 190-11. AR 190-11 requires coordination with the engineer office, the safety office, the provost marshal office (PMO), or the security-force office when definitive drawings and specifications for new construction or upgrades or modifications of AA&E storage structures are proposed. This coordinated effort ensures that safety and physical-security requirements are met. AR 190-11 also addresses waivers and exceptions for AA&E storage structures, as well as the requirements for a tactical (training or operational) or shipboard environment. Waivers and exceptions are not discussed in this manual. The following guidelines are provided for securing AA&E in tactical and shipboard environments:

- The criteria and standards for protecting AA&E will be developed by the major Army command (MACOM) according to AR 190-11.
- The deploying commander will establish and enforce procedures for securing deployed AA&E based on the assessment of the threat, the objectives, the location, and the duration of the deployment.
- The AA&E in the tactical environment will be secured at all times.
- The AA&E will be under continuous positive control.
- Persons charged with the custody of AA&E will have the capability to sound the alarm if a forceful theft is attempted.
- A response force will be available to protect the AA&E.
- A system of supervisory checks will be established to ensure that all personnel comply with security measures. Supervisory checks of the AA&E holding area will be made to ensure that the AA&E being guarded have not been tampered with.
- All officers, noncommissioned officers (NCOs), or civilian equivalents will closely monitor the control of ammunition and explosives during field training or range firing.
- Selection of personnel to perform guard duties at AA&E holding areas will be closely monitored by commanders to ensure that only responsible individuals are assigned duties.

Chapter 5

Physical-Security Lighting

Security lighting allows security personnel to maintain visual-assessment capability during darkness. When security-lighting provisions are impractical, additional security posts, patrols, MWD patrols, NVDs, or other security means are necessary.

OVERVIEW

5-1. Security lighting should not be used as a psychological deterrent only. It should also be used along perimeter fences when the situation dictates that the fence be under continuous or periodic observation.

5-2. Lighting is relatively inexpensive to maintain and, when properly used, may reduce the need for security forces. It may also enhance personal protection for forces by reducing the advantages of concealment and surprise for a determined intruder.

5-3. Security lighting is desirable for those sensitive areas or structures within the perimeter that are under observation. Such areas or structures include pier and dock areas, vital buildings, storage areas, motor pools, and vulnerable control points in communication and power- and water-distribution systems. In interior areas where night operations are conducted, adequate lighting facilitates the detection of unauthorized persons approaching or attempting malicious acts within the area. Security lighting has considerable value as a deterrent to thieves and vandals and may make the job of the saboteur more difficult. It is an essential element of an integrated physical-security program.

5-4. A secure auxiliary power source and power-distribution system for the facility should be installed to provide redundancy to critical security lighting and other security equipment. During deployed operations, primary power may not exist or may be subject to constraints or interruptions due to poor infrastructure or hostile activity. Auxiliary power sources must be available for critical electrical loads and must be secured against direct and indirect fires as well as sabotage. If automatic-transfer switches are not installed, security procedures must designate the responsibility for the manual start of the source.

COMMANDER'S RESPONSIBILITY

5-5. Commanders determine perimeter lighting needs based on the threat, site conditions along the perimeter, surveillance capabilities, and available guard forces. Commanders ensure that security lighting is designed and used to discourage unauthorized entry and to facilitate the detection of intruders approaching or attempting to gain entry into protected areas.

PLANNING CONSIDERATIONS

5-6. Security lighting usually requires less intensity than working lights, except for ID and inspection at entry-control points. Each area of a facility presents its own unique set of considerations based on physical layout, terrain, atmospheric and climatic conditions, and security requirements. Information is available from the manufacturers of lighting equipment and from the installation's director of public works, who will assist in designing a lighting system. This information includes—

- Descriptions, characteristics, and specifications of various lighting fixtures, arc, and gaseous-discharge lamps.
- Lighting patterns of various fixtures.
- Typical layouts showing the most efficient height and spacing of equipment.
- Minimum levels of illumination and lighting uniformity required for various applications.

5-7. In planning a security-lighting system, the physical-security manager considers the—

- Cost of replacing lamps and cleaning fixtures, as well as the cost of providing the required equipment (such as ladders and mechanical buckets) to perform this maintenance.
- Provision of manual-override capability during a blackout, including photoelectric controls. These controls may be desirable in a peacetime situation but undesirable when a blackout is a possibility.
- Effects of local weather conditions on lighting systems.
- Fluctuating or erratic voltages in the primary power source.
- Grounding requirements.
- Provisions for rapid lamp replacement.
- Use of lighting to support a CCTV system.
- Limited and exclusion areas. Specific lighting requirements are referenced in AR 190-59 and TM 5-853-2. TM 5-853-4 provides guidance for facility applications that include CCTV cameras.
  - Lighting in these areas must be under the control of the guard force.
  - For critical areas (such as weapons storage areas), instantaneous lighting with a backup source is required. Any period without lighting in a critical area is unacceptable. Therefore, these areas generally have a requirement for backup power (such as diesel-engine generators, uninterrupted power supplies, and batteries) in case of power loss.
  - Security-lighting systems are operated continuously during hours of darkness.
  - Protective lights should be used so that the failure of one or more lights will not affect the operation of the remaining lights.
- Lighting requirements for adjoining properties and activities.
- Restrike time (the time required before the light will function properly after a brief power interruption).

- Color accuracy.
- Other facilities requiring lighting, such as parking areas.

PRINCIPLES OF SECURITY LIGHTING

5-8. Security lighting enables guard-force personnel to observe activities around or inside an installation while minimizing their presence. An adequate level of illumination for all approaches to an installation will not discourage unauthorized entry; however, adequate lighting improves the ability of security personnel to assess visually and intervene on attempts at unauthorized entry. Lighting is used with other security measures (such as fixed security posts or patrols, fences, and ESSs) and should never be used alone. Other principles of security lighting include the following:

- Optimum security lighting is achieved by adequate, even light on bordering areas; glaring lights in the eyes of an intruder; and little light on security-patrol routes. In addition to seeing long distances, security forces must be able to see low contrasts (such as indistinct outlines of silhouettes) and must be able to detect an intruder who may be exposed to view for only a few seconds. Higher levels of illumination improve these abilities.
- High brightness contrast between an intruder and the background should be the first consideration when planning for security lighting. With predominantly dark, dirty surfaces or camouflage-type painted surfaces, more light is needed to produce the same brightness around installations and buildings than when clean concrete, light brick, and grass predominate. When the same amount of light falls on an object and its background, the observer must depend on contrasts in the amount of light reflected. His ability to distinguish poor contrasts is significantly improved by increasing the illumination level.
- The observer primarily sees an outline or a silhouette when the intruder is darker than his background. Using light finishes on the lower parts of buildings and structures may expose an intruder who depends on dark clothing and darkened face and hands. Stripes on walls have also been used effectively, as they provide recognizable breaks in outlines or silhouettes. Providing broad-lighted areas around and within the installation against which intruders can be seen can also create good observation conditions.

5-9. To be effective, two basic systems or a combination of both may be used to provide practical and effective security lighting. The first method is to light the boundaries and approaches; the second is to light the area and structures within the property's general boundaries. Protective lighting should—

- Discourage or deter attempts at entry by intruders. Proper illumination may lead a potential intruder to believe detection is inevitable.
- Make detection likely if entry is attempted.
- Prevent glare that may temporarily blind the guards.

TYPES OF LIGHTING

5-10. The type of lighting system used depends on the installation's overall security requirements. Four types of lighting units are used for security-lighting systems—continuous, standby, movable (portable), and emergency.

5-11. Continuous lighting is the most common security-lighting system. It consists of a series of fixed lights arranged to flood a given area continuously during darkness with overlapping cones of light. Two primary methods of using continuous lighting are glare projection and controlled lighting.

- The glare security-lighting method is used when the glare of lights directed across the surrounding territory will not be annoying nor interfere with adjacent operations. It is a strong deterrent to a potential intruder because it makes it difficult to see inside of the area. Guards are protected by being kept in comparative darkness and being able to observe intruders at a considerable distance beyond the perimeter.
- Controlled lighting is best when it limits the width of the lighted strip outside the perimeter, such as along highways. In controlled lighting, the width of the lighted strip is controlled and adjusted to fit the particular need. This method of lighting may illuminate or silhouette security personnel.

5-12. Standby lighting has a layout similar to continuous lighting. However, the luminaries are not continuously lit but are either automatically or manually turned on when suspicious activity is detected or suspected by the security force or alarm systems.

5-13. Movable lighting consists of manually operated, movable searchlights that may be lit during hours of darkness or only as needed. The system normally is used to supplement continuous or standby lighting.

5-14. Emergency lighting is a system of lighting that may duplicate any or all of the above systems. Its use is limited to times of power failure or other emergencies that render the normal system inoperative. It depends on an alternative power source such as installed or portable generators or batteries.

FENCED PERIMETERS

5-15. Fenced perimeters require the lighting specifications indicated in TM 5-853-2. Specific lighting requirements are based on whether the perimeter is isolated, semi-isolated, or nonisolated.

- Isolated fenced perimeters are fence lines around areas where the fence is 100 feet or more from buildings or operating areas. The approach area is clear of obstruction for 100 or more feet outside of the fence. Other personnel do not use the area. Use glare projection for these perimeters and keep patrol routes unlit.
- Semi-isolated fenced perimeters are fence lines where approach areas are clear of obstruction for 60 to 100 feet outside of the fence. The general public or installation personnel seldom have reason to be in the area. Use controlled lighting for these perimeters and keep patrol routes in relative darkness.

- Nonisolated fenced perimeters are fence lines immediately adjacent to operating areas. These areas may be within an installation or public thoroughfares. Outsiders or installation personnel may move about freely in this approach area. The width of the lighted strip depends on the clear zones inside and outside the fence. Use controlled lighting for these perimeters. It may not be practical to keep the patrol area dark.

ENTRANCES

5-16. Entrances for pedestrians will have two or more lighting units providing adequate illumination for recognition of persons and examination of credentials. Vehicle entrances will have two lighting units located to facilitate the complete inspection of passenger cars, trucks, and freight cars as well as their contents and passengers. Semiactive and inactive entrances will have the same degree of continuous lighting as the remainder of the perimeter, with standby lighting to be used when the entrance becomes active. Gatehouses at entrances should have a low level of interior illumination, enabling guards to see approaching pedestrians and vehicles.

OTHER

5-17. Areas and structures within the installation's property line consist of yards; storage spaces; large, open working areas; piers; docks; and other sensitive areas and structures.

- Open yards (unoccupied land only) and outdoor storage spaces (material storage areas, railroad sidings, motor pools, and parking areas) should be illuminated. An open yard adjacent to a perimeter (between guards and fences) will be illuminated according to the perimeter's illumination requirements. Where lighting is necessary in other open yards, illumination will not be less than 0.2 foot-candle at any point.
- Lighting units are placed in outdoor storage spaces to provide an adequate distribution of light in aisles, passageways, and recesses to eliminate shadowed areas where unauthorized persons may hide.
- Illuminating both water approaches and the pier area safeguards piers and docks located on an installation. Decks on open piers will be illuminated to at least 1 foot-candle and the water approaches (extending to a distance of 100 feet from the pier) to at least 0.5 foot-candle. The area beneath the pier floor will be lit with small wattage floodlights arranged on the piling. Movable lighting is recommended as a part of the protective lighting system for piers and docks. The lighting must not in any way violate marine rules and regulations (it must not be glaring to pilots). Consult the US Coast Guard (USCG) for approval of protective lighting adjacent to navigable waters.

WIRING SYSTEMS

5-18. The wiring circuit should be arranged so that failure of any one lamp will not leave a large portion of the perimeter line or a major segment of a critical or vulnerable position in darkness. Feeder lines will be placed underground (or sufficiently inside the perimeter in the case of overhead wiring) to minimize the possibility of sabotage or vandalism from outside the perimeter. Another advantage to underground wiring is reduced effects from adverse weather conditions.

MAINTENANCE

5-19. Periodic inspections will be made of all electrical circuits to replace or repair worn parts, tighten connections, and check insulation. Keep fixtures clean and properly aimed.

POWER SOURCES

5-20. Primary and alternate power sources must be identified. The following is a partial list of considerations:

- The primary source is usually a local public utility.
- An alternate source (standby batteries or diesel-fuel-driven generators may be used) is provided where required and should—
  - Start automatically upon failure of primary power.
  - Be adequate to power the entire lighting system.
  - Be equipped with adequate fuel storage and supply.
  - Be tested under load to ensure efficiency and effectiveness.
  - Be located within a controlled area for additional security.

CCTV-CAMERA LIGHTING REQUIREMENTS

5-21. TM 5-853-4 provides a detailed discussion of CCTV-camera lighting requirements and guidelines for minimum lighting levels and lighting uniformity. The following considerations apply when lighting systems are intended to support CCTV assessment or surveillance:

- The camera's field of view.
- Lighting intensity levels.
- Maximum light-to-dark ratio.
- Scene reflectance.
- Daylight-to-darkness transitions.
- Camera mounting systems relative to lighting.
- The camera's spectral response.
- The cold-start time.
- The restrike time.

Chapter 6

Electronic Security Systems

An overall site-security system is comprised of three major subelements—detection, delay, and response. The detection subelement includes intrusion detection, assessment, and entry control. An ESS is an integrated system that encompasses interior and exterior sensors; CCTV systems for assessing alarm conditions; electronic entry-control systems (EECSs); data-transmission media (DTM); and alarm reporting systems for monitoring, controlling, and displaying various alarm and system information. Interior and exterior sensors and their associated communication and display subsystems are collectively called IDSs.

OVERVIEW

6-1. Many Army and DOD regulations specify protective measures, policies, and operations related to security. Although the regulations specify minimum requirements, it is possible that more stringent requirements will be necessary at specific sites. A designer will use a previously performed site survey to determine which regulations apply and to determine whether circumstances require more stringent measures. Refer to TM 5-853-4 for additional detailed information.

6-2. AR 190-13 requires the use of a standardized ESS, if practical and available. The receiving element must determine whether a standardized system can meet the requirements and whether it is available. After coordinating with the product manager for physical-security equipment to verify that a standardized system is available, the associated MACOM can issue approval to procure a commercial system in lieu of a standardized system.

USE OF ESS

6-3. An ESS is used to provide early warning of an intruder. This system consists of hardware and software elements operated by trained security personnel.

6-4. A system is configured to provide one or more layers of detection around an asset. Each layer is made up of a series of contiguous detection zones designed to isolate the asset and to control the entry and exit of authorized personnel and materials.

GENERAL ESS DESCRIPTION

6-5. An ESS consists of sensors interfaced with electronic entry-control devices, CCTV, alarm reporting displays (both visual and audible), and security lighting. The situation is assessed by sending guards to the alarm point or by using CCTV. Alarm reporting devices and video monitors are located in the security center. The asset's importance will determine whether multiple or redundant security centers are required and, ultimately, the required sophistication of all elements in the ESS. Digital and analog data are transmitted from local (field) interior and exterior locations to the security center for processing. Reliability and accuracy are important functional requirements of the data-transmission system.

ESS IMPLEMENTATION PROCESS

6-6. The ESS implementation process is shown in Figure 6-1. Implementing an ESS is based on general requirements tailored to a site-specific mission and physical profile. The process begins with a site survey that includes a top-down view of basic needs and classic configurations that are tailored to such site-specific characteristics as terrain, site geography, climatic conditions, the type of asset, and priorities. This data is used to determine the hardware and software requirements, taking into account the additional capacity that should be factored into the design system for future expansion. Once the requirements for an ESS have been identified, the user must determine whether an existing standardized system is suitable for the application. (AR 190-13 outlines the process for gaining approval to use nonstandard equipment.) The user must also secure funding for the equipment (refer to Appendix J). Depending on the current funding regulations, operation-and-maintenance, procurement, or other funds may be required. For example, operations and procurement, Army (OPA) funds may be required for IDS devices; and operations and maintenance, Army (OMA) funds may be required for installation items. A contract is normally awarded to procure and install the equipment. The procurement or installation must be overseen. This may be accomplished by reviewing submittals, inspecting the contractor's work, or responding to the contractor's requests for information. Once the equipment is installed, the acceptance-testing activities must be witnessed and verified. Site conditions during acceptance testing affect the demonstrated detection capability of an exterior IDS. As feasible, acceptance testing should be designed to determine a sensor system's probability of detection (PD) under a range of conditions. For some types of sensor systems, this may be as straightforward as conducting both daytime and nighttime trials to experience differences in temperature and solar heating. After the ESS has been accepted, it must be operated and maintained throughout the remainder of its life cycle. Planning for manpower to operate the system and forecasting the funding and personnel to properly maintain the system is critical for success.

ESS DESIGN CONSIDERATIONS

6-7. A facility may require interior and exterior ESS elements, depending on the level of protection required. The applicable regulations, threat, and design criteria will define the ESS's general requirements. For an existing ESS, hardware and software may need to be supplemented, upgraded, or completely replaced. A site layout (in which all assets are identified and located) is required. It is a useful design tool for such tasks as configuring the DTM.

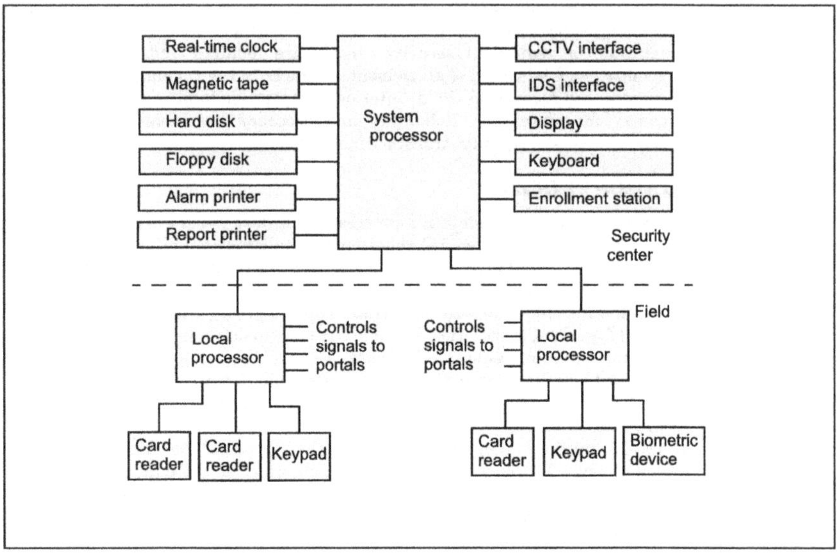

Figure 6-1. Entry-Control System Configured With Distributed Control

6-8. The exterior and interior IDSs should be configured as layers of unbroken rings concentrically surrounding the asset. These rings should correspond to defensive layers that constitute the delay system. The first detection layer is located at the outermost defensive layer necessary to provide the required delay. Detection layers can be on a defensive layer, in the area between two defensive layers, or on the asset itself, depending on the delay required. For example, if a wall of an interior room provides sufficient delay for effective response to aggression, detection layers could be between the facility exterior and interior-room wall or on the interior-room wall. These would detect the intruder before penetration of the interior wall is possible.

RESPONSE AND DELAY

6-9. When dealing with an ESS, the response time is defined as the time it takes the security force to arrive at the scene after an initial alarm is received at the security center. The total delay time is defined as the sum of all of the barriers' delay times, the time required to cross the areas between barriers after an intrusion alarm has been reported, and the time required to accomplish the mission and leave the protected area.

6-10. An ESS's basic function is to notify security personnel that an intruder is attempting to penetrate, or has penetrated, a protected area in sufficient time to allow the response force to intercept and apprehend him. To accomplish this, there must be sufficient physical delay between the point where the intruder is first detected and his objective. This provides delay time equal to or greater than the response time (refer to TM 5-853-1).

6-11. When dealing with interior sensors, boundary sensors that detect penetration (such as structural-vibration sensors or passive ultrasonic sensors) provide the earliest warning of an attempted intrusion. This alarm is usually generated before the barrier is penetrated. This gives the security force advance notification of an attempted penetration, thus allowing the barrier's delay time to be counted as part of the total delay time. Door-position sensors and glass-breakage sensors do not generate an alarm until the barrier has been breached; therefore, the delay time provided by the barrier cannot be counted as part of the total delay time.

6-12. Volumetric motion sensors do not generate an alarm until the intruder is already inside the area covered by the sensors. Therefore, if these sensors are to be used to provide additional response time, additional barriers must be placed between the volumetric motion sensors and the protected asset. Point sensors, such as capacitance sensors and pressure mats, provide warning of attempted penetration only if they detect the intruder before access is gained to the protected area.

BASIC GUIDANCE

6-13. An IDS is deployed in and around barriers (as detailed in TM 5-853-1). Voice communication links (radio, intercom, and telephone) with the response force are located in the security center. Security personnel will man the center and will alert and dispatch response forces in case of an alarm.

6-14. The barrier should always be deployed behind the IDS to ensure that integrity is maintained against intruders. An intruder will then activate the alarm sensor before penetrating or bypassing the barriers, thus providing delay for alarm assessment and response. The delay time is the determining factor in whether an assessment is conducted by dispatching a guard or by observing the CCTV. Normally, an intruder can climb a fence before a guard can be dispatched; therefore, a CCTV is usually required with an exterior IDS. Barriers can be located ahead of an alarm sensor as a boundary demarcation and can serve to keep people and animals from causing nuisance alarms by inadvertently straying into a controlled area. These barriers provide no additional response time because the barrier could be breached before the IDS sensors could be activated.

6-15. Data for monitoring and controlling an ESS are gathered and processed in the security center where the operator interacts with information from the ESS components located at remote facilities. The ESS's alarm-annunciation computer and its DTM line-termination equipment should be located in a controlled area and provided with tamper protection. Supervisory personnel should permit changes to software only, and these changes should be documented. If redundant DTM links connect the central computer to the local processor, diverse paths should be used to route these links.

6-16. The preferred medium for transmitting data in an ESS is a dedicated fiber-optics system. It provides for communications not susceptible to voltage transients, lightning, electromagnetic interference, and noise. Additionally, the fiber optics will provide a measure of communication-line security and wide bandwidth for video signals and increased data-transmission rates.

ESS EFFECTIVENESS

6-17. An ESS has a degree-of-protection effectiveness that is based on its probability of detecting intruders attempting to go over, under, around, or through the physical-security system. The intruder may use forced-entry, covert-entry, or insider-compromise tactics. A well-designed system will minimize the possibility of a successful penetration through covert entry or insider compromise. Interior and exterior alarm sensors have a PD based on the capability to detect an intruder passing through a sensing field. An intruder disturbs the steady-state quiescent condition of a sensor for a finite period. Sensors are designed to detect a person of minimum stature moving within a specific range of speeds and distances from the sensor, and any target outside of those parameters will probably not be detected. The PD for a specific sensor is usually specified at 0.9 or greater, but the designer must be aware that the PD is based on certain constraints and environmental conditions.

6-18. Manufacturer specifications usually do not discuss environmental or nuisance alarms that can be caused by climatic conditions (such as wind or rain) or by the intrusion of animals (including birds). The alarm annunciation is valid because the sensor's thresholds have been exceeded; however, the alarm does not represent a valid penetration attempt. If the assessment system is slow, the operator may not be able to determine the cause of the alarm and must, therefore, treat an environmental or nuisance alarm as real.

6-19. Another type of false alarm is caused by electronic-circuit tolerances being exceeded, resulting in the sensor's actuation. False alarms may also result from improper installation of the sensor or from effects of other equipment in the immediate area.

6-20. After an alarm is sensed and information is displayed in the security center, the console operator must determine the cause of the alarm (intrusion, nuisance, environmental, or false). Timely assessment is required when determining its cause. For example, if an intruder scales a fence in 10 seconds and runs 20 feet per second, the intruder will have overcome the barrier and be 2,200 feet from the point of penetration in 2 minutes. To conduct an accurate assessment of the alarm after 2 minutes, guards will have to search an area of about 200 acres. A fixed-television camera properly located and integrated with the alarm processor can assess the situation while the intruder is still in the controlled area.

6-21. For a CCTV camera to be effective, the area it views must be adequately lighted. To correlate the alarms and cameras in a large system (more than 10 cameras) in a timely manner, a computer-based processing system must be used to select and display alarms and camera scenes for the operator. A complex ESS has the following basic components:

- Intrusion-detection sensors.
- Electronic entry-control devices.
- CCTV.

- Alarm-annunciation system.
- DTM.

6-22. The intrusion-detection sensors are normally deployed in a series of concentric layers. The overall PD improves with each added layer of sensors. The layers (interior and exterior) should be functionally uniform; however, their overall effectiveness and cost are different. The exterior zones significantly differ from the interior zones due to the following considerations:

- The consistency of the PD.
- The PD.
- The cost per detection zone.
- The number of zones.
- The overall sensor coverage.

6-23. Exterior IDSs usually have PDs equal to those of interior IDSs. However, exterior sensors are more likely to experience weather-related situations that cause the system's PD to vary. Sensor phenomenology (passive infrared [PIR], microwave radar, and so forth) determines which environmental factors may alter the system's PD. The frequency of occurrence, severity, and duration of a weather event jointly determine whether it represents security vulnerability with the IDS in use. Typically, sophisticated intruders will attempt their penetration and challenge an ESS under conditions most favorable to themselves. Inclement weather (fog, snow, and rain) affects the usefulness of CCTVs and security lighting such that the capability for remote assessment of alarm events may be lost. Exterior IDSs are not necessarily less likely to detect a penetration attempt during fog, rain and snow; the effect of such site conditions on the IDS depends on sensor phenomenology. For example, fence motion caused by rain impact may drive the response of a fence-mounted sensor closer to satisfying the system's alarm criteria, with the result that the margin of disturbance available to the intruder is less. Also, certain buried sensors are more likely to detect an intruder when the ground is wet because of rain or melting snow. Since interior sensor systems are less influenced by environmental conditions, their PD is typically more consistent than that of some types of exterior sensor systems. Other considerations in comparing an interior and exterior ESS are the cost, the number and size of detection zones required, and the detection height.

- Because of environmental conditions, the exterior electronics must be designed and packaged for extremes of temperature, moisture, and wind. The result is that exterior electronic packages are more costly than equivalent packages for interior applications.
- State-of-the-art exterior sensors do not detect penetration attempts above the height of a fence (typically 8 feet). Fence-mounted sensors are usually limited to this height because the fence fabric or poles are used to support the sensor. For aboveground sensors in the controlled area between the fences, the sensor's mounting brackets and posts limit the detection height. In some applications of field sensors (especially buried sensors), the detection height is no more than 3 feet. For a facility, interior sensors can be deployed on walls, floors, or ceilings, thus permitting complete protection of the asset.

6-24. An interior ESS may be far less costly than that of a comparable exterior ESS. This comparison indicates to the designer the value of selecting and deploying a well-planned, well-designed, layered system. The basic rule in overall design of an ESS is to design from the inside out; that is, layered from the asset to the site boundary.

INTERIOR ESS CONSIDERATIONS

6-25. An interior ESS is typically deployed within a boundary in the immediate vicinity of the asset being protected. If the interior ESS operates in a controlled environment, its PD will be independent of any weather-induced variation in exterior conditions. Also, the physical-security system's effectiveness is enhanced by the interior barriers (walls, ceiling, and floor) that inherently impose a longer delay than exterior barriers (fences and gates).

6-26. Functionally, an interior asset should be viewed as being contained within a cube with sensors protecting all six faces. Interior sensors can be deployed at the cube's perimeter, in its interior space, or in the space immediately outside of the cube.

6-27. If an increased level of protection is dictated by the threat, and if the building is large enough, multiple layers of interior sensors may be deployed for a given asset. A multilayered interior IDS will improve the overall PD. Tamper protection and access-/secure-mode capabilities must be considered when planning and laying out interior sensors.

TAMPER PROTECTION

6-28. To minimize the possibility of someone tampering with circuitry and associated wiring, all sensor-related enclosures must be equipped with tamper switches. These switches must be positioned so that an alarm is generated before the cover has been moved enough to permit access to the circuitry of adjustment controls. In addition, several types of sensors should be equipped with tamper switches to protect against being repositioned or removed. Security screens containing grid-wire sensors and vibration sensors that can be easily removed from a wall are examples of sensors that require tamper switches.

ACCESS/SECURE MODE

6-29. During regular working hours, many of the interior sensors must be deactivated by placing the area in the access mode. For example, door-position sensors and volumetric sensors in occupied areas must be deactivated to prevent multiple nuisance alarms caused by the normal movement of people. This can be done locally or remotely. With local control, a switch is used to bypass or shunt alarm contacts when the sensor is placed in the access mode. When done remotely, the security-center operator usually enters a command that causes the processor software to ignore incoming alarms from those sensors placed in the access mode. However, when a sensor is placed in the access mode, its tamper-protection circuitry must remain in the activated or secure mode. During nonworking hours when the facility is unoccupied, all sensors must be placed in the secure mode. Certain devices (such as duress-alarm switches, tamper switches, grid-wire sensors covering vent openings, and glass-breakage sensors) should never be placed in the access mode. The designer must ensure that selected sensors can be placed in an access mode (if required) and that certain types of sensors (such as duress and tamper switches) are configured so that they cannot be put in the access mode under any condition.

EXTERIOR ESS CONSIDERATIONS

6-30. An exterior ESS is typically deployed at a site's boundary or some other significant boundary such as the demarcation fence for a group of bunkers. An exterior ESS has the advantage that it remains in the secure mode at all times.

6-31. The ideal configuration for an exterior ESS is a rectangle or a polygon, with all sides being straight. The ESS is located in and around barriers that typically include a dual fence. The outside fence is used for demarcation, and the interior fence is used to aid in detection and provide some delay. If dual fences are not used, the sensors should be deployed on the fence or inside it.

DESIGN GUIDELINES

6-32. The general-design criteria of a perimeter IDS involves primarily the selection and layout of exterior sensors that are compatible with the physical and operational characteristics of a specific site. Important factors to consider during the selection process include physical and environmental conditions at the site, the sensor's performance, and the overall cost of the system. Refer to TMs 5-853-1 and 5-853-2 for additional guidance on the requirements for and placement of exterior sensor systems. Since exterior barriers provide very little delay, exterior sensor systems generally do not provide a significant increase in the available response time.

Physical and Environmental Considerations

6-33. Physical and environmental considerations are often the determining factors for selecting exterior sensors. The site's characteristics can significantly affect a sensor's operational performance, both in terms of PD and the susceptibility to nuisance alarms. Exterior sensor systems should be selected on the basis of the frequency and duration of weather-related periods of poor detection capability. An exterior IDS may have an unacceptably low PD during a particular weather event or site condition, yet otherwise be superior to other IDSs in terms of good detection capability and a low nuisance-alarm rate. It may be appropriate to select that IDS in spite of its known vulnerability, precisely because the circumstances of its vulnerability are known and precautionary measures can be taken at those times. The overall performance of that IDS, together with its cost, may justify its selection.

6-34. Weather and climatic conditions at a specific site can significantly influence sensor selection. For example, IR detectors are not very effective in heavy rain, fog, dust, or snow. Deep snow can affect detection patterns and performance of both IR and microwave sensors. High winds can cause numerous false alarms in fence-mounted sensors. Electrical storms can cause alarms in many types of sensors and may also damage the equipment.

6-35. Vegetation can be a significant cause of nuisance alarms. Tall grass or weeds can disturb the energy pattern of microwave and both thermal IR and near-IR beam-break sensors. Vegetation growing near electric-field sensors and capacitance sensors can cause nuisance alarms. Large weeds or bushes rubbing against a fence can produce nuisance alarms from fence-mounted sensors. Large trees and bushes moving within the field of view of video motion sensors can cause nuisance or environmental alarms. A clear area must be established for exterior sensors. This area must be void of vegetation or contain vegetation of carefully controlled growth.

6-36. Topographic features are extremely important. Ideally, perimeter terrain should be flat, although gently sloping terrain is acceptable. Irregular terrain with steep slopes may preclude the use of LOS sensors and make CCTV assessment difficult. Gullies and ditches crossing the perimeter represent a vulnerability to LOS sensors and may be a source of false alarms (from flowing water) for buried line sensors. Large culverts can provide an intruder with an entry or exit route across the perimeter without causing an alarm. Likewise, overhead power and communication lines may permit an intruder to bridge the perimeter without causing an alarm.

6-37. Large animals (such as cows, horses, and deer) can cause nuisance alarms in both aboveground and buried sensors. Sensors sensitive enough to detect a crawling or rolling intruder are susceptible to nuisance alarms from small animals such as rabbits, squirrels, cats, and dogs. To minimize the interference from animals, a dual chain-link-fence configuration may be established around the site perimeter with the sensors installed between the fences.

Sensor Performance

6-38. Exterior sensors must have a high PD for all types of intrusion and have a low unwanted-alarm rate for all expected environmental and site conditions. Unfortunately, no single exterior sensor that is presently available meets both these criteria. All are limited in their detection capability, and all have high susceptibility to nuisance and environmental conditions. Table 6-1, page 6-10, provides estimates of PDs for various types of intrusions. Table 6-2, page 6-10, lists the relative susceptibility of various types of sensors to nuisance and environmental alarms.

Economic Considerations

6-39. Exterior sensor costs are usually given in cost per linear foot per detection zone (typically 300 feet). These costs include both equipment and installation. Fence-mounted sensors (such as strain-sensitive cable, electromechanical, and mechanical) are generally less costly than stand-alone and buried line sensors. Installation costs can vary significantly, depending on the type of sensor. Table 6-3, page 6-11, provides a comparison of relative costs for procuring and installing various types of exterior sensor systems. It should be remembered that the sensor system's cost is only a portion of the total cost for employing a perimeter IDS. Additional costs include fencing, site preparation, CCTV assessment, and perimeter lighting.

Table 6-1. Estimate of PD by Exterior Sensors

Type of Sensor	Intruder Technique											
	Slow walk	Walking	Running	Crawling	Rolling	Jumping	Tunnelling	Trenching	Bridging	Cutting	Climbing	Lifting
Fence mounted	N/A	N/A	N/A	N/A	N/A	VH	VL	L	VL	M/H	H	M/H
Taut wire	N/A	N/A	N/A	N/A	N/A	VH	VL	VL	VL	H	H	H
Electric field	VH	VH	VH	H	VH	VH	VL	L	L	N/A	N/A	N/A
Capacitance	VH	VH	VH	H	H	VH	VL	L	L	N/A	N/A	N/A
Ported cable	H	VH	VH	VH	VH	H	M	VH	L	N/A	N/A	N/A
Seismic	H	VH	H	M	M	M	L	M	L	N/A	N/A	N/A
Seismic/magnetic	H	VH	H	M	M	M	L	M	L	N/A	N/A	N/A
Microwave	H	VH	H	M/H	M/H	M/H	VL	L/M	L	N/A	N/A	N/A
IR	VH	VH	VH	M/H	M/H	H	VL	L	VL	N/A	N/A	N/A
Video motion	H	VH	VH	H	H	H	VL	L/M	M	N/A	N/A	N/A
VL = very low, L = low, M = medium, H = high, VH = very high, N/A = not applicable												

Table 6-2. Relative Susceptibility of Exterior Sensors to False Alarms

Type of Sensor	Intruder Technique											
	Wind	Rain	Standing water/runoff	Snow	Fog	Small animals	Large animals	Small birds	Large birds	Lightning	Overhead power lines	Buried power lines
Fence mounted	H	M	L	L	VL	L	M	L	L	L	VL	VL
Taut wire	VL	VL	VL	VL	VL	VL	L	VL	VL	VL	VL	VL
Electric field	M	L/H	VL	M	VL	M	VH	L	M	M	L	VL
Capacitance	M	M	VL	M	VL	M	VH	L	M	M	L	VL
Ported cable	VL	M	H	L	VL	VL	M	VL	VL	M	VL	L
Seismic	M	L	L	L	VL	L	VH	VL	VL	L	L	M
Seismic/magnetic	M	L	L	L	VL	L	VH	VL	VL	H	M	H
Microwave	L	L	M/H	L/M	L	M/H	VH	VL	M	L/M	L	VL
IR	L	L	L	M	M	M	VH	L	M	L	VL	VL
Video motion	M	L	L	L	M/H	L	VH	VL	M	L	L	VL
VL = very low, L = low, M = medium, H = high, VH = very high												

Table 6-3. Exterior IDS Sensor Cost Comparison

Type of Sensor	Equipment	Installation	Maintenance
Fence mounted	L	L	L
Taut wire	H	H	M
Electric field	H	M	M
Capacitance	M	L	M
Ported cable	H	M	M
Seismic	M	M	L
Seismic/magnetic	H	M	L
Microwave	M	M	L
IR	M	L	M
Video motion	M	L	M
L = low, M = medium, H = high			

PERIMETER LAYOUT AND ZONING

6-40. A protected area's perimeter is usually defined by an enclosing wall or fence or a natural barrier such as water. For exterior sensors to be effective, the perimeter around which they are to be deployed must be precisely defined. In most applications, a dual chain-link-fence configuration will be established around the perimeter. Typically, fences should be between 30 and 50 feet apart; as the distance increases, it is harder for an intruder to bridge the fences. If fence separation is less than 30 feet, some microwave and ported-coax sensors cannot be used. The area between fences (called the controlled area or isolation zone) may need to be cleared of vegetation and graded, depending on the type of sensor used. Proper drainage is required to preclude standing water and to prevent the formation of gullies caused by running water after a heavy rain or melting snow. Cleared areas are required inside and outside of the controlled area. These areas enhance routine observation, as well as sensor-alarm assessment, and minimize the protective cover available to a would-be intruder.

6-41. After the perimeter has been defined, the next step is to divide it into specific detection zones. The length of each detection zone is determined by evaluating the contour, the existing terrain, and the operational activities along the perimeter. Detection zones should be long and straight to minimize the number of sensors or cameras necessary and to aid guard assessment if cameras are not used. It may be more economical to straighten an existing fence line than to create numerous detection zones in accommodating a crooked fence line. If the perimeter is hilly and LOS sensors or CCTV assessment are used, the length of individual detection zones will be commensurate with sensor limitations. Entry points for personnel and vehicles must be configured as independent zones. This enables deactivation of the sensors in these zones; that is, placing them in the access mode during customary working hours (assuming the entry points are manned) without having to deactivate adjacent areas.

6-42. The specific length of individual zones can vary around the perimeter. Although specific manufacturers may advertise maximum zone lengths exceeding 1,000 feet, it is not practical to exceed a zone length of 300 feet. If the zone is longer, it will be difficult for an operator using CCTV assessment or for the response force to identify the location of an intrusion or the cause of a false alarm.

6-43. When establishing zones using multiple sensors, the designer should establish coincident zones where the length and location of each individual sensor will be identical for all sensors within a given zone. If an alarm occurs in a specific zone, the operator can readily determine its approximate location by referring to a map of the perimeter. This also minimizes the number of CCTV cameras required for assessment and simplifies the interface between the alarm-annunciation system and the CCTV switching system.

ESS ALARM-ANNUNCIATION SYSTEM

6-44. Status information from the various intrusion-detection sensors and entry-control terminal devices must be collected from the field and transmitted to the alarm-annunciation system in the security center, where it is processed, annunciated, and acted on by security personnel. The alarm-annunciation system may also interface with a CCTV system. There are typically two types of alarm-annunciation configurations available. The simplest configuration, which is suitable for small installations, is the point-to-point configuration. With this configuration, a separate transmission line is routed from the protected area to the security center (see Figure 6-2). The Joint-Service Interior Intrusion-Detection System (J-SIIDS) is typical of this type of configuration but will not be further discussed in this manual. The second, and more popular type, is a digital multiplexed configuration that allows multiple protected areas to communicate with the security center over a common data line. A block diagram of a typical multiplexed alarm-annunciation system is shown in Figure 6-3.

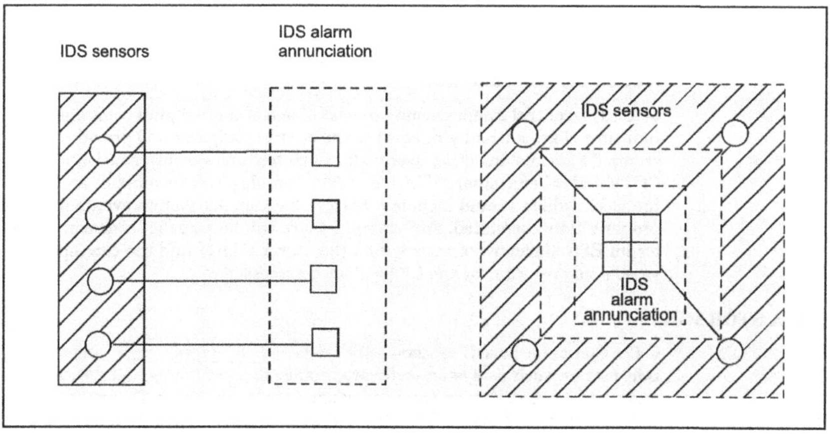

Figure 6-2. Typical Point-to-Point IDS

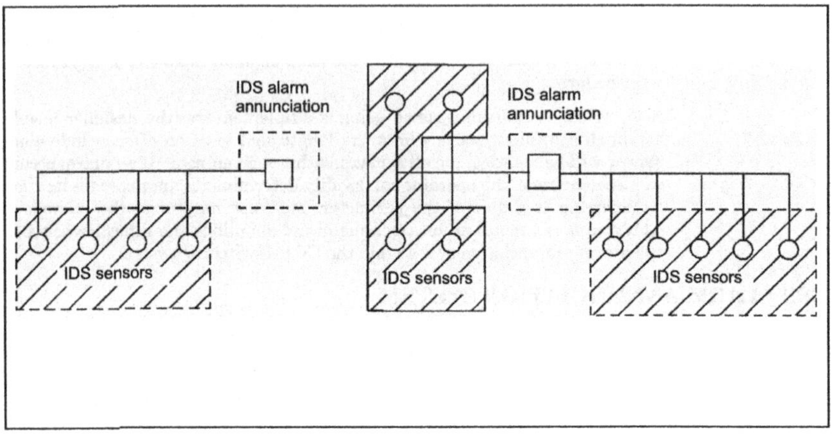

Figure 6-3. Typical Multiplexed IDS

ALARM-ANNUNCIATION CONFIGURATION

6-45. A block diagram of a typical alarm-annunciation system is shown in Figure 6-4, page 6-14. As shown in the figure, the central computer is the hub of the information flow. The central computer receives and displays alarm and device status information and sends operator-control commands to the ESS's local processors. It also interfaces with the CCTV system. For larger facilities, the management of the DTM communications tasks may be delegated to a separate communication processor so that the central computer can turn its full attention to interpreting the incoming information and updating the control and display devices located at the security console (display, logging, control, and storage devices).

6-46. The central computer may consist of one or more digital computers. The real-time clock is usually integral to the central computer and provides a time stamp for alarms and other events. It allows for time synchronization with the CCTV and other systems, if included. The console operator must be able to set the clock, which should include a battery backup. All system events must be properly time-correlated. For example, there will be an exact time correlation for an ESS alarm event reported on the alarm printer and the corresponding video scene recorded by the CCTV's video processor.

DATA STORAGE

6-47. Computer-based systems are required to store large amounts of information such as system software, application programs, data structures, and system events (alarm transactions and status changes). Therefore, a large amount of nonvolatile memory is required. The semiconductor memory provided with a central computer is designed for rapid storage and retrieval

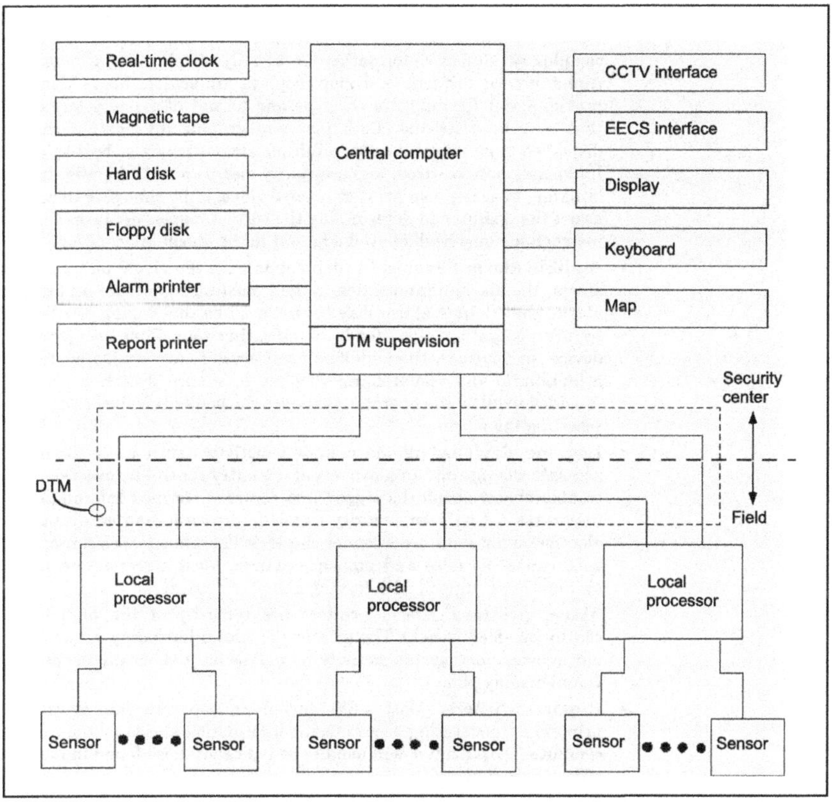

Figure 6-4. Typical IDS Alarm-Annunciation System

and possesses extremely fast access times. The most commonly used media for archival storage are magnetic tape; compact-disk, read-only memory (CD-ROM); and magnetic disk. These media are capable of economically storing large amounts of data.

OPERATOR INTERFACE

6-48. The operator interacts with the alarm-annunciation system through devices that can be seen, heard, or touched and manipulated. Visual displays and printers can be used to inform the operator of an alarm or the equipment's status. Audible devices are used to alert an operator to an alarm or the equipment's failure. Devices such as push buttons and keyboards permit an operator to acknowledge and reset alarms, as well as change operational parameters.

- **Visual displays.** The type of display used to inform the operator visually of the ESS's status is determined primarily by the system's complexity. Status information is usually displayed on monitors. Alphanumeric displays and map displays are seldom used. Monitors provide great flexibility in the type and format of alarm information that may be displayed. Both text and graphic information can be displayed in a variety of colors. Multiple alarms may also be displayed. If alarms are prioritized, higher-priority alarms may be highlighted by blinking, by using bold print or reverse video, or by changing colors. To assist the operator in determining the correct response, alarm-specific instructions may be displayed adjacent to the alarm information.
- **Audible alarm devices.** In conjunction with the visual display of an alarm, the alarm-annunciation system must also generate an audible alarm. The audible alarm may be produced by the ringing of a bell or by the generation of a steady or pulsating tone from an electronic device. In any case, the audible alarm serves to attract the operator's attention to the visual-alarm display. A silence switch is usually provided to allow the operator to silence the bell or tone before actually resetting the alarm.
- **Logging devices.** All alarm-system activity (such as a change of access/secure status, an alarm event, an entry-control transaction, or a trouble event) should be logged and recorded. Logged information is important not only for security personnel investigating an event, but also for maintenance personnel checking equipment performance for such causes as false and nuisance alarms. Most alarm-annunciation systems are equipped with logging and alarm printers.
- **Alarm printers.** Alarm printers are typically of the high-speed, continuous-feed variety. The printer provides a hard-copy record of all alarm events and system activity, as well as limited backup in case the visual display fails.
- **Report printers.** Most ESSs include a separate printer (report printer) for generating reports using information stored by the central computer. This printer will usually be typical of those found in modern office environments.
- **Operator control.** A means is required to transmit information from the operator to the system. The type of controls provided usually depends on the type of display provided. The following are consistent with the controls:
  - Keypads consist of a numeric display system that will generally be provided with a 12-digit keypad and several function keys such as access, secure, acknowledge, and reset. The keypad enables an operator to key in numeric requests for the status of specific zones.
  - Monitor-based systems are usually provided with a typewriter-type keyboard that enables an operator to enter more information using a combination of alphanumeric characters and function keys.
  - An ESS may be equipped with enhancement hardware/devices to help the operator enter information or execute commands quickly. A mouse or a trackball are typical examples.

FIELD-DATA COLLECTION

6-49. Sensor and terminal device data must be transmitted to the central alarm monitor located in the security center using a selected DTM. The following are DTM methods that may be used:

Local Processors

6-50. Multiplexing techniques can be used to minimize the number of data links needed to communicate field-device status to the security center. This is done through devices called local processors. The following is descriptive of a local processor's capabilities:

- A local processor may have very few device inputs, or it may have many (depending on the manufacturer). Rather than having a fixed number of inputs, many local processors are expandable. For example, a basic local processor may be provided with eight device inputs with additional blocks of eight inputs available by using plug-in modules.
- The local processor must provide line supervision for all communication links to sensors, terminal devices, and so forth. Usually, direct-current (DC) line supervision is supplied as the standard with more secure techniques available as options. The data communication links between the local processor and the central alarm monitor must also be supervised.
- Local processors can also provide output signals that can be used for such functions as activating sensor remote test features, light control, or portal control or activating a deterrent (such as a loud horn).
- The local processor contains a microprocessor, solid-state memory, and appropriate software. It has the capability to perform a number of functions locally (such as access-/secure-mode selection, alarm reset, card or keypad electronic-entry control, portal control, and device testing). If the communication link to the security center is temporarily lost, local processors can continue to operate in a stand-alone mode, storing data for transmission after the link is restored.
- The number of local processors required for a specific site depends on the number of protected areas and their proximity to each other and the number of sensors within a protected area. For example, a small building may require one local processor, whereas a large building may require one or more for each floor. An exterior IDS perimeter with two or three different sensors may require one local processor for every two perimeter zones. All local processors may be linked to the central computer using one common DTM link, or the DTM may consist of several links. The designer should note that the temporary loss of a DTM link would render all local processors on that link inactive for the duration of the loss.

Central Computer and Local-Processor Data Exchange

6-51. When the ESS is powered up or reset at the security center, the central computer will download all necessary operational information over the DTM to all local processors. After the download is complete, the central computer will automatically begin polling the local processors for ESS device status. In addition to alarm status, tamper indications, and local-processor status, the DTM may be required to convey security-center console-operator commands to field devices. Examples include security-area access-/secure-mode changes and initiation of the intrusion-sensor self test.

CCTV Interface

6-52. If a CCTV assessment system is deployed with the ESS, an interface between the two is required. This interface allows CCTV system alarms (such as loss of video) to be displayed by the ESS's alarm-annunciation system. The interface also provides IDS alarm signals to the CCTV's video switcher so that the correct CCTV camera will be displayed on the CCTV monitors to allow real-time alarm assessment and video recording as required.

ESS SOFTWARE

6-53. The software provided with computer-based ESS alarm-annunciation systems consists of three types—a standard operating system (such as the Microsoft®-disk operating system [MS-DOS]); vendor-developed application programs; and user-filled, site-specific data structures.

- **System software.** The designer will ensure that system software provided by the vendor conforms to accepted industry standards so that standard, follow-on maintenance and service contracts can be negotiated to maintain the central computer system.
- **Application software.** The vendor-developed application programs are typically proprietary and include ESS monitoring, display, and entry-control capabilities.
- **User-filled data structures.** These data structures are used to populate the site-specific database. Specific electronic address information, personnel access schedules, and normal duty hours are typically included in the site-specific database. The information may include preferred route descriptions for the response force, the phone number of the person responsible for the alarmed area, and any hazardous material that may be located in the alarmed area.

6-54. ESS software functions typically include the following:

- **Alarm monitoring and logging.** The software should provide for monitoring all sensors, local processors, and data communication links and notifying the operator of an alarm condition. All alarm messages should be printed on the alarm printer, archived, and displayed at the console. As a minimum, printed alarm data should include the date and time (to the nearest second) of the alarm and the location and type of alarm.
- **Alarm display.** The software should be structured to permit several alarms to be annunciated simultaneously. A buffer or alarm queue should be available to store additional alarms until they are annunciated and, subsequently, acted upon and reset by the console operator.
- **Alarm priority.** A minimum of five alarm-priority levels should be available. Higher-priority alarms should always be displayed before

lower-priority alarms. This feature permits an operator to respond quickly to the more important alarms before those of lesser importance. For example, the priority of alarm devices may be as follows:

  - Duress.
  - Intrusion detection.
  - Electronic-entry control.
  - Tamper.
  - CCTV alarms and equipment-malfunction alarms.
- **Reports.** The application software should provide for generating, displaying, printing, and storing reports.

PASSWORDS

6-55. Software security will be provided by limiting access to personnel with authorized passwords assigned by a system manager. A minimum of three password levels shall be provided. Additional security can be provided by programmed restrictions that limit the keyboard actions of logged-in passwords to the user ranks of system managers, supervisors, and console operators, as appropriate.

OPERATOR INTERFACE

6-56. The software should enable an operator with the proper password to enter commands and to obtain displays of system information. As a minimum, an operator should be able to perform the following functions through the keyboard or the keypad:

- Log on by password to activate the keyboard.
- Log off to deactivate the keyboard.
- Request display of all keyboard commands that are authorized for the logged-in password.
- Request display of detailed instructions for any authorized keyboard command.
- Acknowledge and clear alarm messages.
- Display the current status of any device in the system.
- Command a status change for any controlled device in the system.
- Command a mode change for any access/secure device in the system.
- Command printouts of alarm summaries, status summaries, or system activity on a designated printer.
- Add or delete ESS devices or modify parameters associated with a device.

INTERIOR INTRUSION-DETECTION SENSORS

6-57. Interior intrusion-detection sensors are devices used to detect unauthorized entry into specific areas or volumetric spaces within a building. These sensors are usually not designed to be weatherproof or rugged enough to survive an outdoor environment. Therefore, this type of sensor should not be used outdoors unless described by the manufacturer as suitable for outdoor use.

6-58. Interior intrusion-detection sensors generally perform one of three detection functions—detection of an intruder penetrating the boundary of a protected area, detection of intruder motion within a protected area, and detection of an intruder touching or lifting an asset within a protected area. Therefore, interior sensors are commonly classified as boundary-penetration sensors, volumetric motion sensors, and point sensors. Although duress switches are not intrusion-detection sensors, they are included in this discussion because they are usually wired to the same equipment that monitors the interior intrusion-detection sensors.

BOUNDARY-PENETRATION SENSORS

6-59. Boundary-penetration sensors are designed to detect penetration or attempted penetration through perimeter barriers. These barriers include walls, ceilings, duct openings, doors, and windows.

Structural-Vibration Sensors

6-60. Structural-vibration sensors detect low-frequency energy generated in an attempted penetration of a physical barrier (such as a wall or a ceiling) by hammering, drilling, cutting, detonating explosives, or employing other forcible methods of entry. A piezoelectric transducer senses mechanical energy and converts it into electrical signals proportional in magnitude to the vibrations. To reduce false alarms from single accidental impacts on the barrier, most vibration sensors use a signal processor that has an adjustable pulse-counting accumulator in conjunction with a manual sensitivity adjustment. The count circuit can be set to count a specific number of pulses of specific magnitude within a predefined time interval before an alarm is generated. However, the circuitry is usually designed to respond immediately to large pulses, such as those caused by an explosion. The sensitivity adjustment is used to compensate for the type of barrier and the distance between transducers. Typically, several transducers can be connected together and monitored by one signal processor. Figure 6-5, page 6-20, shows an example of wall-mounted, structural-vibration sensors.

Glass-Breakage Sensors

6-61. Glass-breakage sensors detect the breaking of glass. The noise from breaking glass consists of frequencies in both the audible and ultrasonic range. Glass-breakage sensors use microphone transducers to detect the glass breakage. The sensors are designed to respond to specific frequencies only, thus minimizing such false alarms as may be caused by banging on the glass.

Passive Ultrasonic Sensors

6-62. Passive ultrasonic sensors detect acoustical energy in the ultrasonic frequency range, typically between 20 and 30 kilohertz (kHz). They are used to detect an attempted penetration through rigid barriers (such as metal or masonry walls, ceilings, and floors). They also detect penetration through windows and vents covered by metal grilles, shutters, or bars if these openings are properly sealed against outside sounds.

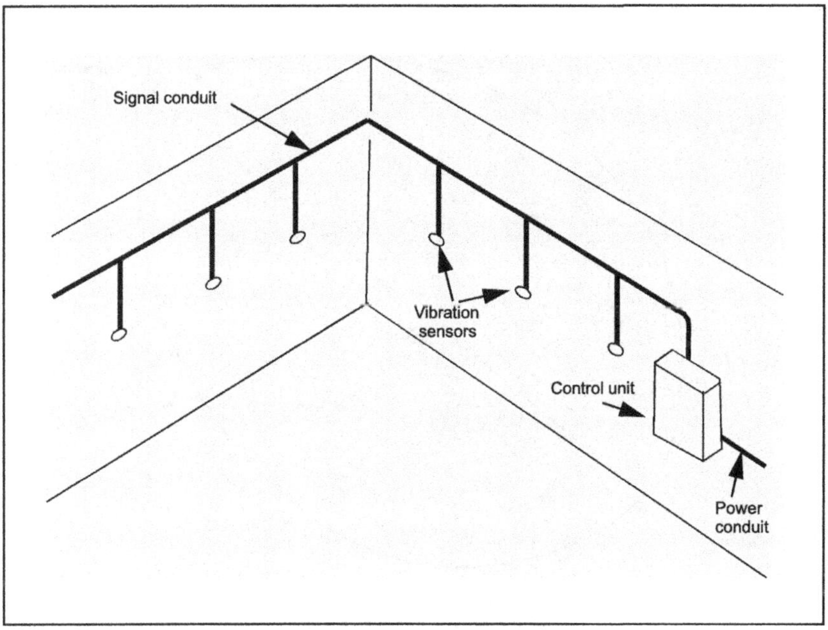

Figure 6-5. Wall-Mounted, Structural-Vibration Sensors

6-63. **Detection Transducer.** The detection transducer is a piezoelectric crystal that produces electrical signals proportional to the magnitude of the vibrations. A single transducer provides coverage of an area about 15 by 20 feet in a room with an 8- to 12-foot ceiling. A typical detection pattern is shown in Figure 6-6. Ten or more transducers can be connected to a signal processor. As with vibration sensors, the signal processor for a passive ultrasonic sensor has manual sensitivity adjustment and an adjustable pulse-counting accumulator.

6-64. **Sensors.** Passive ultrasonic sensors detect ultrasonic energy that results from the breaking of glass, the snipping of bolt cutters on metal barriers, the hissing of an acetylene torch, and the shattering of brittle materials (such as concrete or cinderblock). However, the sensors will not reliably detect drilling through most material nor attacks against soft material such as wallboard. Their effective detection range depends largely on the barrier material, the method of attempted penetration, and the sensitivity adjustment of the sensor. Examples of maximum detection distances for a typical sensor for different types of attempted penetration are shown in Table 6-4.

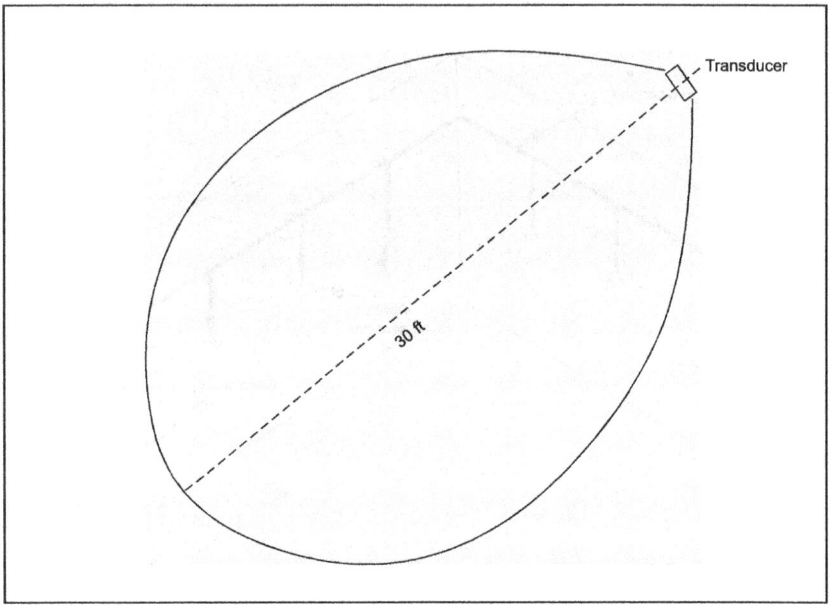

Figure 6-6. Typical Passive-Ultrasonic-Sensor Detection Pattern

Table 6-4. Detection Range for Passive Ultrasonic Sensors

Penetration	Distance (in Feet)
Cut 1/4-inch-thick expanded metal with bolt cutters	55
Cut 5/8-inch reinforcing bar with bolt cutters	45
Use acetylene cutting torch	39
Cut wood with circular saw	30
Cut 5/8-inch reinforcing bar with hacksaw	19
Drill through brick	15
Drill through 1/8-inch steel plate	6
Cut 1/8-inch steel plate with hacksaw	4
Drill through cinderblock	3

6-65. **Balanced Magnetic Switches.** Balanced magnetic switches (BMSs) are typically used to detect the opening of a door. These sensors can also be used on windows, hatches, gates, or other structural devices that can be opened to gain entry. When using a BMS, mount the switch mechanism on the door frame and the actuating magnet on the door. Typically, the BMS has a three-position reed switch and an additional magnet (called the bias magnet) located adjacent to the switch. When the door is closed, the reed switch is held in the balanced or center position by interacting magnetic fields. If the door is opened or an external magnet is brought near the sensor in an attempt to defeat it, the switch becomes unbalanced and generates an alarm. A BMS must be mounted so that the magnet receives maximum movement when the door or window is opened. Figure 6-7 shows several configurations for mounting BMSs.

Grid-Wire Sensors

6-66. The grid-wire sensor consists of a continuous electrical wire arranged in a grid pattern. The wire maintains an electrical current. An alarm is generated when the wire is broken. The sensor detects forced entry through walls, floors, ceilings, doors, windows, and other barriers. An enamel-coated number 24 or 26 American wire gauge (AWG) solid-copper wire typically forms the grid. The grid's maximum size is determined by the spacing between the wires, the wire's resistance, and the electrical characteristics of the source providing the current. The grid wire can be installed directly on the barrier, in a grille or screen that is mounted on the barrier, or over an opening that requires protection. The wire can be stapled directly to barriers made of wood or wallboard. Wood panels should be installed over the grid to protect it from day-to-day abuse and to conceal it. When used on cinder, concrete, and

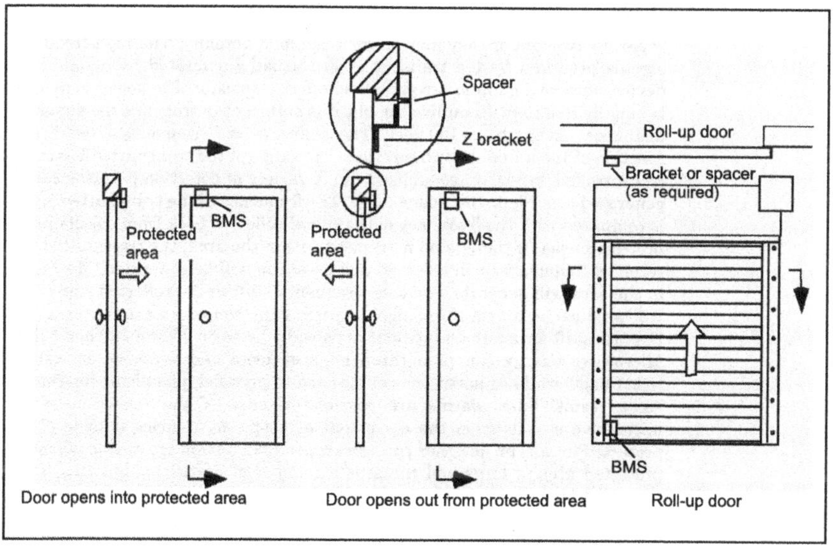

Figure 6-7. BMS Mounting Configurations

masonry surfaces, these surfaces must first be covered with plywood or other material to which the wire can be stapled. An alternative method is to staple the wire grid to the back side of a panel and install the panel over the surface.

VOLUMETRIC MOTION SENSORS

6-67. Volumetric motion sensors are designed to detect intruder motion within the interior of a protected volume. Volumetric sensors may be active or passive. Active sensors (such as microwave) fill the volume to be protected with an energy pattern and recognize a disturbance in the pattern when anything moves within the detection zone. Whereas active sensors generate their own energy pattern to detect an intruder, passive sensors (such as IR) detect energy generated by an intruder. Some sensors, known as dual-technology sensors, use a combination of two different technologies, usually one active and one passive, within the same unit. If CCTV assessment or surveillance cameras are installed, video motion sensors can be used to detect intruder movement within the area. Since ultrasonic motion sensors are seldom used, they will not be discussed here.

Microwave Motion Sensors

6-68. With microwave motion sensors, high-frequency electromagnetic energy is used to detect an intruder's motion within the protected area. Interior or sophisticated microwave motion sensors are normally used.

6-69. **Interior Microwave Motion Sensors.** Interior microwave motion sensors are typically monostatic; the transmitter and the receiver are housed in the same enclosure (transceiver). They may each be provided with a separate antenna or they may share a common antenna. The high-frequency signals produced by the transmitter are usually generated by a solid-state device, such as a gallium arsenide field-effect transistor. The power generated is usually less than 10 milliwatts, but it is sufficient to transmit the signal for distances up to about 100 feet. The shape of the transmitted beam is a function of the antenna configuration. The range of the transmitted beam can be controlled with a range adjustment. A variety of detection patterns can be generated (see Figure 6-8, page 6-24). The frequency of the transmitted signal is compared with the frequency of the signal reflected back from objects in the protected area. If there is no movement within the area, the transmitted and received frequencies will be equal and no alarm will be generated. Movement in the area will generate a Doppler frequency shift in the reflected signal and will produce an alarm if the signal satisfies the sensor's alarm criteria. The Doppler shift for a human intruder is typically between 20 and 120 hertz (Hz). Microwave energy can pass through glass doors and windows as well as lightweight walls or partitions constructed of plywood, plastic, or fiberboard. As a result, false alarms are possible because of the reflection of the microwave signals from the movement of people or vehicles outside of the protected area. The designer can sometimes take advantage of this when the protected area is large and contains a number of partitions, but this is not normally done.

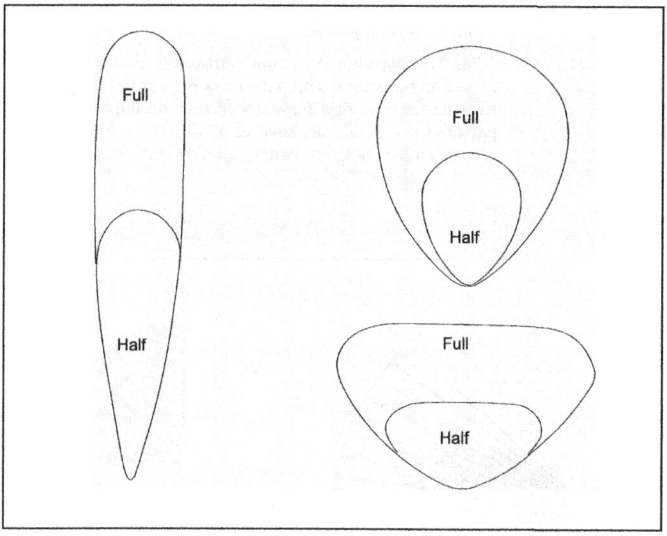

Figure 6-8. Typical Detection Patterns for Microwave Motion Sensors

6-70. **Sophisticated Microwave Motion Sensors.** Sophisticated microwave motion sensors may be equipped with electronic range gating. This feature allows the sensor to ignore the signals reflected beyond the settable detection range. Range gating may be used to effectively minimize unwanted alarms from activity outside the protected area.

PIR Motion Sensors

6-71. PIR motion sensors detect a change in the thermal energy pattern caused by a moving intruder and initiate an alarm when the change in energy satisfies the detector's alarm criteria. These sensors are passive devices because they do not transmit energy; they monitor the energy radiated by the surrounding environment.

6-72. All objects with temperatures above absolute zero radiate thermal energy. The wavelengths of the IR energy spectrum lie between 1 and 1,000 microns. Because the human body radiates thermal energy of between 7 and 14 microns, PIR motion sensors are typically designed to operate in the far IR wavelength range of 4 to 20 microns.

6-73. The IR energy must be focused onto a sensing element, somewhat as a camera lens focuses light onto a film. Two techniques are commonly used. One technique uses reflective focusing; parabolic mirrors focus the energy. The other uses an optical lens. Of the various types of optical lenses, Fresnel lenses are preferred because they can achieve short focal lengths with minimal thickness. Because IR energy is severely attenuated by glass, lenses are usually made of plastic.

6-74. The sensor's detection pattern is determined by the arrangement of lenses or reflectors. The pattern is not continuous but consists of a number of rays or fingers, one for each mirror or lens segment. Numerous detection patterns are available, several of which are shown in Figure 6-9. The PIR is not provided with a range adjustment, but the range can be adjusted

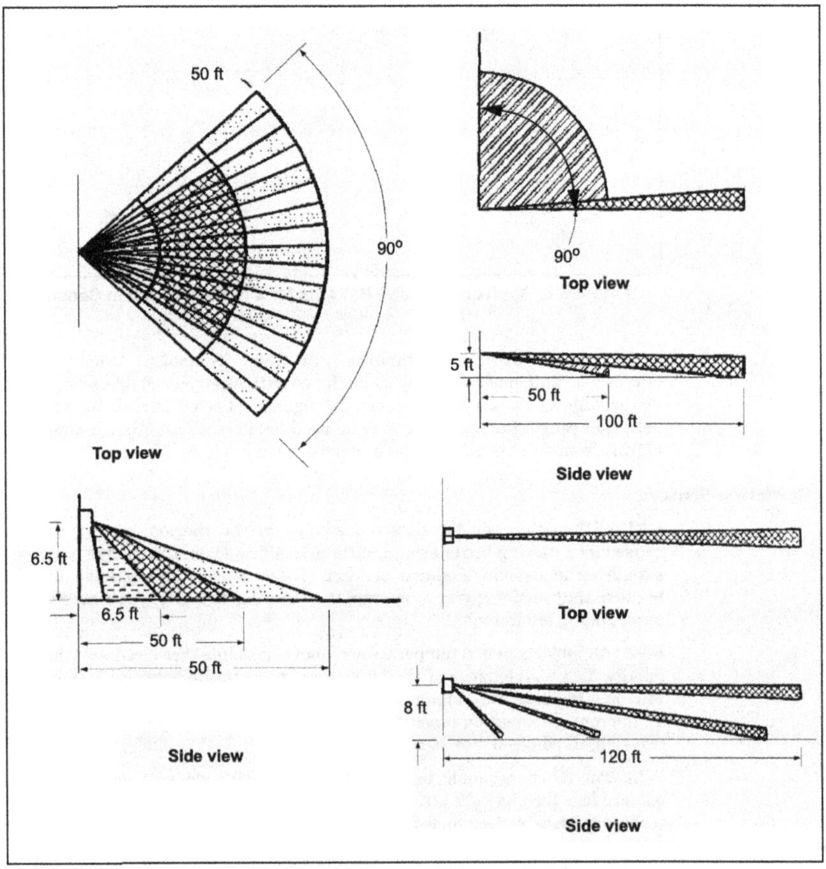

Figure 6-9. Typical Detection Patterns for a PIR Motion Detector

somewhat by manipulating the sensor's position; therefore, careful selection of the appropriate detection pattern is critical to proper sensor performance.

6-75. Most manufacturers use a pyroelectric material as the thermal sensing element. This material produces a change in electric charge when exposed to changes in temperature. To minimize false alarms caused by changes in ambient temperature, most manufacturers use a dual-element sensor. The sensing element is split into halves, one that produces a positive voltage pulse and the other a negative pulse when a change in temperature changes. An intruder entering one of the detection fingers produces an imbalance between the two halves, resulting in an alarm condition. Quadelement sensors that combine and compare two dual-element sensors are also in use. Pulse-count activation, a technique in which a predefined number of pulses within a specific interval of time must be produced before an alarm is generated, is also used.

Dual-Technology Sensors

6-76. To minimize the generation of alarms caused by sources other than intruders, dual-technology sensors combine two different technologies in one unit. Ideally, this is achieved by combining two sensors that individually have a high PD and do not respond to common sources of false alarms. Available dual-technology sensors combine an active ultrasonic or microwave sensor with a PIR sensor. The alarms from each sensor are logically combined in an "and" configuration; that is, nearly simultaneous alarms from both active and passive sensors are needed to produce a valid alarm. Although combined technology sensors have a lower false-alarm rate than individual sensors, the PD is also reduced. For example, if each individual sensor has a PD of 0.95, the PD of the combined sensors is the product of individual probabilities (0.9). Also, ultrasonic and microwave motion sensors have the highest probability of detecting movement directly toward or away from the sensor, whereas PIR motion sensors have the highest probability of detecting movement across the detection pattern. Therefore, the PD of sensors combined in a single unit is less than that obtainable if the individual sensors are mounted perpendicular to each other with overlapping detection patterns. Because of the lower false-alarm rate, the reduced PD can be somewhat compensated for by increasing the sensitivity or detection criteria of each individual sensor.

Video Motion Sensors

6-77. A video motion sensor generates an alarm when an intruder enters a selected portion of a CCTV camera's field of view. The sensor processes and compares successive images between the images against predefined alarm criteria. There are two categories of video motion detectors—analog and digital. Analog detectors generate an alarm in response to changes in a picture's contrast. Digital devices convert selected portions of the analog video signal into digital data that are compared with data converted previously; if differences exceed preset limits, an alarm is generated. The signal processor usually provides an adjustable window that can be positioned anywhere on the video image. Available adjustments permit changing horizontal and vertical window size, window position, and window sensitivity. More sophisticated units provide several adjustable windows that can be individually sized and positioned. Multiple windows permit concentrating on several specific areas of an image while ignoring others. For example, in a scene containing six doorways leading into a long hallway, the sensor can be set to monitor only two critical doorways.

POINT SENSORS

6-78. Point sensors are used to protect specific objects within a facility. These sensors (sometimes referred to as proximity sensors) detect an intruder coming in close proximity to, touching, or lifting an object. Several different types are available, including capacitance sensors, pressure mats, and pressure switches. Other types of sensors can also be used for object protection.

Capacitance Sensors

6-79. Capacitance sensors detect an intruder approaching or touching a metal object by sensing a change in capacitance between the object and the ground. A capacitor consists of two metallic plates separated by a dielectric medium. A change in the dielectric medium or electrical charge results in a change in capacitance. In practice, the metal object to be protected forms one plate of the capacitor and the ground plane surrounding the object forms the second plate. The sensor processor measures the capacitance between the metal object and the ground plane. An approaching intruder alters the dielectric value, thus changing the capacitance. If the net capacitance change satisfies the alarm criteria, an alarm is generated.

6-80. The maximum capacitance that can be monitored by this type of sensor is usually between 10,000 and 50,000 picofarads. The minimum detectable change in capacitance can be as low as 20 picofarads. The signal processor usually has a sensitivity adjustment that can be set to detect an approaching intruder several feet away or to require that the intruder touch the object before an alarm is generated.

6-81. Because air forms most of the dielectric of the capacitor, changes in relative humidity will affect the sensor's sensitivity. An increase in humidity causes the conductivity of the air to increase, lowering the capacitance. Moving a metal object (such as a file cabinet) closer to or away from the protected object can also affect the sensitivity of a capacitance sensor. Figure 6-10, page 6-28, illustrates a typical application using a capacitance sensor.

Pressure Mats

6-82. Pressure mats generate an alarm when pressure is applied to any part of the mat's surface, as when someone steps on the mat. One type of construction uses two layers of copper screening separated by soft-sponge rubber insulation with large holes in it. Another type uses parallel strips of ribbon switches made from two strips of metal separated by an insulating material and spaced several inches apart. When enough pressure is applied to the mat, either the screening or the metal strips make contact, generating an alarm. Pressure mats can be used to detect an intruder approaching a protected object, or they can be placed by doors or windows to detect entry. Because pressure mats are easy to bridge, they should be well concealed, such as placing them under a carpet.

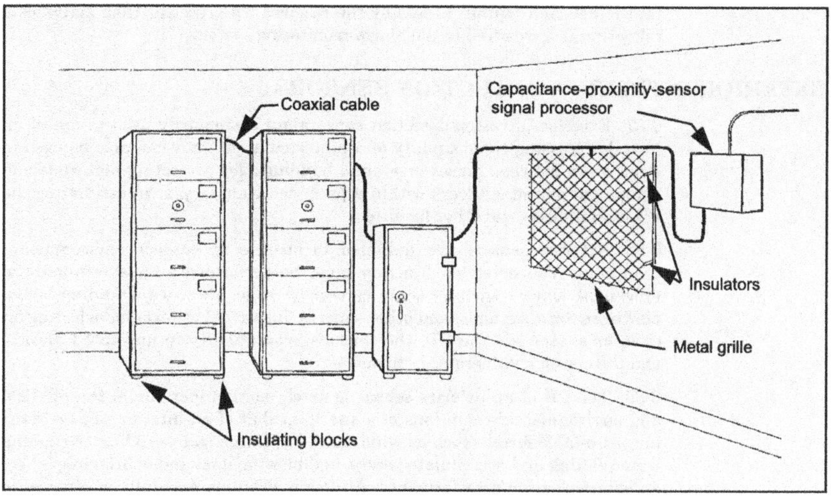

Figure 6-10. Capacitance-Proximity-Sensor Application

Pressure Switches

6-83. Mechanically activated contact switches or single ribbon switches can be used as pressure switches. Objects that require protection can be placed on top of the switch. When the object is moved, the switch actuates and generates an alarm. In this usage, the switch must be well concealed. The interface between the switch and the protected object should be designed so that an adversary cannot slide a thin piece of material under the object to override the switch while the object is removed.

DURESS-ALARM DEVICES

6-84. Duress-alarm devices may be fixed or portable. Operations and security personnel use them to signal a life-threatening emergency. Activation of a duress device will generate an alarm at the alarm-monitoring station. Because of the nature of the alarm, duress devices should never annunciate at the point of threat. These devices are customarily manually operated.

6-85. Fixed duress devices are mechanical switches permanently mounted in an inconspicuous location, such as under a counter or desk. They can be simple push-button switches activated by the touch of a finger or hand or foot-operated switches attached to the floor.

6-86. Portable duress devices are wireless units consisting of a transmitter and a receiver. The transmitter is portable and small enough to be conveniently carried by a person. The receiver is mounted in a fixed location within the facility. Either ultrasonic or RF energy can be used as the communication medium. When activated, the transmitter generates an alarm that is detected (within range) by the receiver. The receiver then activates a relay that is hardwired to the alarm-monitoring system.

EXTERIOR INTRUSION-DETECTION SENSORS

6-87. Exterior intrusion-detection sensors are customarily used to detect an intruder crossing the boundary of a protected area. They can also be used in clear zones between fences or around buildings, for protecting materials and equipment stored outdoors within a protected boundary, or in estimating the PD for buildings and other facilities.

6-88. Exterior sensors are designed to operate in outdoor environmental conditions. The detection function must be performed with a minimum of unwanted alarms such as those caused by wind, rain, ice, standing water, blowing debris, animals, and other sources. Important criteria for selecting an exterior sensor are the PD, the sensor's susceptibility to unwanted alarms, and the sensor's vulnerability to defeat.

6-89. The PD of an exterior sensor is much more vulnerable to the physical and environmental conditions of a site than that of an interior sensor. Many uncontrollable forces (such as wind, rain, ice, frozen soil, standing or running water, falling and accumulated snow, and blowing dust and debris) may affect an exterior sensor's performance. Although attention generally is directed to circumstances that cause a dramatic drop in the PD, environmental factors can also cause short-term increases in the PD. If controlled intrusions (intrusions by security personnel to verify the current detection capability of an IDS) are done while an IDS temporarily has a higher than usual PD as the result of current site conditions, the results may give a false indication of the general effectiveness of that IDS.

6-90. Because of the nature of the outdoor environment, exterior sensors are also more susceptible to nuisance and environmental alarms than interior sensors. Inclement weather conditions (heavy rain, hail, and high wind), vegetation, the natural variation of the temperature of objects in the detection zone, blowing debris, and animals are major sources of unwanted alarms.

6-91. As with interior sensors, tamper protection, signal-line supervision, self-test capability, and proper installation make exterior sensors less vulnerable to defeat. Because signal-processing circuitry for exterior sensors is generally more vulnerable to tampering and defeat than that for interior sensors, it is extremely important that enclosures are located and installed properly and that adequate physical protection is provided. Several different types of exterior intrusion-detection sensors are available. They can be categorized as—

- Fence sensors.
- Buried line sensors.
- LOS sensors.
- Video motion sensors.

FENCE SENSORS

6-92. Fence sensors detect attempts to penetrate a fence around a protected area. Penetration attempts (such as climbing, cutting, or lifting) generate mechanical vibrations and stresses in fence fabric and posts that are usually different than those caused by natural phenomena like wind and rain. The basic types of sensors used to detect these vibrations and stresses are strain-sensitive cable, taut wire, and fiber optics. Other types of fence sensors detect penetration attempts by sensing changes in an electric field or in capacitance. Mechanical and electromechanical fence sensors are seldom used and will not be discussed here.

Strain-Sensitive Cable

6-93. Strain-sensitive cables are transducers that are uniformly sensitive along their entire length. They generate an analog voltage when subject to mechanical distortions or stress resulting from fence motion. Strain-sensitive cables are sensitive to both low and high frequencies. The signal processor usually has a band-pass filter that passes only those signals characteristic of fence-penetration actions. An alarm is initiated when the signal's frequency, amplitude, and duration characteristics satisfy the processor's criteria. Because the cable acts like a microphone, some manufacturers offer an option that allows the operator to listen to fence noises causing the alarm. Operators can then determine whether the noises are naturally occurring sounds from wind or rain or are from an actual intrusion attempt. This feature is relatively costly to implement because it requires additional cable from each signal processor to the security center and, if CCTV is being used, it may be of limited benefit. Strain-sensitive cable is attached to a chain-link fence about halfway between the bottom and top of the fence fabric with plastic ties. One end of the cable is terminated at the signal processor and the other end with a resistive load. The DC through the cable provides line supervision against cutting or electrically shorting the cable or disconnecting it from the processor. A typical installation is shown in Figure 6-11.

Taut-Wire Sensor

6-94. A taut-wire sensor combines a physically taut-wire barrier with an intrusion-detection sensor network. The taut-wire sensor consists of a column of uniformly spaced horizontal wires up to several hundred feet in length and securely anchored at each end. Typically, the wires are spaced 4 to 8 inches apart. Each is individually tensioned and attached to a detector located in a sensor post. Two types of detectors are commonly used—mechanical switches and strain gauges.

- The mechanical switch consists of a specially designed switch mechanism that is normally open. The tensioned wires are mechanically attached to the switch, and movement of the wire beyond a preset limit causes the switch to close. To counteract small gradual movements of a wire (such as that caused by settling of the fence or by freezing or thawing of soil) switches are usually supported in their housing by a soft plastic material. This material allows the switch to self-adjust when acted upon by gradual external forces and wire effects

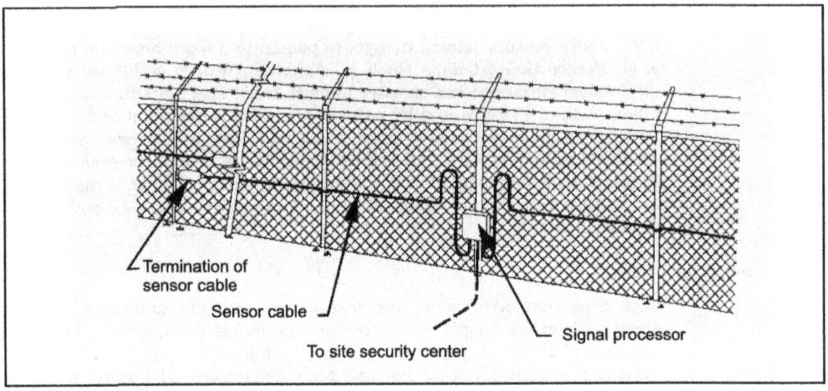

Figure 6-11. Typical Strain-Sensor Cable Installation

such as the relaxation of the wire with time and its thermal expansion or contraction.

- Strain-gauge detectors are attached to the taut wire with a nut on a threaded stud. When a force is applied to the taut wire, the resulting deflection is converted by the strain gauge into a change in electrical output that is monitored by a signal processor.

6-95. With sensors that use mechanical switches as detectors, the switches in a single sensor-post assembly are wired in parallel and are connected directly to the alarm-annunciation system. Pulse-count circuitry is not used because a single switch closure, such as that caused by an intruder moving or cutting one wire, is indicative of an intrusion attempt. Strain-gauge detectors in a sensor post are monitored by a signal processor. When the signal from one or more strain gauges satisfies the processor's criteria, an alarm is initiated.

6-96. The taut-wire sensor can be installed as a freestanding fence or can be mounted on an existing fence or wall. Figure 6-12, page 6-32, shows a freestanding configuration.

Fiber-Optic Cable Sensors

6-97. Fiber-optic cable sensors are functionally equivalent to the strain-sensitive cable sensors previously discussed. However, rather than electrical signals, modulated light is transmitted down the cable and the resulting received signals are processed to determine whether an alarm should be initiated. Since the cable contains no metal and no electrical signal is present, fiber-optic sensors are generally less susceptible to electrical interference from lightning or other sources.

Electric-Field Sensors

6-98. Electric-field sensors consist of an alternating-current (AC) field generator, one or more field wires, one or more sense wires, and a signal

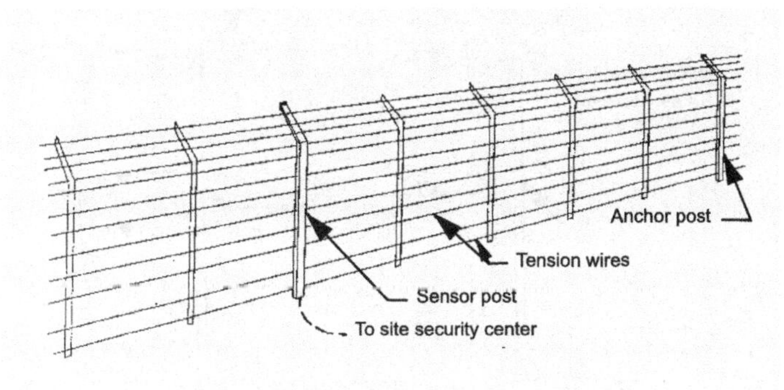

Figure 6-12. Typical Taut-Wire Installation

processor. The generator excites the field wires around which an electrostatic-field pattern is created. The electrostatic field induces electrical signals in the sense wires, which are monitored by the signal processor. Under normal operating conditions, the induced signals are constant. However, when an intruder approaches the sensor, the induced electrical signals are altered, causing the signal processor to generate an alarm.

6-99. Several different field- and sense-wire configurations are available. They range from one field wire and one sense wire to as many as four field wires and one sense wire or four field wires and four sense wires. Figure 6-13 shows the detection pattern produced by vertical three-wire (one field and two sense wires) configurations. The three-wire system has a wider detection envelope and is less costly (one less field wire and associated hardware). However, because of the tighter coupling between wires, the four-wire system is less susceptible to nuisance alarms caused by extraneous noise along the length of the zone.

6-100. A signal processor monitors the signals produced by the sense wires. The processor usually contains a band-pass filter that rejects high-frequency signals such as those caused by objects striking the wires. Additional criteria that must be satisfied before the processor initiates an alarm include signal amplitude and signal duration. By requiring the signal to be present for a preset amount of time, false alarms (such as those caused by birds flying through the detection pattern) can be minimized.

6-101. As with taut-wire sensors, electric-field sensors can be freestanding (mounted on their own posts) or attached by standoffs to an existing fence. They can also be configured to follow contours of the ground. The area under the sensor must be clear of vegetation, since vegetation near or touching sense wires can cause false alarms. These sensors can also be installed on the walls and roof of a building.

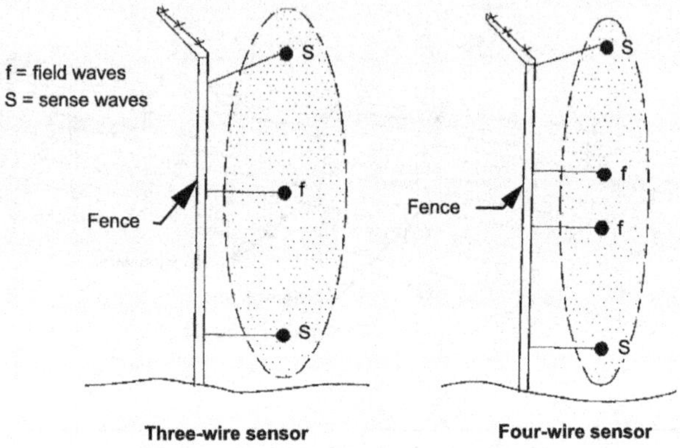

Figure 6-13. Typical Electric-Field-Sensor Detection Patterns

Capacitance Proximity Sensors

6-102. Capacitance proximity sensors measure the electrical capacitance between the ground and an array of sense wires. Any variations in capacitance, such as that caused by an intruder approaching or touching one of the sense wires, initiates an alarm. These sensors usually consist of two or three wires attached to outriggers along the top of an existing fence, wall, or roof edge. Figure 6-14, page 6-34, shows a typical capacitance sensor consisting of three sensor wires attached to the outrigger of a fence. To minimize environmental alarms, the capacitance sensor is divided into two arrays of equal length. The signal processor monitors the capacitance of each array. Changes in capacitance common to both arrays (such as produced by wind, rain, ice, fog, and lightning) are canceled within the processor. However, when changes occur in one array and not the other because of an intruder, the processor initiates an alarm.

BURIED-LINE SENSORS

6-103. A buried-line sensor system consists of detection probes or cable buried in the ground, typically between two fences that form an isolation zone. These devices are wired to an electronic processing unit. The processing unit generates an alarm if an intruder passes through the detection field. Buried-line sensors have several significant features:

- They are hidden, making them difficult to detect and circumvent.
- They follow the terrain's natural contour.
- They do not physically interfere with human activity, such as grass mowing or snow removal.

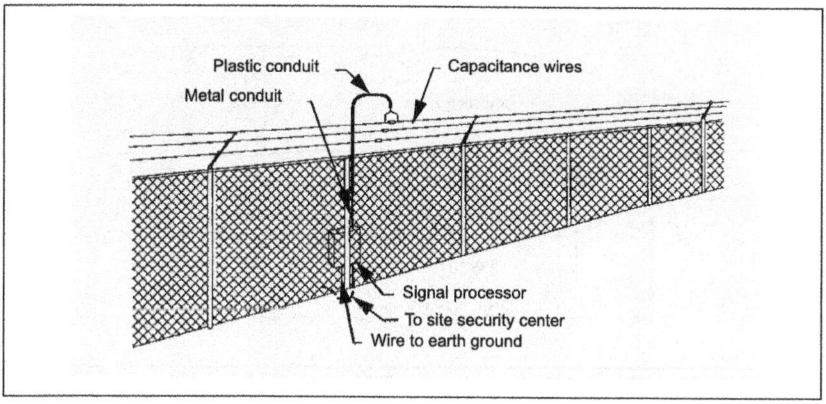

Figure 6-14. Typical Capacitance-Sensor Configuration

- They are affected by certain environmental conditions, such as running water and ground freeze/thaw cycles. (Seismic, seismic/magnetic, magnetic, and balanced pressure sensors are seldom used and will not be discussed here.)

6-104. The ported-coax cable sensor consists of two coax cables buried in the ground parallel to each other. An RF transmitter is connected to one cable and a receiver to the other. The outer conductor of each cable is ported (fabricated with small holes or gaps in the shield). The transmitter cable radiates RF energy into the medium surrounding the cables. A portion of this energy is coupled into the receiver cable through its ported shield. (Because of the ported shields, these cables are frequently referred to as leaky cables.) When an intruder enters the RF field, the coupling is disturbed, resulting in a change of signal monitored by the receiver, which then generates an alarm. Two basic types of ported-coax sensors are available—pulse and continuous wave.

- Pulse-type sensors transmit a pulse of RF energy down one cable and monitors the received signal on the other. The cables can be up to 10,000 feet long. The signal processor initiates an alarm when the electromagnetic field created by the pulse is disturbed and identifies the disturbance's approximate location.
- Continuous-wave sensors apply continuous RF energy to one cable. The signal received on the other cable is monitored for electromagnetic-field disturbances that indicate an intruder's presence. Cable lengths are limited to 300 to 500 feet. Additionally, the sensor is available in a single-cable configuration as well as two separate cables. The pattern typically extends 2 to 4 feet above the ground and can be 5 to 13 feet wide, depending on cable spacing and soil composition. Figure 6-15 represents a typical cross-section of a detection pattern created by a ported-cable sensor.

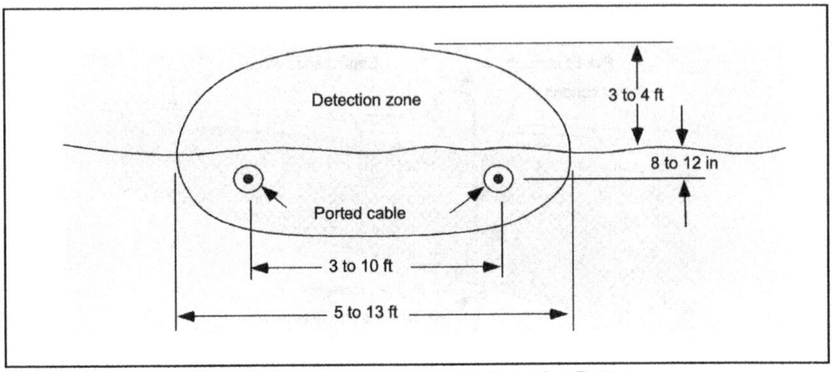

Figure 6-15. Typical Ported-Cable Detection Pattern

6-105. Sensor performance depends on properties of the medium surrounding the cables. Velocity and attenuation of the RF wave that propagates along the cables and the coupling between the cables are functions of the dielectric constant of the soil and its conductivity which, in turn, depends on its moisture content. For example, the velocity is greater and the attenuation is less for cables buried in dry, low-loss soil than in wet, conductive soil. Freeze/ thaw cycles in the soil also affect the sensor's performance. When wet soil freezes, the wave velocity and the cable coupling increase and the attenuation decreases, resulting in greater detection sensitivity. Seasonal sensitivity adjustments may be necessary to compensate for changing ground conditions.

6-106. Although usually buried in soil, ported cables can also be used with asphalt and concrete. If the asphalt or concrete pavement area is relatively small and only a few inches thick (such as a pedestrian pavement crossing the perimeter), the ported cables can be routed under the pavement. However, for the large and deep pavements, slots must be cut into the asphalt or concrete to accept the cable.

6-107. A portable ported-coax sensor is available that can be rapidly deployed and removed. The cables are placed on the surface of the ground rather than buried. This sensor is useful for temporary perimeter detection coverage for small areas or objects (such as vehicles or aircraft).

LOS SENSORS

6-108. The LOS sensors, which are mounted above ground, can be either active or passive. Active sensors generate a beam of energy and detect changes in the received energy that an intruder causes by penetrating the beam. Each sensor consists of a transmitter and a receiver and can be in a monostatic or bistatic configuration. Passive sensors generate no beam of energy; they simply look for changes in the thermal characteristics of their field of view. For effective detection, the terrain within the detection zone must be flat and free of obstacles and vegetation.

Microwave Sensors

6-109. Microwave intrusion-detection sensors are categorized as bistatic or monostatic. Bistatic sensors use transmitting and receiving antennas located at opposite ends of the microwave link, whereas monostatic sensors use the same antenna.

- A bistatic system uses a transmitter and a receiver that are typically separated by 100 to 1,200 feet and that are within direct LOS with each other. The signal picked up by the receiver is the vector sum of the directly transmitted signal and signals that are reflected from the ground and nearby structures. Detection occurs when an object (intruder) moving within the beam pattern causes a change in net-vector summation of the received signals, resulting in variations of signal strength.
  - The same frequency bands allocated by the Federal Communications Commission (FCC) for interior microwave sensors are also used for exterior sensors. Because high-frequency microwave beams are more directive than low-frequency beams and the beam pattern is less affected by blowing grass in the area between the transmitter and the receiver, most exterior sensors operate at the next to highest allowable frequency, 10.525 gigahertz (GHz).
  - The shape of the microwave beam and the maximum separation between the transmitter and the receiver are functions of antenna size and configuration. Various antenna configurations are available, including parabolic-dish arrays, strip-line arrays, and slotted arrays. The parabolic antenna uses a microwave-feed assembly located at the focal point of a metallic parabolic reflector. A conical beam pattern is produced (see Figure 6-16). A strip-line antenna configuration produces a nonsymmetrical beam that is higher than its height. Larger antenna configurations generally produce narrower beam patterns.

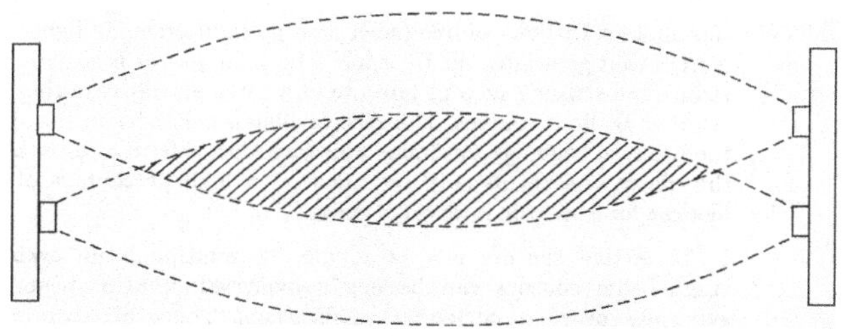

Figure 6-16. Stacked Microwave Configuration

- Monostatic microwave sensors use the same antenna or virtually coincident antenna arrays for the transmitter and receiver, which are usually combined into a single package. Two types of monostatic sensors are available. Amplitude-modulated (AM) sensors detect changes in the net-vector summation of reflected signals similar to bistatic sensors. Frequency-modulated (FM) sensors operate on the Doppler principle similar to interior microwave sensors. The detection pattern is typically shaped like a teardrop (see Figure 6-17). Monostatic sensors can provide volumetric coverage of localized areas, such as in corners or around the base of critical equipment.

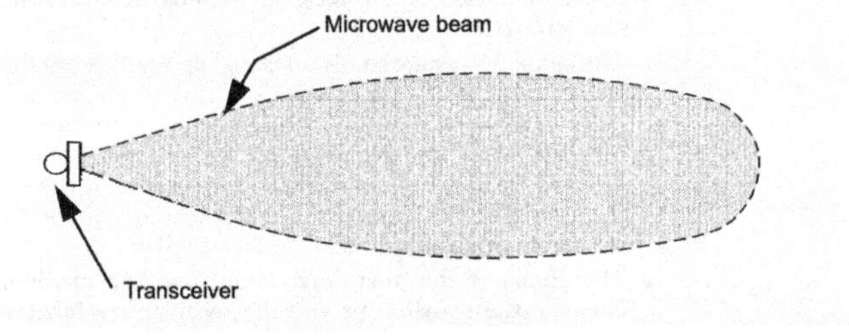

Figure 6-17. Typical Monostatic-Microwave-Sensor Detection Pattern

IR Sensors

6-110. The IR sensors are available in both active and passive models. An active sensor generates one or more near-IR beams that generate an alarm when interrupted. A passive sensor detects changes in thermal IR radiation from objects located within its field of view.

6-111. Active sensors consist of transmitter/receiver pairs. The transmitter contains an IR light source (such as a gallium arsenide light-emitting diode [LED]) that generates an IR beam. The light source is usually modulated to reduce the sensor's susceptibility to unwanted alarms resulting from sunlight or other IR light sources. The receiver detects changes in the signal power of the received beam. To minimize nuisance alarms from birds or blowing debris, the alarm criteria usually require that a high percentage of the beam be blocked for a specific interval of time.

6-112. Active sensors can be single- or multiple-beam systems. Because single-beam sensors can be easily bypassed, multiple-beam systems are generally used in perimeter applications. There are two basic types of multiple-beam configurations—one type uses all transmitters on one post and all receivers on the other post; the second type uses one transmitter and several receivers on each post. Both types are illustrated in Figure 6-18, page 6-38.

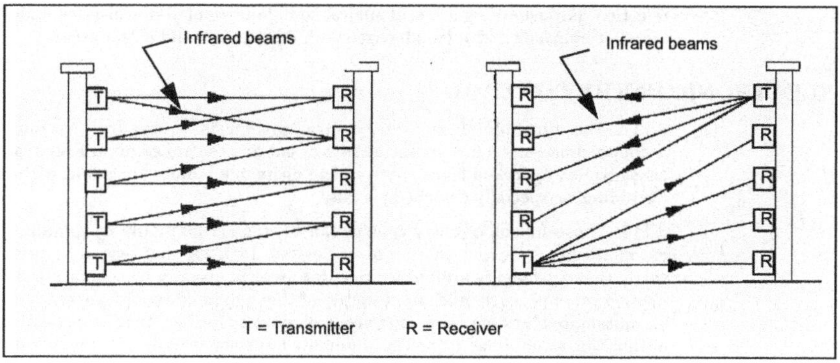

Figure 6-18. Typical IR-Sensor Beam Patterns

6-113. The spacing between transmitters and receivers can be as great as 1,000 feet when operation is under good weather conditions. However, conditions such as heavy rain, fog, snow, or blowing dust particles attenuate the IR energy, reducing its effective range to 100 to 200 feet or less.

VIDEO MOTION SENSORS

6-114. A video motion sensor generates an alarm whenever an intruder enters a selected portion of a CCTV camera's field of view. The sensor processes and compares successive images from the camera and generates an alarm if differences between the images satisfy predefined criteria. Digital devices convert selected portions of the analog video signal into digital data that are compared with data converted previously; if differences exceed preset limits, an alarm is generated.

6-115. The signal processor usually provides an adjustable window that can be positioned anywhere on the video image. Available adjustments permit changing the window's horizontal and vertical sizes, its position, and its sensitivity. More sophisticated units provide several adjustable windows that can be individually sized and positioned. Multiple windows permit concentrating on several specific areas of an image while ignoring others. For example, in a scene that contains several critical assets and multiple sources of nuisance alarms (such as large bushes or trees), the sensor can be adjusted to monitor only the assets and ignore the areas that contain the nuisance-alarm sources.

6-116. The use of video motion-detection systems for exterior applications has been limited, primarily because of difficulties with uncontrolled exterior environments. Lighting variations caused by cloud movement and shadows of slow-moving objects, birds and animals moving within the camera's field of view, camera motion and moving vegetation during windy conditions, and severe weather conditions have traditionally caused a multitude of unwanted alarms in this type of system. Systems using more advanced signal-processing algorithms have improved motion-detection capability and nuisance-alarm rejection; however, they are still subject to high unwanted-alarm rates under certain conditions and should be used with due caution and extreme care.

ELECTRONIC ENTRY CONTROL

6-117. The function of an entry-control system is to ensure that only authorized personnel are permitted into or out of a controlled area. Entry can be controlled by locked fence gates, locked doors to a building or rooms within a building, or specially designed portals.

6-118. These means of entry control can be applied manually by guards or automatically by using entry-control devices. In a manual system, guards verify that a person is authorized to enter an area, usually by comparing the photograph and personal characteristics of the individual requesting entry. In an automated system, the entry-control device verifies that a person is authorized to enter or exit. The automated system usually interfaces with locking mechanisms on doors or gates that open momentarily to permit passage. Mechanical hardware (such as locking mechanisms, electric door strikes, and specially designed portal hardware) and equipment used to detect contraband material (such as metal detectors, X-ray baggage-search systems, explosives detectors, and special nuclear-material monitors) are described in other documentation. Refer to TM 5-853-1 for additional information on determining entry-control requirements and integrating manual electronic-entry control into a cohesive system.

6-119. All entry-control systems control passage by using one or more of three basic techniques—something a person knows, something a person has, or something a person is or does. Automated entry-control devices based on these techniques are grouped into three categories—coded, credential, and biometric devices.

CODED DEVICES

6-120. Coded devices operate on the principle that a person has been issued a code to enter into an entry-control device. This code will match the code stored in the device and permit entry. Depending on the application, a single code can be used by all persons authorized to enter the controlled area or each authorized person can be assigned a unique code. Group codes are useful when the group is small and controls are primarily for keeping out the general public. Individual codes are usually required for control of entry to more critical areas. Coded devices verify the entered code's authenticity, and any person entering a correct code is authorized to enter the controlled area. Electronically coded devices include electronic and computer-controlled keypads.

Electronic Keypad Devices

6-121. The common telephone keypad (12 keys) is an example of an electronic keypad. This type of keypad consists of simple push-button switches that, when depressed, are decoded by digital logic circuits. When the correct sequence of buttons is pushed, an electric signal unlocks the door for a few seconds.

Computer-Controlled Keypad Devices

6-122. These devices are similar to electronic keypad devices, except they are equipped with a microprocessor in the keypad or in a separate enclosure at a different location. The microprocessor monitors the sequence in which the keys are depressed and may provide additional functions such as personal ID and digit scrambling. When the correct code is entered and all conditions are satisfied, an electric signal unlocks the door.

CREDENTIAL DEVICES

6-123. A credential device identifies a person having legitimate authority to enter a controlled area. A coded credential (plastic card or key) contains a prerecorded, machine-readable code. An electric signal unlocks the door if the prerecorded code matches the code stored in the system when the card is read. Like coded devices, credential devices only authenticate the credential; it assumes a user with an acceptable credential is authorized to enter. Various technologies are used to store the code upon or within a card. Hollerith, optically coded, magnetic-spot, capacitance, and electric-circuit cards are seldom used and will not be discussed here. The most commonly used types of cards are described as follows:

Magnetic-Stripe Card

6-124. A strip of magnetic material located along one edge of the card is encoded with data (sometimes encrypted). The data is read by moving the card past a magnetic read head.

Wiegand-Effect Card

6-125. The Wiegand-effect card contains a series of small-diameter, parallel wires about one-half inch long, embedded in the bottom half of the card. The wires are manufactured from ferromagnetic materials that produce a sharp change in magnetic flux when exposed to a slowly changing magnetic field. This type of card is impervious to accidental erasure. The card reader contains a small read head and a tiny magnet to supply the applied magnetic field. It usually does not require external power.

Proximity Card

6-126. A proximity card is not physically inserted into a reader; the coded pattern on the card is sensed when it is brought within several inches of the reader. Several techniques are used to code cards. One technique uses a number of electrically tuned circuits embedded in the card. Data are encoded by varying resonant frequencies of the tuned circuits. The reader contains a transmitter that continually sweeps through a specified range of frequencies and a receiver that senses the pattern of resonant frequencies contained in the card. Another technique uses an integrated circuit embedded in the card to generate a code that can be magnetically or electrostatically coupled to the reader. The power required to activate embedded circuitry can be provided by a small battery embedded in the card or by magnetically coupling power from the reader.

Laser Card

6-127. The optical memory card, commonly called the laser card, uses the same technology developed for recording video and audio disks for entertainment purposes. Data is recorded on the card by burning a microscopic hole (using a laser) in a thin film covering the card. Data is read by using a laser to sense the hole locations. The typical laser card can hold several megabytes of user data.

Smart Card

6-128. A smart card is embedded with a microprocessor, memory, communication circuitry, and a battery. The card contains edge contacts that enable a reader to communicate with the microprocessor. Entry-control information and other data may be stored in the microprocessor's memory.

Bar Code

6-129. A bar code consists of black bars printed on white paper or tape that can be easily read with an optical scanner. This type of coding is not widely used for entry-control applications because it can be easily duplicated. It is possible to conceal the code by applying an opaque mask over it. In this approach, an IR scanner is used to interpret the printed code. For low-level security areas, the use of bar codes can provide a cost-effective solution for entry control. Coded strips and opaque masks can be attached to existing ID badges, alleviating the need for complete badge replacement.

BIOMETRIC DEVICES

6-130. The third basic technique used to control entry is based on the measurement of one or more physical or personal characteristics of an individual. Because most entry-control devices based on this technique rely on measurements of biological characteristics, they have become commonly known as biometric devices. Characteristics such as fingerprints, hand geometry, voiceprints, handwriting, and retinal blood-vessel patterns have been used for controlling entry. Typically, in enrolling individuals, several reference measurements are made of the selected characteristic and then stored in the device's memory or on a card. From then on, when that person attempts entry, a scan of the characteristic is compared with the reference data template. If a match is found, entry is granted. Rather then verifying an artifact, such as a code or a credential, biometric devices verify a person's physical characteristic, thus providing a form of identity verification. Because of this, biometric devices are sometimes referred to as personnel identity-verification devices. The most common biometric devices are discussed below.

Fingerprints

6-131. Fingerprint-verification devices use one of two approaches. One is pattern recognition of the whorls, loops, and tilts of the referenced fingerprint, which is stored in a digitized representation of the image and compared with the fingerprint of the prospective entrant. The second approach is minutiae comparison, which means that the endings and branching points of ridges and valleys of the referenced fingerprint are compared with the fingerprint of the prospective entrant.

Hand Geometry

6-132. Several devices are available that use hand geometry for personnel verification. These devices use a variety of physical measurements of the hand, such as finger length, finger curvature, hand width, webbing between fingers, and light transmissivity through the skin to verify identity. Both two- and three-dimensional units are available.

Retinal Patterns

6-133. This type of technique is based on the premise that the pattern of blood vessels on the human eye's retina is unique to an individual. While the eye is focused on a visual target, a low-intensity IR light beam scans a circular area of the retina. The amount of light reflected from the eye is recorded as the beam progresses around the circular path. Reflected light is modulated by the difference in reflectivity between blood-vessel pattern and adjacent tissue. This information is processed and converted to a digital template that is stored as the eye's signature. Users are allowed to wear contact lenses; however, glasses should be removed.

Device Combinations

6-134. Frequently, an automated entry-control system uses combinations of the three types of entry-control devices. Combining two different devices can significantly enhance the system's security level. In some cases, combining devices results in reduced verification times.

APPLICATION GUIDELINES

6-135. The primary function of an automated entry-control system is to permit authorized personnel to enter or exit a controlled area. Features available to the designer are described below.

- **Enrollment.** All entry-control systems must provide a means of entering, updating, and deleting information about authorized individuals into the system's database files. This is usually accomplished with a dedicated enrollment station for enrolling and disenrolling purposes that is directly connected to the central-processing unit. When credential devices are used, all authorized users must be provided with an appropriate credential. A means should also be provided to disenroll a person quickly without having to retrieve a credential. When using biometric devices, additional enrollment equipment will be required.
- **Entry-control techniques.** Some entry-control functions require additional hardware, while others are accomplished with software. Those features accomplished with software require that the appropriate database be available for every portal affected by them. Typically, these techniques include—
  - Area zones.
  - Time zones.
  - Team zones.
  - Anti-pass back.

  - Antitailgate.
  - Guard tour.
  - Elevator control.
- **Alarms.** Several types of alarms can be used with an entry-control system. These alarms must annunciate audibly and visually in the security center.
- **Entry denial.** Most entry-control devices are configured to permit the user three entry attempts. If more than three unsuccessful entry attempts are made within a specified period, the device generates an alarm. An alarm is also generated if an invalid credential is used or attempted entries are detected that violate specified area, time, or team zoning requirements.
- **Communication failure.** This alarm is generated when a loss of communication between the central processor and the local equipment is detected.
- **Portal open.** If a portal door remains open longer than a predefined time, an alarm is generated.
- **Duress.** This alarm is generated when a special duress code is entered at a keypad.
- **Guard overdue.** This duress alarm is generated when a security guard is determined to be overdue at a checkpoint during a predefined guard tour.
- **Software tamper.** This type of alarm is generated when unauthorized persons are detected attempting to invoke certain system commands or modify database files.

PERFORMANCE CRITERIA

6-136. The overall performance of an entry-control system can be evaluated by examining the verification error rate and the throughput rate. An entry-control system can produce two types of errors—denial of admission of a person who should be admitted or admission of a person who should not be admitted. These are commonly referred to as false-reject errors (type I errors) and false-accept errors (type II errors). Although a false-reject error does not constitute a breach of security, it does create an operational problem that must be handled by an alternative method. False-accept errors constitute a breach of security. Ideally, both false-reject and false-accept error rates should be zero; in practice, however, they are not. In fact, they tend to act in opposition to each other. When the system is adjusted to minimize the false-accept error rate, the false-reject error rate usually increases. Verification error rates are typically measured in percent (number of errors/number of attempts x 100 percent). These error rates are typically very low for coded and credential devices, but many become significant if biometric devices are used.

6-137. The throughput rate is the number of persons that can pass through an entry point in a given unit of time and is usually expressed in persons per minute. It is the time required to approach the entry-control device and for the device to verify information (verification time) and the time required passing through the entry point. Typically, an individual can approach the device and pass through in 3 to 5 seconds. Verification time depends on the type of device and may vary from 3 to 15 seconds. Table 6-5 provides a list of typical verification times for different types of entry-control devices.

Table 6-5. Typical Verification Times of Entry-Control Devices

Device	Verification Time
Keypad	3 seconds
Card reader	3 seconds
Keypad/card reader	6 seconds
Biometric/keypad	6 to 15 seconds
Biometric/card reader	6 to 15 seconds
Biometric	2 minutes

DATA TRANSMISSION

6-138. A critical element in an integrated ESS is the DTM that transmits information from sensors, entry-control devices, and video components to display and assessment equipment. A DTM link is a path for transmitting data between two or more components (such as a sensor and alarm reporting system, a card reader and controller, a CCTV camera and monitor, or a transmitter and receiver). The DTM links connect remote ESS components to the security center. An effective DTM link ensures rapid and reliable transmission media, transmission technique, associated transmission hardware, and degree of security to be provided for the communication system.

6-139. A number of different media are used in transmitting data between elements of an IDS, an EECS, and a CCTV system. These include wire lines, coaxial cable, fiber-optic cable, and RF transmission.

- **Wire line.** Wire lines are twisted pairs that consist of two insulated conductors twisted together to minimize interference by unwanted signals.
- **Coaxial cable.** Coaxial cable consists of a center conductor surrounded by a shield. The center conductor is separated from the shield by a dielectric. The shield protects against electromagnetic interference.
- **Fiber optics.** Fiber optics uses the wide bandwidth properties of light traveling through transparent fibers. Fiber optics is a reliable communication medium best suited for point-to-point, high-speed data transmission. Fiber optics is immune to RF electromagnetic interference and does not produce electromagnetic radiation emission. The preferred DTM for an ESS is fiber-optic cables unless there are justifiable economic or technical reasons for using other types of media.
- **RF transmission.** Modulated RF can be used as a DTM with the installation of radio receivers and transmitters. An RF transmission system does not require a direct physical link between the points of communication, and it is useful for communicating over barriers such as bodies of water and heavily forested terrain. A disadvantage is that the signal power received depends on many factors (including transmission power, antenna pattern, path length, physical

obstructions, and climatic conditions). Also, RF transmission is susceptible to jamming and an adversary with an appropriately tuned receiver has access to it. The use of RF will be coordinated with the communications officer to avoid interference with other existing or planned facility RF systems.

6-140. There are two basic types of communication links—point-to-point and multiplex lines. A point-to-point link is characterized by a separate path for each pair of components. This approach is cost effective for several component pairs or when a number of scattered remote areas communicate with a single central location. The multiplex link, commonly referred to as a multidrop or multipoint link, is a path shared by a number of components. Depending on the number and location of components, this type of configuration can reduce the amount of cabling required. However, the cost reduction from reduced cabling is somewhat offset by costs of equipment required to multiplex and demultiplex data.

6-141. Data links used to communicate the status of ESS devices or other sensitive information to the security center must be protected from possible compromise. Attempts to defeat the security system may range from simple efforts to cut or short the transmission line to more sophisticated undertakings, such as tapping and substituting bogus signals. Data links can be made more secure by physical protection, tamper protection, line supervision, and encryption.

CCTV FOR ALARM ASSESSMENT AND SURVEILLANCE

6-142. A properly integrated CCTV assessment system provides a rapid and cost-effective method for determining the cause of intrusion alarms. For surveillance, a properly designed CCTV system provides a cost-effective supplement to guard patrols. For large facilities, the cost of a CCTV system is more easily justified. It is important to recognize that CCTV alarm-assessment systems and CCTV surveillance systems perform separate and distinct functions. The alarm-assessment system is designed to respond rapidly, automatically, and predictably to the receipt of ESS alarms at the security center. The surveillance system is designed to be used at the discretion of and under the control of the security center's console operator. When the primary function of the CCTV system is to provide real-time alarm assessment, the design should incorporate a video-processing system that can communicate with the alarm-processing system.

6-143. A candidate site for a CCTV assessment system will typically have the following characteristics:

- Assets requiring ESS protection.
- A need for real-time alarm assessment.
- Protected assets spaced some distance apart.

6-144. Figure 6-19, page 6-46, shows a typical CCTV system configuration. A typical site will locate CCTV cameras—

- Outdoors, along site-perimeter isolation zones.
- Outdoors, at controlled access points (sally ports).

- Outdoors, within the protected area, and at viewing approaches to selected assets.
- Indoors, at selected assets within the protected area.

6-145. The security console is centrally located in the security center. The CCTV monitors and the ancillary video equipment will be located at this console, as will the ESS alarm-processing and -annunciation equipment.

CCTV CAMERA COMPONENTS

6-146. An optical-lens system that captures and focuses reflected light from the scene being viewed onto an image target is common to all CCTV cameras. The image target converts reflected light energy into electrical impulses in a two-dimensional array of height and width. An electronic scanning system (reading these impulses in a predetermined order) creates a time-sensitive voltage signal that is a replica of optical information captured by the lens and focused on the target. This voltage signal is then transmitted to a location where it is viewed and possibly recorded. For components and technical information regarding CCTV cameras, see the appropriate TMs.

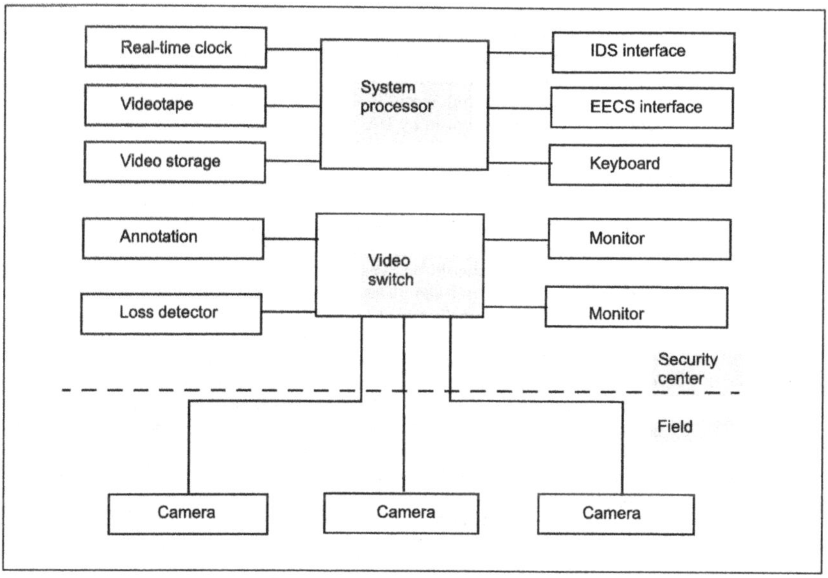

Figure 6-19. Typical CCTV System

VIDEO SIGNAL AND CONTROL LINKS

6-147. A CCTV transmission system is needed to convey video signals from various facility cameras to the security center and to carry commands from the security center to the cameras. Information may be sent via metallic cable, RF, or optical transmission.

Metallic Cable

6-148. Metallic video cables are electrical conductors manufactured specifically for the transmission of frequencies associated with video components. Coaxial cable is a primary example of this type of transmission media. Devices such as video-equalization amplifiers, ground loop correctors, and video-distribution amplifiers may be required.

RF Transmission

6-149. For a system that has widely separated nodes, RF transmission may be a good alternative to metallic cable and associated amplifiers. The information can be transmitted over a microwave link. A microwave link can be used for distances of about 50 miles, as long as the receiver and the transmitter are in the LOS.

Fiber-Optic Cable

6-150. In fiber-optic cable systems, electrical video signals are converted to optical light signals that are transmitted down the optical fiber. The signal is received and reconverted into electrical energy. An optic driver and a receiver are required per fiber. The fiber-optic transmission method provides a low-loss, high-resolution transmission system with usable length three to ten times that of traditional metallic in cable systems. Fiber-optic cable is the transmission media favored by DA.

CCTV-SYSTEM SYNCHRONIZATION

6-151. Timing signals are processed within the image-scan section of the CCTV camera. These signals may be generated internally from a crystal clock, derived from the camera's AC power source, or supplied by an external signal source. The camera should be capable of automatic switchover to its internal clock in case of external signal loss. When CCTV cameras are supplied by a common external (master) signal source or are all powered from the same AC power source, all cameras scan in synchronism. In this case, a console CCTV monitor will display a smooth transition when switched from one video source to another. Without this feature, the monitor display breaks up or rolls when switched between video sources. The rolling occurs for as long as it takes the monitor to synchronize its scan with that of the new video source, typically one second. The resynchronization delay will be experienced by all system components that receive video information, including recorders. To avoid this delay, the designer must specify that all cameras are powered from the AC power phase or must specify master synchronization for the design.

VIDEO PROCESSING AND DISPLAY COMPONENTS

6-152. As shown in Figure 6-19, page 6-46, CCTV camera signals propagate through the video transmission system and through coverage at the security center. In very simple configurations with only a few cameras and monitors, a hardwired connection between each camera and console monitor is adequate. As the number of cameras increases, the need to manage and add supplemental information to camera signals also increases. Psychological testing has demonstrated that the efficiency of console-operator assessment improves as the number of console monitors is reduced, with the optimum number being four to six monitors. Effectiveness is also enhanced by the use of alarm-correlated video. Major components of the video-processor system are the video switcher, the video-loss detector, the alarm-processor communication path, the master video sync generator, video recorders, and monitors.

- **Video switchers.** Video switchers are required when the number of cameras exceeds the number of console monitors or when a monitor must be capable of selecting video from one of many sources. Video switchers are capable of presenting any of multiple video images to various monitors, recorders, and so forth.
- **Video-loss detector.** Video-loss detectors sense the continued integrity of incoming camera signals.
- **ESS interface and communication path.** There must be a means of rapid communication between the ESS alarm-annunciation and video-processor systems. The alarm processor must send commands that cause the video switcher to select the camera appropriate for the sensor reporting an alarm. The video-processor system must report system tampering or failures (such as loss of video) to the alarm processor. The path should also pass date-and-time synchronizing information between processors so that recorded video scenes and printed alarm logs are properly correlated.
- **Master video-sync generation and distribution.** Master video sync includes a crystal-controlled timing generator, distribution amplifiers, and a transmission link to each camera.
- **Video recorders.** Video recorders provide the means to record alarm-event scenes in real time for later analysis. A recorder typically receives its input through dedicated video-switcher outputs. To support recorder playback, the recorder output is connected to a dedicated switcher input and must be compatible with the switcher-signal format. In addition, the recorder receives start commands from the switcher, and compatibility must exist at this interface. Videocassette recorders should be used when alarm events are to be recorded for later playback and analysis. The cassettes can record in time lapse for up to 240 hours (depending on the user-selected speed) and will change to real-time recording on command. The cassettes can be erased and reused or archived if required.
- **Monitors.** Monitors are required to display the individual scenes transmitted from the cameras or from the video switcher. In alarm-assessment applications, the monitors are driven by dedicated outputs of the video switcher and the monitors display video sources selected by the switcher. For security-console operations, the 9-inch monitor is the

smallest screen that should be used for operator recognition of small objects in a camera's field of view. Two 9-inch monitors can be housed side by side in a standard 19-inch console. If the monitors are to be mounted in freestanding racks behind the security console, larger units will be used.

6-153. Video-processor equipment will be specified to append the following alphanumeric information so that it appears on both monitors and recordings. The equipment must allow the operator to program the annotated information and dictate its position on the screen. This information includes—

- Time and date information.
- Video-source or alarm-zone identification.
- Programmable titles.

CCTV APPLICATION GUIDELINES

6-154. Site-specific factors must be taken into consideration in selecting components that comprise a particular CCTV system. The first is the system's size in terms of the number of cameras fielded, which is the minimum number needed to view all ESS sensor-detection fields and surveillance cameras. Another factor is that some CCTV cameras may require artificial light sources. Finally, there are CCTV-system performance criteria and physical, environmental, and economic considerations. Each is discussed in detail in TM 5-853-4.

Scene Resolution

6-155. The level to which video details can be determined in a CCTV scene is referred to as resolving ability or resolution. It is generally accepted that for assessment purposes, three resolution requirements can be defined. In order of increasing resolution requirements, they are detection, recognition, and identification.

- Detection is the ability to detect the presence of an object in a CCTV scene.
- Recognition is the ability to determine the type of object in a CCTV scene (animal, blowing debris, or crawling human).
- Identification is the ability to determine object details (a particular person, a large rabbit, a small deer, or tumbleweed).

6-156. A CCTV assessment system should provide sufficient resolution to recognize human presence and to detect small animals or blowing debris. Given an alarmed intrusion sensor, it is crucial that the console operator be able to determine if the sensor detected an intruder or if it is simply responding to a nuisance condition. (Refer to TM 5-853-4 for detailed design applications.)

Illumination Levels

6-157. For interior applications where the same camera type is used in several different areas and the scene illumination in each area is constant, specify the manually adjustable iris. This allows a manual iris adjustment appropriate for each particular area's illumination level at the time of installation. If the camera must operate in an area subject to a wide dynamic range of illumination levels (such as would be found outdoors), specify the automatically adjusted iris feature.

Cost Considerations

6-158. The cost of a CCTV system is usually quoted as cost-per-assessment zone. When estimating the total system cost, video-processor equipment costs and the video-transmission system's costs must be included. Other potentially significant costs are outdoor lighting system upgrades and the site preparation required to support the CCTV cameras. The CCTV systems are expensive compared to other electronic security subsystems and should be specified with discretion.

DESIGN GUIDELINES

6-159. The design and application of CCTV systems are quite complex and should be left to professionals who are abreast of the current state-of-the-art systems. Some of the general design guidelines include the following:

- **System familiarity.** Before designing an effective CCTV assessment system, the designer must be familiar with the ESS's sensor placement and the detection field's shape.
- **CCTV camera placement and lighting.** The placement of exterior cameras requires more attention than that of interior cameras because of weather and illumination extremes. The field-of-view alignment, illumination range, and balanced lighting are major design factors. Exterior CCTV design considerations include environmental housings, camera mounting heights, system types, and so forth. Indoor design considerations include the mounting location and tamper detection. The layout for indoor alarm-assessment cameras is subject to three constraints—
  - The camera's location should enclose the complete sensor detection field in the camera's field of view.
  - Lighting that is adequate to support alarm assessment will be provided.
  - Protection from tampering and inadvertent damage by collision during normal area operations will be provided.

Chapter 7

Access Control

Perimeter barriers, intrusion-detection devices, and protective lighting provide physical-security safeguards; however, they alone are not enough. An access-control system must be established and maintained to preclude unauthorized entry. Effective access-control procedures prevent the introduction of harmful devices, materiel, and components. They minimize the misappropriation, pilferage, or compromise of materiel or recorded information by controlling packages, materiel, and property movement. Access-control rosters, personal recognition, ID cards, badge-exchange procedures, and personnel escorts all contribute to an effective access-control system.

DESIGNATED RESTRICTED AREAS

7-1. The installation commander is responsible for designating and establishing restricted areas. A restricted area is any area that is subject to special restrictions or controls for security reasons. This does not include areas over which aircraft flight is restricted. Restricted areas may be established for the following:

- The enforcement of security measures and the exclusion of unauthorized personnel.
- Intensified controls in areas requiring special protection.
- The protection of classified information or critical equipment or materials.

DEGREE OF SECURITY

7-2. The degree of security and control required depends on the nature, sensitivity, or importance of the security interest. Restricted areas are classified as controlled, limited, or exclusion areas.

- A controlled area is that portion of a restricted area usually near or surrounding a limited or exclusion area. Entry to the controlled area is restricted to personnel with a need for access. Movement of authorized personnel within this area is not necessarily controlled since mere entry to the area does not provide access to the security interest. The controlled area is provided for administrative control, for safety, or as a buffer zone for in-depth security for the limited or exclusion area. The commander establishes the control of movement.
- A limited area is a restricted area within close proximity of a security interest. Uncontrolled movement may permit access to the item. Escorts and other internal restrictions may prevent access within limited areas.

- An exclusion area is a restricted area containing a security interest. Uncontrolled movement permits direct access to the item.

7-3. The security protection afforded by a restricted area pertains particularly to subversive-activity control; that is, protection against espionage, sabotage, or any such action adversely affecting national defense. Within this context, the designation "restricted area" is not applicable to an area solely for protection against common pilferage or misappropriation of property or material that is not classified or not essential to national defense. For example, an area devoted to the storage or use of classified documents, equipment, or materials should be designated as a restricted area to safeguard against espionage. An installation communications center should also be so designated to safeguard against sabotage. On the other hand, a cashier's cage or an ordinary mechanic's tool room should not be so designated, although the commander may impose controls to access. This may be a simple matter of posting an "off limits to unauthorized personnel" sign. The PM or the physical-security manager acts as an advisor to the commander. In his recommendations, he must consider evaluating the purpose of designating a restricted area and coordinating with the intelligence officer and the staff judge advocate (SJA).

7-4. A restricted area must be designated in writing by the commander and must be posted with warning signs according to AR 190-13. In areas where English is one of two or more languages commonly spoken, warning signs will be posted in English and in the local language (see Figure 7-1).

7-5. An installation may have varying degrees of security. It may be designated in its entirety as a restricted area, with no further restrictions; or it may be subdivided into controlled, limited, or exclusion areas with restrictions of movement and specific clear zones. Figure 7-2 depicts a simplified restricted area and the degrees of security.

CONSIDERATIONS

7-6. There are other important considerations concerning restricted areas and their lines of division. These considerations include the following:

- A survey and analysis of the installation, its missions, and its security interests. This can determine immediate and anticipated needs that require protection. Anticipated needs are determined from plans for the future.
- The size and nature of the security interest being protected. Safes may provide adequate protection for classified documents and small items; however, large items may have to be placed within guarded enclosures.
- Some security interests are more sensitive to compromise than others. Brief observation or a simple act by an untrained person may constitute a compromise in some cases. In others, detailed study and planned action by an expert may be required.
- All security interests should be evaluated according to their importance. This may be indicated by a security classification such as confidential, secret, or top secret.

RESTRICTED AREA

THIS (INSTALLATION OR ACTIVITY) HAS BEEN DECLARED A RESTRICTED AREA BY AUTHORITY OF (TITLE, COMMANDING GENERAL OR COMMANDING OFFICER) IN ACCORDANCE WITH THE PROVISIONS OF THE DIRECTIVE ISSUED BY THE SECRETARY OF DEFENSE ON 20 AUGUST 1954, PURSUANT TO THE PROVISIONS OF SECTION 21, INTERNAL SECURITY ACT OF 1950. UNATHORIZED ENTRY IS PROHIBITED.

ALL PERSONS AND VEHICLES ENTERING HEREIN ARE LIABLE TO SEARCH. PHOTOGRAPHING OR MAKING NOTES, DRAWINGS, MAPS, OR GRAPHIC REPRESENTATIVES OF THIS AREA OR ITS ACTIVITIES ARE PROHIBITED UNLESS SPECIFICALLY AUTHORIZED BY THE COMMANDER. ANY SUCH MATERIAL FOUND IN THE POSSESSION OF UNAUTHORIZED PERSONS WILL BE CONFISCATED.

Figure 7-1. Sample Restricted-Area Warning

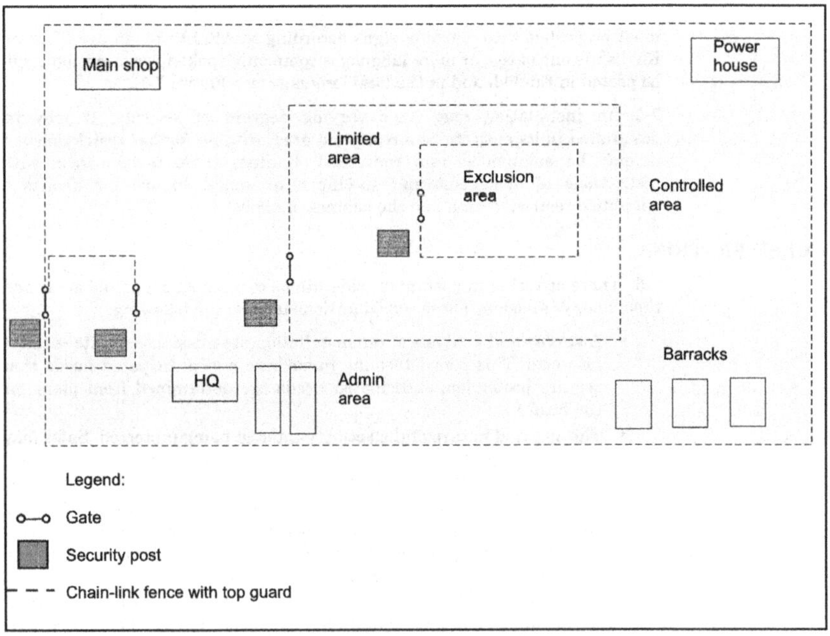

Figure 7-2. Schematic Diagram of a Simplified Restricted Area and the Degrees of Security

- Parking areas for privately owned vehicles (POVs) are established outside of restricted areas. Vehicle entrances must be kept at a minimum for safe and efficient control.
- Physical protective measures (such as fences, gates, and window bars) must be installed.

EMPLOYEE SCREENING

7-7. Screening job applicants to eliminate potential acts of espionage and sabotage and other security risks is important in peacetime and is critical during a national emergency. Personnel screenings must be incorporated into standard personnel policies.

7-8. An applicant should be required to complete a personnel security questionnaire, which is then screened for completeness and used to eliminate undesirable applicants. A careful investigation should be conducted to ensure that the applicant's character, associations, and suitability for employment are satisfactory. The following sources may be helpful in securing employment investigative data:

- State and local police (including national and local police in overseas areas).
- Former employers.
- Public records.
- Credit agencies.
- Schools (all levels).
- References. (These references should include those names not furnished by the applicant. These are known as throw offs, and they are obtained during interviews of references furnished by applicants.)
- Others as appropriate. (These may include the FBI, the US Army Criminal Records Repository, and the Defense Investigative Agency).

7-9. Medical screening considerations should be made (based on an applicant's position [such as a guard]) to evaluate physical and mental stamina. Once an applicant has been identified for employment, he is placed on an access-control roster.

IDENTIFICATION SYSTEM

7-10. An ID system is established at each installation or facility to provide a method of identifying personnel. The system provides for personal recognition and the use of security ID cards or badges to aid in the control and movement of personnel activities.

7-11. Standard ID cards are generally acceptable for access into areas that are unrestricted and have no security interest. Personnel requiring access to restricted areas should be issued a security ID card or badge as prescribed in AR 600-8-14. The card's/badge's design must be simple and provide for adequate control of personnel.

7-12. A security ID card/badge system must be established for restricted areas with 30 or more employees per shift. Commanders may (at their discretion) authorize a card/badge system in restricted areas for less than 30 people.

ID METHODS

7-13. Four of the most commonly used access-control ID methods are the personal-recognition system, the single-card or -badge system, the card- or badge-exchange system, and the multiple-card or -badge system.

Personal-Recognition System

7-14. The personal-recognition system is the simplest of all systems. A member of the security force providing access control visually checks the person requesting entry. Entry is granted based on—

- The individual being recognized.
- The need to enter has been established.
- The person is on an access-control roster.

Single-Card or -Badge System

7-15. This system reflects permission to enter specific areas by the badge depicting specific letters, numbers, or particular colors. This system lends to comparatively loose control and is not recommended for high-security areas. Permission to enter specific areas does not always go with the need to know. Because the ID cards/badges frequently remain in the bearer's possession while off duty, it affords the opportunity for alteration or duplication.

Card- or Badge-Exchange System

7-16. In this system, two cards/badges contain identical photographs. Each card/badge has a different background color, or one card/badge has an overprint. One card/badge is presented at the entrance to a specific area and exchanged for the second card/badge, which is worn or carried while in that area. Individual possession of the second card/badge occurs only while the bearer is in the area for which it was issued. When leaving the area, the second card/badge is returned and maintained in the security area. This method provides a greater degree of security and decreases the possibility of forgery, alteration, or duplication of the card/badge. The levels of protection described in TM 5-853-1 require multiple access-control elements as the levels of protection increase. In the case of the badge exchange, this system counts as two access-control elements.

Multiple-Card or -Badge System

7-17. This system provides the greatest degree of security. Instead of having specific markings on the cards/badges denoting permission to enter various restricted areas, the multiple card/badge system makes an exchange at the entrance to each security area. The card/badge information is identical and allows for comparisons. Exchange cards/badges are maintained at each area only for individuals who have access to the specific area.

MECHANIZED/AUTOMATED SYSTEMS

7-18. An alternative to using guards or military police (MP) to visually check cards/badges and access rosters is to use building card-access systems or biometric-access readers. These systems can control the flow of personnel entering and exiting a complex. Included in these systems are—

- Coded devices such as mechanical or electronic keypads or combination locks.
- Credential devices such as magnetic-strip or proximity card readers.
- Biometric devices such as fingerprint readers or retina scanners.

7-19. Access-control and ID systems base their judgment factor on a remote capability through a routine discriminating device for positive ID. These systems do not require guards at entry points; they identify an individual in the following manner:

- The system receives physical ID data from an individual.
- The data is encoded and compared to stored information.
- The system determines whether access is authorized.
- The information is translated into readable results.

7-20. Specialized mechanical systems are ideal for highly sensitive situations because they use a controlled process in a controlled environment to establish the required database and accuracy. One innovative technique applied to ID and admittance procedures involves dimension comparisons. The dimension of a person's full hand is compared to previously stored data to determine entry authorization. Other specialized machine readers can scan a single fingerprint or an eye retina and provide positive ID of anyone attempting entry.

7-21. An all-inclusive automated ID and access-control system reinforces the security in-depth ring through its easy and rapid change capability. The computer is able to do this through its memory. Changes can be made quickly by the system's administrator.

7-22. The commercial security market has a wide range of mechanized and automated hardware and software systems. Automated equipment is chosen only after considering the security needs and the environment in which it operates. These considerations include whether the equipment is outdoors or indoors, the temperature range, and weather conditions. Assessment of security needs and the use of planning, programming, and budgeting procedures greatly assist a security manager in improving the security posture.

CARD/BADGE SPECIFICATIONS

7-23. Security cards/badges should be designed and constructed to meet the requirements of AR 600-8-14. Upon issuing a card/badge, security personnel must explain to the bearer the wear required and the authorizations allowed with the card/badge. This includes—

- Designation of the areas where an ID card/badge is required.

- A description of the type of card/badge in use and the authorizations and limitations placed on the bearer.
- The required presentation of the card/badge when entering or leaving each area during all hours of the day.
- Details of when, where, and how the card/badge should be worn, displayed, or carried.
- Procedures to follow in case of loss or damage of the card.
- The disposition of the card/badge upon termination of employment, investigations, or personnel actions.
- Prerequisites for reissuing the card/badge.

VISITOR IDENTIFICATION AND CONTROL

7-24. Procedures must be implemented to properly identify and control personnel. This includes visitors presenting their cards/badges to guards at entrances of restricted areas. Visitors are required to stay with their assigned escort. Guards must ensure that visitors stay in areas relating to their visit; an uncontrolled visitor, although conspicuously identified, could acquire information for which he is not authorized. Foreign-national visitors should be escorted at all times.

7-25. Approval for visitors should be obtained at least 24 hours in advance (if possible). Where appropriate, the installation should prepare an agenda for the visitor and designate an escort officer. Measures must be in place to recover visitor cards/badges on the visit's expiration or when they are no longer required.

7-26. Physical-security precautions against pilferage, espionage, and sabotage require the screening, ID, and control of visitors. Further information about visiting requirements and procedures are found in ARs 12-15 and 381-20. Visitors are generally classed in the following categories:

- Persons with whom every installation or facility has business (such as suppliers, customers, insurance inspectors, and government inspectors).
- Individuals or groups who desire to visit an installation or facility for personal or educational reasons. Such visits may be desired by educational, technical, or scientific organizations.
- Individuals or groups specifically sponsored by the government (such as foreign nationals visiting under technical cooperation programs and similar visits by US nationals). Requests for visits by foreign nationals must be processed according to AR 380-10.
- Guided tours to selected portions of the installation in the interest of public relations.

7-27. The ID and control mechanisms for visitors must be in place. They may include the following:

- Methods of establishing the authority for admitting visitors and any limitations relative to access.

- Positive ID of visitors by personal recognition, visitor permit, or other identifying credentials. Contact the employer, supervisor, or officer in charge to validate the visit.
- The use of visitor registration forms. These forms provide a record of the visitor and the time, location, and duration of his visit.
- The use of visitor ID cards/badges. The cards/badges bear serial numbers, the area or areas to which access is authorized, the bearer's name, and escort requirements.

7-28. Individual groups entering a restricted area must meet specific prerequisites before being granted access. The following guidance is for group access into a restricted area:

Visitors

7-29. Before allowing visitors into a restricted area, contact the person or activity being visited. After verifying the visitor's identity, issue a badge, complete the registration forms, and assign an escort (if required). Visitors may include public-utility and commercial-service representatives.

Very Important Persons

7-30. The procedures for admitting very important persons (VIPs) and foreign nationals into restricted areas are contained in AR 12-15. Special considerations and coordination with the protocol office are necessary. A 24-hour advance notice is desirable for these requests, along with an agenda for the visit and the designation of an escort, if appropriate.

Civilians Working on Jobs Under Government Contract

7-31. To allow these personnel to conduct business in restricted areas, the security manager must coordinate with the procurement office. The security manager must also identify movement-control procedures for these employees.

Cleaning Teams

7-32. Supervisors using cleaning teams must seek technical advice from the physical-security office on internal controls for each specific building. This may include providing escorts.

DOD Employees in Work Areas After Normal Operating Hours

7-33. Supervisors establish internal controls based on coordination with the security manager. They also notify security personnel of the workers' presence, type, and duration of work.

ENFORCEMENT MEASURES

7-34. The most vulnerable link in any ID system is its enforcement. Security forces must be proactive in performing their duties. A routine performance of duty will adversely effect even the most elaborate system. Positive enforcement measures must be prescribed to enhance security. Some of these measures may include—

- Designating alert and tactful security personnel at entry control points.
- Ensuring that personnel possess quick perception and good judgment.
- Requiring entry-control personnel to conduct frequent irregular checks of their assigned areas.
- Formalizing standard procedures for conducting guard mounts and posting and relieving security personnel. These measures will prevent posting of unqualified personnel and a routine performance of duty.
- Prescribing a uniform method of handling or wearing security ID cards/badges. If carried on the person, the card must be removed from the wallet (or other holder) and handed to security personnel. When worn, the badge will be worn in a conspicuous position to expedite inspection and recognition from a distance.
- Designing entry and exit control points of restricted areas to force personnel to pass in a single file in front of security personnel. In some instances, the use of turnstiles may be advisable to assist in maintaining positive control.
- Providing lighting at control points. The lighting must illuminate the area to enable security personnel to compare the bearer with the ID card/badge.
- Enforcing access-control measures by educating security forces and employees. Enforcement of access-control systems rests primarily with the security forces; however, it is essential that they have the full cooperation of the employees. Employees must be instructed to consider each unidentified or improperly identified individual as a trespasser. In restricted areas where access is limited to a particular zone, employees must report unauthorized individuals to the security force.
- Positioning ID card/badge racks or containers at entry control points so that they are accessible only to guard-force personnel.
- Appointing a responsible custodian to accomplish control procedures of cards/badges according to AR 600-8-14. The custodian is responsible for the issue, turn in, recovery, and renewal of security ID cards/badges.

7-35. The degree of compromise tolerable in the ID system is in direct proportion to the degree of security required. The following control procedures are recommended for preserving the integrity of a card/badge system:

- Maintenance of an accurate written record or log listing (by serial number) all cards and badges and showing those on hand, to whom they are issued, and their disposition (lost, mutilated, or destroyed).
- Authentication of records and logs by the custodian.
- A periodic inventory of records by a commissioned officer.
- The prompt invalidation of lost cards/badges.
- The conspicuous posting at security control points of current lists of lost or invalidated cards/badges.
- The establishment of controls within restricted areas to enable security personnel to determine the number of persons within the area.
- The establishment of the two-person rule (when required).

- The establishment of procedures to control the movement of visitors. A visitor-control record will be maintained and located at entry control points.

SIGN/COUNTERSIGN AND CODE WORD

7-36. This method of verifying identity is primarily used in a tactical environment. According to the local SOP, the sign/countersign or code-word procedures must be changed immediately if compromised.

DURESS CODE

7-37. The duress code is a simple word or phrase used during normal conversation to alert other security personnel that an authorized person is under duress. A duress code requires planning and rehearsal to ensure an appropriate response. This code is changed frequently to minimize compromise.

ACCESS-CONTROL ROSTERS

7-38. Admission of personnel to a restricted area is granted to those identified and listed on an access-control roster. Pen-and-ink changes may be made to the roster. Changes are published in the same manner as the original roster.

7-39. Rosters are maintained at access control points. They are kept current, verified, and accounted for by an individual designated by the commander. Commanders or their designated representatives authenticate the rosters. Admission of persons other than those on the rosters is subject to specific approval by the security manager. These personnel may require an escort according to the local SOP.

METHODS OF CONTROL

7-40. There are a number of methods available to assist in the movement and control of personnel in limited, controlled, and restricted areas. The following paragraphs discuss the use of escorts and the two-person rule:

ESCORTS

7-41. Escorts are chosen because of their ability to accomplish tasks effectively and properly. They possess knowledge of the area being visited. Escorts may be guard-force personnel, but they are normally personnel from the area being visited. Local regulations and SOPs determine if a visitor requires an escort while in the restricted area. Personnel on the access list may be admitted to restricted areas without an escort.

TWO-PERSON RULE

7-42. The two-person rule is designed to prohibit access to sensitive areas or equipment by a lone individual. Two authorized persons are considered present when they are in a physical position from which they can positively detect incorrect or unauthorized procedures with respect to the task or operation being performed. The team is familiar with applicable safety and security requirements, and they are present during any operation that affords access to sensitive areas or equipment that requires the two-person rule. When application of the two-person rule is required, it is enforced constantly by the personnel who constitute the team.

7-43. The two-person rule is applied in many other aspects of physical-security operations, such as the following:

- When uncontrolled access to vital machinery, equipment, or materiel might provide opportunity for intentional or unintentional damage that could affect the installation's mission or operation.
- When uncontrolled access to funds could provide opportunity for diversion by falsification of accounts.
- When uncontrolled delivery or receipt for materials could provide opportunity for pilferage through "short" deliveries and false receipts.
- When access to an arms or ammunition storage room could provide an opportunity for theft. Keys should be issued so that at least two people must be present to unlock the locks required under the provisions of AR 190-11.

7-44. The two-person rule is limited to the creativity of the PM and the physical-security manager. They should explore every aspect of physical-security operations in which the two-person rule would provide additional security and assurance and include all appropriate recommendations and provisions of the physical-security plan. An electronic-entry control system may be used to enforce the two-person rule. The system can be programmed to deny access until two authorized people have successfully entered codes or swiped cards.

SECURITY CONTROLS OF PACKAGES, PERSONAL PROPERTY, AND VEHICLES

7-45. A good package-control system helps prevent or minimize pilferage, sabotage, and espionage. The local SOP may allow the entry of packages with proper authorization into restricted areas without inspection. A package-checking system is used at the entrance gate. When practical, inspect all outgoing packages except those properly authorized for removal. When a 100 percent inspection is impractical, conduct frequent unannounced spot checks. A good package-control system assists in the movement of authorized packages, material, and property.

7-46. Property controls are not limited to packages carried openly, but they include the control of anything that could be used to conceal property or material. Personnel should not be routinely searched except in unusual situations. Searches must be performed according to the local SOP.

7-47. All POVs on the installation should be registered with the PM or the installation's physical-security office. Security personnel should assign a temporary decal or other temporary ID tag to visitors' vehicles to permit ready recognition. The decal or the tag should be distinctly different from that of permanent-party personnel.

7-48. When authorized vehicles enter or exit a restricted area, they undergo a systematic search, including (but not limited to) the—

- Vehicle's interior.
- Engine compartment.
- External air breathers.
- Top of the vehicle.
- Battery compartment.
- Cargo compartment.
- Undercarriage.

7-49. The movement of trucks and railroad cars into and out of restricted areas should be supervised and inspected. Truck and railroad entrances are controlled by locked gates when not in use and are manned by security personnel when unlocked. The ID cards/badges are issued to operators to ensure proper ID and registration for access to specific loading and unloading areas.

7-50. All conveyances entering or leaving a protected area are required to pass through a service gate manned by security forces. Drivers, helpers, passengers, and vehicle contents must be carefully examined. The examination may include—

- Appropriate entries in the security log (including the date, operator's name, load description, and time entered and departed).
- A check of the operator's license.
- Verification of the seal number with the shipping document and examination of the seal for tampering.

7-51. Incoming trucks and railroad cars must be assigned escorts before they are permitted to enter designated limited or exclusion areas. Commanders should establish published procedures to control the movement of trucks and railroad cars that enter designated restricted areas to discharge or pick up cargo (escorts will be provided when necessary).

7-52. The best control is provided when all of these elements are incorporated into access-control procedures. Simple, understandable, and workable access-control procedures are used to achieve security objectives without impeding operations. When properly organized and administered, access-control procedures provide a method of positively identifying personnel who have the need to enter or leave an area.

TACTICAL-ENVIRONMENT CONSIDERATIONS

7-53. Access-control procedures during tactical operations may establish additional challenges for the commander. In some instances, the commander cannot provide a perimeter barrier (such as a fence) based on METT-TC. Commanders are still required to provide security measures for restricted areas, although they may not always have the necessary assets. Early-warning systems and the use of guards become crucial. A restricted area may become a requirement without prior notice during an operation. Figure 7-3 and Figure 7-4, page 7-14, are examples of temporary tactical restricted and exclusion areas.

7-54. Commanders must plan for these considerations when developing their budget. Funding must be requested and set aside to support physical-security requirements during tactical operations. Resources will not always be available; therefore, commanders must implement procedures that support access-control measures. Improvising will become common practice to overcome shortfalls concerning access-control equipment in the field.

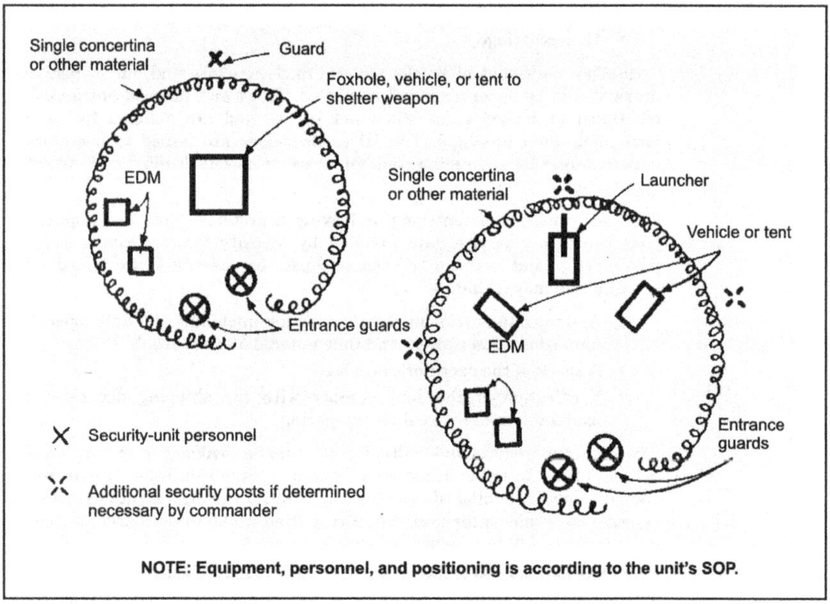

Figure 7-3. Sample Layout of Temporary Tactical Restricted Areas

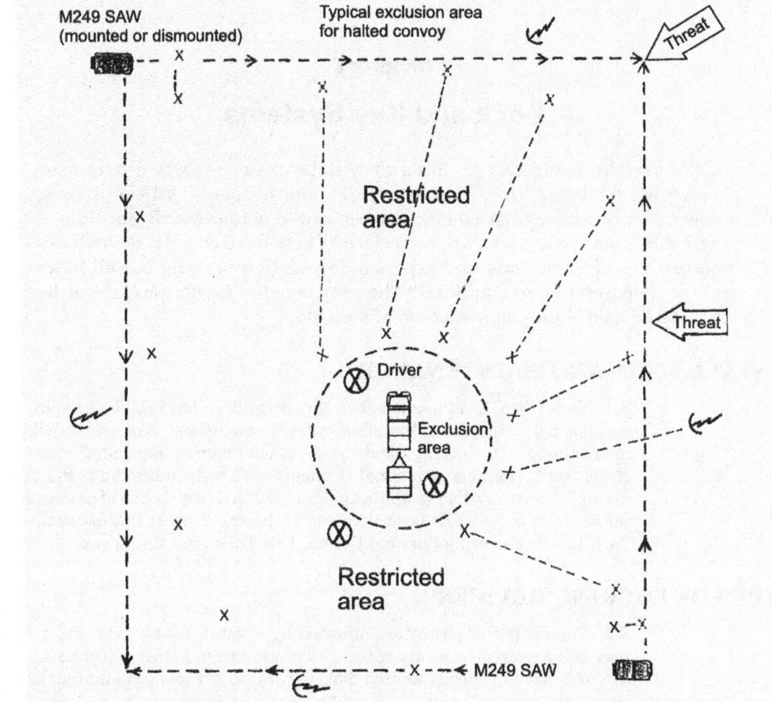

NOTE: Other vehicles and personnel, to include the LPs, form a defensive perimeter inside the primary directions of fire of the M60 machine guns. During attack, alternate machine-gun crews should be placed in the perimeter away from primary machine guns and be spaced so that they can be used as replacements and so that large gaps do not appear in the perimeter.

Legend:

X — MP rifleman

◄– – – Principal direction of fire

( ) Temporary exclusion area (size depends on terrain and threat force)

Listening post

Figure 7-4. Sample Layout for a Temporary Tactical Exclusion Area

Chapter 8

Lock and Key Systems

Locks are the most acceptable and widely used security devices for protecting facilities, classified materials, and property. All containers, rooms, and facilities must be locked when not in actual use. Regardless of their quality or cost, locks are considered delay devices only. Some locks require considerable time and expert manipulation to open, but all locks can be defeated by force and with the proper tools. Locks must never be considered as a stand-alone method of security.

INSTALLATION AND MAINTENANCE

8-1. The USACE is responsible for installing locking devices in newly constructed facilities. Installation-level engineers are responsible for maintaining the locking devices. Physical-security personnel must work closely with engineer personnel to ensure that locks meet the standards and are installed according to applicable regulations. One source of assistance and information is the DOD Lock Program Technical Support Hotline at the Naval Facilities Engineering Services Center, Port Hueneme, California.

TYPES OF LOCKING DEVICES

8-2. The degree of protection afforded by a vault, a safe, or a filing cabinet may be measured in terms of the lock's resistance. Locking devices are listed in TM 5-805-8. Types of locking devices include key and combination locks.

8-3. ARs 190-11, 190-51, 50-5, and 50-6 prescribe specific types of locks for specific types of facilities. AR 380-5 prescribes standard facilities for storing classified material and contains guidance for different storage requirements.

KEY LOCKS

8-4. Key locks consist of, but are not limited to, the following:

- Cylindrical locksets are often called key-in-knob or key-in-lever locks. They are normally used to secure offices and storerooms. The locking cylinder located in the center of the doorknob distinguishes these locks. Some cylindrical locksets have keyways in each of the opposing knobs that require a key on either side to lock and unlock them. Others unlock with a key, but may be locked by pushing or rotating a button on the inside knob. These locks are suitable only for very low-security applications. Using these locks may require compensatory measures in the form of additional locks on containers within the room.
- Dead-bolt locks are sometimes called tubular dead bolts. They are mounted on the door in a manner similar to cylindrical locksets. The primary difference is in the bolt. When the bolt is extended (locked),

the dead bolt projects into the doorframe at least one inch, and it cannot be forced back (unlocked) by applying pressure to the end of the bolt. The dead-bolt lock has the potential for providing acceptable levels of protection for storerooms and other areas where more security is desired. It is recommended for use in military housing as an effective security measure in the installation's crime-prevention program. In situations where there is a window in or adjacent to the door, a double-cylinder dead-bolt lock (one that requires a key to open from either side) should be used.

- Mortise locks are so named because the lock case is mortised or recessed into the edge of the door. The most common variety of mortise locks has a doorknob on each side of the door. Entrance doors often have an exterior thumb latch rather than a doorknob. The mortise lock can be locked from inside by means of a thumb turn or by a button. Mortise locks are considered low-security devices since they weaken the door in the mortised area.
- Drop-bolt locks (often referred to as jimmy-proof locks) are normally used as auxiliary locks similar to dead bolts. Both the drop-bolt lock body and the strike have interlocking leaves similar to a door hinge. When closed, locking pins in the lock body drop down into the holes provided in the strike and secure the locking system. Since the lock body and the strike are interconnected with locking pins when closed, the lock essentially becomes a single unit and is extremely difficult to separate.
- Rim-cylinder locks are mounted to the door's inside surface and are secured by screws in the door face. These locks are generally used with drop-bolt and other surface-mounted locks and latches. They consist of an outer barrel, a cylinder and ring, a tailpiece, a back mounting plate, and two mounting screws. The tailpiece screws are usually scored so that the lock can be tailored to fit varying door thicknesses.
- Unit locks are ideal in heavily traveled facilities (such as hospitals or institutional buildings). These locks are a complete, one-piece unit that slides into a notch cut into the door's latch edge. The one-size cutout of the door edge simplifies the door preparation for the lock.
- Mechanical, push-button combination locks are digital (push buttons numbered 1 through 9) combination door-locking devices used to deny area access to any individual not authorized or cleared for a specific area. These locks are normally used for access control and should be backed up by door locking devices when the facility is unoccupied.
- Padlocks are detachable locks that are typically used with a hasp. Low-security padlocks, sometimes called secondary padlocks, are used to deter unauthorized access, and they provide only minimal resistance to force. Low-security locks are made with hardened steel shackles. Precautions must be taken to avoid confusing these locks with similar brass or bronze locks. The brass or bronze locks are commonly used but do not meet the security requirements of the hardened shackled locks. High-security padlocks may be used to secure AA&E. They provide the maximum resistance to unauthorized entry when used with a high-security hasp.

8-5. Some locks have interchangeable cores, which allow the same key system to include a variety of locks. Padlocks, door locks, cabinet locks, and electrical key switches can all be operated by the same key system. Because these cores are removable by a special key, this system allows for rapid rekeying of locks in the event that the key is compromised.

8-6. Locks are keyed in several different ways. When several locks are keyed differently, each is operated by its own key. When they are keyed alike, one key will open them all. Locks that are master-keyed are keyed differently, yet have one key that will open them all. Master-keying is done for convenience and represents the controlled loss of security. Master-keying is not used unless permitted by regulation.

COMBINATION LOCKS

8-7. Combination locks are available as padlocks or as mounted locks. They are low-security padlocks with combinations that are either fixed or changeable. Combination locks may be either mechanical or electronic. They are operated by entering a particular sequence of numbers. When the correct combination is entered, the lock's bolt is retracted. Combination locks used for securing classified material must meet Federal Specification FF-L-2740.

8-8. Although the lock is the most accepted and widely used security device, it is only a delay device and should never be considered as a positive bar to entry. A lock can (and will) be defeated. The best defense for locking devices is a good key-control program. Refer to AR 190-51, Appendix D, for standard key and lock procedures. Additional key and lock procedures for AA&E can be found in AR 190-11, Chapter 3.

Chapter 9

Security Forces

The security force for an installation or a facility provides the enforcement element in the physical-security program. This force consists of personnel specifically organized, trained, and equipped to protect the command's physical-security interests. It is a commander's most effective tool in a comprehensive, integrated, physical-security program. Vulnerability tests are periodically conducted to determine and ensure the state of readiness of security forces (see Appendix K).

TYPES OF SECURITY FORCES

9-1. On installations, security forces may be MP forces, security police, DOD civil-service security guards, or contract guards. Interior guard duties are performed by installation unit troops on a roster basis. MP forces normally perform security duties that require higher degrees of training and experience. These include—

- Security of restricted areas.
- Security of specific sensitive gates.
- Supervisory or coordinated roles with other military or DOD civil-service security guards.
- Responsibility for monitoring and responding to intrusion alarms.

9-2. An MP unit may perform the entire physical-security function alone based on METT-TC, the area, and the facilities. When an MP unit cannot assume responsibilities for all of the physical-security requirements in the command, other forces may be required. Additional forces may consist of the following:

- Personnel furnished by units of the installation's command on a daily or weekly basis. While this method has the single advantage of providing additional manpower, it has the disadvantages of rapid turnover and the lack of training. If this manpower is used, personnel should be assigned the least sensitive posts or patrols. For extended augmentation, units may be attached to MP units. The MP unit may also be augmented by reserve units or units in rotation.
- The combat-arms branches (especially the infantry) may attach their forces to MP units and may be designated as security guards assisting in the required operations.
- Military or paramilitary units of the host country may also be attached to or operate in coordination with MP forces. They may also be supplemented with national police of their own country.

- The installation's band may be a source of military force during wartime. (The band is assigned enemy-prisoner-of-war [EPW] duty as a wartime duty.) The band is doctrinally capable of providing security at the division tactical operations center (DTOC) and ASPs, assisting in the perimeter defense of the command post (CP), and operating the dismount point for the CP. It is capable of providing access control at the DTOC and the ASPs and augmenting or relieving security personnel on the defensive perimeter.

9-3. Civil-service security guards are uniformed civilian employees from a government agency. They are customarily trained and organized along military lines. The organization may be completely civil service or may be composed of civil-service personnel under military supervision. In either case, they are under the operational control of the PM or the security officer.

9-4. Labor-service personnel (local civilian personnel) have been organized and used successfully in theaters of operation. These types of units were organized after World War II and since that time have established enviable records in the physical-security field. They are distinctively uniformed, organized, and equipped. They have set and maintained the highest security standards, resulting in a minimal loss of property. While not military organizations, these units have successfully developed a high sense of duty and esprit de corps that has been reflected in their outstanding contributions to the physical security of installations in overseas commands.

AUTHORITY AND JURISDICTION

9-5. It is most important that the PM or the security officer determine (and instruct his security force in) the extent and limitations of the commander's jurisdiction in the field of law enforcement and investigations. Those jurisdictions include—

- Jurisdiction of place.
  - Military installations and facilities. Whether state or federal law or both are applicable on a military installation or facility depends largely on the nature of jurisdiction over the land involved. The amount of federal jurisdiction may vary between different areas of the installation or facility. The legal formalities of acquiring jurisdiction over land under the control of the Secretary of the Army are accomplished at DA level and according to the provisions of AR 405-20. Information and advice relating to jurisdictional questions should be referred to the local SJA.
  - Areas outside of military installations. Areas outside of military installations are generally subject to state and local laws; however, there are exceptions. Information and advice in this regard should be obtained through the local SJA.
  - Overseas areas. In overseas areas, jurisdiction varies according to the military situation and existing international treaties, contracts, and agreements. Guidance should be obtained in each instance from the commander and the SJA and set forth in appropriate command directives.

- Jurisdiction of personnel.
  - Jurisdiction of personnel generally follows the limitations of jurisdiction of the installation.
  - MP forces have jurisdiction and authority over personnel as described in AR 190-14 and related publications.
  - Authority for federal civilian employees assigned to security, police, and guard duties is derived from the installation's commanding officer. These personnel can have no more authority than he possesses and are subject to any limitations imposed thereon.
  - Security-force personnel may enforce all offenses under the Uniform Code of Military Justice (UCMJ), military regulations, federal laws and regulations, and state laws where applicable.
  - Security force personnel may be given the same authority as MP forces over all personnel subject to military jurisdiction, including apprehension, detention, and search.
  - Civilian security-force personnel have no specific grant of authority over civilians other than the right of citizen's arrest.
  - The commander is the source of jurisdiction and authority for all other personnel assigned to security-force duties.

PERSONNEL SELECTION

9-6. Regardless of the use of structural, mechanical, or electronic equipment, the human element in security operations makes the difference between success or failure. Commanders and supervisors have a responsibility to ensure that security personnel who control access to restricted areas and classified activities are qualified based on criteria in AR 380-67. Personnel who perform physical-security duties must be disciplined and alert, have sound judgment, be confident and physically fit, and possess good interpersonal communication skills.

SECURITY CLEARANCE

9-7. Security-clearance criteria for security positions must be based on the security classifications of the information to which access will be granted. Security positions are normally designated as sensitive and require a secret security clearance. ARs 381-20 and 380-67 describe criteria and procedures governing security clearances. Appropriate civilian-personnel regulations should also be consulted when civilians are involved.

9-8. Positive evaluation of the reliability of all personnel must be made before they are entrusted with classified or sensitive information. (The Individual Reliability Program is prescribed in AR 190-56.) Follow-up action must be made on all personnel who are granted a security clearance to ensure that they continue to meet the criteria for their clearance. Personnel not meeting or adhering to the prescribed standards must have their security clearances revoked and thereby lose their access to areas containing classified information or material (see AR 380-67).

ORGANIZATION AND EMPLOYMENT OF FORCES

9-9. The organization of a security force will vary, depending on circumstances and the forces available. Forces consist of—

- Mobile patrols. A mobile detachment of ground, sea, or air forces dispatched to gather information or carry out a security mission.
- The response force. A mobile force with appropriate fire support (usually designated by the area commander) to deal with Level II threats in the rear area (Army). This is normally an MP function.
- Reserves. That portion of a force withheld from action or uncommitted to a specific course of action so as to be available for commitment at the decisive moment. Its primary purpose is to retain flexibility throughout an offensive action.
- Any combination of these three.

9-10. Instructions to the security force should be issued in writing. These instructions are normally in the form of general, special, or temporary orders. They should be carefully and clearly worded and include all phases of each assignment. They should be reviewed at least monthly to ensure that they are current. Categories of instructions of each are as follows:

- General orders are those orders that concern the security force as a whole and are applicable at all posts and patrols.
- Special orders pertain to a permanent post or patrol. Each permanent post or patrol should have special orders issued concerning the location, duties, hours manned, arms, ammunition, and other equipment required and the instructions for using force in enforcement and apprehension activities.
- Temporary orders are issued for a short period and cover a special or temporary situation. If it can be predetermined, such orders should indicate the period of time for which they are valid.

9-11. A security-force SOP that outlines policies, organization, authority, functions, and other required information should be prepared for required reading. Each security-force member should be held responsible for full knowledge and understanding of the contents of the SOP. Each installation PM, physical-security officer, or chief of a guard force should conduct periodic inspections and tests to determine each individual's degree of understanding of these instructions. Instructions should be provided in writing regarding the safeguarding and control of the SOP. Its contents may not be classified; however, the information could assist an intruder in breaching security.

HEADQUARTERS AND SHELTERS

9-12. The location of the security force's headquarters will depend on the size and layout of the installation or facility. The objectives are the efficient control of the security force and the adequate security of vital activities. On a small installation, there is frequently only one full-time entrance that may be supplemented by several part-time entrances. At these installations, the logical location of the headquarters would be at or near the main entrance. On larger installations, it might be better to locate the headquarters near the center of the cantonment area.

9-13. The security force's headquarters should be the control point for all physical-security matters for the installation and the monitoring point for protective alarm and communication systems. This office should have a reliable and independent means to contact nearby civil authorities. A list of key telephone numbers should be available for use in emergency operations.

9-14. Personnel shelters should be available to protect the guards from the elements. The design can be temporary or hardened and include adequate space for guard-force personnel only. The facility should have heat, ventilation, storage space for essential accessories, lighting that will not expose the occupant, and good visibility in all directions.

EXECUTION OF SECURITY ACTIVITIES

9-15. Security personnel must exercise good interpersonal communication skills when carrying out their duties with other employees. Bad employee relations can result if security personnel become impertinent and assume powers not rightfully theirs. Security personnel must understand the methods and techniques that will detect security hazards and assist in identifying violators and intruders.

9-16. Written reports or journals are recommended for security activities. These should be prepared by either the security force's supervisor or the personnel at the security post. These reports should record all activities, actions, and visits at the security post.

9-17. It must be strongly emphasized that security personnel will be used for security duties only and should not be given other routine functions except as directed by the commander or his representative. Security personnel should have no fire-fighting or similar duties regularly assigned. Such emergencies offer an excellent diversion to cover an intruder's entrance. Consequently, during such times, security personnel must be exceptionally alert when performing their duties. However, the security force may be cross-trained in other areas (such as fire fighting) so that they may be used when required and when circumstances permit (such as when they are off duty).

9-18. Personnel who are assigned to fixed posts should have a designated method of relief. The security force's shift supervisor should establish a relief schedule (about every two hours) according to local policies and the SOP. A simple but effective plan of operation should be worked out for the security force to meet every foreseeable emergency. Practice alarms should be conducted frequently to test the plan's effectiveness. Such plans should be designed to prevent a diversion at one point on the installation, drawing off the guards or distracting their attention from another section of the installation where unauthorized entry may be made. Routes and times for security patrols should also be varied at frequent intervals to preclude establishing a routine that may be observed by potential intruders.

TRAINING REQUIREMENTS

9-19. The extent and type of training required for security forces will vary according to the importance, vulnerability, size, and other factors affecting a particular installation or facility. The training program's objective is to ensure that all personnel are able to perform routine and emergency duties competently and efficiently.

BENEFITS OF PROPER TRAINING

9-20. Efficient and continuing training is the most effective means of obtaining and maintaining maximum proficiency of security-force personnel. Regardless of the selection process, new personnel seldom have all of the qualifications and experience necessary to do the job. In addition, new or revised job requirements frequently mean that personnel must be retrained. Training can bridge the void between ability and job requirement.

9-21. Supervisors need to remember that all personnel do not have the same training needs. It is a waste of valuable time to train an individual in a subject that he has already mastered. Past experience, training, acquired skills, and duty assignments should be evaluated for each person as an aid in planning an effective training program.

9-22. A good training program benefits both the installation and the security force. The task of supervising the security force is made easier, there is much less wasted time, fewer mistakes are made, and there is less friction with other agencies. A good training program helps to instill confidence through developing increased skill proficiency. The training program provides for more flexibility and better physical protection, fewer required personnel, and less time to learn duties. Training establishes systematic and uniform work habits.

BASIC TRAINING

9-23. As a minimum, personnel (including civil-service security personnel) who have not had security training should receive training in their security duties. This training includes—

- The care and use of weapons, if required. No person should be placed on security duty unless weapons training has occurred within the past 12 months. Weapons training must be according to AR 190-14.
- Areas of responsibility and authority of security personnel, particularly on apprehension, search and seizure, and the use of force.
- The location and use of first aid and fire-control equipment and electrical switches.
- Duties in case of emergencies such as alerts, fires, explosions, and civil disturbances.
- Common forms of sabotage and espionage activity.
- The location of hazardous and vulnerable equipment and material.

IN-SERVICE TRAINING

9-24. All newly assigned individuals are given special instructions for each post. When possible, their first assignment should be with an experienced person. Additional in-service training and periodic retraining to review basic material and procedures are continuous requirements.

9-25. Scheduling in-service training and classes to enable all of the security force or a complete shift to participate is often difficult. Therefore, the supervisor must exercise good judgment when scheduling training to ensure that each person has the opportunity to receive the training.

EVALUATION OF TRAINING

9-26. Testing designed to evaluate performance is a necessary step in the training program. These tests may be oral or written or may be a type of performance test. They should be administered annually to ensure that the entire force maintains high standards of proficiency. A testing program also helps to improve training by—

- Discovering gaps in learning.
- Emphasizing main points.
- Evaluating instructional methods.

9-27. Security training received by personnel at their units must be entered in unit training charts or records. The record serves to—

- Indicate individual degrees of skill.
- Establish priorities of instruction.
- Present a consolidated picture of the security force's training status.
- Help certify guard personnel.

SUPERVISION

9-28. A security supervisor is tasked with overseeing and directing the work and behavior of other members of the security force. To obtain maximum performance from each member of his force, the supervisor must have a complete understanding of leadership principles and be capable of applying them.

9-29. The supervisor is responsible for understanding the operations of all posts. Additionally, he is often responsible for selecting, inducting, training, and ensuring the productivity, safety, morale, and advancement of guard-force members.

9-30. To ensure an alert, presentable, and efficient security force, the leadership must provide consistent and intelligent supervision. To earn the respect and cooperation of the guard force, supervisors must be professional in their conduct. The security force's morale and efficiency is a direct reflection of the quality of its supervision.

9-31. The ratio of supervisory personnel to security personnel should be determined by the individual characteristics of each installation. At small installations, the ratio may be higher than at large installations.

9-32. There must be sufficient supervision to enable the inspection of each post and patrol. It is also essential that supervisors be in contact with security headquarters to control emergencies that may arise. Specific duties of a supervisor include the inspection and briefing of the relief shift and the inspection of posts, vehicles, and equipment during visits to posts and patrols.

SUPPLEMENTS TO SUPERVISION

9-33. Various means and devices may be used as supplements to personnel supervision. These include the following:

- **Recorded tour systems.** Personnel record their presence at strategic points throughout an installation by using portable watch clocks or similar devices. These are effective means of ensuring that such points are regularly covered. This system provides an after-the-fact type of supervision.
- **Supervisory tour systems.** A signal is transmitted to a manned central headquarters at the time the post is visited. These systems provide instantaneous supervision and a means of detecting interference with normal security activities and initiating an investigation or other appropriate action.

9-34. All personnel on security duty should be required to report periodically to headquarters by the usual means of communication. The frequency of such reports will vary, depending on a number of factors. Regularity should be avoided to preclude setting a pattern by which an intruder can gauge an appropriate time for entrance.

MANAGEMENT

9-35. The physical-security supervisor is responsible for managing and developing the security organization. A physical-security program is greatly enhanced by a well-developed educational program.

9-36. The physical-security supervisor acts as an advisor and assists in formulating policies for the installation's physical-security measures. The goal should be the best security within the restrictions of the commander's budget guidance. Physical-security planners must remember that anyone can provide adequate security with unlimited funds; however, this is not a realistic approach. There must be a constant endeavor to effect justifiable economy where possible without jeopardizing the physical-security program.

UNIFORMS

9-37. All security-force personnel are required to wear the complete prescribed uniform as outlined in their special orders. Deviations from the prescribed uniform should not be made except for items to protect the guard force's health, comfort, and safety. The duty uniform will be worn during all tours of duty and may be worn during off-duty hours only between the place of residence and the place of duty. Each member of the security force is required to maintain high standards of appearance.

VEHICLES

9-38. The security force should be furnished with sufficient and reliable vehicles to maintain patrol standards established by the installation commander. Vehicles assigned to the force should be equipped with two-way radios to obtain the greatest possible use of all personnel and vehicles.

FIREARMS

9-39. Before issuing weapons, the security force will be briefed on the use of force. Security-force personnel will be issued weapons as prescribed by AR 190-11 and the unit's SOP. The commander may prescribe other weapons for the security force based on needs and requirements. Weapons normally are loaded with live ammunition, except where prohibited for safety reasons. The use of privately owned weapons while on duty is not authorized. Weapons and ammunition issued to security-force personnel will not be removed from the installation except in the course of official duty. When not in use, weapons are secured in arm racks in storage rooms as prescribed by AR 190-11.

9-40. Weapons are inspected as necessary to ensure proper maintenance. A written report is prepared and filed on the discharge of any weapon except for authorized and supervised training. The patrol supervisor or an MP investigator prepares the report (DA Form 3975).

9-41. Ammunition supplies for the security force's use must be maintained in secured storage containers according to AR 190-11. Ammunition must be issued only under proper supervision for authorized purposes. Ammunition issued to members of the security force must be accounted for by individual members immediately upon completion of duty. Any ammunition unaccounted for will be the subject of a report of its disposition by the individual.

COMMUNICATIONS

9-42. The security force should be equipped with two-way radios. These may be vehicle-mounted and portable, or they may be telephones. A secure-voice capability should be used where possible. This equipment is considered essential for the efficient operation of the security force and the accomplishment of its assigned mission. Proper use and care by security personnel will enhance the equipment's usefulness and capability.

MISCELLANEOUS EQUIPMENT

9-43. Security managers or supervisors should obtain other equipment necessary to accomplish their security mission. Items in this category may include (but are not limited to) warning lights; sirens; spotlights; portable lights; flashlights; first aid kits; traffic-control devices; and items of wear for the health, comfort, and safety of security personnel. Some of this equipment may require local purchase.

MILITARY WORKING DOGS

9-44. The requirements for physical protection of installations or facilities within the US and overseas theaters of operation continue to increase. Manpower available for this purpose has always been (and probably will continue to be) limited. The MWD, properly trained and properly used, can enhance a physical-security program. See AR 190-12 and DA Pam 190-12 for information regarding the use of MWDs.

SUMMARY

9-45. A security force is the critical element of a successful physical-security program. It is as strong as its weakest member. A comprehensive training program is essential to a knowledgeable, disciplined, and alert security force. A well-trained security force will be prepared to respond to a security breach.

Chapter 10

In-Transit Security

In-transit security subjects the movement of cargo to different, and frequently, more demanding aspects of physical security. Cargoes may be moved via port, rail, pipeline, or convoy. Regardless of the mode of movement, commanders must aggressively apply the principles of physical security to their protection. Security forces must be provided at the most vulnerable areas of each cargo movement.

IN-PORT CARGO

10-1. Ports and harbors are prime targets for enemy and criminal activities. Perimeter areas of these facilities are more vulnerable because of the extensive distance and exposed beach or pier areas. Terminal areas may include fully developed piers and warehouses or may be an unimproved beach where logistics-over-the-shore (LOTS) or roll-on/roll-off (RORO) operations are conducted.

10-2. If a Theater Army Area Command (TAACOM) MP unit must provide security for cargo in a port, the main effort is to provide security from the perimeter of the port outward. Security measures focus on aggressive patrolling to detect, report and, if need be, combat enemy threats. Measures may include—

- Conducting route and area reconnaissance patrols.
- Developing police intelligence in the area of operations (AO).
- Controlling traffic in the area surrounding the port.
- Conducting mounted or dismounted patrols (with MWDs, if available) around the port's perimeter.
- Establishing an access-control/ID section.
- Watching for diversions of supplies out of the port.
- Providing a response force to react to incidents inside the port's perimeter.
- Providing observation and early warning of threat ground and air attacks.

10-3. When providing security for cargo, the focus is on providing a security overwatch for the cargo as it moves from the port to the combat area. Inside a port's perimeter, access to cargo is limited by—

- Operating random mounted or dismounted patrols (with MWDs, if available).
- Using combined patrols as a response force for incidents inside the perimeter.

- Controlling access to the most restricted areas.

10-4. On occasion, the MP may have to safeguard highly critical cargo inside a port's perimeter. The type and degree of security provided is based on logistical security information. Some examples are the—

- Types and values of the cargo stored.
- Vulnerability of the cargo to a land threat.
- Likelihood of theft, diversion, pilferage, or sabotage by military personnel, local workers, black marketers, or enemy agents.
- Location and nature of the port facilities.
- HN agreements.
- Degree of entrance and exit controls.

10-5. Safeguarding the most critical cargo waiting to be transferred to land transport is the priority. The following measures help to safeguard stored cargo:

- Establishing access-control procedures.
- Searching bundles and packages being taken from the area.
- Examining trip tickets and documentation of cargo vehicles.

10-6. If the restricted area is a pier or other maritime environment, access from the water must be controlled as well as from the land. Entry on the landward side of a pier can be limited with fencing, pass control, and aggressive patrolling; but the part of the pier that protrudes over the water is accessible from the sides and from below. Methods for securing the pier along its water boundaries include—

- Patrols (both walking on the pier and in small boats).
- Protective lighting.
- Log booms.
- Nets.
- Buoys or floats.
- Anchored or pile-mounted navigational aids and signaling devices.
- Barges.

10-7. While most of the barriers described above will stop or impede access to facilities from boats or swimmers, nets are among the most effective. Well-marked, partially submerged objects are also effective; however, there may be legal prohibitions against placing barriers that may constitute a hazard to navigation. These barriers should be placed only after coordination with and approval by the appropriate legal and HN authorities. Sometimes it is best to close off the waterside of a pier. A floating boom will keep small boats out. Suspending a cable or a chain-link net from the bottom of the boom will deny access underwater.

10-8. At least two security zones must be established on a facility's waterside—the reaction zone and the keep-out (exclusion) zone. Security forces in these zones notify vessels, craft, and swimmers that they are entering restricted waters and should alter their course. Security forces may stop and search intruders if necessary. Security zones should extend at least 1,000 meters from the nearest protected asset; however, in some port areas this large security zone is not possible. In such cases, other measures (such as boat patrols) must be increased to mitigate the possibility of attack.

10-9. A reaction zone extends from the high-water mark to a distance beyond the maximum range of anticipated waterborne threats. Security forces will stop and challenge intruders inside the reaction zone.

10-10. The keep-out zone is the zone closest to the protected assets. It extends from the asset to the maximum range of anticipated threat weapons. Security forces should prevent the entry of all unauthorized craft or vessels into this zone. The tactical response force (in this case, a boat) may be used. In addition to organic security, forces may be provided by HN or contracted personnel.

10-11. To keep the cargo secured while transferring from one transport method to another, the traffic moving in and out of cargo-handling areas must be controlled. MP forces can—

- Set up a single access-control point.
- Erect field-expedient barriers. Truck trailers or other large vehicles can be used to constrict the traffic flow if permanent barriers are not in place.
- Limit entry to mission-essential personnel, vehicles, and equipment (as designated by the port authority).

10-12. A holding area should be provided if gates are used by vehicles other than cargo vehicles. Cargo vehicles can pull into the holding area while they are being checked. The holding area should be large enough to handle the volume and size of traffic. A wooden deck or platform at, or slightly higher than, the level of the truck bed can be used to facilitate checking. The platform must be at least as long as the vehicle (such as an empty flatbed trailer). Such a platform makes it quicker and easier to observe and check cargo.

10-13. Cargo is less likely to be diverted if a close watch is kept on cargo documentation and container safety. Containerized cargo is less likely to be stolen or sabotaged. However, containers must be watched closely as they are filled and sealed. Cargo can be pilfered before the seal is applied. An unsealed container can be moved to a stacking area; or someone may apply a false seal, break the seal later, remove the cargo, and then apply a legitimate seal.

10-14. At access-control points—

- Inbound and outbound containers should be inspected. Signs of damage or unserviceability should be observed.
- Containers must be inspected for the presence of seals or locks and hinges. Their serviceability should also be checked.
- The document's transport number, container number, and seal number should be checked to ensure that they match those numbers on the transportation control-and-movement document. (Check the seals by handling them, not simply by a visual check.)
- Containers with valid documents only should be allowed to pass inbound or outbound through the control point.

RAIL CARGO

10-15. Because a train's movement is determined directly by the condition of the tracks, cargo moving by rail is particularly vulnerable to attack. The destruction of switches, signals, or the track may be a delaying harassment; or it could trigger a major catastrophe. Since railroads can be such high-value targets, the commander may task MP or other US forces to provide on-board security for critical cargo.

10-16. Most train crews consist of four or five people who control the train—the engineer, a conductor, a fireman, a senior brakeman, and a brakeman or a flagman. The conductor is the train commander unless a transportation railway service officer is assigned to the train. The train commander is responsible for the train's operation and security. He makes all decisions affecting the train. The security force's commander is responsible for the cargo's security. The train crew and the security force watch for and report any discrepancies or interruptions to normal procedures at any time during the movement. Information about the movement is usually sent along the movement route by the chief dispatcher through a telephone circuit.

10-17. A four- to six-person security force is usually enough to secure railway shipments of sensitive freight, but additional security forces may be needed for moving critical cargo. In addition to a military security force, the shipper or loading agency may send specially trained personnel with highly sensitive cargo. The number of MP in a train security force depends on the—

- Sensitivity of the freight.
- Priority of need for the freight.
- Terrain over which the train will pass.
- Length of the train.
- Duration of the trip.
- Degree of threat.

10-18. Security forces prepare and maintain a record (by car number) of guarded cars in the train. Security forces can ride in—

- A specific car that requires protection.
- The caboose.
- A security-force car. (If only one security car is used, it should be near the center of the train; if more than one is used, cars should be spaced to provide the best protection for the train.)

10-19. The security force on a train must keep a constant check on car doors, seals, wires, and locks to detect tampering. The following instances must be noted and reported immediately:

- Irregularities in procedures.
- The presence or actions of unauthorized persons.
- Deficiencies or incidents that occur.

10-20. When planning rail-cargo security, the time schedule for the rail movement must be obtained. A map reconnaissance of the route should be provided, detailing bridges and tunnels that are especially vulnerable.

10-21. Security-force actions should be planned at scheduled stops or relief points, and forces should be deployed according to these plans. Locations of MP units and other friendly forces should be plotted along the route, and their radio frequencies and call signs should be noted. An intelligence report covering the route should also be obtained. This report should indicate sites where sabotage may occur, attacks may be expected, or thefts and pilferage are likely.

10-22. The shipper is responsible for the security of all carload freight until it is turned over to the Transportation Railway Service and the loaded cars are coupled to a locomotive for movement. The shipper or field transportation officer should complete the freight waybill or the government bill of lading. This report shows the car number, a brief description of contents, the weight of the load, the consignor, the consignee, the origin, and the destination. In addition, it may show special instructions for the movement or security of the car and its contents. Careful documentation is essential for—

- Securing the shipment.
- Locating cars with critical cargo.
- Ensuring that priority movement is authorized.

10-23. Transportation officers are responsible for the completeness, correctness, and proper handling of waybills. Each car must have a waybill; this allows cars to be detached or left behind should they become defective en route. If this occurs, a team from the security force must remain with the cargo until they are relieved.

10-24. Railway cars are sealed after loading. A seal shows that a car has been inventoried and inspected. The standard method of sealing a railway boxcar door (in addition to padlocks or wires) is with a soft metal strap or a cable seal that contains a serial number. Maintaining rigid accountability of all seals is necessary to prevent the undetected replacement of an original seal with another. While sealing does not prevent pilferage, a broken seal is a good indicator that the car and its contents have been tampered with. Train security forces or operating crews can easily check the seals on cars when the train stops. Broken seals should be reported immediately to help pinpoint the time and place of a possible theft. When vehicles are shipped by railcar, sensitive and high-value items must not be secured in the vehicles. Container-express (CONEX) and military-van (MILVAN) containers are ideal for shipping these and other small items on flatcars since they greatly reduce the chance of pilferage. These containers must be locked and sealed and, if possible, placed door to door for additional security.

10-25. When operations permit, cars containing highly pilferable freight, high-priority cargo, or special shipments are grouped in the train to permit the most economical use of security forces. When flatcars or gondolas are used to transport sensitive or easily pilfered freight, security forces should be placed where they can continuously observe and protect these cars.

10-26. When the train is stopped, security forces should dismount and check both sides of the train, verifying that seals, locks, and wires are intact. They must report a broken seal immediately to help pinpoint the time and place of the theft.

10-27. If the security force is relieved by another security force while en route, a joint inspection of the cars is conducted. The relief force signs the record being kept on the guarded cars. Consignees assume responsibility for the security of loaded freight cars at the time they arrive at their destination. When the trip is complete, the receiver or his agent will inspect the cars. The security force obtains a receipt for the cars, which is then attached to the trip report. The trip report should include—

- Dates and times the trip started and ended.
- Any additional information required by the local SOP or command directive.
- Recommendations for correcting deficiencies or for improving future security on trains.

10-28. Because unloading points are highly vulnerable to pilferage and sabotage, cars should be unloaded as soon as possible to reduce the opportunity for loss. MP forces are normally not available for the security of freight in railway yards. For more information regarding rail cargo, see FM 55-20.

PIPELINE CARGO

10-29. Pipeline systems are widely used in a theater of operation to transport bulk petroleum products or other liquids. Such systems are open to a number of security threats from the point of entry to the point of final delivery. Pipeline systems are composed of storage and dispersing facilities, pump stations, and extended pipelines. They also include discharging facilities for tankers at ports or other water terminals.

10-30. The type and extent of risk to a pipeline varies with the level of conflict in the AO. In a communications zone, the chief hazard is likely to be pilferage. Pipelines can be tapped by loosening the flange bolts that join sections of pipe or by cutting holes in the hose line. The risk rises if gasoline is scarce and expensive on the civilian market. Sabotage is a security hazard during all levels of conflict. It is committed by any method such as simply opening pipe flanges, cutting a hose line, or setting fires and causing explosions to destroy portions of the line.

10-31. In areas of conflict, the likelihood of sabotage and interdiction increases. Pipeline systems are vulnerable to air attacks, especially aboveground sections of the pipeline, pump stations, and storage facilities.

10-32. Security forces should be deployed in the best manner to provide coverage to the most vulnerable portions of the pipeline that are at the greatest risk to enemy, terrorist, partisan, and ground attack. Patrols should be set up to screen isolated areas and remote pumping stations. Sensors should also be considered, along with aerial security. Security patrols will—

- Detect, report, and respond to attacks on or sabotages of the pipeline.
- Monitor critical parts of the pipeline on a routine but random basis.
- Monitor ground sensors and other intrusion-detection devices. These are often used at pump stations and elsewhere along the pipeline to detect and identify threats to the system.

- Check line-pressure devices in pipeline and pumping facilities. These devices monitor the flow and detect breaks in the line, which may indicate pilferage of gasoline (or other petroleum products).

10-33. Dedicated security forces are rarely sufficient in number for the surveillance of an entire pipeline system. All available supporting forces (in the course of their normal duties) should observe and report items of intelligence for further investigation. Examples of suspicious activities in the pipeline area might include the unusual presence of commercial tanker trucks, the appearance of gasoline drums or cans, or an increased use of motor vehicles in fuel-scarce areas. Other resources available to the commander for coordination and support include HN and MP elements responsible for the AO, as well as the security officer of the petroleum group or battalion.

CONVOY MOVEMENT

10-34. As convoy movements are tactical in nature and are discussed in detail in FM 19-4, they will be briefly discussed here. When moving by convoy, consideration should be made for the following:

- Congested traffic areas.
- Travel during night hours when traffic is reduced rather than travel during daylight hours when traffic congestion is heaviest.
- National holidays. Traffic may be three times heavier than on a normal day. Also, if you are moving a convoy overseas on a national holiday, the HN people may not be receptive to your action and the result may be unwanted reactions on their part.
- The use of a marked HN police vehicle in conjunction with the convoy. The HN people are more receptive to an activity when it is represented by one of their own. Additionally, the HN police may be able to diffuse a potential crisis.
- Security of the convoy. Security of the convoy is foremost important both during movement and stops. During extended or overnight stops, special consideration must be given to securing the loaded vehicles.

Chapter 11

Inspections and Surveys

Inspections and surveys are valuable tools to a commander's physical-security program. These tools collectively measure and identify the readiness of a commander's physical-security program. The survey provides the installation commander with an overall security posture of the installation.

INSPECTIONS

11-1. Physical-security inspections are conducted at DA installations, activities, and facilities by trained physical-security inspectors. Some facilities on an installation may be exempt from inspection due to their mission. These facilities are inspected under the guidance of regulations and directives unique to those activities. Inspection personnel will be trained and will conduct inspections according to AR 190-13.

11-2. A physical-security inspection is a recorded assessment of physical-security procedures and measures implemented by a unit or an activity to protect its assets. The inspection is recorded on DA Form 2806-1-R (see AR 190-13 for using the form).

COORDINATION

11-3. Liaison and coordination should be established with other agencies on the installation before an inspection. The director of facility engineers can provide information to benefit the overall security program. Other agencies, such as MI (threat analysis) and local law-enforcement agencies, may have input essential to the security program.

SECURITY LIBRARY

11-4. A security library is necessary to help personnel prepare for and conduct an inspection. This library may include—

- The mission and history of each activity to be inspected.
- Previous inspection reports.
- A copy of the most current risk analysis.
- The SOPs and ARs specific to physical security.

ENTRANCE INTERVIEWS

11-5. Entrance interviews are usually required before the actual inspection. During the interview, the inspector establishes a rapport with the unit representative. The inspector identifies the following during the interview:

- All members of the inspection team.

- An overview of the last inspection.
- Areas to be inspected and the order of inspection.
- A review of waivers, work orders, and exceptions.
- Changes to the unit's mission (if any).

CONDUCTING INSPECTIONS

11-6. The inspection should be conducted from the outside to the inside of the facility, activity, or area with regard to the following:

- Observation of the facility will be conducted during all hours of the day.
- Interviews of managerial and operational personnel will be performed.
- Security forces should be inspected so as not to disrupt the mission (if possible).
- An assessment should be made of security-force training, especially if security-force knowledge proves inadequate.
- Inspection of entry and movement control by the guard force should not hinder operations.
- All communications systems used by the guard force should be thoroughly inspected. The guard force should have two reliable and efficient means of communication, one of which is a radio.
- Inspections should be conducted according to regulations appropriate for the facility.

EXIT INTERVIEWS

11-7. Exit interviews should be conducted as soon as possible after the inspection. The commander should be informed of any deficiencies or compliments noted. A rating on the inspection's results will not be provided during the exit interview. The approving authority, not the inspector, will determine the inspection rating. The rating will be forwarded to the unit along with the final report.

11-8. Recommendations will be made according to regulations. Written reports should be forwarded through channels in a timely manner according to the PM's SOP. The commander's report of actions taken will be required and reviewed by the PM's staff.

SURVEYS

11-9. A physical-security survey differs from an inspection in that a survey covers a formal assessment of an installation's physical-security program. Each survey includes a complete reconnaissance, study, and analysis of installation property and operations. The survey provides the commander with an assessment of the installation's overall security posture. It consists of the threat and the mission, and it advises the commander on the installation's physical-security program's strengths and weaknesses.

PHYSICAL-SECURITY SURVEY

11-10. The physical-security survey is a formal recorded assessment of an installation's physical-security program. See AR 190-13 for further information on this type of survey.

SECURITY-ENGINEERING SURVEYS

11-11. While a security-engineering survey is largely an engineer function, it must be coordinated with physical-security personnel to be successful. A security-engineering survey is the process of identifying (by means of an on-site survey) engineering requirements associated with facility enhancements for physical security and antiterrorism, including an IDS installation. This type of survey should be conducted when planning new construction, renovations, or upgrades to existing facilities where there are likely to be physical-security requirements. A security-engineering survey may also be requested by the PM or an equivalent security officer to evaluate existing security. This survey—

- Identifies assets to be protected.
- Identifies threats to these assets and the level of protection required to protect them.
- Identifies the protective measures.
- Determines the cost of the protective measures.

Appendix A

Metric Conversion Chart

This appendix complies with current Army directives which state that the metric system will be incorporated into all new publications. Table A-1 is a conversion chart.

Table A-1. Metric Conversion Chart

Metric to English			English to Metric		
Multiply	**By**	**To Obtain**	**Multiply**	**By**	**To Obtain**
Length					
Centimeters	0.0394	Inches	Inches	2.54	Centimeters
Meters	3.28	Feet	Feet	0.0305	Meters
Meters	1.094	Yards	Yards	0.9144	Meters
Kilometers	0.621	Miles (stat)	Miles (stat)	1.5609	Kilometers
Kilometers	0.540	Miles (naut)	Miles (naut)	1.853	Kilometers
Millimeters	0.039	Inches	Inches	25.40	Millimeters
Area					
Square centimeters	0.1550	Square inches	Square inches	6.45	Square centimeters
Square meters	10.76	Square feet	Square feet	0.0929	Square meters
Square meters	1.196	Square yards	Square yards	0.836	Square meters
Volume					
Cubic centimeters	0.610	Cubic inches	Cubic inches	16.39	Cubic centimeters
Cubic meters	35.3	Cubic feet	Cubic feet	0.0283	Cubic meters
Cubic meters	1.308	Cubic yards	Cubic yards	0.765	Cubic meters
Milliliters	0.0338	US liq ounces	US liq ounces	29.6	Milliliters
Liters	1.057	US liq quarts	US liq quarts	0.946	Liters
Liters	0.264	US liq gallons	US liq gallons	3.79	Liters
Weight					
Grams	0.0353	Ounces	Ounces	28.4	Grams
Kilograms	2.20	Pounds	Pounds	0.454	Kilograms
Metric tons	1.102	Short tons	Short tons	0.907	Metric tons
Metric tons	0.984	Long tons	Long tons	1.016	Metric tons

Appendix B

Sample Installation Crime-Prevention Handbook

This appendix provides guidance on planning, organizing, directing, and controlling installation crime-prevention programs. It provides guidance on developing an installation program, criminal analyses to identify crimes, guidance on which crimes to address, command and individual countermeasures for particular crimes, and program-evaluation procedures.

SECTION I — INSTALLATION CRIME-PREVENTION PROGRAMS

B-1. In the past few years, the Army has shifted an increasingly larger percentage of its manpower from combat-service-support activities to combat organizations. This change means that fewer MP personnel are available to support a larger number of units. To meet this challenge, it is necessary to reevaluate the way we do business and to emphasize those programs or procedures that have the greatest impact on our installation crime rates. Crime prevention is one program that can have a major impact on installation crime rates at a relatively minor cost in both dollars and manpower. It takes less effort to discourage a criminal from perpetrating a crime or to teach a soldier to avoid becoming a victim than it does to investigate a crime, identify the offender, prosecute him, and punish him. In addition, a proactive approach to law enforcement can help maintain the high quality of service life that can improve the retention of first-term soldiers.

B-2. The Army is a large organization that performs a variety of activities in many different environments. Crimes that are major problems on one installation may be totally absent from others. For example, most military installations have a significant number of robberies while most depots have none. Because of this, any rigid, centrally controlled program—no matter how carefully thought out—is bound to be inappropriate in many locations. Therefore, DA has elected to provide only the most general guidance and to allow commanders to develop crime-prevention programs that address their local problems.

CRIME-PREVENTION WORKING GROUPS

B-3. The installation is the smallest practical level for implementing crime-prevention programs. If these programs are developed and implemented at a lower level, then crime is often not eliminated but is merely displaced from units with good programs to units with less effective programs. Also, crime does not affect personnel only when they are in their place of duty. In many cases, a company commander's troops are victimized in areas over which he has little control. Unit commanders are responsible for implementing many anticrime measures; however, the selection of overall program goals, the ID of appropriate countermeasures, and quality control should be done at the installation level.

B-4. Crime prevention must always be recognized as a commander's program rather than as an MP program. MP personnel have the expertise to analyze data, identify major problems, and develop lists of possible countermeasures. They should perform these functions in support of an installation crime-prevention council appointed by the installation commander and composed of representatives of all of the installation's major organizations and activities. The advantages of using this type of system are—

- It provides representatives of all major segments of the post population with a forum where they can identify criminal problems that are of the greatest concern to them.
- It allows the representatives of all major commands to review the available options to counter a crime and to select the level of resource commitment that is compatible with their missions and internal priorities.
- It helps ensure that the resources of the entire community, rather than only those of the MP force, are mobilized to attack the problem.
- It is easier to obtain the support of the whole population if its representatives are instrumental in the development of the program.

CRIME-PREVENTION OFFICERS

B-5. The installation's crime-prevention officer is normally a senior NCO or an officer who has a solid background as an MP investigator or a physical-security inspector (PSI). He supports the installation council by performing a crime-data analysis to identify problem areas, drafting programs for the council's consideration, inspecting the implementation of council-mandated measures, and coordinating the efforts of unit crime-prevention officers in the implementation of the crime-prevention program.

B-6. As a member of the PM's staff, the crime-prevention officer develops the law-enforcement section of the crime-prevention program, develops and maintains the written crime-prevention plan, and coordinates crime-prevention programs with civilian police agencies and community groups.

B-7. Crime-prevention officers are also appointed in each organization down to the company level. At this level, written crime-prevention plans are not required; however, SOPs are established. The crime-prevention officers serve as their organizations' focal points for coordinating installation crime-prevention plans; they supervise the implementation of the installation's program within their organizations.

CRIME-PREVENTION PROGRAM DEVELOPMENT

B-8. The starting point for developing a crime-prevention program must be a thorough analysis of criminal activity on the installation. This identifies significant criminal problems that are susceptible to crime-prevention efforts. Crimes that are most susceptible to crime-prevention measures are those for which a high probability of reoccurrence exists. Crimes such as murder normally are not repetitive and are poor candidates for inclusion in the crime-prevention program. Since it is seldom practical to attack all criminal problems simultaneously, they should be prioritized based on their impact on the command's ability to perform its mission and their impact on installation personnel. Next, the whole range of countermeasures that can be used to combat each problem must be identified (see Figure B-1). Table B-1, page B-4, identifies (by offense) programs that have been successful in countering specific criminal problems. Sections III and V of this appendix contain discussions of the strengths, weaknesses, and applicability of the countermeasures listed in Table B-1. Once developed and prioritized, the list of criminal problems and possible countermeasures must be presented to the installation crime-prevention council for action. The council should decide which crimes will be addressed and which countermeasures will be used for each crime. The council must then identify specific objectives for its anticrime campaigns.

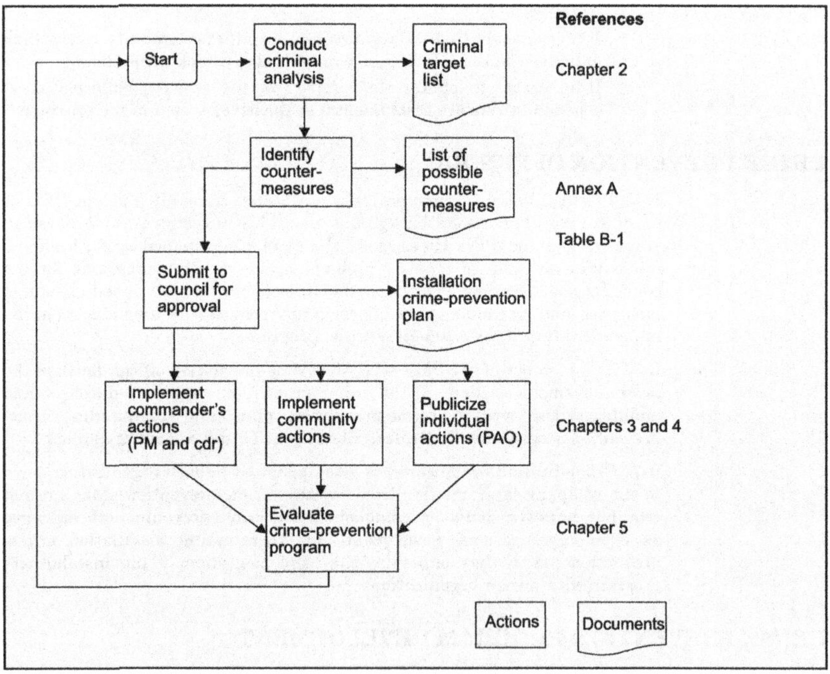

Figure B-1. Crime-Prevention Program-Development Cycle

Table B-1. Offenses Countermeasure Matrix

Crime	Command/Law-Enforcement Countermeasures												Community Programs					
	Crime Hot line	Lighting	Environmental Changes	Patrol/Surveillance	Publicity Campaigns	Residential Security Survey	Juvenile Programs	Fraud Programs	Employee Theft	Arson Programs	CCTV	Warning Signs	Neighborhood Watch	Operation ID	Neighborhood Walks/Escorts	Mobile Patrols	Project Lock	Education
Arson				X	X					X			X			X		X
Auto theft	X	X			X								X		X	X	X	X
Burglary/ housebreaking	X	X	X	X	X	X							X	X	X	X		X
Employee theft	X				X				X		X	X		X				X
Fraud	X				X			X				X						X
Larceny	X		X			X			X					X				X
Rape	X	X	X	X	X	X							X		X	X		X
Robbery	X	X	X	X	X						X		X		X	X		X
Juvenile delinquency	X			X	X		X						X		X	X		X
Vandalism	X	X		X	X		X						X		X	X		X

B-9. Objectives must identify—

- What crime will be reduced.
- What target population will be addressed.
- What specific changes and behaviors on the part of the victims or perpetrators will be encouraged.
- What actions the command must take to reduce the opportunity for the crime to occur.

B-10. Once objectives have been clearly defined, specific areas of responsibility should be assigned to each council member (based on their organization's primary area of responsibility) and major milestones should be identified for developing the campaign against each targeted crime.

TRAINING

B-11. The prerequisite skills for successful performance as an installation crime-prevention officer are best developed through on-the-job experience as the supervisor of MP investigations or physical-security inspections. More important than any technical skill is the cultivation of a frame of mind that instinctively examines each case to determine not only what occurred, but also how the crime could have been prevented. Technical skills (such as criminal-data analysis) that may not have been developed as an MP investigator or physical-security supervisor are presented in courses taught by several civilian agencies. These classes should be used to the fullest extent possible.

CIVILIAN CRIME-PREVENTION ORGANIZATIONS

B-12. There are many civilian crime-prevention organizations at the national, state, and local levels. Many of these organizations have produced crime-prevention material (including posters, radio spots, and leaflets). Material and programs sponsored by civilian agencies should be used to support Army crime-prevention efforts. However, when material from a source outside of DOD is used, a copyright release must be obtained. Normally, it is necessary to get a release for each separate item that is used. If there is any doubt as to the necessity of securing a copyright release, the crime-prevention officer should refer the matter to the local SJA.

SECTION II — CRIMINAL ANALYSIS

B-13. Criminal analysis is a system for identifying trends and patterns where they may exist. It is a routine, ongoing function for the PM and battalion- and brigade-level staffs. Criminal analysis is the foundation upon which the installation force-protection program is based. Moreover, criminal analysis is an integral component of the police intelligence-operations function and is applicable across the operational continuum. An effective criminal analysis establishes the following:

- Crimes having a significant impact on the installation.
- The segments of the population being victimized.
- The ID of criminals/perpetrators.
- The most common time of occurrence.
- The areas that experience the highest number of incidents.
- Offense information (such as types of weapons or victims' actions that contribute to the offense).
- Information critical to an installation's VA.
- Information essential in formulating a successful patrol-distribution plan.
- Police information and criminal intelligence fused with tactical intelligence.

B-14. With this type of information, specific countermeasures are developed to reduce the opportunity for a crime to occur or to remove the incentives for perpetrators. Without an effective criminal analysis, the overall security effort is unfocused. Moreover, the installation/base patrol-distribution plan may be skewed. When this occurs, broad countermeasures are implemented and meaningful results are lost.

B-15. A professional analysis of criminal data is essential for protecting soldiers, units, and installations. The Military Police Management Information System (MPMIS) provided the initial software capable of assisting in the development of criminal-analysis information. As this system is modernized and replaced by the Military Police Automated Control System (MPACS), the MP Corps and the CID will have interface with the Army Battle Command System (ABCS). The analysis of police, criminal, and tactical information results in the development of police intelligence (PI). The PI, which is coordinated with the Assistant Chief of Staff, G2 (Intelligence) (G2) or the Intelligence Directorate (Joint Command) (J2), provides relevant information and intelligence (RII) that greatly contribute to the commander's critical information requirements (CCIR). The RII provided vertically and horizontally ensures that the common operational picture is properly shared.

SOURCES OF INFORMATION

B-16. The basic source for identifying crimes that warrant examination is the installation's Law Enforcement and Discipline Report (DA Form 2819). The information in this report can be used to plot the amount of and seasonal variations for each major type of crime. This will tell little about conditions that produce the crime, but it is useful in identifying crimes that can be eliminated from detailed analyses due to low frequency rates.

B-17. The Law Enforcement and Discipline Report identifies which major category of crime should be targeted, but it does not tell which type of crime is causing the most problems. For example, It may indicate that robbery is a problem, but it will not discriminate between robbery of commercial establishments (such as the post exchange [PX], the Class VI store, or banks) and muggings of individuals. Since countermeasures for these two types of robbery are different, it is necessary to collect additional information. The best sources for this information are the military-police report (MPR) and the CID report of investigation (ROI). If there are fewer than 200 cases in the past year for a particular type of crime, they should all be examined. With a larger annual caseload, a random sample that is large enough to give results of plus or minus 5 percent accuracy should be examined.

B-18. General factors must be identified for any type of crime. These are the types of victims, the perpetrators, geographical data, and chronological data.

VICTIMS

B-19. It is essential to determine (as precisely as possible) the segment of a post population that is being victimized. Junior soldiers live in different areas on post than officers and senior NCOs. They patronize different clubs and, for the most part, work in different areas. Information programs must use multiple sources to ensure that the same message reaches the different portions of the population. If the specific population segment that is being victimized is not identified, it is possible to spend a large amount of resources and have little impact on the targeted crime because the message is not getting to the people who need it.

PERPETRATORS

B-20. As with the victim profile, data on perpetrators is essential to ensure that countermeasures are targeted against the correct population. When on-post auto thefts are committed by civilians with legitimate reasons for entering the installation, it may not make sense to increase security (as a countermeasure) at post entrances. Countermeasures are developed based on specific profiles identified by case type and the information described above. Successful countermeasures depend on correct criminal analyses. An analysis and profile of perpetrators will help to focus MP/CID countermeasures.

GEOGRAPHICAL DATA

B-21. The MPR and the ROI should contain a street address or a description of the incident's location (such as "in a parking lot on the east side of building 1409"). This type of description fulfills requirements for identifying the location of the incident in court. It is further enhanced by identifying the normal duty hours and the type of activity that takes place in the location (for example, "in a parking lot on the east side of the NCO club (building 1401), normal operating hours 1800-0400" or "in a parking lot on the east side of building 1408, a troop billets for Company A, 15th Cav"). This type of information may help to develop a list of specific types of areas where a particular crime is occurring, making it easier for the PM to provide increased MP patrols in these areas. For example, if there have been several incidents in which vehicles have been broken into and stereos removed, it would help to know that 90 percent of them occurred in the parking lot by troop billets.

CHRONOLOGICAL DATA

B-22. As with geographical data, the more specific the time-of-occurrence pattern is for a crime, the easier it is to apply sufficient resources to affect the crime rate. For each crime having a significant impact, determine the following:

- Major seasonal variations.
- Monthly variations. Is there a concentration of crimes immediately before or after payday?
- Weekend occurrences. Is there a concentration of incidents on weekends? Each day represents about 14.25 percent of the week. Concentrations of crime higher than that for any particular day may be significant.
- Time of day. Is there a particular period that accounts for a disproportionate share of the incidents?

CRIME-SPECIFIC FACTORS

B-23. In addition to the factors that should be examined for all crimes, the following crime-specific factors are useful in analyzing specific offenses:

- Housebreaking and burglary.

- The type of building that was attacked (family-housing unit, troop barracks, PX, and so forth).
- Whether the facility was occupied or unoccupied.
- The point of entry (door, window, and so forth).
- The method of entry (unsecured door, forced door, forced window, and so forth).
- The property that was stolen. (Was it marked?)

- Robbery.
  - The number of perpetrators.
  - The perpetrator's method of operation.
  - The type of weapon used.
  - The type of robbery (street mugging, residential robbery, or commercial robbery).
  - The victim's injuries.
  - Actions by the victim that contributed to his being targeted.
- Larceny.
  - The type of property taken.
  - Whether the property was secured or unsecured.
  - The perpetrator's method of operation.
- Auto theft.
  - The type of vehicle stolen (POV, motorcycle, and so forth).
  - Whether the vehicle was secured or unsecured.
  - Whether the vehicle was recovered and where it was recovered.
  - Whether the vehicle was stripped of parts.
  - The perpetrator's method of operation.
- Forgery.
  - The type of document that was forged.
  - How the document was obtained.
  - The type of ID used in passing the forged document.
  - The con games or techniques used.
- Rape and sex offenses.
  - The perpetrator's method of operation.
  - The relationship between the victim and the perpetrator (blood relatives, acquaintances, or strangers).
  - The degree of force used.
- Aggravated assault or murder.
  - The relationship between the victim and the perpetrator.
  - The motivation.
  - The weapon used.

B-24. Personnel who have tried to use MPRs and ROIs as a sole source of data know that in many cases the required information is not contained in sufficient detail to be useful. To correct this situation, investigators and inspectors must be trained to recognize conditions that contributed to the crime as well as the information needed to identify and prosecute the offender. As a minimum, the PM ensures that all law-enforcement personnel know general and specific crime factors that are required to analyze each type of crime. He must also provide feedback through the law-enforcement operations staff when incomplete reports are received. If this is done, high-quality data can be collected without generating additional reports to collect criminal information for analysis.

B-25. The MPRs and ROIs are not the only sources of information on which to base criminal analyses. Physical-security inspection results, summaries of common deficiencies noted by the inspector general, data from the SJA claims section, and information summaries from reports of survey on lost government material can all produce worthwhile information on conditions that lead to the commission of crimes. Additionally, a review of civilian criminal statistics and civilian police information may contribute to the pattern analysis. The reviewing authority must understand which information is relevant. The MP commanders ensure that the CCIR are published in operations orders (OPORDs), operation plans (OPLANs), and SOPs. This effort ensures that MP forces at every level recognize relevant information.

INDIVIDUAL CRIMINAL ANALYSIS

B-26. A criminal analysis is most effective when it is applied to the class of criminal offenses with a high probability of recurrence. Single-incident crimes do not lend themselves to analyses. Most crimes against persons do not usually benefit from analyses, with the notable exceptions of rape, robbery, and related combinations of offenses (such as kidnapping and rape; robbery and attempted murder; burglary and rape; and burglary, robbery, and kidnapping). Analyzing isolated criminal offenses has some value (such as gaining knowledge of where these offenses are most likely to reoccur). However, this knowledge is usually difficult to use effectively for preventive purposes.

B-27. There are universal factors available for analysis for most crimes. The availability of these factors varies greatly between criminal types and specific reported offenses.

B-28. In addition to the universal factors, there are an almost infinite number of factors that may be considered specific to a particular criminal class or type. These crime-specific factors are data elements that are usually recorded during the reporting of a particular type of offense and are used for analysis purposes.

B-29. Crime-specific factors provide information that can be used by the analyst to connect crimes with similar characteristics. Information regarding physical evidence may have considerable value in the analysis of several criminal types (such as burglary and auto theft). Therefore, the suitability of different crimes to analyses depends on the general factors, specific modus operandi (MO) factors, and physical evidence. With these considerations in mind, different criminal types and classifications are examined to determine their applicability.

RESIDENTIAL HOUSEBREAKING AND BURGLARY

B-30. Although housebreaking and burglary are two of the most difficult crimes to prevent, they are the most suitable for analyses. The typical housebreaker or burglar establishes an MO pattern based on successful past offenses. Usually, a burglar will continue to commit similar crimes until he is apprehended.

B-31. The available information for evaluation must consider the specificity, accuracy, and value of the information received. Informational factors are presented in the order of relative importance. It should be emphasized that the factors listed in Figure B-2 are for analysis purposes; the order for investigative solutions may vary.

COMMERCIAL HOUSEBREAKING

B-32. The analysis of commercial housebreaking is in many respects easier than that of residential housebreaking or burglaries. More pertinent information is available to the analyst for analyzing commercial housebreaking (which is more specialized and will exhibit more specific MO characteristics). The analyst normally has specific information regarding the point and method of entry, the victim/target description, and the property-loss description. The time factors for commercial housebreaking may be of less value in a commercial-housebreaking analysis than in a residential-burglary analysis since most commercial housebreakings occur at night or during weekends.

B-33. The commercial housebreaker is generally more mobile than the residential burglar. An analysis of commercial housebreaking is not as

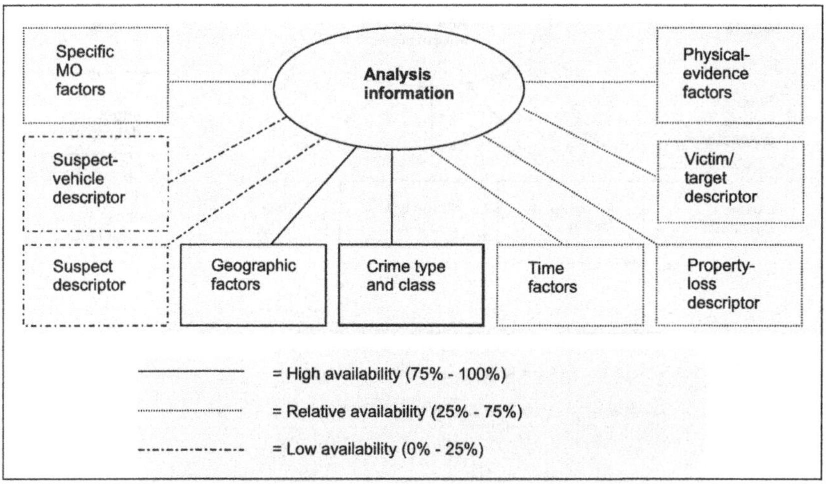

Figure B-2. Residential Housebreaking and Burglary

restricted in geographical area as an analysis of residential burglaries. The commercial burglar may travel a considerable distance to attack a particular type of business. The residential burglar is less discriminating because he has a greater number of potential targets.

B-34. The analysis of commercial housebreaking should be systematically directed to examining the information factors shown in Figure B-3. A commercial housebreaker may be a juvenile criminal, an addict, a professional thief, a soldier, or an enemy agent. Each of these types provides a different set of factors to analyze.

ROBBERY

B-35. Robberies are well suited for analysis. This class of criminal offender usually operates in a given geographical area. The commercial robber generally seeks a sizeable amount of cash; while targets of the street robber frequently include credit cards, checks, and other valuables in addition to cash.

B-36. The presence of physical evidence is more probable in street robberies than in commercial robberies because the criminal offender frequently discards evidence (such as purses and wallets) after completing the offense. The mugger frequently uses surprise as part of his MO in order to reduce his chance of apprehension by physical ID.

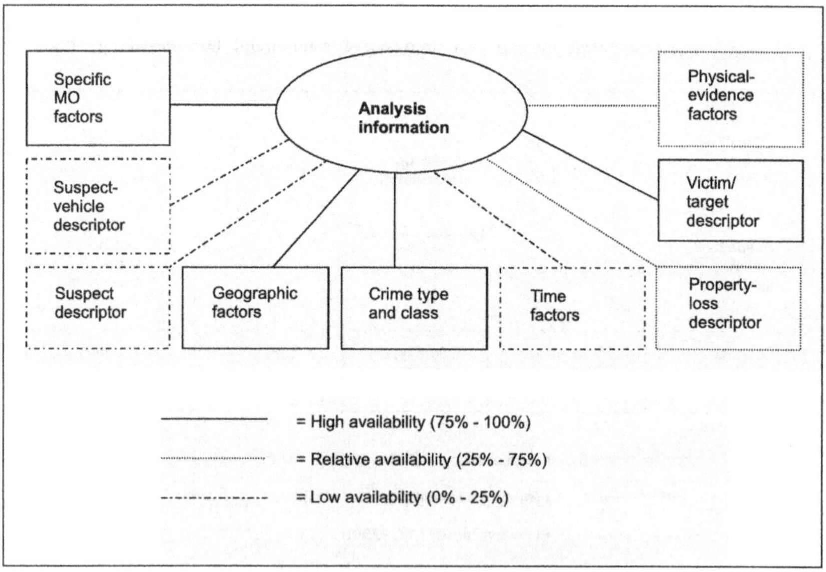

Figure B-3. Commercial Housebreaking

B-37. The probability of universal and crime-specific informational factors for a typical robbery is a systematic method for examining the case. See Figure B-4 for the analysis of a robbery case.

AUTO THEFT

B-38. Auto theft is probably the best-suited crime for analysis. The analyst frequently has more information available on any given auto theft than any other crime. Suspect descriptive information is usually lacking in auto larceny cases. However, the greater availability of crime-specific information makes auto theft extremely receptive to an analysis.

B-39. Auto-theft offenders include joyriders, professional car dismantlers, and wholesalers of stolen vehicles. In many cases, these offenders will establish and maintain a particular MO until apprehended.

B-40. Helpful elements in analyzing auto-theft cases are the availability of information, the presence of physical evidence, and the ability of the stolen property to be traced. Of special importance are the geographic factors, suspect-vehicle descriptors, and property-loss descriptors. In analyzing an auto theft, the analyst usually has information concerning two geographic locations for analysis—where the vehicle was stolen and where it was recovered. The suspect-vehicle descriptors, property-loss descriptors, and victim/target descriptors are the same (see Figure B-5). This greatly augments the analysis of auto thefts. In an examination of specific MO factors in an auto

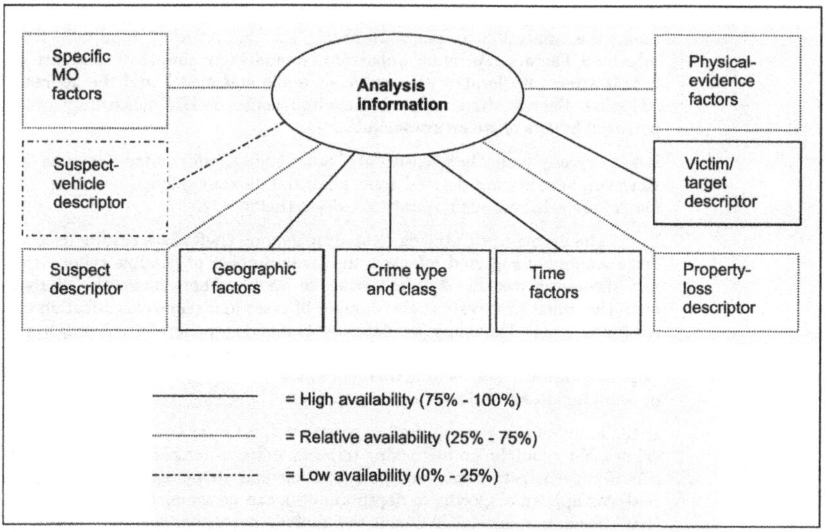

Figure B-4. Robbery

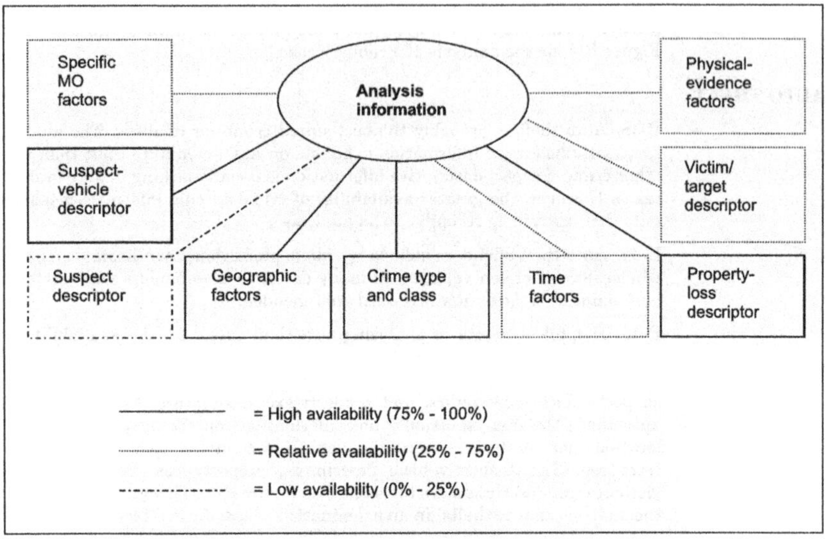

Figure B-5. Auto Theft

theft, the analyst will place emphasis on the vehicle's condition when recovered. Factors of primary importance in analyzing auto-theft cases are the vehicle's recovery location, the vehicle's make and model, and the degree of stripping. Coordination with local civilian police or HN police may reveal potential buyers of stolen automobiles.

B-41. Larceny and theft cases are not usually well suited for analysis. However, benefits are derived from analyzing larceny offenses when specific factors are selected, such as auto accession thefts.

B-42. The problem with trying to analyze general theft cases results from the large volume of reported offenses, the large number of possible crime types, and the many possible MO patterns. To analyze theft cases, the analysis operation must first restrict the number of cases and crime classification to a workable level. This may be done by classifying general theft cases into special categories and analyzing them by considering only special MO factors. Logical classifications include thefts of autos, auto accessories, bicycles, items on shipping docks, and tools and equipment.

B-43. Many of the general theft cases are helpful in analyzing other types of crimes. For example, an increasing trend in stolen automobile parts may have common perpetrators involved in auto theft and stripping. When particular patterns appear, a specific in-depth analysis can be conducted. It is essential that information developed during the analysis process be shared with the G2/ J2, the SJA, the CID, and others (as authorized).

RAPE AND SEX CRIMES

B-44. Rape and sex crimes fall into two distinct categories—those in which the offender is known to the victim and those in which he is unknown (stranger to stranger). Those cases in which the victim knows the offender are of limited informational value to the analyst. The in-depth analysis of rape and sex crimes is restricted to those cases in which there is no apparent relationship between the suspect and the victim. Rape and sex offenses in which the victim knows the suspect are often crimes of opportunity. Usually, these offenders do not establish a particular MO. On the other hand, the rapist or the sex offender committing stranger-to-stranger crimes will usually provide definite MO patterns, making this crime well suited for analysis. For example, the burglary/rape motive is particularly well-suited for analysis, as is the kidnapper/child molester. The advantages to analyzing stranger-to-stranger rape and sex offenses lies in the seriousness of the crime, its relative rareness, and the availability of information for a particular offense or group of offenses.

B-45. As with robbery, the suspect descriptors are important to the analyst for examining rape and sex offenses (see Figure B-6). The victim descriptors are also important in analyzing rape and sex offenses. In many cases, the perpetrator will restrict his attacks to victims of a certain age, a particular race, or a particular occupation grouping.

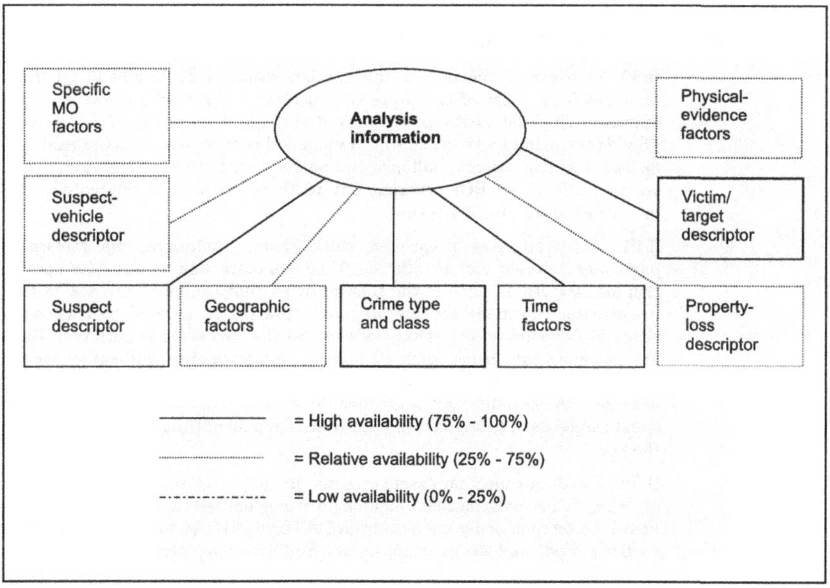

Figure B-6. Rape and Sex Crimes

FORGERY AND FRAUD

B-46. Forgery and fraud cases can provide correlative information for other reported criminal offenses such as burglary and strong-armed robbery. The analysis of forgeries is also aided by the fact that most forgers are repeaters who have established definite MO patterns.

B-47. An analysis of forgery and fraud cases can generate information that is disseminated to area merchants in the form of check-warning bulletins. The use of these bulletins can enhance community relations.

B-48. The primary problem in analyzing forgery and fraud cases is the immediate availability of information. In some cases, a delay of between three or four days to several months occurs between the time the offense was committed and the time it was reported.

ASSAULT AND MURDER

B-49. Most criminal offenses constituting assault and murder do not lend themselves to analysis. The comparative rarity of these crimes involving complete strangers makes it difficult to plot and predict assaults and murders. A tremendous amount of time and research is necessary to predict assaults and murders effectively. This data may identify specific areas with high rates of violent crime. This allows the commander to identify unsafe areas and warn soldiers to stay away from them.

CRIMINAL-ANALYSIS PROCEDURES

B-50. To organize the flow of information within a PMO and to facilitate analyses in support of crime-prevention efforts, the staff reviews a list of offenses. This list contains offenses that lend to analyses. The analyses provide the azimuth for crime suppression and countermeasures (for example, focused investigation/surveillance and patrol distribution). When final action on an MPR or an ROI covering one of these offenses is completed, it is forwarded to the staff for review.

B-51. The staff uses preprinted work sheets containing the factors of particular interest for an offense. They annotate and review the specific criminal factors. If some of the information is not available, the MP or CID investigator continues the investigation. The period covered by each work sheet will depend on the volume of cases at the particular installation. For a low-incidence-rate crime (such as robbery), one work sheet will suffice for the entire calendar year. For a more frequent crime (such as larceny), it may be necessary to use a different work sheet for each month or quarter. A summary sheet can be used to keep track of the data from all of the reporting periods in the year.

B-52. The data-collection sheets give specific information on individual types of crime. This information, together with general crime trends that are identified by comparing the data from DA Form 2819 for the current period to earlier periods and the geographic data from the crime-occurrence map, gives most of the information that is required to develop tightly targeted crime-prevention programs. While this system requires additional forms (the data-collection sheets and the crime-occurrence map) to be developed for use by the

MP Intelligence Officer (US Army) (S2), it does not increase the patrol's workload.

B-53. In addition to the information recorded in the forms, handwritten notes should be kept on items or trends that become apparent to the analyst during the review. For example, as a result of reading the cases on robbery to extract information for the collection form, the analyst may notice that a high proportion of the victims were attacked while crossing an unlighted athletic field near a troop billet area. This should be recorded so that it may be used in publicity campaigns and so that it may be passed along to patrols who routinely patrol the troop billet areas.

B-54. A geographic analysis is performed to determine information not available in the original data-element source (for example, a crime report). When a pin is used to indicate a crime location on a map, the relationship to other reported crimes of the same type becomes apparent to the analyst. Mapping is usually the source for a geographic analysis.

B-55. Mapping-analysis techniques involve the use of a map to depict the actual geographical relationships between particular criminal events according to prescribed data elements. A pin map displaying the actual locations of burglaries over a period of time is an example of such a technique.

B-56. In selecting a mapping technique, several things must be considered. These include the number of data elements that are to be recorded in the mapping, the retrievability of stored data, and the number of maps to be maintained.

B-57. The number of data elements to be recorded on a map can dictate the technique to be used. For example, if the case number is to be recorded in conjunction with the location of the offense, a dot (color-coded, self-adhesive paper disk) map or a flag-pin map can be used, whereas a typical color-coded pin map cannot. If particular MO factors are recorded, various color-coded and marked pins can be applied to identify specific MO patterns.

B-58. In addition, the length of time a particular map or set of maps is maintained should be considered. No set rules are established regarding a map's maintenance time; however, a number of agencies develop annual statistics based upon maps. Generally, maps are maintained for short-term (quarterly) and long-term (annually) periods for each type of crime. This decision should be based on the volume of criminal activity, physical-space limitations, the specific mapping technique adopted, and the number of maps to be maintained. As geospatial-terrain data becomes available, mapping techniques can be fully integrated into the MPACS.

B-59. Stored map data can be easily retrieved for comparing data. Pin maps can be photographically recorded and digitized for software applications. Computers and scanning equipment can be used to store the data for later retrieval. Color-coded paper dots placed on acetate overlays covering area maps facilitate the retrieval of recorded data. Digitized photographs of the maps are excellent sources of evidence for trials and court-martials. Moreover, the developed maps may contribute to the CCIR.

B-60. The number of maps to be maintained must be based on physical-space limitations, staffing, the types of crimes selected for analysis, and the maintenance period for each map. There are five types of crime suited to geographical analysis (burglary, robbery, housebreaking, auto theft, and sex crimes). Other crimes (such as theft of government property) may be depicted as a geographical analysis and contribute to RII. Housebreaking and burglaries can be placed on the same map by using different colored pins for each type of crime. The incorporation of numbered pins of different colors can be used to designate the type of premise attacked. Auto-theft and recovery maps may be kept on the same map with recorded information on thefts of vehicle parts.

B-61. The number of maps maintained should reflect the needs of the agency according to the volume of reported crimes and the different data elements or informational factors to be recorded. Therefore, the more detailed the analysis, the more maps are required. The actual number of maps maintained must be left to the discretion of the PM or the commander.

B-62. Periodically, a summary of the information that is collected, countermeasures, and operational/tactical procedures are briefed to the installation commander. Whether this is done monthly, quarterly, or semiannually will depend on the volume of reported crime and the commander's preference.

CRIMINAL-ANALYSIS SUMMARY

B-63. Criminal analysis is an important element to the MP function of PI. The techniques mentioned in this appendix are not all-inclusive. The PM/CID commander establishes formal procedures for integrating data and developing PI. It may require the formation of an ad hoc team, or it may be accomplished with computers and the MPACS. Criminal analysis requires close coordination with the G2/J2, the public affairs office (PAO), civilian police, and others based on METT-TC.

SECTION III — COMMAND AND LAW-ENFORCEMENT COUNTERMEASURES

B-64. This section discusses actions that commanders or PMs can take to reduce crime on Army installations. In many cases, the actions described have applications other than strictly crime prevention. However, only the crime-prevention aspects are discussed.

B-65. Most of the data available on the effectiveness of law-enforcement measures in reducing crime comes from studies conducted by civilians. While most of the findings of these studies are applicable to crime-prevention programs on Army installations, differences in population, the degree of control that can be exercised by authorities, and other environmental factors may dictate that the civilian recommendation be modified before implementation on military installations.

CRIME HOT LINES

B-66. A crime hot line is a dedicated crime-reporting telephone number located at the MP or security desk. This hot line allows anyone in the community to make an immediate report of an observed crime or suspicious activity. An effective crime-reporting program is a deterrent to crime and enhances law-enforcement responses to such incidents. Considerations in implementing a reporting program include the following:

- It should be publicized that personnel reporting incidents are allowed to remain anonymous if they desire. Some individuals will report their observations only if they know they can remain anonymous. If deemed feasible, neighborhood-watch blocks could be provided designated block numbers that could be used when reporting crimes or other suspicious activities. This system would allow neighborhoods to receive feedback on the disposition of the reported incident.
- The hot line's phone number should be easy to remember. This could include a number where extension digits are all the same, are in ascending or descending order, or spell out a word (such as 4357 [HELP]). Sticker labels listing the number could be placed on the phone with other emergency numbers for quick reference.
- The program should be well-publicized.
- Members on the installation should be educated on the desired procedures for reporting incidents.
- Each call should be documented. Records should be maintained concerning the results of these calls to evaluate the program's effectiveness.

CRIME PREVENTION THROUGH ENVIRONMENTAL DESIGN

B-67. Crime-prevention practitioners are recognizing the importance of considering design and physical planning in crime reduction. The crime-prevention officer has an opportunity to influence the design of facilities through the installation planning board. However, to be effective, he must understand a number of concepts about the relationship between the physical design of buildings and crime occurrences. These include the concepts of territoriality, natural surveillance, and defensible space.

TERRITORIALITY

B-68. Historically, a single-family home on its own piece of land and somewhat isolated from its neighbors (but often by as little as a few feet) has been considered to be the family's territory. The single-family home sits on a piece of land buffered from neighbors and the public street by intervening grounds. At times, symbolic shrubs or fences reinforce a boundary. The positioning of lights in windows that look out on the grounds also act to reinforce the claim.

B-69. Unfortunately, as the population has grown and the need for housing has increased, the trend toward developing single-family units has been paralleled, if not surpassed, by the development of row houses, apartment buildings, and various high-rise structures. Architects, planners, and designers involved in developing structures have not paid a great deal of attention to crime control or the need for an individual or a family group to identify with its home in a manner that might affect crime. Therefore, most families living in apartment buildings consider the space outside their apartment door to be distinctly public. In effect, they relegate responsibility for all activity outside the immediate confines of their apartment to public authorities. A question is whether environmental design can be used to extend the boundaries of these private realms, subdividing public space outside quarters so that more of the common space comes under the resident's influence and responsibility.

B-70. Through extensive research of efficiently functioning housing developments, a number of mechanisms have been identified that may be used in the design process (or may be added after construction). These mechanisms encourage the residents of multifamily dwellings to identify more with the ground or area around their immediate home site and to assume responsibility for its protection. Presented below is a brief discussion of a number of the mechanisms.

- **Site design.** If the grounds around a set of quarters can be directly identified with a particular building and the residents of that building take a personal interest in the use or upkeep of that area, they will play a role in protecting it. Through proper site design, a recreational area adjoining a building may be used as a buffer zone by providing play equipment for young children and seating areas for adults. The fact that children play and adults sit in these areas serves to increase the residents' concerns with the activities taking place there. Strangers are usually recognized and their activities come under observation and immediate questioning.
- **Street design.** Research has shown that by the placement, enclosure, or rerouting of streets and traffic the nature of a particular area can be changed and the crime rate reduced. For example, a particular portion of a street might be closed to vehicular traffic, and play equipment and seats may be added. In a number of areas where this technique has been used, it has been found that most residents know or at least recognize people up and down the block and strangers on the street are identified. Similar approaches that involve rerouting traffic, using one-way streets, or blocking off streets has lowered the crime rate in some areas.
- **Symbolic barriers.** The types of barriers that planners may use in laying out an area include open gateways, light standards, low walls, and plantings. Both physical and symbolic barriers serve the same purpose—to inform an individual that he is passing from a public to a private space. Symbolic barriers identified by residents as boundary lines serve as defining areas of comparative safety. Many places warrant the use of symbolic barriers, including transition points between a public street and the semipublic grounds of a building; an area between a building's lobby and its corridors; or hallways on particular floors of a building.
- **Internal design.** Although economics may sometimes enter the picture, a building's interior may be designed for specific groupings of apartment units and shared entrances. These factors may cause the residents of these apartments to develop a concern for the space immediately adjacent to their dwelling. For example, on each floor of an apartment building, two to four families might be required to share

a common corridor area. The apartment doors would be grouped around that common corridor, and access to elevators or stairs might be screened by a glazed partition. The net effect would be that the floor's residents would adopt the corridor as a collective extension of their dwelling unit and would take an increased interest in its maintenance and use.

- **Facilities and amenities.** The location of particular facilities (such as play and sitting areas and laundry facilities) will tend to give an area a high intensity of use and support the idea of territoriality. The presence of residents involved in various activities (children at play and people chatting or engaged in other types of activities) allows for casual surveillance by concerned members of the family and screens out possible intruders.

B-71. Reducing the number of apartment units grouped together to share a collectively defined area and limiting the number of buildings that comprise a housing project are important factors for creating an environment that residents will help to protect. Research has documented the fact that housing projects comprised of fewer high-rise buildings (two to four) have lower crime rates than projects containing a larger number of buildings. Based on this finding, it is argued that there appears to be much less freedom of movement in the public spaces of the smaller high-rise projects. Unlike buildings and large developments, every building of a small grouping usually has an entrance directly off a public street. These dwellings more closely resemble middle-income, high-rise developments and look more private.

B-72. As a crime-prevention officer, you may not be in a position to directly use these techniques. However, your familiarity with these approaches and the value of their use in the crime-prevention process are important elements in your arsenal of tools to create public involvement in reducing crime. In particular, the purpose of outlining these tools is not to equip you to be a designer, but rather to equip you to communicate with those who are involved in that profession. The discussion that follows will further enhance your ability to converse with designers.

NATURAL SURVEILLANCE

B-73. Experience has shown that the ability to observe criminal activity may not be adequate to stimulate an observer to respond with assistance to the person or property being victimized. The decision to act depends on the presence of motivational conditions, including—

- The degree to which the observer has developed a sense of personal and property rights that are being violated by the criminal act.
- The degree to which the observer feels that the event is within his area of influence.
- The observer's ability to clearly identify whether the act is unusual for the particular area.
- The observer's identification with either the victim or the property being vandalized.
- The degree to which the observer believes he can effectively alter the course of events he is observing.

B-74. Based on these conditions, a number of mechanisms have been identified that can be used to design the grounds and internal areas of apartment units, housing developments, and other residential areas to facilitate natural monitoring of activities taking place. By providing opportunities for surveillance through the positioning of windows in relation to stairs, corridors, or outside areas, continual natural observation will be maintained and crime will be deterred. If such steps are taken, the security of observed areas will be understood by the potential criminal, making him think twice before committing a crime.

B-75. The first of these natural surveillance mechanisms involves the positioning of service areas and access paths leading to apartment buildings to facilitate surveillance by residents and authorities. For example, buildings might be designed so that their entries face and are within 50 feet of a street, so that well-lit paths lead to the front door or the lobby, and so that the lobby is arranged to afford good visibility from the street. Other related steps focus on the strategic placement of windows, fire stairwells, lobby lights, and mailboxes so that they can be easily viewed from the street. Elevator waiting areas on each floor can also be designed so that they can be seen from the street level. Research has proven that if steps such as these are taken, residents will be more likely to become involved with protecting the facility, MP patrols will be in a better position to observe what is going on, and criminals will be discouraged from vandalizing the site.

B-76. A second technique that might be used to increase surveillance is to design facilities so that people within them will naturally view commonly used paths, entries, and play and seating areas during their normal household activities. This concept also focuses on the strategic placement of windows, lighting, and open areas so that natural surveillance by residents is improved.

B-77. Another mechanism involves the subdivision of housing areas into small, recognizable, and identifiable groupings that improve visual surveillance possibilities. Research has shown that in housing developments where the surveillance of a neighbor's outside activities was possible, residents were found to be very familiar with everyone's comings and goings. The overall effect was to cement collective identity and responsibility through social pressure.

DEFENSIBLE SPACE

B-78. Defensible space is a term for a range of combined security measures that bring an environment more under the control of its residents. A defensible space is a residential environment that can be used by inhabitants for the enhancement of their lives while providing security for their families, neighbors, and friends. The physical mechanisms suggested to create safety and improve upkeep (as part of the defensible-space concept) are self-help tools wherein design catalyzes the natural impulses of residents rather than forcing them to surrender their shared social responsibilities to any formal authority.

B-79. Research has revealed investigative techniques that might be used to modify existing housing areas to make them more secure. The following methods may require alteration or adaptation to the particular situation on your installation:

- Widening major pathways and using colored decorative paving.
- Differentiating small private areas (front lawns) outside each dwelling unit from the public path with low, symbolic walls.
- Adding public-seating areas in the center of public paths far enough from private-dwelling units to eliminate conflicts of use but close enough to be under constant surveillance by residents.
- Designing play areas as an integral part of open space.
- Adding new and decorative lighting to highlight various paths and recreation areas at night and extending the residents' surveillance potential and feeling of security.
- Adding seats and path networks to recreational facilities where large, central court areas exist. This increases the interest and usability of the areas.
- Redesigning parking and play areas around buildings to create the illusion that the buildings are grouped where natural opportunities exist.
- Modernizing building entrances to create breezeways into building courts and to accommodate a telephone intercom for opening entry doors to the lobby.
- Providing video surveillance of public grounds and central paths by security of public monitors.
- Installing audio surveillance capabilities in elevators and at the doors of residences.

CRIME-PREVENTION MODEL

B-80. The model for crime prevention through environmental design is based on the theory that action must be taken to counter crime before it occurs. The critical element in this model is the environmental-engineering component. It provides both direct and indirect controls against criminal activity by reducing the opportunity for crime through science and technology and the use of various urban planning and design techniques. The model explains what environmental engineering is and how it supports crime prevention. With this information, you may be in a better position to understand and respond to questions and discussions on how urban design and planning can have an impact on the installation's criminal element.

THE ENVIRONMENTAL INFLUENCE ON CRIMINAL BEHAVIOR

B-81. The basic theory that supports crime prevention through environmental design is that urban environments can influence criminal behavior in two ways. First, the physical surroundings in which people live have an effect on each individual. These physical characteristics include noise, pollution, overcrowding, and the existence and unmonitored spreading of refuse and other unsightly waste. The second element that must be dealt with in the environmental-engineering formula concerns the social characteristics of the community that provide individuals with social relationships to which they must respond. Characteristics such as alienation, loneliness, anxiety, and dehumanization are seen as keys to criminal behavior.

B-82. In terms of these environmental characteristics, buildings are all too often constructed to be dangerous, with corridors and passageways hidden from public view. Elevators, basements, and storage and washroom areas are also laden with danger due to their design. Various large-scale housing developments are not secure in that they are often isolated from the main flow of traffic (both human and automobile) and are closed to public use and public view.

B-83. With regard to altering the social characteristics of the community and their relationship to criminal behavior, it should be recognized that behavior is future-oriented, not past-oriented. A man steals so that he can have a car or money in the future, not because in the past he experienced psychic trauma, a broken home, poverty, or delinquent associates. Criminal behavior can be explained directly in terms of the consequences of behavior and in terms of noncriminal variables such as poverty, race, or social class. Criminal behavior is viewed as a problem to be dealt with and not symptomatic of other problems (such as poverty, mental conflict, class conflict, unemployment, or undereducation). To change criminal behavior, it must be dealt with directly by removing the environmental reinforcement that maintains the behavior. The approach advocated is to change the environment to which the individual responds.

ACTION APPROACHES TO CRIME PREVENTION THROUGH PHYSICAL PLANNING

B-84. The primary focus of crime prevention has been on what architects, planners, and other nonpolice professionals can do in terms of various physical-planning strategies to reduce criminal opportunity. Experienced MP personnel have long recognized that certain physical conditions can contribute to the rate and nature of crime. They have also developed a capability to identify high crime-risk locations by noting such factors as poor lighting and weak points of entry as potential targets. The critical job is to identify specific areas concerning physical planning and design that can be responded to and actions can be taken against on the installation.

B-85. Attempting to reduce crime or the fear of crime by regulating physical environments is easier said than done. In fact, although crime prevention can be built into almost every aspect of community planning, it is often ignored for a number of reasons. For example, fragmentation of responsible agencies is a key problem. In addition, crime has historically been looked upon as the exclusive responsibility of MP forces; not of those in charge of education, housing, or health and welfare. Yet, with an understanding of crime prevention combined with the knowledge that design techniques can change the opportunity for criminal behavior, PSIs will be able to talk the language of the planner and the designer and to be able to advise them from a police perspective.

B-86. It is notable that a number of civilian police agencies have become involved in the physical-planning process and have achieved notable results from their work. For example, the Fremont, California, Police Department has been involved in a planning process and maintains that law enforcement should become an integral part of the master- or comprehensive-plan review to screen all redevelopment plans for safety and crime hazards. Working with other units of municipal government as well as architects and designers, the department drew up a set of model guidelines for the evaluation of projects. The model included evaluation criteria dealing with such subjects as the accessibility of buildings to patrol units; traffic flow and off-street parking provisions; and the location and regulation of cul-de-sacs, playgrounds, common greens, fences, and security entrances. In addition, working with agencies such as the American Institute of Architects, the National Public Works Association, the Association of Public Utilities, and others, the department identified a number of subjects that are of specific concern to police officers and that should be considered in the design and planning stage. As a result of these efforts, the following list of design concerns was developed by the department:

- Building setbacks (front, side, and rear).
- Wall construction, interior and exterior (industrial, commercial, and residential).
- Door construction, setbacks and security (industrial, commercial, and residential) (including carports, garages, and sliding-glass doors).
- Windows and skylights, setbacks, heights (from ground), show-window displays, and the type of frame or pane.
- Stairs (stairwells and staircases).
- Balconies.
- Utility boxes.
- Fences, walls, hedges, screens, setbacks, heights, and louvers.
- Parking (public and private).
- Lighting (industrial, commercial, and residential).
- Streets, sidewalks, and walkways (locations, slopes, curvature, grades, and the length of a block).
- Alleys (blind and through alleys).
- Visibility of valuables (people, safes, cash registers, and personal property).
- Signs (street signs and signals, traffic signs and signals, and advertising signs).
- Accessibility; approach, entrance, and exit (pedestrian, vehicular, services, residential, commercial, and industrial).
- Public utilities and easements (gas, water, telephone, and electrical).
- Public areas and facilities (public restrooms, parks, bus stops and shelters, playgrounds, recreation halls, and so forth).
- Street trees and shrubbery (types, heights, and locations).

B-87. With this information, you can improve the security aspects of the community's physical-planning process. There is a probability that the work and recommendations of the National Advisory Commission on Criminal Justice Goals and Standards may help you in your efforts to get involved in the design process. More specifically, the Commission noted that "every police agency should participate with local planning agencies and organizations, public and private, in community physical planning that affects the rate or nature of crime or the fear of crime."

B-88. The future role of MP forces in crime prevention through physical planning will depend on the PSI's initiative. The perspectives and knowledge of what is happening in this field, combined with a working knowledge of the language, should equip PSIs to sell this approach as part of the overall program. It is important to point out that others who represent professions other than crime prevention are aware of the relationships between urban planning and crime. Remember, they deal in concepts, approaches, and ideas that have not involved the realities that law-enforcement personnel face.

SPECIALIZED PATROL TACTICS AND SURVEILLANCE

B-89. Specialized patrol operations use a variety of tactics in attempting to control identified crime problems. The most common tactics include uniformed tactical patrols and suspect and area surveillances. The following paragraphs discuss these tactics in terms of their target crimes, operation requirements and characteristics, and established or perceived levels of effectiveness.

B-90. The appropriateness of a given tactic depends on the characteristics of a particular crime problem. The selection of specialized patrol tactics should be made on the basis of a careful and continuous analysis of crimes. Most crimes can be addressed by more than one tactic. Several tactics might be tried in an effort to find the best one, and it is quite possible that the most effective approach to a given crime problem will include the combination of several tactics.

UNIFORMED TACTICAL PATROLS

B-91. A uniformed tactical patrol is the most traditional and widely used form of specialized patrol. It is a simple, straightforward approach to specialized patrol that involves the same procedures and techniques used by MP officers on routine patrol. These include constant visible movement throughout an area to generate a sense of police presence, careful observation of street activity, vehicle and pedestrian stops, and citizen contacts. The difference between uniformed tactical patrols and routine patrols are that uniformed tactical patrols use these tactics in an intense, concentrated fashion. MP officers are relieved of the responsibility for responding to routine calls for services so that they can devote their full time and attention to patrol, thus intensifying its impact. In addition, uniformed tactical operations typically deploy a number of MP officers in target areas, thereby increasing the level of patrol in these areas.

B-92. Uniformed tactical patrols can be used to control virtually any type of suppressible crime (for example, crimes that can be viewed from locations where the police have a legitimate right to be and those that can be potentially affected by police operations). These suppressible crimes include street robberies, purse snatches, vehicle thefts, burglaries, and housebreakings. Uniformed tactical patrols can also have an impact on other types of crime as officers use observation, field interrogation, and citizen contacts to develop information on the locations, activities, vehicles, and associates of suspects.

B-93. The primary purpose of uniformed tactical patrol is deterrence. This tactic is based on the assumption that highly visible, active patrols will deter potential offenders. By increasing the perceived probability of apprehension, conspicuous patrol is thought to reduce the likelihood that crimes will occur. If the deterrence should fail, heightened patrol coverage is believed to increase the probability of the immediate apprehension of the suspects.

B-94. Uniformed tactical patrols are often used to saturate an area that is experiencing a particularly serious crime problem. Although it has been widely used for years, saturation patrol has never been clearly and adequately defined. Exactly what level and intensity of patrol constitutes saturation has never been determined, nor have the effects of different levels of patrol been clearly established. It is difficult to prescribe the level of uniformed tactical patrols that should be used to disrupt a crime pattern in a particular area. This should be determined through an analysis of the size and characteristics of the area of concentration of each potential target crime pattern coupled with an assessment of manpower availability.

B-95. Some patterns can be effectively handled by a very small increase in the patrol level. For example, in Portland, Oregon, Operation Crime Reduction Involving Many People (CRIMP) had a significant impact on street robberies and assaults in the city's skid-row area by initiating two-officer, uniformed foot beats in the area. This was sufficient to saturate the primary locations of the target crimes during the high-crime hours, and it led to a substantial reduction in these crimes with little apparent spillover into adjacent areas.

B-96. Other departments have used saturation patrols on a much larger scale. For example, in the mid-1950s, the New York City Police Department attempted to saturate an entire precinct by assigning over 200 additional officers to the precinct's patrol force. Foot beats covering extremely small areas were arranged in straight lines so that an officer could keep the entire street area of his beat in view at all times. The four-month experiment led to a dramatic reduction in crime in the precinct. Compared with the same period in the previous year, street muggings fell by 89.9 percent, burglaries where entry was made from the front of the building dropped by 78 percent, and so on. The only crime category that was not affected was the relatively private crime of murder. Since crime displacement was not examined in the experiment, its true impact remains unknown. The experiment strongly suggests that massive additions of police manpower can have a substantial effect on crime. The problem is that most departments do not have the ability to conduct even a limited version of this experiment.

B-97. The amount of resources required for saturation patrols can vary tremendously, and there is no definite way of determining how much additional patrol is needed. This can best be decided on a problem-by-problem basis, using experience and evaluations of past efforts as a guide. As a rule, patrol augmentation should be sufficient to affect rather quickly the perceptions of would-be offenders concerning the level of police activity in a particular area. It appears that for such a strategy to be effective, sufficient resources should be applied in a limited area to ensure a true saturation effect.

B-98. Uniformed tactical patrols can use various modes of transportation. Patrol cars are most often used; however, foot patrols can be effective in congested service districts and bikes have been used to good advantage in high-density residential areas. The mode of transportation should be selected based on visibility, mobility, and access to citizens.

B-99. Some specialized units deploy MP officers in unmarked police cars. This is done in an effort to effect a balance between overt and covert operations, hopefully gaining many of the advantages of both. Unmarked cars may also be readily available since in many departments investigators work primarily during the day, which leaves their vehicles free for specialized patrol in the evening and early morning watches. This approach has serious drawbacks. First, unmarked MP cars are somewhat less visible than marked cars, yet they are still easily recognizable as police vehicles to large segments of the public (especially when the officers in them are in uniform). Second, the use of these cars in uniformed tactical patrols could lead to the sacrifice of some of the deterrent effects of high visibility without realizing the apprehension benefits of truly covert patrols.

B-100. Several patrol techniques have been tried to increase the visibility of uniformed patrol and to enhance the sense of police presence. A tandem patrol uses two marked patrol cars that follow each other at intervals of one-half to several blocks. Two units can also patrol on parallel streets one block over or in an alley. Another approach combines foot and vehicle patrols in an effort to increase visibility. Officers park their marked cars in conspicuous locations in high-crime areas and then are transported to other high-crime areas where they patrol on foot. The frequent repetition of this procedure is seen as a way of multiplying patrol visibility.

B-101. Unless a target area is more or less completely saturated (as in the New York experiment), MP patrols should move in a random, unpredictable pattern. This will make it difficult for potential offenders to plan their crimes on the basis of observation of patrol activities. One department found that its uniformed-tactical-patrol operation was actually helping burglars to commit break-ins. Interviews with confessed burglars indicated that they were aware of a visible patrol passing through the neighborhoods at regular intervals and they planned their crimes accordingly.

B-102. In addition to increasing the level and visibility of patrol in high-crime areas, uniformed tactical patrols often use aggressive patrol tactics involving frequent vehicle and pedestrian stops. MP patrols stop, question, and perhaps search citizens when there are reasonable grounds for suspecting that the citizens may have committed, may be committing, or may be about to commit a crime. Since the concept of reasonable suspicion is vague, MP officers have a considerable amount of discretion in conducting field interviews. Field interviews that do not result in either immediate arrest of the citizen or in alleviation of the officer's suspicions are usually documented in field-interview cards or spot reports. Field interviews serve to generate information about the activities of probable suspects; more importantly, they make the suspects aware that the police know of their presence in a given area, regard them as suspicious, and are watching them closely. This is expected to reduce the likelihood that they will commit crimes, at least in the area in which the tactical force is working.

B-103. The extensive use of field interviews is often highly controversial. Tactical units that emphasize field interviews have been accused of being storm troopers who constantly harass citizens and abuse their rights. There is sometimes a certain amount of truth to the allegations of harassment, especially when the activities of a particular type or group of suspects are being closely monitored. Not surprisingly, individuals who are stopped and questioned frequently are likely to complain, particularly if they have reason to be concerned about close police scrutiny. A recent study of field interviews in San Diego found that when conducted at moderate levels, field-interview activities do not have a major effect on police and community relations. The majority of citizens in all three study areas accept field interviews as legitimate and properly conducted police activity. The majority of all citizens who were the subjects of contacts felt that the contact was justified and properly conducted. The study found that the suspension of field interviews in a particular area was associated with a significant increase in the level of suppressible crime and when reinstated in the area, these crimes declined. The results of the San Diego study provide confirmation for the widely held belief that field interviews can have an important deterring effect on suppressible crime without doing irreparable damage to police and community relations.

B-104. The potentially negative impact of field interviews on certain segments of the community can be held to a minimum if the interviews are conducted in a professional manner. Citizens should be informed of the reasons why they are being stopped. They should be detained for as little time as possible and should not be subjected to verbal or physical abuse. There is also no need to stop everything that moves.

B-105. While experience and a limited amount of research indicate that uniformed tactical patrols can have a positive impact on the level of suppressible crime in areas, the overall effectiveness of this tactic is a controversial and much-debated issue. The principal concern is that rather than reducing crime, uniformed tactical patrols may simply lead to its displacement from one area to another or from one period to another.

B-106. When conducting uniformed tactical patrols, a PM activity should carefully monitor crime trends for indications of possible displacement effects. This is an important aspect of evaluating the impact of tactical operations, and it will provide information to guide future deployments and tactical decisions.

AREA SURVEILLANCE

B-107. Covert patrol and surveillance of high-crime areas can be used to make apprehensions for crime problems. These problems include those for which there are no suspects who warrant personal surveillance, the suspects are too numerous to permit personal surveillance, and there are too many potential targets to conduct either physical or electronic stakeouts. Examples of these types of problems would be a rash of residential burglaries or auto thefts in a particular area.

B-108. This tactic simply involves the covert patrol of a particular area and the observation of suspicious or unusual activities and occurrences that might indicate the likelihood of a crime. Suspicious individuals are not stopped but are watched until they either commit an offense or the officers' suspicions are removed.

B-109. The list of various techniques that can be used in an area surveillance is virtually endless. The following are some techniques that have been effectively used by specialized patrol units:

- Mingling with citizens at the crime scene to pick up information on possible suspects.
- Maintaining rooftop surveillance of a shopping center's parking lot to locate larcenies from vehicles.
- Surveilling housing areas by posing as maintenance workers.
- Following likely crime victims such as elderly citizens leaving a bank.
- Surveilling rooftops for unusual activity from aircraft or higher buildings. Binoculars are used to facilitate surveillance, and rooftops are marked so that street units can be dispatched to check out suspicious circumstances.

B-110. As in all types of plain or uniformed patrol, care should be taken to ensure that area surveillance is truly covert. Rental vehicles that can be changed frequently provide an excellent, though expensive, means of covert transportation.

B-111. Police on covert patrol should be dressed to blend in with the environment in which they are working, and they should have apparently legitimate, nonpolice-related reasons for being where they are. Several specialized units have found that surveillance teams composed of one male and one female officer can work in many situations without arousing suspicion. An apparently married or romantically involved couple lingering in a park, meandering slowly down the street, or sitting together in a parked car would generally appear less suspicious than two male MP officers doing the same things. Finally, it should be noted that in some small neighborhoods where residents know each other well, covert surveillance may be difficult, if not impossible, since the presence of any stranger arouses immediate curiosity and suspicion.

SUMMARY

B-112. These tactics represent the basic approaches that specialized patrol operations normally take in trying to control suppressible crime. Some of the tactics (such as uniformed tactical patrols) are directed primarily at crime deterrence, while others (such as suspect surveillance) are used to achieve apprehensions for target crimes. The tactics are most commonly used independently of one another. However, there are some indications that the combined use of several tactics in an integrated operation might be an effective way of coping with particular types of crime. Especially promising is the coordinated use of highly visible and covert patrols. A visible patrol force could be deployed to a particular area to deter crime there and direct it toward other areas in which MP forces using covert tactics are working. To date, efforts to direct criminal activity to areas or targets where MP forces are set up to make apprehensions have only been tried on a sporadic basis. However, this appears to be a promising approach to crime control and warrants greater attention in the future. It can be viewed as the creative use of crime displacement.

B-113. Many of the tactics mentioned previously could not be effectively used by MP patrols that are responsible for handling routine calls for service. It is obvious that the effectiveness of decoy operations, physical stakeouts, and suspect surveillance would be destroyed if the MP officers had to interrupt these activities to handle calls for service. It is also generally unwise to have MP officers in civilian clothes respond to calls for service.

B-114. The importance of crime analyses in identifying crime problems and suggesting appropriate tactics deserves repetition. Specialized patrol units should be deployed to address clearly identified crime patterns, and they should choose their tactics on the basis of an analysis of these patterns and a comprehensive knowledge of the area of occurrence. The nature and characteristics of a particular crime pattern should be the driving force behind the selection of tactics. None of the tactics of specialized patrol can be effectively used unless crime patterns have been identified and analyzed.

PUBLICITY CAMPAIGNS

B-115. Public-information campaigns are an essential part of every crime-prevention effort. The installation PAO can provide assistance in—

- Informing the public of the magnitude of the local crime problem.
- Disseminating information on crime circumstances.
- Generating the interest and enthusiasm necessary to initiate and sustain community crime-prevention programs.

B-116. The PAO should be a member of the installation's crime-prevention council and must be involved from the start in planning crime-prevention efforts. There are many information vehicles at the installation level available to carry the message to the public, including—

- Digital marquees.
- Radio.
- CCTV.
- Installation and unit newspapers.
- Posters and leaflets.
- Commanders calls and similar meetings.
- Town-hall meetings.

B-117. Not all media are equally effective in reaching a particular segment of the post population. To ensure that the media campaign is effective, identify the segment of the population requiring the information and the exact message to be communicated. The more specific the target audience, the more effective the media campaign will be.

B-118. Whatever media is used, it is important to provide coordinated input to the PAO well in advance of the desired publicity campaign date to allow for writing, rewriting, and publishing. For material to be published in a magazine, a minimum lead time of six months is normally required. For installation newspapers, a lead time of a week may be sufficient. Check with the PAO to determine the correct lead times for local publications.

B-119. When addressing civic groups or school audiences, films are often more effective than straight lectures because of the animation and other special techniques used to illustrate critical points. Several Army films are available as are a large number of commercial films.

RESIDENTIAL-SECURITY SURVEYS

B-120. A security survey is an in-depth, on-site examination of a physical facility and its surrounding property. The survey is conducted to determine a residence's security status, to identify deficiencies or security risks, to define the protection needed, and to make recommendations to minimize criminal opportunity. Because of several common characteristics, one expert has likened the security survey to the traditional criminal investigation. This comparison hinges primarily on the facts that both techniques are systematic in nature; are aimed at identifying the method of a criminal act; and are, in effect, more an art than a science. It should be recognized, however, that the survey has two other advantages. First, it can be conducted before the commission of a crime; second, it can offer protection against, rather than just remedial action after, criminal victimization.

THE RESIDENTIAL-SECURITY SURVEY AS A CRIME-PREVENTION TOOL

B-121. A residential survey is not a substitute for an aggressive neighborhood watch. It supplements these efforts and should be established after Neighborhood Watch and Operation ID programs are established.

B-122. The security survey is the primary tool used in crime prevention to recognize, appraise, and anticipate potential loss in residential areas. It is often defined as the backbone of a local crime-prevention program. In practice, it combines the security experience, training, and education of the local crime-prevention officer and focuses on a single element—the analysis of a residential facility.

B-123. The inherent value of surveys has been proven by nearly 300 civilian police departments across the country that have established crime-prevention bureaus or units. An even broader endorsement of the survey technique, however, was provided by the National Advisory Commission on Criminal Justice Standards and Goals, which stated that, "Every police agency should conduct, upon request, security inspections of businesses and residences and recommend measures to avoid being victimized by crime."

B-124. In short, the security survey is a tool that informs a homeowner of the particular areas in which his home is susceptible to criminal victimization along with steps that can be taken to reduce and minimize that potential. Further, the survey is a tangible document that reflects the efforts of the MP force not only to be responsive to community needs, but to get the community more directly involved in the criminal-justice process.

THE CRIME-PREVENTION SURVEY—A PROCESS OF INVESTIGATION

B-125. An important factor in understanding the residential-security survey is that it must be considered as an ongoing process. While a particular survey will result in specific recommendations, each survey will provide a foundation for future action. In combination, these surveys will provide a database that can be used in the analysis of the community's crime problems and guide action toward the resolution or reduction of the problem on a community-wide basis.

B-126. As a starting point, five steps must be used in carrying out the actual survey, while four additional steps must be remembered afterward. These steps are—

- Analyzing the overall environment (neighborhood, block, and so forth).
- Assessing the general vulnerability of the premises.
- Defining the specific points of vulnerability.
- Recommending specific security procedures.
- Including specific remedial hardware recommendations.
- Urging the implementation of the recommendations.
- Conducting a follow-up to ensure that recommendations have been implemented.
- Keeping crime statistics to evaluate the survey's effect and the implementation of recommendations.
- Conducting a second survey of the premise's statistical analysis to determine the alteration in criminal activity in the areas surveyed.

B-127. This process is the continued involvement and participation of the police with the community. Once the survey is completed, the job is not finished. In fact, if this posture is assumed, it may later be learned that the recommendations were not implemented and that the work was done in vain. This can easily lead to a loss of community-wide public confidence in the programs. Similarly, even in those cases when recommendations are implemented, additional crime might be experienced. Attention to this fact, coupled with immediate follow-up, will be essential to avoid losing the confidence gained from the original survey. Through prompt actions, the proposal of additional tactics may become the final step needed to substantially reduce criminal opportunity. In short, the security-survey process is not a one-shot operation and it is not to be looked upon as a simple, straightforward task. It is a continuing, difficult, rigorous, yet effective approach to reducing crime that has not been systematically applied by police agencies in this country.

THE ROLE OF POLICE IN IMPLEMENTING CRIME-REDUCTION PROGRAMS THROUGH SECURITY SURVEYS

B-128. At a minimum, there are two ways to encourage people to improve their personal security. First, you can organize, conduct, and participate in broad-based public-information and -education programs that make use of such media as radio, television, and the press. Second, you can organize, undertake, and follow up on a series of person-to-person security surveys. Clearly, both of these techniques have their advantages. The security survey has a unique quality that does not exist in the public-education program. It provides an added incentive on the part of citizens to implement recommendations because of the personal relationships and respect established during the actual survey by a crime-prevention officer.

B-129. For example, consider a homeowner listening to someone on television recommending improvements of security by installing better locks or alarm systems. You might ask yourself these questions: How secure am I? Are the locks that I have on my doors adequate? If they are inadequate, what kind of locks should I install? Do I really need an alarm system? What can the MP officers do for me in terms of making recommendations about security? In short, you would be aware of the possible need for improved security, but you would not know enough of the specifics to warrant real action. Because home security is a complicated matter requiring careful analysis and the installation of carefully tailored systems, implementation must be approached in a personal way. Certainly, public-education programs can assist in some aspects of crime prevention (such as prevention against schemes, auto theft, and personal protection), but they are not as effective in causing improved security as is personal contact.

B-130. Moreover, for the first time, police are placed in a position where they can actually provide a different kind of advice or service that can be offered in an environment where a crisis has not yet occurred. This is a truly unique opportunity for crime-prevention officers.

CONDUCTING THE SURVEY

B-131. To develop a proper perspective of the types of crimes that a PSI will most frequently be trying to reduce, a review of the cases is necessary to get a broader feel for the actual conditions in your area. During this review process, pay specific attention to photographs in the files. Study crime scenes in an effort to identify the type of security device that was defeated. In particular, if a door was used as a point of entry, note whether it was of adequate construction; if the door frame was broken or separated; if hardware, such as strike plates and door trim was inadequate; and so forth. In addition, review photos to determine if lock cylinders were pulled or if door chains fastened to trim moldings were simply pulled away to permit easy entrance. If photographs of the crime scenes are not available, visit as many crime scenes as possible. While doing so, photograph security risks you can study later and use as examples in future presentations to community groups.

B-132. By becoming familiar with the MO of persons committing such crimes as burglary and larceny, you will be better-equipped to understand potential risk situations and to point them out to potential victims. Quite surprisingly, many of the cases investigated were invited by some obvious crime-risk hazard that was overlooked by a resident. Also, many additional crime risks existed at a particular crime scene that a burglar could have exploited. Such vulnerability might be an indication of other crime targets within the community to which you should pay particular attention in your survey work.

B-133. To be an effective security surveyor, it is necessary to develop an intimate knowledge of the crime factors in your community. You do not have to become a statistician; however, the more you know and understand about crime problems, the methods used in your community, and security devices or systems that were defeated, the better you will be equipped to analyze the potential crime when surveying a home.

B-134. In addition to the general types of crimes that occur in your community, it is necessary to develop an understanding of the details of particular types of offenses. For example, with regard to residential housebreaking, you should be familiar with the types of burglaries and approaches used in particular sections of the community.

B-135. On the surface, it might appear as if this would be a monumental task. However, in terms of housebreaking, you might pull the files for the last quarter and carry out the following steps:

- Tally the number of times entry was made through the front door, rear door, or through a window.
- Identify the approach used for entry (kicking the lock, throwing a shoulder through the door, jimmying the lock, and so forth).
- Determine how often a window was broken or removed or a mechanism was used to force the latch (when entry was made through a window).
- Attempt to determine whether security devices were used in residences (such as alarms, special lighting, or other systems).
- Identify the burglar's general escape route (down a back alley, through a schoolyard, and so forth).

B-136. In developing an understanding of the MO of crimes, statistics that illustrate exactly what is happening in your community will be valuable tools. Not only will you be able to use this information in explaining crime risks while you are surveying the site, but it will be invaluable in making public presentations.

B-137. Only when you have developed the ability to visualize the potential for criminal activity will you have become an effective crime-scene surveyor. It is important that when you arrive on a survey site, you are prepared to give a property owner sound advice on the type of security precautions he should consider.

B-138. In summary, to be a good crime-prevention surveyor, you have to be a good investigator. You must understand the criminal's method of operation and the limitations of standard security devices. In addition, you must be knowledgeable about the type of security hardware necessary to provide various degrees of protection. (Specific information on conducting surveys is contained in Chapter 11).

JUVENILE CRIME PREVENTION

B-139. Law enforcement has long recognized the importance of establishing personal contact with youth. This need was first met with the "Officer Friendly" concept. The goal of this concept was to enhance the image of police among younger children and attempt to negate the unfavorable image other segments of society provided children. This concept has evolved into today's Drug Abuse Resistance and Education (DARE) program which has spread to most law-enforcement agencies.

B-140. Crime-prevention officers can have an impact on juvenile crime by establishing a positive interaction with juveniles. Many areas suffer a high crime rate of which over 50 percent is attributable to juveniles. Nationally, it is known that youth 17 years of age and under (who make up just 16 percent of the population) commit 42 percent of the crimes that cause injury or loss of property. As the punitive aspects of the juvenile justice system today are of questionable value, it is logical that crime-prevention officers should develop programs aimed at juvenile crime.

B-141. The basic problem in establishing any juvenile crime-prevention program is the destruction of preconceived ideas by both youths and law-enforcement personnel. This obstacle can only be overcome by the building of an honest rapport with individual youths. Just as a crime-prevention officer must sell crime-prevention procedures to the public, he must also sell himself and his program to the youth.

B-142. One method of establishing the needed rapport with youth reverts to a similar system used in the Officer Friendly concept. Classroom presentations in high schools and junior high schools can be used to destroy the preconceived ideas held by youth regarding police and law enforcement in general. Once an MP officer enters the classroom, he discovers that the majority of students obtain their ideas of police one of two ways—

- From television shows that portray every officer as a slow-witted moron (such as the old "Car 54" show).
- From their peers at school who relate every negative aspect of law enforcement.

B-143. However the students gain their information, it seldom resembles fact and must be changed to effectively establish a juvenile crime-prevention program.

B-144. In making classroom presentations for crime-prevention purposes, the crime-prevention officer must display honesty at all times. When confronted by uncomfortable questions, the MP officer should render the law-enforcement's point of view and make a truthful explanation. Obviously, some answers will not always be well received, but they will establish the credibility of his presentation. Further, when making these presentations, many areas dealing with specific areas of crime prevention will arise (such as rape prevention, property engravement, residential and apartment security, vehicle theft, and burglary prevention).

B-145. In dealing with schools, the crime-prevention officer can establish many positive contacts with individuals inside the school system that will prove to be of great value. Individuals such as administrators, counselors, and faculty members offer the crime-prevention officer assistance in many different ways. Perhaps the most valuable assistance is in the area of establishing a juvenile crime-prevention program that is based on peer influence.

B-146. A crime-prevention program aimed at peer influence is essential to a crime-prevention unit. Peer influence is directly related to the Determination Theory of Delinquency Causation. Simply stated, the theory holds that delinquency occurs as a result of external influences on youth. While attempting to control all external influences on youths is an impossible task, the control of peer influence is possible and perhaps the most important external influence to be controlled.

PEER-INFLUENCE CRIME-PREVENTION PROGRAM

B-147. A peer-influence crime-prevention program can be established after the crime-prevention officer establishes himself in the junior/senior high schools as described earlier. The basic concept of the program is the negation of undesirable peer influence on predelinquent youths and the establishment of a positive peer influence.

B-148. The peer-influence program functions within the school district. A problem youth is detected at the junior high school level by his teacher, counselor, or the crime-prevention officer. When it is believed that the peer-influence program could serve a youth, his counselor arranges a conference with him. During the conference, the counselor attempts to determine the basic problem experienced by the youth and has him complete an application for admittance into the program.

B-149. When the application is complete, the counselor contacts the youth's parent or guardian and explains the program. If the parent agrees to allow participation, the application is sent to the parent for his signature.

B-150. When the completed application is returned to the counselor, it is forwarded to a senior high school counselor for review. If necessary, the senior high school counselor may contact the junior high school counselor personally to discuss the youth in more detail. The high school counselor then selects a suitable senior high school student who has been accepted for the peer-influence program and matches the student with the junior high school youth.

B-151. Once the match has been made, the senior high school student is given a briefing on the junior high school student by the crime-prevention officer. The senior youth is instructed to contact the junior youth (preferably at the junior youth's home) and introduce himself to the youth and to the youth's parent or guardian. After the initial contact has been made, the senior youth is instructed to plan activities with the junior youth and spend as much time as possible with him.

B-152. Each senior youth that volunteers for the program is required to complete an application. This application gives personal background information and references that are thoroughly checked by the crime-prevention officer. The youth's academic record is examined, but prime consideration is given to citizenship factors rather than grades. The crime-prevention officer personally interviews each applicant.

B-153. When a senior youth is selected for the program, he is given a basic outline to follow. He is warned not to preach to the junior youth but to lead him by allowing him to observe the senior youth's actions at different activities. The senior youth must devote the necessary amount of time to limit the junior youth's opportunity to continue the negative associations he had in the past.

B-154. While the peer-influence program is simple in nature, the results can be remarkable. During a three-year period in Irvin, Texas, approximately 600 students participated in the program. From this number, seven arrests were made from the group with one juvenile accounting for four of the seven arrests. As about 30 percent of the juveniles participating were legally judged delinquent, the relapse rate was officially established as 0.1 percent.

B-155. While juveniles judged as delinquent can participate in the program, the crime-prevention officer should concentrate his efforts on those juveniles identified as predelinquent by counselors or faculty members. Again, the personal contact of the crime-prevention officer and the school faculty is essential to the program's success. A personal rapport should be cultivated and maintained.

B-156. Operationally, the peer-influence program can be administered by the collective efforts of school personnel and MP officers who are committed to the program. Once the program is established, it will probably not require the services of an officer on a full-time basis (depending on the number of students involved). However, constant contact is necessary; the more time that can be devoted, the more success will be apt to follow.

CRIME-PREVENTION PRESENTATIONS FOR ELEMENTARY AND JUNIOR HIGH SCHOOLS

B-157. As the traditional American family structure has experienced a significant decline over recent years, more and more emphasis is being placed on schools to provide guidance for children. As a crime-prevention officer who is frequently invited to speak to students, you can play an important role in assisting the school to provide this guidance. To stress the importance of the above statement, many experts in child behavior believe that the school is second only to the family in molding a child's socialization and behavior. Many people feel that the school is increasingly replacing much of the family socialization process. For this reason, schools are the most logical environment to begin a juvenile crime-prevention program.

B-158. Crime-prevention officers can have an impact on the most common types of juvenile crime by preparing specific lectures for particular grade levels in the elementary and junior high schools. Classes discussing shoplifting, vandalism, and bicycle theft are well received by students and have a wide-reaching effect on the community as a whole due to the children relaying the information to their parents.

B-159. A number of films are currently on the market that assist in making crime-prevention presentations in the classroom. In addition to specifically addressing juvenile crimes, the films directly relate to the peer-influence causation of crime.

OVERVIEW OF SELECTED JUVENILE CRIME-PREVENTION PROGRAMS

B-160. In addition to the programs described above, there are a variety of other juvenile programs. You will notice that some of the programs are designed to offer activities for the juvenile while at the same time attempting to provide one or more of the following:

- Positive peer influence.
- Education.

- Police/juvenile cooperation.
- An understanding of law-enforcement and MP duties.

B-161. One program deals with the enforcement aspect, with the goal being truancy reduction. Also mentioned are programs designed for child safety and victimization reduction.

DARE Program

B-162. DARE is a drug-abuse educational program that gives children and parents the skills to recognize and resist the subtle and overt pressures that cause them to experiment with drugs, alcohol, and violence. This unique program uses uniformed law-enforcement officers to teach a formal curriculum to students ranging from kindergarten to twelfth grade. DARE gives special attention to fifth through ninth graders to prepare them for entry into middle/junior high and high school where they are most likely to encounter pressures to use drugs or alcohol.

Police Explorer Programs

B-163. Explorer programs are run in conjunction with the Boy Scouts of America and a sponsoring police agency. The programs offer a wide range of flexibility in that they may be structured along the lines of a police cadet program with all activities centered around law-enforcement training. This type of explorer program usually seeks juveniles who have an active interest in pursuing a law-enforcement career. Being an explorer in this type of program will offer the student the opportunity to view the duties of a police officer and to take part in minor police activities (such as traffic direction and observation riding in patrol cars). The selection and training of the students for this type of program strictly depend on the sponsoring agency.

B-164. Other police-sponsored explorer programs may be designed to provide the members with other nonpolice activities (such as hiking, camping, and canoeing). At the same time, this type of program will offer the student the chance to meet with police officers and learn more about police functions, though this is a secondary purpose. Selection for this program would not be as stringent as the above-mentioned program. A local council office of the Boy Scouts can provide additional information on this program.

Alateen

B-165. The Alateen program is run in cooperation with the Alcoholics Anonymous organization. This program's intended purpose is to provide assistance for a child's emotional needs if he or she is faced with family problems caused by alcoholism of one or both parents. Your local chapter of Alcoholics Anonymous can provide more information on this program.

Truancy-Enforcement Program

B-166. Various cities have successfully adopted truancy-enforcement programs that focus attention on identifying and apprehending youths that are on unexcused absence from school. The primary goal behind this program is to prevent or reduce the incidence of daytime residential burglary. The San Angelo, Texas, Truancy Enforcement Program selected a target area in an effort to see how effective the program would be in reducing daytime burglaries. The initial evaluation of the program showed positive results. The San Angelo program uses uniformed crime-prevention officers in marked police cars patrolling around the school and known hangouts. All street personnel in the patrol division are also encouraged to check school-age individuals. Those students under 17 years of age who are suspected of truancy are taken to the school office while those over 17 years of age are detained only long enough to obtain their name and other identifying information, which is then turned into the school office. The assistant principal in charge of attendance completes the investigation and takes the appropriate action. The action varies (depending on the circumstances) but can include counseling with the student, the teacher, and the parents or, in extreme cases, suspension for a short period.

B-167. The emphasis behind these programs is to inform youths of the facts of the law. It was shown that many of the problems with the youthful offenders would be eliminated if they were presented the facts of the law to which they are subject. The courses can be adapted to fit the needs and desires of the individual school system, and the extent of the areas covered will depend on the commitment and availability of the local MP officers. Various civilian school districts have incorporated these law-enforcement courses as part of their regular curriculum.

Criminal Victimization Reduction

B-168. Programs designed toward this goal normally are structured to reduce the possibility of a child becoming a victim of a child molester. Most commonly, officers give a verbal presentation to groups of children using visual aids that help to act as a positive reinforcement to learning. Some effective methods have been the use of "talking traffic lights," "talking bicycles," and "talking motorcycles." These items are usually equipped with a tape-recorded message that can be controlled by the officer conducting the program. The talking traffic light uses a full-size traffic light that flashes red or green according to the type of person that is being described.

VANDALISM

B-169. Today's vandals often attack their own territory. School vandalism—the illegal and deliberate destruction of school property—is committed by students themselves. They break so many windows that in large districts the funds spent annually on replacing broken windows could pay for a new school. Vandals destroy about $3 million worth of school-bus seats annually, and they commit enough arson to account for 40 percent of all vandalism costs.

B-170. School vandalism outranks all other assaults on private and public property. At the end of the 1973 school year, the average cost of damages from vandalism was estimated at $63,031 per school district. That figure could have paid the salaries of eight reading specialists or could have financed a school breakfast program for 133 children for one year. A typical school's chance of being vandalized in a month are greater than one in four, and the average cost of each act of vandalism is $81. Yet, these figures do not include the hidden costs of school vandalism—increased expenses for fencing, intrusion and fire detectors, special lighting, emergency communications equipment, and vandal-resistant windows. Every dollar spent in replacing a window or installing an alarm cannot be spent on education.

B-171. School vandalism can also have enormous social cost. The impact of an 89-cent can of spray paint used to cover a wall with racial epithets far exceeds the monetary cost of removing the paint. An abusive word scrawled across a hallway wall can destroy student morale, disrupt intergroup relations, undermine the authority of an administration, or even close the school. Incidents with high social costs damage the educational process as much as those with high monetary costs. Today's vandal is not a hardened, war-scarred veteran. Instead of grizzled whiskers, he sports peach fuzz. He is almost literally the boy next door. In fact, the typical vandal differs quite dramatically from the typical juvenile delinquent.

B-172. It is significant that vandals fall into a well-defined and relatively narrow age group. They are usually early adolescent males who are highly subject to group pressures and transitory impulses. It is not at all unusual for adolescents to act out whatever is controlling them at the moment—rage, boredom, pent-up energy, or the sheer joy of "wreckreation." While there are conditions that may predispose or provoke a youth toward vandalism, the problem seems to be almost human nature. Few among us have never written on a sidewalk or scrawled initials on a school desk. Vandalism cuts across all strata of society, all geographic regions, and all racial lines.

B-173. The causes of vandalism remain obscure. Though research addressing the "why" of vandalism is growing, it has yet to yield clear-cut answers. Among the motivating factors often cited are anger, frustration, hostility, bitterness, alienation, futility, inequality, restricted opportunity, emotional pain, failure, prejudice, revenge, and the need for attention. Although much of the research is convincing, the fact remains that many vandals do not appear to be among the most angry, frustrated, hostile, alienated, or needy youth. Only a small fraction of the youngsters who fall into that category actually commit acts of vandalism. So, while most experts agree that vandalism is not totally senseless, they do not claim to fully understand its causes. In fact, vandalism is often not understood by vandals themselves. Many vandals report that they do not know why they did it. Many others, according to case reports, offer the unsolicited observation that destruction is fun. Still others express satisfaction and exhilaration. Few consider themselves criminals. For the time being, we can conclude only that motives for vandalism are diverse. But the whys notwithstanding, the vandal profile suggests that our task is, in large part, to anticipate and redirect the impulses of young teenagers.

B-174. Schools are by no means the helpless victims of early adolescence. Many school factors, most of which are amenable to change, influence the amount of vandalism that schools experience. The following characteristics are typical of schools that suffer high property damage or loss:

- Vandalism is higher when there is poor communication between the faculty and the administration (such as when the principal fails to define policy or makes policy decisions unilaterally).
- Hostility and authoritarian attitudes on the part of teachers toward students often result in students "taking it out" on the school.

- Limited contact between teachers and students reduces student involvement with the school and increases the likelihood of vandalism.
- Schools characterized by intense competition for leadership positions suffer greater property damage and loss.
- The chances for vandalism increase when the students do not value their teachers' opinions of them.
- Schools at which students strive to get good grades experience more vandalism.
- Parents of students in high-damage schools express less favorable attitudes toward their schools than do other parents.
- The school is a convenient target for vandalism when it is close to students' homes.
- The chances for vandalism are increased when grades are used as a disciplinary tool.
- Damage is greater in larger schools where there is more property to destroy. This correlation between school size and vandalism prevails regardless of whether the school is located in an urban, suburban, or rural setting.
- Fewer offenses occur when rules are well understood by students and are consistently and firmly enforced by teachers and administrators.

Vandalism Prevention

B-175. If the special problems of early adolescence, often intensified by social or personal pressures, interact with school conditions to produce vandalism, then preventive measures must address the nature of both the child and the school. Furthermore, prevention must include both physical and human responses. At present, most vandalism-prevention or -reduction programs rely on physical security—bigger and better electronic alarm systems, patrol guards, dogs, tamperproof locks, and window grilles. These techniques help, but they address only 20 percent of the problem—those incidents involving breakage. These incidents usually occur when school is not in session and in the absence of witnesses. The techniques have little effect on the day-to-day trashing of the school or on the disruptive acts aimed at the school's routine (bomb threats, the setting of fires, and false fire alarms) that are committed during school hours. The most sophisticated physical and electronic barriers are not sufficient to keep vandals from what they consider an attractive target. In fact, it has been argued that alarms and armed guards, besides lowering student and staff morale, often themselves become a challenge, inviting rather than deterring vandals. Vandalism prevention requires not a narrow or piecemeal approach, but a varied and comprehensive effort that includes both physical and human components geared to the school's specific problems. Furthermore, an effective long-term program must involve the community, parents, neighbors, police, and civic groups as well as students, teachers, and school administrators.

B-176. Schools are an easy target for vandals. Most are public, secular, and often unoccupied. Most will remain public and secular; but they need not remain unoccupied, unprotected, or unobserved. The following are techniques that have made some schools less vulnerable to vandals. These are especially effective against problems occurring during nonschool hours.

B-177. **Occupy the School.** Employ a custodial force around the clock. In most schools, the entire custodial force works at one time, leaving the school at night. As an alternative, custodians can be assigned staggered shifts so that the school is always occupied. Twenty-four-hour custodians are particularly appropriate in schools suffering sporadic property damage that demand more than a roving patrol but less than permanent security guards.

B-178. Invite police to use the school buildings at night. Police can be issued keys to the schools in their patrol areas so that they can use school offices to write their reports. This places a police officer in the school when it might otherwise be unoccupied, and it places a police car in front of the school.

B-179. Bring the community into the school. The school is an excellent place for recreational programs; health clinics; adult-education classes; counseling centers; community gatherings; and Boy Scout, Girl Scout, and Parent-Teacher Organization (PTO) meetings. The presence of people in the school building not only reduces the opportunity for vandalism, but also stimulates community and student interest in the school.

B-180. **Watch the School.** Use school neighbors as eyes and ears. Ask nearby homeowners to watch the school and report suspicious activities. Emphasize careful observation and rapid reporting, but discourage direct involvement in any situation observed. Such programs work best if they are organized but based on informal involvement, if they are accompanied by overall involvement of parents and community with the school, and if they offer some sort of prestige to participants.

B-181. **Employ Roving Patrols.** A uniformed patrol used instead of or in conjunction with an alarm system can deter vandalism. The individuals who patrol should establish rapport with neighborhood youths and open communication with community leaders. They should also vary their patrol patterns.

B-182. Hire student patrols during the summer and on weekends. The school district or community can provide its youth with part-time or summer employment and, at the same time, curb vandalism by paying students to patrol the school grounds during weekends, holidays, and summer vacations. These students should be paid an adequate wage and considered an integral part of the school's security force.

B-183. **Control Access to the School by Using an Alarm System.** Alarms are the most expensive vandalism control measure a school can use. While they can detect vandals, they cannot apprehend them; they can merely signal the alarm-system monitor, which may be miles away. They can, however, be an efficient deterrent and should be considered as part of any comprehensive plan to control vandalism. If alarm systems are linked with a surveillance camera, the chances of apprehending intruders are greatly increased.

B-184. **Design the School With Vandalism Prevention in Mind.** The following designs for preventing vandalism can be implemented when building a school:

- Limit ground-to-roof access.
- Eliminate low, overhanging roofs.
- Avoid unnecessary exterior fixtures.
- Plant trees that cannot be climbed near buildings.
- Consider raising as much of the school building as possible from ground level.
- Build the school at some distance from residential areas. While it should be located near the homes of most of those it serves, it will suffer less property damage if there is a buffer zone between it and surrounding residential areas.
- Design the school with plenty of defensible space so that the normal flow of school traffic allows continuing, casual surveillance of the premises.
- Use vandal-resistant surfaces. Use harder surfaces in damage-prone areas. For walls, use epoxy paint or glazed tiles that are easily and inexpensively replaced or repaired; use small wall panels and keep replacement panels in stock; and place permanent signs, building names, and decorative hardware at a level that cannot be reached from the ground. Replacing damaged areas immediately shows a sense of pride in the appearance and helps to eliminate copycat acts of vandalism.
- Plan windows carefully. Avoid windows that are vulnerably placed. Use small panes of glass to simplify replacement; use thick, tempered glass, thick acrylic, or Plexiglas® for windows in heavily traveled or hangout areas. Avoid useless windows in student stores, administrative offices, and industrial-arts storage areas.
- Plan entries with multiple uses in mind. Install flexible internal gates to block off specific areas or corridors when necessary. Provide separate exterior entries for community and student use. Inside the building, create areas for informal gatherings near entrances and exits by installing soft-drink machines and benches.
- Locate or relocate the playground's access roads to provide better surveillance by roving patrols.
- Consider outdoor lighting. Opinions on this issue are divided. Many schools report a decline in vandalism after installing hardened exterior night lighting. Others report that elimination of all night lighting reduces vandalism, presumably because young adolescents are afraid of the dark. If lighting is used, it should be directed away from windows to keep vandals from seeing the process of destruction or its outcome.
- Channel graffiti. Graffiti artists will usually select light, smooth surfaces rather than dark, rough surfaces. Therefore, school officials can channel graffiti onto one or two walls designed to withstand such treatment. Students or maintenance staff can paint most walls at regular but not too frequent intervals. One wall can be officially designated a "legitimate" graffiti wall, though this approach removes some of the challenge inherent in informal graffiti.

B-185. **Make the Target Less Attractive.** The school is not only an easy target but also an attractive one. It represents enforced learning, discipline, and mandatory attendance to students who are, simply by virtue of their age, rejecting adult standards and striving to achieve independence from adult control. Students are additionally provoked if the school functions in an impersonal, undemocratic, repressive, or manipulative manner, further increasing the likelihood of vandalism.

B-186. **Revise the Curriculum.** The list of characteristics associated with vandals and vandalized schools indicates that property damage and loss are higher when competition for rewards is intense, the availability of rewards is limited, or the distribution of rewards is unfair. All of this suggests that school policy and atmosphere have a direct bearing on school vandalism. Below are changes in school governance that can help remove the features that make a school an attractive target for vandals. These procedures, especially in combination, have proven effective against all forms of vandalism, including those that most commonly occur while school is in session.

- **Alternative schools.** Though originally designed to perform educational functions, alternative schools have proven effective in reducing school violence and vandalism. They provide an option to students who are not benefiting from the regular academic classroom. These schools operate within or alongside the traditional school. They are characterized by low student-to-teacher ratios, intense individual counseling, and extensive use of the community as a learning resource. They offer an alternative to suspension, personalize the learning environment, and provide successful experiences to students who have performed poorly in the past.
- **Law-related education and police-school liaison programs.** In many communities, the police department has assigned school resource or liaison officers to local junior and senior high schools. These officers may on occasion assume policing duties, but their primary function is to counsel students and teach law-related courses. These activities acquaint students with the positive role that law plays in our society and personalize the relationship between the "cop on the beat" and "the kid on the corner."
- **Action learning.** This term refers to apprenticeship programs as well as training in practical aspects of adult life. The former allows students academic credit for community work (such as tutoring and assisting physicians, lawyers, or other professionals). The latter provides instruction in skills such as checkbook balancing, comparative shopping, and applying for a job. Both address the boredom and frustration that are linked to truancy, violence, and vandalism.
- **Reward distribution.** The school's reward structure is related to school crime. Although the school may offer attractive incentives, most students receive very little in the way of rewards. Many of those who lose out still care about the competition. They become frustrated, and they vent their anger on the apparent source of their problems, the school. It should be possible to spread the rewards around without compromising performance standards, perhaps by recognizing improvement as well as achievement or by acknowledging forms of achievement other than scholastic, athletic, and social.

B-187. **Change Administrative Policy and Organizational Structure.** According to the National Institute of Education's Safe-School Study, the principal's leadership is a critical factor in reducing or preventing school crime. Strong principals who are visible, available, and responsive to students and staff appear to be most successful in eliminating violence and vandalism. Those who are less successful are often described as unavailable and ineffective.

B-188. The Safe-School Study also found that the exercise of discipline through clear enunciation and even-handed enforcement of rules is associated with a low incidence of school crime. To increase the likelihood that students will find school a fulfilling experience, many districts are establishing minischools—schools small enough to allow the individual student to feel significant. Similarly, several large schools are currently functioning on a house basis—the school is divided programmatically into several smaller units with which students can more easily identify.

B-189. **Involve the Students.** Establishing a vandalism fund uses peer pressure to the school's advantage. The community or the school district puts aside a certain amount of money and announces that the funds will be used to cover the costs of vandalism. Any money left over reverts to the students to be used as they choose. This plan works because it educates students about the costs of vandalism, allows them to see the positive results of curbing property damage and, most important, gives them full responsibility for the problem.

B-190. Several school districts have established voluntary student security advisory councils that conduct workshops and small group discussions focusing on vandalism and violence prevention. These committees increase awareness of the school's problems, generate recommendations for action, and give students opportunities to participate in school decision making.

B-191. Older students can be helpful in influencing younger students. In several communities, junior and senior high school students visit fourth- and sixth-grade classrooms where they show a film about vandalism and then lead a discussion on the causes and consequences of vandalism.

B-192. School beautification projects involve students in the care of the school building and grounds in an attempt to increase their pride in the responsibility for the school. The more effective projects are selected by the students themselves and continue throughout the year. The projects focus on marginal students rather than school leaders.

B-193. **Involve the Parents.** An open-door policy allows parents to go directly to their child's classroom whenever they wish without securing a special visitor's permit from the office. This simple policy offers parents concrete evidence that they are indeed welcome at school.

B-194. In some schools, parents serve on antitruancy committees along with teachers and students. They visit youngsters and their families in an effort to resolve truancy problems. In other schools, parents serve as hall monitors, supervise extracurricular activities, and help in classrooms. Parents are also the school's best source of guest speakers and contacts for work-study or apprenticeship programs. Some school districts have initiated faculty men's clubs to acquaint fathers with the male teaching staff and to encourage them to assume more responsibility for their children's progress in school.

B-195. **Involve the Community.** In some communities, students and law-enforcement, school-district, and city personnel sponsor daylong forums on vandalism. The forums introduce citizens to the causes and costs of vandalism and give them an opportunity to voice their concerns and initiate preventive programs.

B-196. Police departments can initiate public-relations programs within schools and youth-service agencies. In addition, they can enlist the help of youth in preventing vandalism through police-sponsored groups, such as the Police-Youth Service Corps, which recruits adolescents from high-crime areas to work as public-safety aides. Similarly, law students from neighboring universities can be brought into the schools at minimal or no cost to discuss the legal implications of vandalism.

Picking up the Pieces

B-197. The preventive measures listed on the preceding pages can function as part of a long-range, proactive response plan. However, they do not address the question of immediate response. What should the school do, right away, about 20 broken windows, a cherry bomb in the toilet, or recurring racial graffiti on the wall? The answer is to repair the damage immediately. Quick repair keeps perpetrators from admiring their handiwork, lessens the epidemic effect of vandalism, and minimizes any social impact the act may have.

B-198. Initiate formal record-keeping procedures, and ensure that they are followed. Schools faced with serious problems should begin recording all acts of vandalism. They should also consult law-enforcement personnel about when police should and should not be called. When a school begins to have problems, it should work with the juvenile justice system so that the two institutions can coordinate their efforts with regard to school-age offenders.

B-199. Careful record keeping allows a school to plot the incidence of vandalism to find out precisely where and when each type of offense is occurring. For example, using incident analyses, the National Institute for Education's Safe-School Study found that—

- Fire and bomb threats most often occur on Tuesdays.
- School-property offenses tend to occur with greater frequency toward the end of each semester, especially in November and December.
- Break-ins and school-property offenses occur most often on weekends and Mondays.

B-200. This type of information is invaluable in planning a vandalism reduction and prevention plan.

Restitution

B-201. Restitution is a set of legal and administrative procedures through which the school receives payment from vandals for damages they cause. While it seems reasonable to require payment for damages, restitution does not appear to be worth the effort. In the first place, most vandals are not caught. In the Los Angeles School District (which has an aggressive restitution program), only 30 percent of the offenders are ever identified.

From this 30 percent, most restitution is paid before matters get to court. Going through lengthy legal processes to obtain the rest is simply not cost-effective. However, a parent faced with the possibility of a court case may make a greater effort to keep his or her child out of trouble.

CONCLUSION

B-202. The goal of crime prevention is to reduce crime through public awareness and education. The skills that crime-prevention officers acquire can eventually benefit all segments of our communities. It is logical then for crime-prevention officers to attack one of the major crime problems in our society—juvenile delinquency. Efforts of crime-prevention units will certainly be lacking if the juvenile-crime problem is not given a high priority and if juvenile-crime programs are not established.

FRAUD

B-203. Two types of fraud can be affected by crime-prevention efforts—fraud against the government and fraud against individuals in the Army. Since the countermeasures for each of these categories of fraud are different, they will be discussed separately.

FRAUD AGAINST THE GOVERNMENT

B-204. Fraud of this type is a loss to the Army due to manipulation of systems from within the government with criminal intent. Typical examples of fraud are—

- The diversion or theft of government property by falsifying documents such as purchase orders, shipping documents, and so forth.
- False claims for temporary-duty (TDY) pay, for reimbursement for losses due to the movement of household goods, or for reimbursement for material reported as stolen.
- Overcharges or underproduction on contracts with the Army.
- Bribery.
- Kickbacks to secure purchase orders.
- Use of one's official position for personal gain.

B-205. This list is not all-inclusive. There are many more ways to defraud the government. Some of the factors that make it easy to perpetrate a fraud are—

- The lack of independent verification of records, transactions, and reports.
- The lack of adequate supervision.
- Unrealistic budgeting and acquisition requirements.
- Failure to correct deficiencies identified by existing systems.
- Concentration, at the operational level, of responsibility and authority for an entire process in one individual.

B-206. There are built-in measures to discourage fraud for most Army material-control systems. These programs are routinely examined by the inspector general (IG) along with other command inspectors. A crime-prevention officer's main function is to encourage the reporting of fraud by ensuring that the community understands what type of activities should be reported and by ensuring that the numbers are posted in work areas where personnel are in a position to detect fraud.

FRAUD AGAINST INDIVIDUALS

B-207. The second type of fraud is con games or consumer fraud. This crime is perpetrated by professional con men, unscrupulous businesses, and amateurs who see a chance to make a fast buck. Most of the time, the operator offers a "something for nothing" deal. By the time the victim realizes that he has been duped, the con man is long gone. Frequently, the victim is so embarrassed that he has been duped that he does not report the crime. The only way that a crime-prevention officer can help prevent this crime is to publicize common con games and illegal business practices and to encourage victims to report them. When a case of fraud occurs, the specific con technique that was used should be advertised widely to prevent others from being fooled by the same ploy.

Repair Fraud

B-208. Repair frauds are simple to execute but difficult to detect. Some unscrupulous repairmen do not fix the problem, but they charge you anyway. Some use inferior parts; others charge you for work that you did not expect or need. Some also do "insurance" work; they will repair one thing but ensure that something else will soon go wrong. The following are methods that you can use to protect yourself from repair fraud:

- Shop around. Ask friends, neighbors, or coworkers for references. When you find repairmen you trust, stick with them.
- Do not try to diagnose the problem yourself unless you are an expert. The mechanic may take your advice, even if it is wrong. If you know exactly what the problem is, do not tell the mechanics; wait and see if their recommendations agree with your diagnosis. That way you will know whether needless repairs are suggested.
- Try to get several detailed written estimates before any work is done. Compare job descriptions and materials to be used. Be sure to ask if there is a charge for an estimate.
- Ask for the old parts to ensure that replacements were really installed.
- Ensure that you get a guarantee on any work done.
- Make sure that the work was done before you pay. Take your car for a test drive. Plug in the refrigerator. Test the television.
- Ask your local consumer affairs office about the laws in your state and what specific protection it gives regarding professional services to be licensed or certified.

Home-Improvement Fraud

B-209. Home repairs and improvements can be costly. Be cautious if somebody offers to do an expensive job for an unusually low price, if a firm offers to make a free inspection, or if the workers "just happened to be in the neighborhood." These are the favorite tricks of dishonest home-repair firms.

Some firms offer a price you just cannot resist. Once you sign the contract, you learn why—they never deliver the service. Others send door-to-door inspectors to do a free roof, termite, or furnace inspection. These free inspections will turn up plenty of expensive repairs. Some fly-by-night companies will offer to do the work on the spot. When they leave, you may be left with a large bill and a faulty repair job. To avoid home-improvement fraud, try the following:

- Get several estimates for every repair job. Compare prices and terms. Check to see if there is a charge for estimates.
- Ask your friends for recommendations, or ask the firm for references and check them.
- Check the ID of all inspectors.
- Call the local consumer affairs office or the Better Business Bureau to check the company's reputation before you authorize any work.
- Be suspicious of high-pressure sales tactics.
- Pay by check, never with cash. Arrange to make payments in installments—one-third at the beginning of the job, one-third when the work is nearly completed, and one-third after the job is done.

Land Fraud

B-210. Real estate can be a great investment. The enterprising real estate salesperson knows how anxious you are to find just the right property, especially for an investment or a retirement home—a nice warm climate, not too crowded, or a new development. Some dishonest agents will promise you anything—a swimming pool, country club, or private lake—to get your name on the contract. Even if the sales agent promises you luxury, they may not guarantee the basics such as water, energy sources, and sewage disposal.

B-211. Most land developers offering 50 or more lots (of less than 5 acres each) for sale or lease through the mail or by interstate commerce are required by law to file a Statement of Record with the Housing and Urban Development Administration (HUD). This document tells you almost everything you need to know about your future home—legal title, facilities available in the area (such as schools and transportation), availability of utilities and water, plans for sewage disposal, and local regulations and development plans. All of this information must be given to you in a property report prepared by the developer. Always ask to see this report before you sign anything.

B-212. If the developer does not give you a copy of the property report for the lot you are considering, you can obtain it from HUD for a fee. Send requests to: Department of Housing and Urban Development, Office of Interstate Land Sales Registration, 451 Seventh Street South West, Washington, DC 20410.

Door-to-Door Sales

B-213. Beware of the following door-to-door sales tactics:

- "Nothing like it in the stores!" This is a true statement. The vacuum cleaners in the stores are probably of better quality and come with a better warranty.

- "Won't find this price anywhere." This is also a true statement. The prices in the stores are probably lower.
- "Easy credit!" This is another true statement. They do not care what your credit rating looks like. Once you sign for the purchase, paying for it is your problem. Be wary of low monthly payments. Find out the total amount you will pay over the life of the loan, then subtract the actual cost of the item itself. The difference is what you will pay in interest. Your bank, your credit union, or a local legal-aid society can tell you if the interest rate is fair.

B-214. Be cautious of these words, and be firm if the salesperson pressures you to buy. If you do get trapped, you are protected by a Federal Trade Commission regulation. When you make a purchase in your home totaling $25 or more, the salesperson must give you a written contract and two Notice of Cancellation forms. You have three days to change your mind and use one of these forms to cancel your contract.

Charity Fraud

B-215. Charity fraud does a lot of harm. The swindler takes advantage of a person's good will and takes his cash—money that was meant for people in need. You can ensure that any money you give gets into the right hands. Remember these pointers when somebody asks you for a donation:

- Ask for ID, either the organization's or the solicitor's. Find out what the charity's purpose is and how the funds are used. Ask if contributions are tax deductible. If you are not satisfied with the answers, do not give money.
- Give to charities that you know. Check out those you have never heard of or others whose names are similar to a well-known charity.
- Do not fall for high-pressure tactics. If solicitors will not take no for an answer, tell them no anyway, but do not give them your money.
- Be suspicious of charities that only accept cash. Always send a check made out to the organization, not to an individual.
- If a solicitor reaches you by telephone, offer to mail your donation. Shady solicitors usually want to collect quickly.

Self-Improvement Fraud

B-216. Con artists know that everyone wants to look better, feel better, and be a better person. Selling worthless plans and cures is one of the easiest ways for them to make a "quick buck." The following ads can look tempting:

- "Miracle reducing plan."
- "Look like a model in only five days."
- "Learn to speak Spanish while you sleep."
- "You can have a new, dynamic personality."

B-217. What can you do? Be careful! Read the small print. Know what the product contains. You should check with your doctor before you embark on any diet or exercise program.

Medical and Health Fraud

B-218. Most of us do not know much about medicine; that is why we go to doctors. It is also why we fall for miracle cures and other phony health products and services. Patent medicines, health spas, and mail-in lab tests should be warning signs for the potential consumer. A laboratory in Texas advertised nationally that it had perfected a fail-safe urine test for cancer. More than 15,000 tests were made at $10 each before authorities stopped this fraudulent company.

Unsolicited Merchandise

B-219. Cagey con artists will send you a gift in the mail—a tie, a good luck charm, or a key chain. What do you do with it if you did not order it? If you are the kind of person they are looking for, you will feel guilty and pay for it, but you are not obligated to. If you—

- Have not opened the package, mark it "return to sender." The post office will send it back at no charge to you.
- Open the package and do not like what you find, throw it away.
- Open the package and like what you find, keep it, free of charge. This is a rare instance where the rule of "finders, keepers" applies unconditionally.

B-220. Whatever you do, do not pay for it. Look at your gift as an honest-to-goodness way of getting something for nothing. Do not get conned if the giver follows up with a phone call or a visit—by law the gift is yours to keep.

Mail Fraud

B-221. The following are examples of mail fraud:

- **The contest winner.** "You've won! This beautiful brand-name sewing machine is yours for a song. To claim your prize, come to our store and select one of these attractive cabinets for your new machine. Bring this letter with you and go home with a new sewing machine for next to nothing." Treat an offer like this carefully. Shop around before you claim your prize. Chances are, the cost of the cabinet will be more than the machine and cabinet are worth.
- **The missing heir.** You have just received a very official looking document. The sender is looking for the rightful heirs to the estate of someone with your last name. It could be you. To find out, just send $10 for more information. There may be thousands of people with your last name, and letters like these are often mailed nationwide. Even if there really was claimed estate, it is highly unlikely that you would be an heir. Save your money; why help a swindler get rich?

B-222. These are just two examples of mail fraud. Many of the other frauds described in this section can be handled through the mail. When they are, the US Postal Service can launch a full-scale investigation. If you think you have been cheated in a mail-fraud scheme—

- Save all letters, including envelopes.

- See if your neighbors or business associates received the same material.
- Contact your local postmaster who can direct you to your regional Postal Inspector's Office.

SUMMARY

B-223. The following are recommendations to avoid being duped:

- Do not believe something-for-nothing offers. You get what you pay for.
- Be suspicious of high-pressure sales efforts.
- Take your time. Think about the deal before you part with your money.
- Get all agreements in writing. Insist that agreements are made in "plain English," not "legalese."
- Read all contracts and agreements before signing. Have a lawyer examine all major contracts.
- Compare services, prices, and credit offers before agreeing to a deal. Ask friends what their experiences have been with the firm or service in question.
- Check the firm's reputation with your local consumer affairs office or the Better Business Bureau.

INTERNAL THEFT

B-224. Employee theft is a major problem in most large organizations. It is estimated that one-third of all business bankruptcies are a result of theft. Of course, in the Army there is no danger of the organization going out of business; however, waste and internal theft can divert significant amounts of critical resources from mission-essential activities. Unlike many businesses, the Army has recognized that internal losses can be a problem, and most areas have adequate controls mandated by regulation. As crime-prevention officers, we should monitor reports of survey, physical-security inspection reports, and results of crime-prevention inspections to ensure that control measures are being followed. When it is apparent that the correct measures are not being followed, the problem must be identified for the senior commander's action.

B-225. Research has shown that an organization's atmosphere is just as important as management controls and physical security in preventing employee theft. To control this problem, the following items are essential:

- Employees, both civilians and soldiers, must believe they are part of a professional organization that expects superior performance from all of its members. When second-rate work is accepted, the lax attitude carries over into property-control procedures and losses increase. Standards must be high, but fair.
- Leadership must set the example. If leaders, supervisors, and managers take advantage of their positions to use government material or services for their own benefit (even in minor ways), their employees will also feel justified in diverting Army resources to their own use.

- The organization must show a genuine concern for the problems of its personnel. If an employee feels he has been treated unfairly, it is easy for him to justify stealing from the organization. He believes that he is only taking what he would be getting if the organization was fair.
- The organization must take appropriate disciplinary or administrative actions in cases of theft. There can be no acceptable level of internal theft. If employees believe that an activity considers a certain level of loss to be acceptable, then material theft will grow rapidly. Everyone will consider the material he takes to be within the 2 percent that the activity expects to lose.
- Policies on internal thefts must be enforced. Frequently, a thief is a long-term employee whose honesty has always been above question. In these cases, the temptation is very strong to let the offender go with a "slap on the hand." However, this is a clear signal to other employees that diversion of Army resources for personal use is acceptable to management. If an employee chooses to steal and is caught, he should be prosecuted.
- Publicity campaigns using posters and other media should be used to disseminate command policies on internal theft. Crime hot lines are also useful in increasing reports of employee thefts.

PILFERAGE

B-226. The protection of property, including the prevention of pilferage of government supplies and equipment, is one of the primary functions of the MP and civil service security forces. This function may include protecting supplies and equipment while in storage areas, during the issue process, or while they are in transit.

B-227. Pilferage is probably the most common and annoying hazard with which security personnel are concerned. It can become such a financial menace and detriment to operations that a large portion of the security force's efforts may have to be devoted to its control. Pilferage, particularly petty pilferage, is frequently difficult to detect, hard to prove, and dangerous to ignore.

B-228. Military property loss throughout the world would increase millions of dollars each year if subjected to uncontrolled pilferage. However, the risks incurred cannot be measured in terms of dollars alone. Loss of critical supplies for tactical units could result in loss of life or a danger to national defense. In some areas, losses could assume such proportions as to jeopardize an installation's mission. All installations and facilities can anticipate loss from pilferage. Actual losses will depend on such variable factors as the type and amount of materials, equipment, and supplies produced, processed, and stored at the facility; the number of persons employed; social and economic conditions in surrounding communities; command attitude; and physical-security measures used. Because these factors differ greatly in various types of installations and in different geographical locations, each must be considered separately.

B-229. To determine the severity of this hazard at any given installation or facility, it is necessary to determine the amount of loss that may be occurring.

Unfortunately, this is not always an easy task. Accounting methods may not be designed to pinpoint thefts; consequently, such losses remain undisclosed or they are lumped together with other shrinkages, thus effectively camouflaging them.

B-230. Inventory losses may be inaccurately labeled as pilferage for the following reasons:

- Failure to detect shortages in incoming shipments.
- Improper stock usage.
- Poor stock accounting.
- Poor warehousing.
- Improper handling and recording of defective and damaged stock.
- Inaccurate inventories.

B-231. In some cases, inventory losses may be impossible to detect because of the nature and quantities of materials involved. Stock inventory records may not be locally maintained, or there may be no method for spot checks or running inventories to discover the shortages.

PROFILE OF PILFERERS

B-232. Physical-security personnel must be able to recognize and counteract two types of pilferers—casual and systematic. A casual pilferer is one who steals primarily because he is unable to resist the temptation of an unexpected opportunity and has little fear of detection. There is usually little or no planning or premeditation involved in casual pilferage, and the pilferer normally acts alone. He may take items for which he has no immediate need or foreseeable use, or he may take small quantities of supplies for the use of family or friends or for use around his home. The degree of risk involved in casual pilferage is normally slight unless a very large number of persons are involved.

B-233. Casual pilferage occurs when the individual feels the need or desire for a certain article, and the opportunity to take it is provided by poor security measures. Though it involves unsystematic theft of small articles, casual pilferage is nevertheless very serious. It may have a great cumulative effect if permitted to become widespread, especially if the stolen items have a high cash or potential value.

B-234. There is always the possibility that casual pilferers, encouraged by successful theft, may turn to systematic pilferage. Casual pilferers are normally employees of the installation and usually are the most difficult to detect and apprehend.

B-235. A systematic pilferer is one who steals according to preconceived plans. He steals any and all types of supplies to sell for cash or to barter for other valuable or desirable commodities. He may work with another person or with a well-organized group of people, some of whom may be members of a cleaning team or even be in an advantageous position to locate or administratively control desired items or remove them from storage areas or transit facilities.

B-236. The act of pilferage may be a one-time occurrence, or such acts may extend over a period of months or even years. A large quantity of supplies with great value may be lost to groups of persons engaged in elaborately planned and carefully executed systematic-pilferage activities. Systematic pilferers may or may not be employees of the installation; if they are not, they frequently conspire with employees.

TARGETS FOR PILFERAGE

B-237. Both casual and systematic pilferers have certain problems to overcome to accomplish pilferage objectives. These problems include the following:

- A pilferer's first requirement is to locate the item or items to be stolen. For the casual pilferer, this may be accomplished through individual search or even accidental discovery. In systematic pilferage, more extensive means are generally used. These may consist of surveilling by members of the group or checking shopping and storage areas or documents by those who have access to them.
- The second requirement is to determine the manner in which he can gain access to and possession of the desired item. This may involve something as simple as breaking open a box, or it may be as complex as surveying security factors (such as physical safeguards or security procedures) for weaknesses. It may also involve attempting to bribe security forces, altering or forging shipping documents or passes, or creating disturbances to divert the attention of security personnel while the actual theft is taking place.
- The third requirement is to remove the stolen items to a place where the thief may benefit from his act. Articles of clothing may be worn to accomplish this. Small items may be concealed in many possible places on the thief's body or in vehicles. Through falsification of documents, whole truckloads of supplies may be removed from their proper place without immediate discovery.
- Finally, to derive any benefit from his act, the pilferer must use the item himself or dispose of it in some way. The casual pilferage of supplies is intended primarily to satisfy the need or desires of the thief. The systematic pilferer usually attempts to sell the material through "fences," pawnbrokers, or black-market operations.

B-238. Detection of use or disposal can help prevent similar pilferage through investigation and discovery of the means used to accomplish the original theft. Similarly, each of the problems faced by would-be pilferers offers an opportunity for constructive preventive measures. Careful study of the possible opportunities for the pilferer to solve his problem is essential in security work.

B-239. The primary concern of a systematic pilferer in selecting a target is its monetary value. Since he steals for personal profit, the systematic pilferer looks for items from which he can realize the greatest financial gain. This means he must also have or be able to find a ready market for items he may be able to steal. He pilfers small items of relatively high value (such as drugs, valuable metals, or electronic items). However, we cannot discount the possibility that a systematic pilferer may, if the profit is substantial, select a target of great size and weight. As a rule, bulk-storage areas contain most of the material that may be selected by systematic pilferers.

B-240. The casual pilferer is likely to take any item easily accessible to him. Since he normally will remove the item from the installation by concealing it on his person or in his POV, size is also an important consideration. Monetary value and available markets are not of any great concern to the casual pilferer, because he usually does not have any idea of selling the property he steals. Storage areas containing loose items are more likely to tempt casual pilferers than bulk-storage areas.

METHODS OF PILFERAGE

B-241. There are many ways that pilfered items may be removed from installations. Because the motives and targets likely to be selected by systematic and casual pilferers are very different, the methods of operation for each are very different.

B-242. As stated above, the casual pilferer steals whatever is available to him and generally removes it from the installation by concealing it on his person or in his automobile. The methods of the systematic pilferer are much more varied and complex. The means he may use are limited only by his ingenuity.

B-243. Shipping and receiving operations are extremely vulnerable to systematic pilferage. It is here that installation personnel and truck drivers have direct contact with each other and readily available means of conveyance. This offers a tempting opportunity for collusion.

B-244. One individual must not have control of all shipping and receiving transactions. Obviously, this procedure invites manipulation of government bills of lading and inaccurate storage and movement procedures through failure of one activity to compare bills and invoices with another activity. The opportunities for monetary kickbacks increase without a sound system of checks and balances.

B-245. Railway employees assigned to switching duties on the installation can operate in a similar manner. However, this operation is more difficult because a railway car normally cannot be directed to a location where stolen property can be easily and safely removed.

B-246. Tanker trucks used for shipping petroleum products may be altered to permit pilferage of the product. Trash- and salvage-disposal activities offer excellent opportunities to the systematic pilferer to gain access to valuable material. Property may be hidden in waste materials to be recovered by a accomplice who removes trash from the installation.

B-247. Other methods that may be used by systematic pilfers to remove property from military installations include—

- Throwing items over fences to be retrieved later by themselves or by accomplices.
- Packaging property and sending it to outside addresses through mail channels.
- Conspiring with security personnel.

- Wearing loose-fitting clothing to conceal small items.
- Removing items by using vehicles belonging to outside contractors and vendors.

CONTROL MEASURES FOR CASUAL PILFERAGE

B-248. Specific measures for preventing pilferage must be based on careful analyses of the conditions at each installation. The most practical and effective method for controlling casual pilferage is to establish psychological deterrents. This may be accomplished in a number of ways, such as—

- Searching individuals and vehicles leaving the installation at unannounced times and places.
- Conducting spot searches occasionally to detect attempts of theft.
- Making employees aware that they may be apprehended if they attempt to illegally remove property.

B-249. Care must be taken to ensure that personnel are not demoralized nor their legal rights violated by oppressive physical control or unethical security practices. An aggressive security-education program is an effective means of convincing employees that they have much more to lose than to gain by engaging in acts of theft. Case histories may be cited where employees were discharged or prosecuted for pilferage. Care must be taken in discussing these cases to preclude the identification of individuals because of possible civil suits for defamation of character. Also, it is generally poor policy to publicize derogatory information pertaining to specific individuals. It is important for all employees to realize that pilferage is morally wrong no matter how insignificant the value of the item taken.

B-250. It is particularly important for supervisory personnel to set a proper example and maintain a desirable moral climate for all employees. All employees must understand that they have a responsibility to report any loss to the proper authorities. Adequate inventory and control measures should be instituted to account for all material, supplies, and equipment. Poor accountability, if it is commonly known, provides one of the greatest sources of temptations to the casual pilferer.

B-251. Identifying all tools and equipment by some mark or code (where feasible) is necessary so that government property can be identified. Installation tools and equipment have counterparts on the civilian economy and cannot otherwise be identified as government property. Another control method requires individuals to sign for tools and equipment. The use of the signature control method reduces the temptation to pocket the item.

B-252. In establishing any deterrent to casual pilferage, physical-security officers must not lose sight of the fact that most employees are honest and disapprove of theft. Mutual respect between security personnel and other installation employees must be maintained if the facility is to be protected from other more dangerous forms of human hazards. Any security measure that infringes on the human rights or dignity of others will jeopardize rather than enhance the overall protection of the installation.

CONTROL MEASURES FOR SYSTEMATIC PILFERAGE

B-253. Unlike the casual pilferer, the systematic thief is not discouraged by psychological controls. Nothing short of active physical-security measures are effective in eliminating loss from this source. Some of these measures include—

- Establishing security surveillance at all exits from the installation.
- Establishing an effective package- and material-control system.
- Locating parking areas for POVs outside of the activity's perimeter fencing.
- Eliminating potential thieves during the hiring procedure by careful screening and observation.
- Investigating all losses quickly and efficiently.
- Establishing an effective key-control system.
- Establishing adequate security patrols to check buildings, grounds, perimeters, and likely locations for clandestine storage of property removed from its proper location.
- Installing mechanical and electrical intrusion-detection devices where applicable and practical.
- Coordinating with supply personnel to establish customer ID to authenticate supply-release documents at warehouses and exit gates.
- Establishing appropriate perimeter fencing, lighting, and parking facilities and effective pedestrian, railway, and vehicle gate-security control.

HOW TO STOP EMPLOYEE THEFT

B-254. No matter what it is called—internal theft, peculation, embezzlement, pilferage, inventory shrinkage, stealing, or defalcation—thefts committed by employees are behind at least 60 percent of crime-related losses. So many employees are stealing so much that employee theft is the most critical crime problem facing businesses today.

B-255. Although employee theft results in part from factors beyond control, the extent of employee theft in any business is a reflection of its management—the more mismanagement, the more theft. An effective stop-employee-theft policy must include at least the following:

- Pre-employment screening.
- Analysis of opportunities for theft.
- Analysis of how employees steal.
- Management-employee communication.
- Prosecution of employees caught stealing.

B-256. Each employer must reduce losses as much as possible. A police state need not be created. Large monetary expenditures need not be made.

B-257. The best way to stop employee theft is to simply not hire those employees inclined to steal. The best way is also impossible. The employer must set up a screening process that will weed out obvious security risks. Many experts believe that personnel screening is the most vital safeguard against internal theft. The following are some basic guidelines for the employer:

- Have the applicant fill out a written application. Ensure that the written application does not discriminate and that it conforms to any applicable laws.
- Solicit references, but keep in mind that those contacted will give favorable opinions. Ask primary references for secondary references. In contacting the latter, make it clear that the applicant did not refer them.
- Interview. During an interview, assess the applicant's maturity and values. Observe his gestures.
- Use psychological deterrents. Inform the applicant that your business routinely runs a background security check or that fingerprints will be taken. The hope is that the dishonest applicant will not be back.
- Obtain credit-bureau reports, but only after following guidelines set forth in the Fair Credit Reporting Act.

Opportunities, Methods, and Control

B-258. Cases of employee theft have been documented in almost every conceivable phase of business operations, from theft of petty cash to theft of railroad cars. An infinite variety of methods have been used. Some of the areas that are most vulnerable are—

- Shipping and receiving.
- Inventory.
- Accounting and record keeping.
- Cash, check, and credit transactions.
- Accounts payable.
- Payroll.
- Facility storage units.

B-259. Some of the methods used include—

- Pilfering (one item at a time).
- Stealing from cash registers or altering cash-register records.
- Issuing false refunds.
- Using the back door or trash containers.
- Taking advantage as a supervisor.
- Avoiding package controls.
- Embezzling.
- Forging checks.
- Stealing credit cards.
- Manipulating computers and stealing computer time.
- Conspiring with night cleaning crews.
- Duplicating keys or using a master key that is not properly controlled.
- Conspiring with outsiders (such as inflating insurance claims).

B-260. Too many opportunities exist for employees to exploit. Reduce these opportunities and the losses will be reduced. Reduce opportunities by using the following controls:

- Perform random spot checks on all phases of business in addition to regular, comprehensive audits.
- Check the payroll. Make sure that you are not paying a fictitious or dead employee.
- Take physical inventory seriously.
- Know what you own. Be able to identify it.
- Do not allow one employee to perform all functions. Separate receiving, purchasing, and accounts payable. Separate accountants from cash.
- Control payment authorizations.
- Keep blank checks secured. Do not presign or use uncoded, unnumbered checks.
- Reconcile cancelled checks with original invoices or vouchers.
- Secure exits. Restrict employees to one exit, preventing exit from the rear of buildings.
- Establish strict package control.
- Inspect cash-register receipts daily. Inspect the tape and ensure that the employee is identified on the slip. Deposit money daily.
- Issue ID badges to decrease employee presence in unauthorized areas.
- Simplify red tape; make it harder for the employee to disguise theft.
- Locate employee parking away from the business area.
- Establish a usage schedule of supplies to isolate irregularities.

Management-Employee Communication

B-261. Leadership must be firm, yet reasonable. Most employees pattern their values after their leaders, so a good example must be set. If you expect employees to remain honest, do not take home office supplies or goods.

B-262. Train new employees, advising them of the company's values and the standards by which they will be expected to perform. Explain all security procedures, stressing their importance. Emphasize that any deviations will be thoroughly investigated.

B-263. Establish grievance procedures; give your employees an outlet for disagreement and be receptive to all grievances submitted. Ensure that employees are aware of grievance procedures and that no reprisals are taken.

B-264. Regularly evaluate employee performance and encourage employees to evaluate management. Unrealistic performance standards can lead either to desperation and anger (resulting in dishonesty) or to get-even attitudes. Regularly review salaries, wages, and benefits—do not force employees to steal from you.

B-265. Delegate responsibility. Unless decision making exists among lower and middle levels, there is a tendency for development of an "it's-us-against-them" attitude. Delegate accountability as well; no decision is valid if it is lost in a "pass-the-buck" routine.

SECTION IV — ARMY PROPERTY AT THE LOCAL LEVEL

B-266. Proper accountability by commanders and subordinate personnel cannot be overemphasized. To ensure accountability of property, commanders must establish, implement, and supervise an installation's, activity's, or organization's security program.

B-267. Weaknesses in security procedures at the installation, activity, or organizational levels involving military property create vulnerability supported by criminal activity. Criminal activity includes—

- Theft.
- Fraud.
- Property diversion.
- Property manipulation.

B-268. Commanders and subordinate personnel must conduct a risk analysis; identify military property and; in the interest of monetary value and mission accomplishment, design mandatory security measures for specific property.

B-269. Security doctrine (as outlined in this manual) should be used to the maximum extent in securing Army property vulnerable to theft, destruction, or manipulation. Certain categories of property (such as in Table B-2, page B-62) must be assessed for security vulnerability and protective treatment. Security protective measures addressing this military property should be documented in the installation's physical-security plan. If the security measures recommended in Table B-2 are implemented using established doctrine, they should eliminate or reduce the property's vulnerability. This will reduce the incidents of theft, pilferage, and manipulation at the installation.

MOTOR VEHICLES

B-270. Security of tactical vehicles should be based on a uniform and cost-effective approach. For example, to ensure that a tactical vehicle without a locking device is properly secured, install a clamp, chain, and locking device as illustrated in Figure B-7, page B-63. To install the security device properly while maintaining safety, refer to Technical Bulletin (TB) 9-2300-422-20. Army motor-vehicle security should also incorporate the use of the following:

- Key/lock security and accountability.
- Protective lighting.
- Fencing.
- Walking patrols, as appropriate.
- Frequent observation and visits by mobile patrols or unit personnel (such as the charge of quarters [CQ], the staff duty officer [SDO], and the staff duty noncommissioned officer [SDNCO]).

Table B-2. Recommended Security Measures

Property	Inventoried by—					Inspected by—							
	Security Plan	Hand Receipt/Property Book	Secured by IDS	PS Plan	Regulation	Unit Leaders	PS Officer	SDO	Maint Officer	Supply Officer	Dining-Facility Officer	CQ	SDNCO
Arms/ammunition	x	x	x	x	x	x	x	x	x	x			x
Small arms	x	x	x	x	x	x	x	x	x	x			x
Explosives	x	x	x	x	x	x	x	x	x	x			x
Communications/electronic equip	x	x	x	x	x	x	x	x	x	x			x
Hand tools, tool sets/kits, and shop equip	x	x	x	x	x	x	x		x	x			x
Subsistence items	x	x		x	x	x				x	x		x
Controlled substances, precious metals, and tax-free items	x	x	x	x	x	x	x	x		x			x
Accounts	x	x	x	x	x	x							
POL products	x	x		x	x	x	x	x	x	x		x	
Repair parts	x	x	x	x	x	x	x		x				
Aircraft	x	x	*	x	x	x	x	x	x				
Vehicles	x	x		x	x	x	x	x	x	x		x	x
Towed weapon systems/components	x	x	x	x	x	x	x	x	x				
Carriage-mounted weapon systems	x	x	x	x	x	x	x	x	x				x
Construction material	x		*	x		x	x			x			x
Special-issue clothing (CTA)	x	x	x	x	x	x	x	x			x		
Individual clothing and equipment	x	x	x	x	x	x		x		x			
Organizational equip/components	x	x	*	x	x	x	x	x	x	x		x	x
Compasses, binoculars, and flashlights	x	x		x	x	x	x			x			x
Medical-unique items	x	x	x	x	x	x	x	x		x			x
Housekeeping supplies and equipment	x	x	x	x	x	x				x		x	
Housing furniture	x	x		x	x	x				x		x	
Mess equipment	x	x	x	x	x	x	x			x	x		x
Office machines	x	x	x	x	x	x	x			x			x
Expendable/consumable supplies	x		x	x	**	x		x		x	x		x

*** Depends on facility availability and cost effectiveness**
**** Depends on local policy**

Figure B-7. Typical Clamp and Chain Installation

CONSUMER OUTLETS

B-271. The lack of initiative at the management level within operational consumer outlets does little to prevent or reduce pilferage. Such shortcomings are identified as—

- Failure to present a professional image.
  - Lack of continuing interest, motivation, and direction.
  - No alertness to internal control of pilferage.
- Failure to institute and implement methods of operational effectiveness and efficiency. These methods include—
  - A clearly defined delegation of responsibility.
  - Insistence on stringent accountability
  - Orientation and training programs for subordinate supervisors and current and new employees.
- Failure to emphasize and enforce established criteria for continual employment.
  - Rules of conduct.
  - Standards of job performance. (Officially request appropriate action for employees guilty of criminal acts or infractions conducive to criminal acts.)
  - Inattentive job attitudes of subordinate supervisors.
  - Inadequate personal checks of established accounting and inventory procedures. **NOTE: Checks on both a regular and unannounced basis tend to control access to official stock records and to ensure careful and organized storage or stocking of merchandise.**
  - Infrequent observation of an employee's job performance.

- Failure to report misconduct, criminal or otherwise, to superiors or responsible law-enforcement personnel in the activity.
- Failure to implement recommendations made during physical-security inspections or crime-prevention surveys.

PILFERAGE

B-272. Pilferage may be accomplished by individual employees, by a team of employees, or by employees and patrons in collusion. These actions can be greatly reduced by tightening supervision and security in the following areas:

- **Merchandise display or dispensing areas.** The following measures can reduce merchandise pilferage in display areas:
  - Detecting unauthorized price reductions.
  - Preventing or making it difficult to alter price tags.
  - Checking procedures for declaring merchandise old, shopworn, damaged, or salvage.
  - Providing more unpackaged items for personal consumption.
  - Discouraging the careless waste of foods and other perishable items.
- **Cash registers.** The following are methods of pilferage from cash registers:
  - Stealing directly from an unattended register.
  - Rerunning register tapes at lower figures (this is preventable if the reset key is maintained by the supervisor).
  - Clearing the register at a lower total figure than actual receipts for the operational period.
  - Reporting overrings and refunds falsely.
- **Theft of merchandise.** The following methods enable merchandise to be pilfered:
  - Underringing.
  - Reusing cash-register tapes. This occurs when employees fail to provide patrons with tapes or when patrons allow employees to retain tapes. The tapes allow employees to package merchandise and remove it from the premises.
  - Removing items from bags or containers carried out by employees.

SHOPLIFTING

B-273. Shoplifting is usually confined to sales areas and is committed by casual and systematic pilferers. Items that are most frequently pilfered—

- Are relatively small in size.
- Have a high degree of consumer desirability.
- Are easily carried in pocketbooks or secreted on the person.

B-274. Amateur, adult shoplifters share the following characteristics:

- Theft is from a sudden temptation (impulse theft). There is success in the initial theft, which leads to more temptation, stronger impulses, and more thefts.
- They rarely have a genuine need for the item.

- They usually have enough money to pay for the item.
- They display symptoms of nervousness and uneasiness.

B-275. Juvenile shoplifters have the following traits:

- They act on a dare or "to belong."
- They may be coached or directed by an adult.

B-276. Professional shoplifters share these characteristics. They—

- May be talkative and are usually polite and deliberate.
- Look for opportunities continually.
- Do not take many chances.
- Are very capable of spotting security personnel.
- Steal for resale.
- Have "fences" (usually).
- Steal "to order" (often). They may have a list describing the items to be pilfered.
- Use innovative techniques.

B-277. Genuine cases of kleptomania are rare. Kleptomaniacs share the following characteristics. They—

- Take items without regard to their value or use.
- Steal compulsively and often openly.
- Are nervous and shy.

B-278. Narcotic addicts as shoplifters have the following characteristics. Only MP/security personnel should attempt apprehension. These types of shoplifters—

- Are desperate for money and they fear imprisonment.
- Take big chances.
- Take the merchandise and exit the premises quickly.
- Steal when they are at their lowest physical and/or psychological ebb.
- Are dangerous if you try to apprehend them (sometimes violent).

B-279. Alcoholics and vagrants as shoplifters share the following traits. They—

- Usually steal because of need.
- Are often under the influence of liquor at the time of theft.
- Take the merchandise quickly, then exit the premises.
- Are less likely to steal regularly at a single location.

B-280. The most pilferage occurs when employee coverage is low or when employees are untrained, inexperienced, or indifferent to the issue. The ineffective use of floor space aids shoplifters by creating congestion in the patron's traffic flow. Allowing an emphasis on small rooms or partitioned areas causes congestion that clusters, isolates, or partially hides displays. Shoplifting involves the use of one or more of the following means to obtain items:

- Palming or placing an open hand on a small article, squeezing the muscles of the hand over the article to grasp it, and lifting the still open and apparently empty hand.
- Using fitting rooms to put on tight or close-fitting garments under clothing worn into the store.
- Trying on hats, gloves, sweaters, and jackets, then exiting the store.
- Stepping around counters and removing items from unlocked showcases.
- Handling several items at once and replacing all except the item pilfered.
- Using accomplices to create a diversion of employee attention when secreting items on the person. Such items include:
  - Clothing.
  - Pocketbooks or handbags.
  - Umbrellas.
  - Various items placed in packages or paper sacks containing merchandise paid for at other departments.

ARSON

B-281. On an installation there is little incentive for a professional arsonist to operate since the government owns the buildings and insurance fraud in collusion with the property owner is not possible. However, arson can still be a problem. In the civilian community, most deliberate fires are not set for profit. They are set to "get even" with the property owner or just for the excitement of watching something burn. Military installations are susceptible to this type of crime. On many Army posts, there is a large number of empty, wooden structures that are ideal targets for revenge seekers or vandals.

B-282. Arson is an easy crime to perpetrate, and it is relatively difficult to collect the information needed to convict an arsonist. However, this does not mean we are helpless in combating this crime. Several major cities, including Seattle, Denver, Houston, and Philadelphia, have developed successful programs to reduce the number of arsons. Some of the successful, proactive measures that have been developed are—

- Securing or disposing of materials that could be used to start fires. Ensuring that the regulations on the storage of gasoline, paint, and solvents are enforced and that paint lockers are locked.
- Enforcing command policies on the police of the post. Removing piles of trash, scrap lumber, and other material that burns easily.
- Securing empty buildings (especially empty wooden structures) and posting them as off-limits areas.
- Patrolling areas susceptible to arson.
- Encouraging participation in neighborhood watches, taxi patrols, and other community programs that increase surveillance.
- Encouraging the reporting of suspicious activities through the establishment of crime hot lines.

- Establishing a close working relationship between fire fighters and law-enforcement personnel to ensure that fires of suspicious origin are reported and thoroughly investigated.
- Offering rewards for information leading to the apprehension of the arsonist (if there is an outbreak of arson).

SECTION V — COMMUNITY CRIME-PREVENTION PROGRAMS

B-283. There are a number of crime-prevention programs set up in communities to help deter and detect crime. These programs are supported and often headed by members of the community in conjunction with local law-enforcement agencies. The programs are as diverse in nature as they are in number, yet they complement one another.

NEIGHBORHOOD WATCH PROGRAM

B-284. A neighborhood watch program is an organized network of citizens interacting with other neighbors and the police in preventing and detecting crime in their neighborhood. Law-enforcement efforts to reduce crime cannot be accomplished effectively without the support and cooperation of all citizens. A strong community involvement with neighbors helping themselves and other neighbors in becoming more alert to activities in the neighborhood, protecting their property, and reporting suspicious activities is essential to an effective crime-prevention program.

B-285. The Army Neighborhood Watch Program is designed to encourage Army service members and their families to actively participate in protecting their own property and the property of their neighbors, joining community crime-prevention programs, and reporting suspicious activities to MP officers. The program is designed to develop the following:

- The awareness of community crime trends and prevention efforts.
- The knowledge of quarters' security procedures.
- A cooperative system of surveillance over each neighbor's property.
- Accurate observation and reporting of suspicious activities.
- The establishment of a reliable, two-way information link between the community and MP forces.

B-286. Most neighbors know the routines of the other families that live near them. They know what cars are normally parked in the neighborhood and when families are on vacation or out of the area. Neighbors are in a very good position to recognize burglars and other intruders. Also, residents are in a good position to recognize safety hazards and crime-conducive conditions near their homes.

B-287. To capitalize on these advantages, neighborhood watch programs organize blocks in family-housing areas or floors in troop billets to improve police and community interaction. They also disseminate information on crime problems and countermeasures.

BLOCK CLUBS

B-288. Block clubs are the basic components of an installation-wide neighborhood watch program. The geographical size of a block club may vary widely depending on the population's density and the nature of the terrain. The key factor is that the terrain organized into a single block club should promote a feeling of unity and mutual assistance. In a troop billet area, block clubs could be organized along company lines, by individual barracks, or by floor. In a family-housing area, a block club could cover one high-rise apartment building, one block, or one or more streets that are so situated that the residents identify themselves as a subcommunity within the housing area.

Organization

B-289. Individual residents, community-service organizations, or MP units can initiate block clubs. Existing organizations (such as the PTO) have frequently established contacts within the community and have sponsored the organization of block clubs. Regardless of the approach used to organize the installation, every family or resident in the area should be contacted and encouraged to attend a block club (see Figure B-8).

B-290. A block captain and a deputy block captain should be elected at the initial block meeting. These individuals serve as "spark plugs" to sustain interest in their geographic area. They also represent their block club at district meetings. The effectiveness of the presentation at the initial block-club meeting is critical. A PMO representative should explain the crime problems on the installation and clearly outline the functions of a block club and how it can affect the crime problem. The main objective should be to generate enthusiasm and build a foundation upon which an effective neighborhood watch can be built. The use of camcorders or other audiovisual programs can generate interest. Association with programs in the civilian community may also help to generate interest.

B-291. The functions of a block club should be clearly identified. Block clubs—

- Serve as the "eyes and ears" of the police.
- Encourage the implementation of individual countermeasures, such as marking personal property.
- Disseminate crime-prevention information.
- Improve police and community relations

Block-Club Districts

B-292. A block-club district should be organized as an intermediate link between individual block clubs and the installation's crime-prevention council. These districts should cover at least one housing area and should be composed of the block captains and deputies from each block club in the district's area. District leaders should be elected from the block captains and should serve as members of the installation's crime-prevention council. The primary functions of districts are to transmit information from the crime-prevention council to the block clubs and to develop incentive awards to recognize effective participation by the clubs.

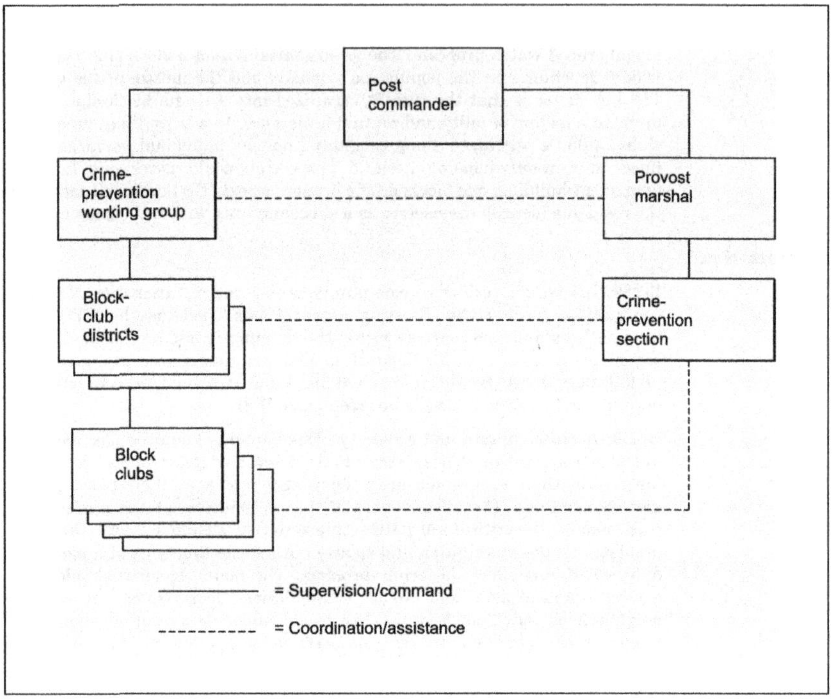

Figure B-8. Block-Club Organization

Maintaining Interest

B-293. Maintaining interest in neighborhood watch programs is a major problem. Typically, a particular incident generates the level of interest required to initially organize a system of block clubs and districts; but as the particular problem is overcome, interest wanes and the clubs gradually dissipate. A review of programs that have maintained their effectiveness for extended periods indicates that successful programs have the following characteristics:

- They are a formal organization with elected block captains and district leaders.
- They address a wide range of problems.
- They feature an incentive award program to recognize individuals and clubs that participate effectively.

- Their members attend periodic workshops to train leaders on various aspects of crime prevention.
- They have clear-cut, attainable objectives. The members need to see accomplishments.
- They are actively supported by the local police.

Block-Club Operations

B-294. Block clubs may become umbrella organizations that conduct the entire range of community crime-prevention programs, including foot and mobile patrols. However, each of these programs will be discussed in a separate section. In their most basic form, block clubs disseminate information on residential security and report suspicious activity in their neighborhood. Members of a neighborhood watch program should exchange names, addresses, and telephone numbers to enhance communication among neighbors. They must learn how to record and report suspicious activities. Some activities that block-club members should be alert for and, when observed, report to the MP officers are—

- A stranger entering a neighbor's house when it is unoccupied.
- Someone screaming.
- An offer of merchandise at a ridiculously low price (it could be stolen).
- Persons entering or leaving a place of business after duty hours.
- The sound of breaking glass or an explosion.
- A person going door to door and then going into a back or side yard, or a person trying a door to see if it is locked.
- A person loitering around a school, a park, a secluded area, or in the neighborhood.
- A person carrying property at an unusual hour or in an unusual place.
- A person exhibiting unusual mental or physical symptoms (he may be in need of medical help).
- A vehicle being loaded with valuables when parked by a closed business or untended residence.
- A business transaction conducted from a vehicle.

ARMY NEIGHBORHOOD WATCH MEETINGS

B-295. A PMO representative should attend the initial neighborhood block meeting to explain the relationships of crime prevention and the neighborhood watch programs. The meeting should be publicized with handout invitations announcing the time, location, and purpose of the meeting. If possible, the meeting should be held in the neighborhood (such as the training room in the troop billets, a local school, or a residence in family quarters). In addition to explaining the program concepts, the initial meeting could include the following:

- Installation and community criminal statistics concerning the nature and volume of housebreakings, larcenies, and other crimes.
- An exchange of names and telephone numbers (if applicable) of attendees. This information should also be placed on a neighborhood block sheet that is a geographical diagram of the block showing the

location of each room, apartment, and building address number in the block. The names and phone numbers of participants would be added to the address number of each residence drawn on the diagram. This block sheet should be distributed to block members at a subsequent meeting.

- Determination of the second meeting date and location.

B-296. Other agenda items and activities for subsequent meetings could include the following discussions:

- Residential security procedures and the conduct of quarters security inspections.
- Operation ID—marking and recording of personal property.
- The observation and report of suspicious activities.
- Other crime-prevention measures and procedures being implemented at the installation (such as rape prevention and citizen escort patrols).
- Fire prevention, personal safety, and other related activities.

OPERATION ID

B-297. Operation ID, which entails marking of property to make it identifiable and traceable to its owner if lost or stolen, was initiated by the Monterey Park, California, Police Department in 1963 and has been adopted by more than 80 percent of the police departments in the US. Operation ID is a low-cost, highly effective crime-prevention program. However, its success is contingent upon the willingness of individuals and communities to actively participate in marking and identifying their personal property.

B-298. Operation ID is designed to encourage Army service members and their families to mark their personal property with a standard Army-wide, owner-applied number. This numbering system permits the positive ID of the property and determines the location of the owner in case of theft or loss.

B-299. The principal advantages of Operation ID are theft deterrence and recovery of personal property. Marked stolen property is more difficult to dispose of, and illegal possession can result in the prosecution of a thief. Recovered lost or stolen property can only be returned if there is some means of identifying and locating the rightful owner.

METHODS OF IDENTIFYING PROPERTY

B-300. Various methods of establishing positive ID and ownership of property in case of loss are available for individuals. Each method has advantages and limitations, and a combination of these methods would be required to ensure the ID of all high-value personal property.

- Inscribing the owner's applied number with an etching or engraving tool would allow the recovering agency to visually identify the number inscribed on the property for notification and subsequent return of the property. Electrostatic markers are available for use at no cost to the individual. However, some personnel are reluctant to use this method since it can mar the property. The inscription should be made in a location that can be readily seen by the recovering agency but which would not deface the property's appearance or reduce its value. Some high-value personal items such as coins, jewelry, and silver cannot be inscribed with an owner's applied number. Another method of identifying these items would be required.

- Using invisible fluorescent ink, powder, or paste to mark the property can make it easier for the agency recovering the property to use an ultraviolet light and identify the owner. This marking will not mar the property. However, the fluorescent markings and ultraviolet light are an additional cost, and many agencies do not have ultraviolet lights to inspect recovered property.
- Using a laser photographic process to identify diamonds. Every diamond emits a unique reflection when penetrated by a low-level laser light. A laser photographic process has been developed to record a diamond's pattern of light reflection on film. Several jewelers throughout the country have the laser photographic equipment available. To register a diamond, two photographs are taken. One is provided to the owner and the other to a central registry. If a diamond is lost or stolen, the recovering agency can take the diamond to a jeweler that maintains the laser photographic equipment for print. This photograph would then be forwarded to the registry for owner ID.
- Photographing the personal item. Individuals can photograph personal high-value items that cannot be engraved. Although the agency recovering the lost or stolen items could not identify the owner, use of photographs could assist in verifying ownership if it is known that the item has been recovered. In addition, a photograph would assist in submitting claims against the government or private insurance companies, as appropriate.

RECORDING PERSONAL PROPERTY

B-301. Individuals should record identifying data (such as brand name, model, serial number, and value of the personal items), even if they use other methods of identifying property. This information would assist in determining what items may be lost, stolen, or damaged through fire, explosion, or other hazards. This information can also be used in claims against the government or private insurance companies, as appropriate.

IDENTIFICATION NUMBERING

B-302. There are various types of owner-applied numbering systems used to mark personal high-value items. Criteria that should be considered in determining which owner-applied numbering system to incorporate include—

- Uniqueness, where no two people have the same identifier.
- Permanence, so that the owner-applied number will not change.
- Ubiquity, so that an identifying number is available to any individual who desires one.
- Availability, where the identifying number can be easily obtained and remembered.
- Indispensable, so that there are incentives requiring an individual to have the number.

- Privacy, so that the number is not a means of infringing upon an individual's right to privacy.
- Uniformity, so that the owner-applied number would be readily recognized by law-enforcement agencies who handle or come into contact with the recovered property.
- Traceability, so that the property owner can be identified and located.

B-303. The most commonly used owner-applied numbering systems include—

- Driver's license number with the issuing state abbreviation prefix.
- Social security number.
- Personal numbers assigned to individuals by a local law-enforcement agency.
- Personal numbers with the marking agency's National Crime Information Center originating-agency ID number.
- A private numbering system maintained by a commercial organization.

STANDARD ARMY NUMBERING SYSTEM

B-304. To ensure that criteria for property ID numbering are met, the standard Army-wide, owner-applied numbering system is designated as the service member's social security number with a "USA" prefix. Upon recovery of lost or stolen property by other military or civilian law-enforcement agencies, the USA prefix, owner-applied number would alert the recovering agency that the property belongs to a member of the Army. The recovering agency can then contact the nearest Army installation's PMO/security office concerning the property and the owner-applied number inscribed. The service number can then be identified and located through the Army worldwide locator system. Since this locator system lists only social security numbers and locations of active Army service members, family members should also use the service member's social security number with the USA prefix when marking their personal items.

IDENTIFYING AND LOCATING OWNERS OF RECOVERED PROPERTY

B-305. The installation's PMO/security officer should be the initiator of tracer actions to identify and locate the owner of recovered property marked by the standard Army Operation ID numbering system. Continuous liaison should be maintained with local civilian law-enforcement agencies and other military installations to ensure that they are cognizant of the standard Army owner-applied numbering system and will contact the PMO/security office upon recovery of private property marked with this system. The PMO/security officer should readily accept custody of the private property if the recovering agency is willing to release the property and it is not required as evidence for criminal prosecution.

B-306. Upon notification of recovered property and the inscribed Army standard owner-applied number, the PMO/security officer should contact the servicing military personnel office (MILPO) for assistance in determining the name and location of the property owner by using the Army's worldwide locator microfiche.

B-307. If the owner-applied social security number is not listed on the Army's worldwide locator microfiche and the servicing MILPO is unable to provide the requested information, the PMO/security office should contact the following offices:

- For enlisted members: USA Enlisted Records and Evaluation Center, ATTN: PCRE-RF-L, Fort Benjamin Harrison, Indiana 46249.
- For officers: PERSCOM, ATTN: TAPC-MSR-S, 200 Stovall Street, Alexandria, Virginia 22332-0444.
- For warrant officers: PERSCOM, ATTN: TAPC-OPW, 200 Stovall Street, Alexandria, Virginia 22332-0444.

B-308. In case the owner-applied number cannot be identified by either the servicing MILPO or the USA Enlisted Records and Evaluation Center, the Army Reserve Personnel Command (AR-PERSCOM) should be contacted for assistance. The AR-PERSCOM retains files on separated, retired, and Reserve component Army members. Written requests should be forwarded to: Commander, AR-PERSCOM, ATTN: ARTC-PS, 1 Reserve Way, St. Louis, Missouri 63132.

B-309. When the name and location of the service member associated with the owner-applied number inscribed on the property is verified, the PMO/security office should notify the service member in writing that the property has been recovered. The notification should ascertain if the recovered property belongs to the service member, if the service member ever reported the property as lost or stolen, and if a claim was submitted to the SJA claims service for loss or theft of the property. Notification should also state where the property is located and a point of contact that the owner can deal with for the return of the property. If the property is to be retained as evidence for legal proceedings, the owner should be informed that the property will be returned upon completion of the proceeding. A copy of this letter should be provided to the US Army Claims Service, Fort Meade, Maryland 20755. The Army Claims Service will advise the PMO/security office if a claim was or was not submitted.

B-310. If the owner cannot be located, recovered property in the custody of the PM/security offices should be disposed of according to AR 190-22 and DOD 4160.21-M.

USE OF THE NATIONAL CRIME INFORMATION CENTER

B-311. Stolen articles may be entered into the National Crime Information Center (NCIC) file if a theft report has been made, if the item is valued at $500 or more, and if it has a unique manufacturer's assigned serial number and/or owner-applied number. Entering stolen personal property items meeting the above criteria into the NCIC or other police information systems as outlined in AR 190-27 is encouraged.

NEIGHBORHOOD WALKS

B-312. While most people are unwilling to participate more actively than as observers in a neighborhood watch, there are some individuals who want to become more actively involved in securing their neighborhoods. For this segment of the population, the organization of "neighborhood walks" provides a welcome opportunity to make a more active contribution. The basic idea is simple; residents patrol on foot through their own neighborhoods to observe and report crime. In practice, it is a little more complicated; however, neighborhood walks can have a dramatic impact on the crime rate, so the effort expended is worthwhile. The points that must be considered are—

- **Patrol composition.** Both adults and teenage children can volunteer to participate in neighborhood walks. However, there should be at least one adult in each party. There should be at least two people in each patrol group. Groups of four to six individuals are desirable since, by their numbers alone, they discourage attacks on the walkers. Larger groups are also more fun, and this is important when volunteers are providing the manpower.
- **Times/patrol duration.** As with other crime-prevention programs, maintaining a high level of interest can be a problem. A successful walk program in Philadelphia schedules groups for one 2-hour patrol per month. More frequent tours caused high drop-out rates among participants. Neighborhood walks should be conducted only during those times when the crime rate is the highest. Normally, there are not enough volunteers to conduct walks at times other than peak crime periods.
- **Functions.** Members of neighborhood-walk groups must understand that they are to observe only and not actively intervene in criminal acts. Participants and the government are legally liable for their actions during walks. When a crime or a suspicious activity is spotted, the neighborhood walkers should report it to the police. In Philadelphia, the walkers are equipped with horns. When a crime is spotted, they activate their horn and go to the nearest house to call the police. When residents hear the walker's horn, those in the immediate area turn on all of their lights and sound their own horns. The noise and increased lighting invariably causes the criminal to flee.
- **Neighborhood escorts.** In addition to observing and reporting criminal activity, neighborhood patrols can escort children and older persons between community-service facilities and residences. They can also request that owners secure property when they find it unsecured; for example, when there are unsecured bicycles parked on a front lawn.

VIGILANTISM

B-313. While active community participation is essential, vigilantism must be discouraged at all costs. Both formal law codes and US common law offer few protections for private citizens who take the law into their own hands. Block captains and installation crime-prevention officers must be alert for indications that neighborhood patrols are doing more than observing and take swift remedial action when required. Of course, all nonpolice participants must be prohibited from carrying weapons of any type while engaged in crime-prevention programs. Experience in neighborhoods having much higher violent crime rates than found on Army installations has demonstrated that passive devices like horns or whistles were adequate to discourage attacks.

These devices, plus the assignment of four to six individuals to a neighborhood patrol, provide sufficient protection.

MOBILE PATROLS

B-314. Some communities with high street-crime rates have been successful in organizing private citizens into mobile-patrol programs. Like neighborhood foot patrols, these mobile patrols serve as the "eyes and ears" of the police but do not actively intervene when they spot a crime in progress.

B-315. In a typical program, block captains or police crime-prevention officers assign specific patrol areas to each private mobile patrol. In addition, each patrol receives training on the patrol's functions, communications procedures, and emergency actions. Normally, patrols are instructed to blow their auto horns steadily when they observe a crime in progress. This is usually sufficient to drive off the criminal.

B-316. Most private citizen patrols use cellular telephones or citizen's band (CB) radios as direct communications links with the supporting police department. Installation CB radio clubs are often willing to sponsor anticrime patrols under police supervision. Commercial taxi companies that operate on military installations are also excellent candidates to organize into patrols. The cab drivers normally cover most of the high-crime-rate areas on the installation. Because of the frequency with which they cover them, they are familiar with the routine conditions in each area and are quick to spot suspicious activity.

B-317. The government cannot provide gasoline for POVs used for anticrime patrols; however, it may often provide magnetic signs to affix to the vehicle for identifying it as part of the police-sponsored, neighborhood-patrol program. As in the case of foot patrols, the installation crime-prevention officer must be alert for signs of vigilantism and must take positive action to discourage it if it appears.

PROJECT LOCK

B-318. Nearly one million automobiles are stolen in the US every year. The total value of cars stolen is around the billion dollar mark, making auto theft the nation's costliest crime involving property. Of even greater importance is the social impact of auto theft. For a growing number of young people each year, stealing cars represents the first step toward a life of crime.

B-319. Police agencies have been diligent in the apprehension of auto thieves and the recovery of stolen vehicles, but the auto-theft problem seems to be more amenable to improvement through prevention rather than punishment. Barring strict security, auto theft is one of the easier crimes to commit. All vehicles left unattended are vulnerable, and widespread prevention by police surveillance is a physical impossibility.

B-320. The problem has grown to serious proportions despite determined law enforcement because motorists continue to be negligent or unaware of their responsibility. As long as people invite theft by leaving their cars unlocked or leaving the key in the ignition, auto thefts will continue to climb. Almost half of all stolen cars each year had been left with keys in the ignition; nine out of ten of the stolen vehicles had been left unlocked.

B-321. If a significant reduction is to be made, the motorists themselves must make it. Widespread adoption of accepted and effective prevention practices by motorists presents the most logical and immediate improvement to this growing problem.

B-322. In 1963, the Boston police department conducted a broad information campaign with the assistance of the National Automobile Theft Bureau and the Insurance Information Institute. Since then, more than 525 "Lock Your Car" campaigns have been held in about 400 communities in 49 states.

B-323. In the months following these campaigns in such cities as Denver, Chicago, Atlanta, and San Francisco, significant reductions in the number of auto thefts (ranging from 9 to 54 percent) have been recorded. Undoubtedly, the "Lock Your Car" campaigns contributed to reducing auto-theft statistics.

B-324. Project Lock is designed to permit sponsoring groups to conduct one-day or one-week "Lock Your Car" campaigns. Its purpose is twofold; it—

- Alerts the public on the importance of locking cars and removing keys as a deterrent to auto theft.
- Contributes to the welfare of youths by preventing the commission of a first crime.

B-325. The following materials can be ordered to support Project Lock:

- Windshield flyers.
- Identifying insignia to be worn by inspectors.
- Tally cards for noting cars left with keys in ignitions or unlocked doors.

B-326. Groups considering sponsorship of this campaign should consider installation areas known to have car-theft problems. The PMO should assign patrolmen to accompany the teams on inspection day. Usually, the teams' routes can be arranged to fit the regular patrols of the MP officers. If uniformed police will not be available, the campaign should not be held.

COMMUNITY ORIENTATION

B-327. If the commander desires, a community orientation meeting might be held a month or so in advance of the campaign. This would be more desirable for a week-long rather than a day-long campaign. If such a meeting is planned, the PMO should issue invitations to representatives of the installation crime-prevention council, service clubs, women's clubs, PTOs, high schools, and churches.

B-328. The meeting should be opened with an introduction of the installation commander. After appropriate comments, the commander would read a proclamation setting the date for Lock-Your-Car Week. If possible, a representative of the National Automobile Theft Bureau should be asked to address the meeting. An alternative would be to have the PM review national and local trends in auto thefts stressing the importance of the forthcoming campaign. In conclusion, a representative of the sponsoring group might review the schedule of activities for the campaign.

PROGRAM ACTIVITIES

B-329. If the campaign is conducted in one day, a kick-off breakfast might be substituted for the orientation meeting. The program and the attendance might be similar if a weeklong campaign is planned and the orientation meeting were held. The kick-off breakfast might have a simple program with attendance limited to the sponsoring group and police representatives. An alternative to the breakfast would be a luncheon at which results to the moment are reported. Whether the campaign will last a week or a day, the general activities will be similar.

B-330. With the assistance of the MP officers, the installation should be zoned according to the established MP patrol areas where possible. The size and number of the zones will depend on the number of police and volunteer personnel that will be available. However, the zones should cover most of the installation's service areas, troop billet areas, and family-housing areas.

B-331. The inspection teams are each composed of a uniformed patrolman and three to five sponsoring group members, cover their assigned zones and place flyers under the windshield wipers of all cars found to have keys in the ignition or to be unlocked. Under no circumstances should the flyers be placed inside the cars, even through open windows.

B-332. It is suggested that flyers be ordered early enough to allow a local printer to inscribe an overleaf statement such as "This public service is provided as a courtesy of the MP force." The tally cards should be used to record the number of cars inspected, the number that were unlocked, and the number with keys in the ignition.

SCHOOL PROGRAM

B-333. The crime-prevention office might sponsor a poster contest for art students. It should be announced at least a month before the campaign to allow the entries to be placed in public in advance of the date. The PMO should present awards in an office ceremony. The ceremony should take place at noon on the campaign day or midweek if the campaign is longer.

PUBLIC ADDRESSES

B-334. Close to or during the campaign date, addresses by crime-prevention professionals should be scheduled for programs of service clubs, women's clubs, PTOs, and other civic groups. Well in advance of the campaign, the PAO should be visited by the PM and the other sponsors to develop comprehensive internal-information programs in support of the installation's Project Lock. Plans for the public-affairs program should include advance publicity for the campaign, coverage of events during the campaign, and wrap-up coverage following its completion. During the inspection days, findings on how many cars were unlocked and how many had keys in the ignition should be reported regularly to a headquarters (preferably to the MP unit).

CONCLUSION

B-335. Project Lock has been outlined to provide basic suggestions for a "Lock Your Car" day or week. No procedure can be designed to fit all needs or circumstances, and variations often will be desirable. However, if the suggestions and the materials contained in this section are used, Project Lock will not be difficult to organize and conduct. It provides better publicity opportunities than most public-service projects. It presents an opportunity for active participation by a number of crime-prevention groups. It has been field-tested and found to be an outstanding success.

SECTION VI — EVALUATION

B-336. Evaluations of installation crime-prevention programs (see Figure B-9, page B-80) do not rely on control groups, a tight control of variables, or elaborate statistical analyses to produce worthwhile results. Most often, the resources to conduct a formal evaluation that will stand up to rigorous academic scrutiny are not available.

CRIME-PREVENTION PROGRAMS

B-337. Several inherent difficulties in data collection on crime-prevention programs make it difficult to determine with 100 percent accuracy that a particular reduction in the crime rate was a result of a particular crime-prevention measure (unless a very elaborate analysis and control system are used). These difficulties include the control of variables, the displacement effect, and unreported crime.

CONTROL OF VARIABLES

B-338. In its simplest form, the type of evaluation most commonly used in academic or scientific settings seeks to determine the relationship between two variables. By varying the independent variable (for example, the dosage of a drug), the effect on the dependent variable (for example, a pulse rate) is determined while all other variables (such as food intake) are held constant.

B-339. Although more complex in form, the same model can be used to evaluate complex programs. Experimental and control groups can be selected. The "treatment" can be administered by researchers or those taught by researchers. The results can be analyzed for their statistical significance. However, since crime-prevention programs deal with human subjects, certain complications arise. The degree of success may have nothing to do with the efficacy of the program, but only with the way it was introduced or with the personal predilections of the groups involved. There is no standard population; human beings are not standardized as mice are for laboratory purposes. A program found successful at one installation may be a failure in another.

B-340. These considerations also apply in the evaluation of crime-control programs. This evaluation is further complicated by another problem—the people whose behavior is to be modified (the offenders) cannot be treated directly or separated into experimental and control groups; they will not stand up and be counted. Although public-health programs often encounter this problem, they often deal with physical cause-and-effect links between treatment and improvement. The same is not true for crime-control programs.

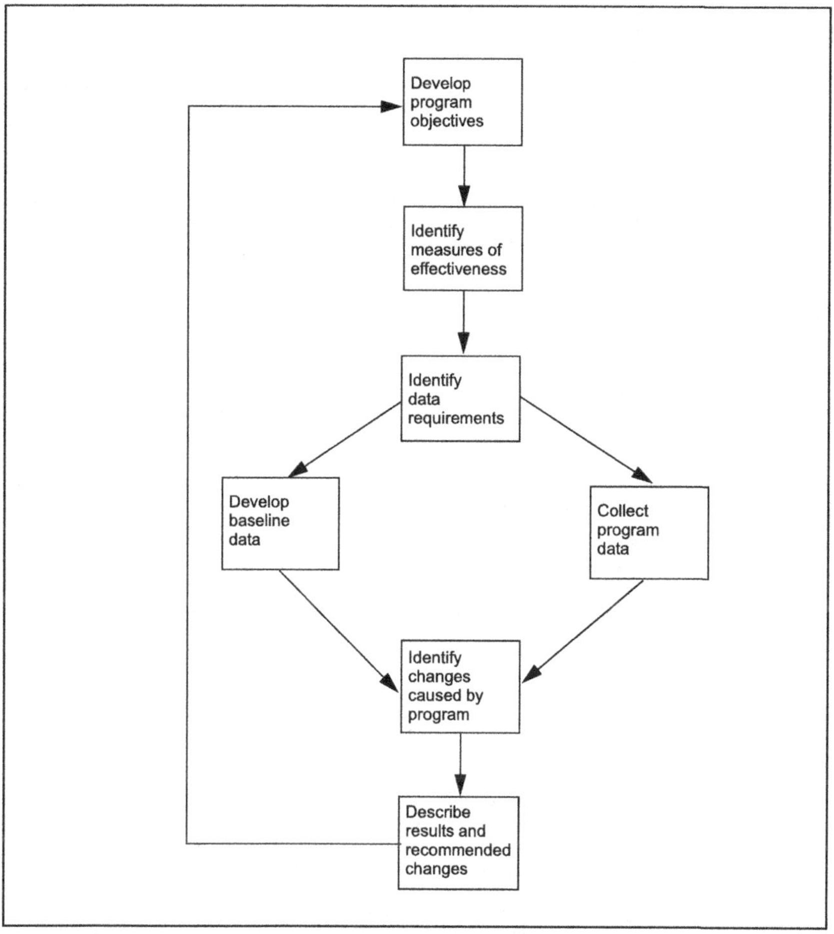

Figure B-9. Crime-Prevention Program Evaluation

The effectiveness of these programs is normally determined by looking at statistics of reported crime and arrests, which are more indirect indicators.

B-341. In a crime-control program, it may be impossible to classify variables as dependent and independent; they may all affect and be affected by each other. Furthermore, because of the difficulty in determining why people behave the way they do, a number of intervening and antecedent variables may go unnoticed. Police programs designed to reduce crime may have their most direct effect on the behavior of the general public toward the police, which in turn affects the crime rate.

B-342. Evaluations are not necessarily restricted to the analysis of objective crime data; they can also include subjective considerations and perceptions. These subjective evaluations can be of significant benefit in augmenting the statistical analyses of the program's results. They are especially helpful in assessing why and how a program worked and whether a statistical outcome is actually evidence that the program was successful. Interviews of participating agency personnel and residents of the program's area are usually used to supply this information. They can give the evaluator new insight into the actual program operation.

DISPLACEMENT EFFECT

B-343. In many cases where crime reductions have been measured and attributed to programs, it is unclear whether there has been an actual reduction in crime or whether the crime has been displaced. The amount of displacement depends to an extent on the offender's characteristics. An opportunistic offender can be pictured as having a relatively elastic demand—if the risk is too high, he will forgo the crime. An addict offender is typically pictured as having a relatively inelastic demand for the product because of his inelastic demand for drugs—despite the risks, he needs the product.

B-344. Deterrents may have little effect on perpetrators of expressive crimes. These are crimes in which the perpetrator is emotionally involved and is expressing these emotions. Most assaults and homicides fit this category. On the other hand, deterrents may have a strong effect on instrumental crimes, those that are seen by the offender only as a means to an end (usually money). If alternative avenues to the same end are made more attractive by comparison, the offender may well be deterred. Deterrence may produce a diversion to legal alternatives to crime; it also may cause displacement to illegal alternatives.

Displacement to Other Crimes

B-345. There is no immutable law that says a burglar cannot hold up a liquor store and a robber cannot burglarize a warehouse. If a specific crime or a set of crimes is the target of a crime-control program, offenders may decide to avoid the target crimes and ply their trade in other ways. Some offenders will be deterred from all crime if their crime specialty is the object of a crime-control program, but the extent of this deterrence should not be overestimated. The statutory categories of crime should not be confused with categories that serve to classify offenders.

B-346. In some cases, the result of displacing offenders to other crimes is beneficial. If the targeted crimes are more serious than the ones to which offenders are diverted, the net effect on the program may be the reduced danger to society. Of course, the converse may also be true; closing off the vulnerable and more easily protected targets of crime may cause an offender to commit more serious crimes with a net increase in the danger to society.

B-347. In some instances, the individual effect may be substantial but the overall effect may be negligible. Protecting a small fraction of premises against burglary will reduce the number of crimes committed against them, but the burglary rate against unprotected premises may go up.

Displacement to Other Tactics and Targets

B-348. Offenders can change their manner of committing a crime when a new program is established to counter their activity. One example of this took place in 1969 in a section of the Bronx that was having a rapid increase in outdoor crime. The crimes took place primarily in the evening hours when people were returning from work. The program instituted by the police consisted of intensive sweeps of randomly selected city blocks, coupled with plainclothes police officers patrolling the streets. It succeeded in reducing the number of offenses committed during the evening hours, but at the expense of increasing the number taking place in the late afternoon when patrolmen were taking their lunch hours or were occupied with school crossings or shift changes.

Displacement to Other Areas

B-349. The most frequently discussed type of crime displacement is from one area to another. For instance, it has been cynically suggested that the goal of the New York subway police is to chase crime into the streets where it belongs. More seriously, some recent police-helicopter-program evaluations have been questioned because they did not consider possible area displacements.

B-350. One type of boundary of interest to crime displacement is the jurisdictional boundary between the installation and surrounding cities. It has been conjectured that the crime reduction experienced in some central cities has been at the expense of the surrounding suburbs that have experienced increased crime rates.

B-351. An initial study of crime displacement was performed for the Washington, DC, area. It concluded that, although the decrease in Washington's crime rate was concurrent with an increase in the suburban crime rate, there is no evidence that the reduction in reported crime in Washington, DC, has resulted in a corresponding crime increase in the nearby suburbs.

B-352. The area-displacement effect can be measured with some degree of reliability. Three zones can be defined for the purposes of the measurement—the area containing the crime-control program (zone 1), a border around the area (zone 2), and the area chosen as the control area (zone 3). The width of the border may depend on the type of program implemented. If the program involves police helicopters, a quarter-mile-wide border may be necessary; for a patrol car, one or two blocks may suffice.

B-353. Crime rates before program initiation should be determined for all three zones. If zone 2 records a greater increase in crime than zone 3 while zone 1's crime rate decreases, then the increase in zone 2 can be attributed to two factors—

- The general increase in crime rate verified by any increase in zone 3.
- The increase caused by a displacement of crime from zone 1.

B-354. A displacement of this crime does not mean that the program is ineffective. It may suggest that the program should be expanded for all three zones.

UNREPORTED CRIME

B-355. Crime statistics are based on crimes reported to the police. It is well known that many crimes go unreported. Victimization studies can determine the extent of unreported crime and its change from year to year by area of the country or the reasons for failure to report them. These victimization studies are best suited to determining long-term effects. They are not well suited to most crime-control evaluations in which short-term changes must be assessed.

B-356. The amount of unreported crime is important but not for planning crime-control programs affecting police activities. The extent of unreported crimes is of little significance unless a program affects it. If a program encourages reported crime, the reported-crime rate may increase despite the program's effectiveness.

B-357. Ironically, a lowered reported-crime rate may be the direct result of an increase in the actual crime rate. Taking reports from victims of crimes occupies a substantial amount of a patrolman's time. Many of these crimes are minor and have no potential for solution. In an effort to increase the police department's time on patrol, some police chiefs have stopped the practice of sending a patrolman to get reports from the victim of a minor crime. This requires the victim to travel to the police station to report the crime. If the crime is minor or is seen by the victim to be unsolvable or the theft is not covered by insurance, the victim may decide not to inconvenience himself by going to the police station to report the crime. Therefore, the number of crimes reported to the police may drop. This may lead to a larger number of unreported crimes and prevent a complete picture to the crime-prevention counsel. Conversely, an actual decrease in crime due to the increased effectiveness of the police may produce an increase in the reported crime rate.

CRIME RATES

B-358. It has been pointed out that the crime rates, as presently calculated, do not reflect the true situation. For example, the rape rate should be calculated by dividing the annual number of rape cases by the number of women (since they are the population at risk). You would expect that the rate of commercial burglaries would be less in a residential area than a commercial area. When calculated on the basis of "per thousand people," this would be true; however, these rates should be obtained by dividing the number of cases by the number of commercial establishments (the population at risk) in each area.

B-359. The victim or the target is only one aspect of the crime. The offender can also be calculated into the rate. For example, the potential offenders in stranger-to-stranger crimes are usually considered to be males between 16 and 25 years of age. Therefore, one would expect fewer of these crimes in a city full of pensioners and retirees than in a city of the same population but with a higher proportion of young men. This fact is of minor importance in evaluating crime-control programs, since the age distribution of people in a city or a section of a city does not normally change greatly over the evaluation period. However, the former factor (the population at risk of becoming victimized) can be misleading if it is not taken into account. If possible, crime rates in experimental and control areas should be compared to the population that risks becoming victims of the target crimes.

MEASURES OF EFFECTIVENESS

B-360. The goals of the program determine the criteria that are used to measure its effectiveness. These goals and criteria should not be seen as confining; the evaluator should be amenable to broadening the criteria, especially if the program to be evaluated is a new one. For example, the program might be beneficial in some unforeseen way, wholly outside the original criteria. Conversely, the program may be an overall failure but a success according to the evaluation. It may be that the specified measures were the wrong ones to use for the program or should not have been used alone.

B-361. Programs aimed at controlling crime should not be evaluated solely for their effect on crime. Most programs cannot, by their very nature, focus on one specific objective alone. They normally are multifaceted in their effect and should be evaluated with respect to all of their facets. Similarly, the measures of effectiveness discussed in this section may not be adequate for every crime-control program, but they comprise some of the more useful measures that can be used.

B-362. This section concentrates on the two evaluation types—internal and external. Internal and external refer to whether the evaluation is conducted on the program's inner workings and logic or the external effect of the program (which depends on the program type). An internal evaluation of a crime-control program involving the use of new police patrol techniques would include the analysis of police response time and how it was effective in controlling crime or why it was successful in one area and not in another. The external evaluation would focus only on the effectiveness of the program in reducing crime rates or solving crimes, not on how or why or the conditions under which the results were achieved.

B-363. Evaluating how well a program achieved its goals is not the only purpose of an evaluation; how and why the results were achieved are of equal importance. External measures relate to the former evaluation; internal measures are concerned with the latter. The following examples will further serve to highlight the differences between these measures:

- Many crime-control programs depend on good community relations in order to achieve their goals. In these cases, a public-affairs campaign is often instituted concurrent with the crime-control program. The success of the public-affairs campaign should not be interpreted as program success. It may be a necessary part of the program, but it does not substitute for the results of the program in controlling crime. Testimonials from people involved in the program should also be considered only as a supplement to the evaluation based on external measures.

- A study undertaken for the President's Crime Commission showed that for certain types of incidents, the probability of arrest increased as the response time decreased. As a result of this finding, many police departments purchased new equipment or tried novel techniques to reduce response time without first determining whether their workloads included enough of the incidents for which quick response is useful. If this measure (response time) is to be used, it should be recognized as an internal measure and not substituted for the external evaluation.

INTERNAL MEASURES

B-364. Each program will have its own internal measures of effectiveness based on the logical elements of which it is constituted. This section covers only the internal measures of effectiveness that are common to most crime-control program evaluations. The measures covered include the crime rate, the clearance rate, the arrest rate, the crime-seriousness index, and the fear of crime.

CRIME RATE

B-365. The crime rate (the number of a specified type of crime committed per resident in a specified time period) is normally considered to be a measure of deterrence. If the crime rate decreases, it is presumed that potential offenders have modified their behavior to some extent and have committed fewer crimes. This is based on the assumption that the program has made the target crimes unattractive by increasing the actual or perceived risk of apprehension, by reducing the expected return from the crime, or by making alternative forms of behavior more attractive than the target group of offenses.

B-366. These deterrent effects use different means for their accomplishment. Most crime-control programs are police-oriented and concentrate on the risk-related aspects of deterrence. Victim-oriented programs focus on reducing the expected return. Many social and recreational programs deal with making alternatives more attractive. Regardless of the orientation of the programs, their deterrent effects are determined by measuring reported crime rates.

B-367. Reported crime rates can be changed by a number of factors, some of which are misleading. The public may feel that the police are becoming less effective in dealing with crimes and, therefore, report them less often. Conversely, if the public perceives that the police are becoming more effective, they may begin to report crimes that previously would have gone unreported. Another apparent crime-rate reduction may be due to the police not recording crimes that have been reported to them. Displacement effects that can produce misleading crime-rate reductions were discussed earlier.

B-368. There may also be an actual reduction in crime due to a program's deterrent effect. In some cases, the reduction in crime can be attributed to psychological deterrence. That is, the police department may have instituted some change (such as painting all police cars canary yellow) in preexisting patterns of operation that may cause a change in the behavior patterns of potential offenders. This type of deterrence is rarely long-lived.

B-369. On the other hand, there may have been a change instituted by the police that had the desired effect of increasing the actual risk of apprehension and, therefore, reducing the number of target offenses. An example of this is the police-operated burglar alarm of commercial establishments. In the experimental program, the number of alarms were increased almost tenfold compared to the (nonalarmed) control establishments. There was 1 capture in 36 control-group burglaries (2.8 percent), while there were 12 captures in 46 experimental-group burglaries (26 percent). Crime displacements to other crimes, tactics, targets, and areas reduced the actual effectiveness of the program, but this example shows that a significant change can be made in the actual risk of apprehension. Preliminary results indicate that the rate of increase of commercial burglaries has been decreased from about 15 percent per year to about 0 percent (at the expense of a greater increase in residential burglaries).

B-370. It is difficult but useful to distinguish between actual deterrence (due to an actual increase in risk) and deterrence that is purely psychological in nature (due to a perceived increase in risk). If it is suspected that part of the deterrent effect may be transient, a long-term study would be of benefit. In this way, the half life of the psychological deterrence can be gauged, which can give some indication of the extent to which resources should be committed to the program.

B-371. Some forms of psychological deterrence are almost entirely counterproductive. They may appear effective to those who would not commit a crime and ineffective to those who are "in the business" and study the presumed deterrent more closely. For example, a tear-gas pen may give a person a sense of security that is entirely without foundation. It may be dangerous to him if he actually attempts to use it when faced with an assailant.

B-372. One investigator has pointed out that for given criminal situations, nondelinquents perceive a higher risk of apprehension than do delinquents; in all probability, the delinquents have a more realistic assessment of the situation. A purely psychological deterrent may have the unfortunate effect of making only a cosmetic improvement. This gives the general population the impression that there has been a change for the better, while in reality the situation may not have changed or may have changed for the worse because of the division of resources to a nonexistent solution.

B-373. The crime rate can be used as a measure of effectiveness. However, the evaluator should delve into the determination of the crime rate to see if any change in the rate reflects a change in reporting procedures or the deterrent effect (with tangible evidence).

CLEARANCE RATE

B-374. The clearance rate is normally considered to be a measure of the ability of police to solve crimes. A cleared crime is one in which the police have identified the offender and have sufficient evidence to arrest him. The clearance rate is the percentage of total crimes that were cleared.

B-375. This measure of effectiveness should be used with care. A decreasing clearance rate may not mean that a police department is becoming less effective, and an increasing clearance rate may not mean that it is becoming more effective. This is due to a number of factors, primarily the public's conception of the role of the police with respect to crime and the present method of collecting crime data.

B-376. Often overlooked in discussion about crime is the role of the public in assisting the police. Police rely on community support to legitimize their authority as well as to help them carry out their work. If a segment of the community becomes alienated from the police (for whatever reason) and offers them little assistance in pursuing offenders, crime rates in these areas may rise. However, it is not only alienation of community groups that reduces the ability of the police to deal with crime; the profit motive is also to blame. Many store owners that have been robbed refuse to give their clerks time off (with pay) to help the police in their investigation. They absorb the loss of a robbery easily (it rarely comes close to the amount lost from shoplifting, employee theft, and damaged goods) and are unwilling to increase it by helping the police. They may feel that the chances of apprehending the offender are too slim, or they may be afraid of retribution if the offender discovers their assistance. They may also be afraid that their insurance will be cancelled.

B-377. If a police department begins a drive to increase its clearance rate, the increase may be forthcoming without any real change in police effectiveness. A survey of three police departments found that arrests for felonies were not made by the police in about 43 percent of the cases, in which there was probable cause, while the police were accompanied by witnesses. Making arrests in such instances would inflate the clearance rate quite easily. However, it should be noted that the police officer has a great deal of discretion in the exercise of his power of arrest. He may feel that the arrest charges will not hold up. One measure of the arrest quality is the percentage of arrests that leads to prosecutions.

B-378. In summary, clearance rate can be a measure for determining the effectiveness of crime-control programs. Its use can be increased by careful selection and specification of the crime categories that are studied, by determining the manner in which the crimes were cleared, and by determining if there has been a change in where the police draw the line in the exercise of their discretion.

ARREST RATE

B-379. Another measure of effectiveness that is often used as a determinant of crime-control effectiveness is the arrest rate, calculated either per police officer or per resident for a specified time period. Most of the considerations concerning the clearance rate (discussed above) also apply to the arrest rate. However, the arrest rate is distinguished from the clearance rate by an additional factor—it is not related to the total number of offenses. For example, the number of arrests for drug violations has risen considerably over the past few years. However, this increase is indicative of the extent of the problem, not of the effectiveness of the solution. It has been described how drug arrests may be traded off against arrests for other offenses and vice versa, especially when informal arrest quotas are established. Therefore, the use of the arrest rate by itself does not appear to be appropriate as a measure of the effectiveness for most crime-control programs.

CRIME-SERIOUSNESS INDEX

B-380. Among the many criticisms of crime statistics is the contention that, even if the data were reliable and complete, we would still have only a count of the number of incidents without an indication of their relative seriousness. The crime-seriousness index was proposed to include some of the major disutilities of crime typically committed by juveniles. Crimes are weighted according to the degree and nature of injury to the victims—whether they were intimidated and the nature of the intimidation or whether premises were forcibly entered, and the kind and value of property stolen. The weights were determined by requesting a sample of people to estimate the relative seriousness of various crimes.

B-381. All of the factors used to determine the weight are (or should be) included in offense reports. It would not be difficult to calculate an incident-seriousness score based on these reports, either for a specific evaluation or as a matter of course. Use of the seriousness index has also been proposed to measure the relative performance of law-enforcement agencies.

B-382. The crime-seriousness index is not the ultimate weighting scheme. The seriousness appears to be calculated more from the offender's viewpoint and the event than from the victim or society's viewpoint. For example, most people would consider the murder of a robbery victim by his assailant to be more serious than the murder of one spouse by the other. With regard to property loss, there is a difference between loss suffered by an individual who is insured and one who is not covered.

B-383. The loss relative to the individual's income is also an important factor; the theft of a $100 television from a low-income family has a much greater impact than the loss of $10,000 of jewels from a wealthy family. Perhaps a better index of the relative value of property loss to the victim would be the value of the loss in relation to the amount of the individual's discretionary income (that is, income not used for the basic necessities of life). Of course, such information is not available on police crime reports.

FEAR OF CRIME

B-384. It has been pointed out that the perceived risk of crime is greater than the actual risk of crime, and that this perceived risk does not seem to be correlated with the actual crime rate. Unless the public feels safer in proportion to its increased actual safety, the full potential of improvements will not have been reached. Therefore, the goal of a crime-control program can be broadened to include not only improved public safety (deterrence), effectiveness (clearance rate), and reduced crime impact (seriousness); but also improved, more accurate, public perceptions of safety as well.

B-385. Measurements of perceived safety can be both direct and indirect. Public-opinion surveys with regard to perceptions about crime and safety have been made frequently. It is also possible to gauge the effect of this fear using indirect measures by observing what people do rather than what they say. The number of patrons of movie theaters and restaurants at night (or other observations of this type of activity) could be used to gauge the fear of crime.

B-386. A reliable measure of the public's perception of public safety has not been developed. Additional research is being done and needs to be done before this type of measure of effectiveness can be used with confidence.

Appendix C

Intelligence, Counterintelligence, and Threat Analysis

Intelligence and counterintelligence make up the first line of defense in an antiterrorism program. A well-planned, systematic, all-source intelligence and counterintelligence program is essential. The role of intelligence and counterintelligence in antiterrorism is to identify the threat. Additionally, counterintelligence provides a warning of potential terrorist attacks and provides information for counterterrorism operations. This appendix provides the elements of the intelligence cycle that have particular importance in a viable antiterrorism program. Effective intelligence and counterintelligence support requires effort, planning and direction, collection and analysis, production, investigation, and dissemination. The entire process provides decision makers with information and timely warnings upon which to recommend antiterrorism actions.

INFORMATION SOURCES

C-1. The primary sources of intelligence and counterintelligence for the antiterrorism program are open-source information, criminal information, government intelligence and counterintelligence, and local information.

- **Open-source information.** This information is publicly available and can be collected, retained, and stored without special authorization. The news media is an excellent open source of information on terrorism. The media reports many major terrorist incidents and often includes in-depth reports on individuals, groups, or various government counterstrategies. Government sources include congressional hearings; publications by the Defense Intelligence Agency (DIA), the FBI, the Central Intelligence Agency (CIA), the Department of State (DOS), and the national criminal justice reference services. Additionally, there are private data services that offer timely information on terrorist activities worldwide. Terrorist groups and their affiliates may also have manuals, pamphlets, and newsletters that reveal their objectives, tactics, and possible targets.
- **Criminal information.** Both military and civil law-enforcement agencies collect criminal information. Because terrorist acts are criminal acts, criminal information is a major source for terrorist intelligence. Commanders must work through established law-enforcement liaison channels because the collection, retention, and dissemination of criminal information are regulated. Local military criminal investigative offices of the CID; the Naval Investigative Service Command (NISCOM); the Air Force Office of Special Investigations (AFOSI); and Headquarters, US Marine Corps,

Criminal Investigations Division, maintain current information that will assist in determining the local terrorist threat.

- **Government intelligence and counterintelligence.** The Community Counterterrorism Board (CCB) is responsible for coordinating with the national intelligence agencies concerning combating international terrorism. These agencies include the CIA (the lead agency), the DIA, the National Security Agency (NSA), the DOS, the Department of Justice (DOJ), the FBI, the Department of Energy (DOE), the Federal Aviation Administration (FAA), the Department of Transportation (DOT) (including the USCG), and the DOD. Service intelligence and counterintelligence production organizations include the US Army Counterintelligence Analysis Center; the Navy Antiterrorism Analysis Center (NAVATAC); Headquarters, US Marine Corps, Counterintelligence; and the US AFOSI Operations Center. These organizations compile comprehensive intelligence and counterintelligence for distribution on a need-to-know basis throughout the services. In combatant commands, the J2 is responsible for the intelligence fusion center. The Counterintelligence Support Officer (CISO) provides interface between the combatant command, the component commands, and the joint staff.
- **Local information.** Other valuable sources of information are the soldiers, civil servants, family members, and individuals with regional knowledge (such as college faculty or members of cultural organizations). Local crime or neighborhood watch programs can be valuable sources of information and can serve as a means to keep individuals informed in dispersed and remote areas. Intelligence exchanges with local government agencies through cooperative arrangements can also augment regional information.

RESPONSIBILITIES OF US GOVERNMENT LEAD AGENCIES

C-2. The FBI is responsible for collecting and processing domestic terrorist information. Overseas, terrorist intelligence is principally a CIA responsibility; but the DOS, the DIA, and the HN are also active players. The MI activities are conducted according to Presidential executive orders, federal law, status of forces agreements (SOFAs), memorandums of understanding (MOUs), and applicable service regulations. Responsibilities of intelligence activities include the following:

- The combatant commander (through the commander's J2 and the CISO in coordination with the DIA, the CIA, the embassy staff, the country team, and applicable HN authorities) obtains intelligence and counterintelligence specific to the AO. The commander issues intelligence and counterintelligence reports, advisories, and assessments to the units within his command or those operating within his command's AO. This network is the backbone for communicating intelligence and counterintelligence information and advisories and for warning of terrorist threats throughout the region.
- The Secretaries of the military departments were asked (in DOD Directive 2000.12) to ensure that a capability exists to receive and evaluate data from a service perspective and that the capability exists

to disseminate all relevant data on terrorist activities, trends, and indicators of imminent attack. To accomplish this task, each Secretary appoints an MI agency (INSCOM, NISCOM, or AFOSI) to conduct intelligence and counterintelligence activities directed against terrorists and to detect, neutralize, or deter terrorist acts. To accomplish this mission, the military department intelligence agency establishes counterintelligence offices on an area basis to collect and disseminate information to combatant commanders. Each military department's intelligence agency—

  - Coordinates with appropriate US and HN agencies.
  - Provides overall direction and coordination of the service counterintelligence effort.
  - Operates a 24-hour operations center to receive and disseminate worldwide terrorist threat information to and from the combatant command's J2, applicable service staff elements, subordinate commands, and national agencies.
  - Provides service commanders with information on terrorist threats concerning their personnel, facilities, and operations.
  - Investigates terrorist incidents for intelligence, counterintelligence, and force-protection aspects (with the FBI or HN authorities).
  - Provides terrorist threat information in threat briefings.
  - Conducts liaison with representatives from federal, state, and local agencies (as well as HN agencies) to exchange information on terrorists.
  - Provides international terrorism summaries and other threat information to supported commanders. On request, provides current intelligence and counterintelligence data on terrorist groups and disseminates time-sensitive and specific threat warnings to appropriate commands.

- Service criminal investigative services (such as the CID, the NISCOM, and the AFOSI) collect and evaluate criminal information and disseminate terrorist-related information to supported installation and activity commanders as well as to the service lead agency. As appropriate, criminal investigative elements also conduct liaison with local military or security police and civilian law-enforcement agencies.
- Intelligence staff elements of commanders at all echelons—
  - Report promptly all actual or suspected terrorist incidents, activities, and early warnings of terrorist attacks to supported and supporting activities, to the local counterintelligence office, and through the chain of command to the service lead agency.
  - Initiate and maintain liaison with the security police or PMO; local military criminal investigative offices; local counterintelligence offices; security offices; HN agencies and; as required, other organizations, elements, and individuals.
  - Develop and present terrorism threat-awareness briefings to all personnel within their commands (in cooperation with the local counterintelligence offices).
- Law-enforcement staff elements will—

  - Report all actual or suspected terrorist incidents or activities to their immediate commander, supported activities, and service lead agency through established reporting channels.
  - Initiate and maintain liaison with local counterintelligence offices and military criminal investigative offices.
  - Maintain liaison with federal, HN, and local law-enforcement agencies or other civil and military antiterrorism agencies as appropriate.
- Installation, base, unit, and port security officers—
  - Report all actual or suspected terrorist incidents or activities to their immediate commander, supporting military law-enforcement office, other supported activities, local counterintelligence office, and local military criminal investigation office.
  - Conduct regular liaison visits with the supporting military law-enforcement office, counterintelligence office, and local criminal investigation office.
  - Coordinate with the supporting military law-enforcement office and counterintelligence office on their preparation and continual updating of the threat assessments.
  - Assist in providing terrorism threat-awareness training and briefings to all personnel and family members as required by local situations.

INFORMATION REQUIREMENTS

C-3. To focus the threat analysis, intelligence and counterintelligence officers develop information requirements to identify targets using the following terrorist considerations—

- Organization, size, and composition of group.
- Motivation.
- Long- and short-range goals.
- Religious, political, and ethnic affiliations.
- International and national support (moral, physical, and financial).
- Recruiting methods, locations, and targets (students).
- Identity of group leaders, opportunists, and idealists.
- Group intelligence capabilities and connections with other terrorist groups.
- Sources of supply and support.
- Important dates, such as religious holidays.
- Planning ability.
- Internal discipline.
- Preferred tactics and operations.
- Willingness to kill.
- Willingness for self-sacrifice.
- Group skills, demonstrated or perceived (for example, sniping, demolitions, masquerade, industrial sabotage, airplane or boat

operations, tunneling, underwater, electronic surveillance, poisons, or contaminants).

- Equipment and weapons (on hand and required).
- Transportation (on hand and required).
- Medical-support availability.
- Means and methods of C^2.
- Means and method of communication.

THREAT ANALYSIS AND ASSESSMENT

C-4. The preparation of the terrorist threat analysis is a continual process of compiling and examining all available information to identify terrorist targeting of US interests. A vulnerability analysis is a continual process of compiling and examining information on a facility's security posture. The threat analysis is then paired with the facility's vulnerability analysis to create the threat and vulnerability assessment. Threat analysis is an essential step in identifying the probability of a terrorist attack. To enhance the capability to collect and analyze information from many sources, the DIA maintains a terrorism database. The combatant command's J2 and CISO (in coordination with the DIA) focus this database information and regional information toward the intelligence and counterintelligence needs specific to the security of the command. However, this terrorism database is limited to foreign terrorist groups because of limitations on US intelligence-collection operations. A country's threat assessments, information and biographies about terrorist organizations, and incidents in the database can be disseminated to commands. Commands at all echelons then augment or refine the DIA's threat analysis to focus on their area of interest. This process, operative across the full range of military operations, promotes coordination between all levels of the intelligence, counterintelligence, and law-enforcement communities; broadens acquisition channels; and enhances timely distribution of information to the supported commander.

C-5. Several factors complicate intelligence and counterintelligence collection and operations. The small size of terrorist groups, coupled with their mobility and cellular organization, make it difficult to identify the members. Unlike other criminals, terrorist cadres often receive training in counterintelligence and security measures from foreign intelligence agencies or other terrorists. Additionally, the traditional orientation of police organizations is toward individual criminals, while MI organizations focus on conventional forces. Terrorist activity, therefore, requires some degree of reorientation for police and MI and counterintelligence collection and operations.

C-6. An intelligence system's ability to provide critical and timely information to the user depends not only on efficient collection and processing, but also on the ability to organize, store, and retrieve this information rapidly. This capability, coupled with early warning, careful observation, and assessment of threat activity, enhances the probability of accurately predicting the types and timing of terrorist attacks.

C-7. Commanders must carefully exercise judgment in estimating both the existing terrorist threat and the need for changes in antiterrorism measures. Key questions are—

- What has changed (mission, political climate, installation and unit personnel or equipment, terrorist capabilities)?
- What affect will the changes have on the security posture?

C-8. Extraordinary security measures, unless part of a deliberate deception during critical or high-threat situations, draw attention and detract from mission accomplishment. Sound physical security, personnel who are aware, accurate threat and vulnerability assessments, and well-rehearsed response plans reduce the probability of a successful terrorist venture. The goal is to make an attack too difficult or the level of risk unacceptable to the terrorist.

DETERMINATION OF THE THREAT LEVEL

C-9. This threat-analysis methodology is used by the DIA, the joint staff, and the unified and specified commands for selecting the level of threat for an installation. It is applicable in an overseas setting, but fails to address issues unique to the sustaining base within CONUS. In CONUS there is a lack of intelligence and counterintelligence information from the CIA, the DIA, or the military services for the CONUS-based threat. That information must be extracted from law-enforcement channels at the local, regional, and national levels. In that context, the factors mentioned in this appendix are not as clear as they are within the intelligence process in place overseas. A modified version of this appendix should be considered for assessing threat levels in CONUS.

C-10. Threat levels within CONUS historically have been either low or negligible. This trend will most likely continue at least through the next decade. Domestic groups not covered by DOD intelligence reporting pose the greatest threat to the CONUS-based military. Due to the lack of reporting and information on these groups, the domestic terrorist groups are not currently factored into the current terrorist threat program. Therefore, the law-enforcement community must become a key player in establishing the threat levels in the context of the recommended model. A possible methodology would be the establishment of an additional threat level between low and medium—one that allows local commanders more flexibility in implementing additional security measures. Table C-1 shows the procedure for determining the threat level.

Table C-1. Threat Levels

Threat Level	
Critical	Factors 1, 2, and 5 are present. Factors 3 or 4 may be present.
High	Factors 1, 2, 3, and 4 are present.
Medium	Factors 1, 2, and 4 are present.
Low	Factors 1 and 2 are present. Factor 4 may be present.
Negligible	Factors 1 and/or 2 may be present.
Explanation of Factors	
Factor 1. Existence. A terrorist group is present, assessed to be present, or able to gain access to a given locale.	
Factor 2: Capability. The acquired, assessed, or demonstrated level of capability to conduct terrorist attacks.	
Factor 3: Intentions. Recent demonstrated anti-US terrorist activity or stated and/or assessed intent to conduct such activity.	
Factor 4: History. Demonstrated terrorist activity over time.	
Factor 5: Targeting. Current credible information on activity indicative of preparations for specific terrorist operations and/or specific intelligence that shows an attack is imminent.	

Appendix D

Crisis-Management Plan

The following pages highlight areas of concern in crisis-management planning. This plan is not meant to be all-inclusive or rigidly followed. Figure D-1 is a sample format only. It does not reflect a format developed and approved for use with OPLANs or contingency plans (CONPLANs) prepared by the CINCs to fulfill tasks assigned in the Joint Strategic Capabilities Plan (JSCP) or as otherwise directed by the Chairman of the Joint Chiefs of Staff. Figure D-2, page D-4, is a sample of the Crisis-Management-Plan Checklist, which is Annex A or Appendix H to the crisis-management plan. This checklist will help ensure that the plan is sound.

Copy No. ________
Issuing Headquarters
Place of Issue
Date of Issue

Crisis-Management Plan

Ref: Maps, charts, and other relevant documents.

Time Zone: X

Task Organization: List units organized to conduct antiterrorism operations. Include attachments, supporting roles, and the delegation of operational control as necessary.

1. Situation. Identify essential information to understand ongoing events.

a. Terrorist force. Identify the terrorist's composition, disposition, methods of operation, and estimated strengths and capabilities that could influence the crisis-management operation. Refer to an appropriate annex.

b. Response force. Explain the response force's abilities and responsibilities. The response force's abilities can influence the crisis-management mission.

c. Attachments and detachments. Address here or refer to an annex.

d. Assumptions. Provide assumptions used as a basis for this plan (for example, the strength of the response force to be supported and the support available from other agencies).

Figure D-1. Sample Crisis-Management Plan

(1) Tactical-situation possibilities. Obtained from the commander's planning guidance.

(2) Personnel situation. Provided by the personnel officer.

(3) Logistics situation. Provided by the logistics officer.

(4) Legal-situation possibilities. Provided by the SJA.

2. Mission. Identify the antiterrorism mission (for example, detect, deter, contain, and neutralize terrorist threats and actions aimed at the disruption of the installation).

3. Execution.

a. Concept of operations. State the commander's tactical plan. The purpose is to inform. It may address how the commander will conduct combat-terrorism operations. It provides enough detail to ensure proper action by subordinates in the absence of specific instructions. If the required details are extensive, address them in an annex. If an operation involves two or more distinct phases, designate each phase and use subparagraphs (for example, Phase I and Phase II).

b. Tasks. Identify specific tasks for each command element charged with executing a crisis-management mission. When giving multiple instructions, itemize and indicate the priority or sequence.

c. Coordinating instructions. Include coordination and control measures applicable to two or more command elements.

4. Service Support. Provide a statement of service-support instructions and arrangements supporting the crisis-management operation. Use the following subparagraphs as required:

a. General. Outline the general plan for service support.

b. Materiel and services. Address supply, transportation, labor, and services required.

c. Medical evacuation and hospitalization. Provide the plan for evacuating and hospitalizing sick, wounded, or injured personnel. Address evacuation responsibilities and the air-evacuation policy.

d. Personnel. Provide required information and instructions to supporting unit personnel.

(1) Maintenance of unit strength.

(a) Strength reports. Provide instructions for submitting status reports. Include requirements for routine and special reports.

(b) Replacements. Address validating existing personnel requisitions, instructions for submitting requisitions, and instructions for processing and removing replacements.

(2) Personnel management. Address military and civilian personnel and civilian detainee management procedures.

(3) Development and maintenance of morale.

(a) Morale and personnel services. Provide postal and financial services, religious activities, personal hygiene, and special services activity information.

(b) Graves registration. Include evacuation procedures and handling personal effects.

Figure D-1. Sample Crisis-Management Plan (continued)

(4) Maintenance of discipline, law, and order. Obtain this guidance from the PMO/security officer.

(5) Miscellaneous. Include personnel administrative matters not specifically assigned to another coordinating staff section or included in preceding subparagraphs.

e. Miscellaneous. Provide special instructions or special reports not covered in preceding paragraphs.

5. Command and Signal. Provide instructions for the command and operation of communications-electronics equipment. Communications-electronics instructions may refer to an annex but should list the index and issue number of the command, control, and communications (C^3) operation instructions in effect. If not already issued, give instructions for the control, coordination, and establishment of priorities in the use of electromagnetic emissions. Command instructions include subordinate and higher unit CP locations and designated alternate CPs.

6. Acknowledgement Instructions.

/s/

Commander

Annexes as applicable

Distribution:

Figure D-1. Sample Crisis-Management Plan (continued)

Annex A

Crisis -Management Plan Checklist

Yes	No	
		1. Intelligence and/or Counterintelligence.
___	___	Does the plan allow for the threat-analysis process (collection, analysis, production, and dissemination) to help identify the local threat?
___	___	Does the plan consider restrictions placed on the collection and storage of information?
___	___	Does the plan indicate an awareness of sources of information for the threat-analysis process (MI, counterintelligence, federal agencies, and state and local authorities)?
___	___	Does the plan allow for liaison and coordination of information (such as establishing a committee)?
		2. Threat Assessment.
___	___	Does the plan identify the local threat (immediate or long term)?
___	___	Does the plan identify other threats (such as national and international groups that have targeted or might target US installations)?
___	___	Does the installation incorporate factors for assessing the threat? Does it address —
___	___	Geography of the area concerned?
___	___	Law-enforcement resources?
___	___	Population cultural resources?
___	___	Communication capabilities?
___	___	Does the plan establish a priority of identified weaknesses and vulnerabilities?
___	___	Is the threat assessment periodically updated?
		3. Security Countermeasures.
___	___	Does the plan have specified THREATCONs and recommended actions?
___	___	Do security countermeasures include a combination of physical operations and sound-blanketing security measures?
___	___	Do the THREATCONs correspond to DOD 0-2000.12-H, Appendix BB?

Figure D-2. Sample Crisis-Management-Plan Checklist

Yes	No	
		4. OPSEC.
____	____	Have procedures been established that prevent terrorists from readily obtaining information about plans and operations (for example, not publishing the commanding general's itinerary and safeguarding classified material)?
____	____	Does the plan allow for in-depth coordination with the installation's OPSEC program?
____	____	Has an OPSEC annex been included in the CONPLAN?
		5. Personnel Security.
____	____	Has the threat analysis identified individuals vulnerable to terrorist attacks?
____	____	Has a training program been established to educate both military and civilian personnel in the proper techniques of personnel protection and security commensurate with the local threat and the type of position held?
		6. Physical Security.
____	____	Are special-threat plans and physical-security plans mutually supportive?
____	____	Do security measures establish obstacles to terrorist activity (such as guards, HN forces, lighting, and fencing)?
____	____	Does the special-threat plan include the threats identified in the threat statements of higher headquarters?
____	____	Does the physical-security officer assist in the threat analysis and corrective action?
____	____	Does the installation have and maintain detection systems and an appropriate assessment capability?
		7. Security Structure.
____	____	Does the plan indicate that the FBI has primary domestic investigative and operational responsibility in the US and US territories?
____	____	Has coordination with the SJA been established?
____	____	Does the plan allow for close cooperation between principal agents of the military, civilian, and HN communities and federal agencies?
____	____	Does the plan clearly indicate parameters for the use of force, including briefing any elements augmenting MP assets?
____	____	Is there a mutual understanding between all local agencies (military, local, FBI resident or senior agent-in-charge, HN forces, and local law enforcement) that might be involved in a terrorist incident on the installation regarding authority, jurisdiction, and possible interaction?

Figure D-2. Sample Crisis-Management-Plan Checklist (continued)

Yes	No	
____	____	Has the SJA considered the ramifications of closing the post (such as possible civilian union problems)?
____	____	Does the plan identify the DOS as having primary investigative and operational responsibilities overseas?
		8. Operations-Center Training.
____	____	Has the operational command and coordination center been established and exercised?
____	____	Is the operations center based on the needs of the installation while recognizing manpower limitations, resource availability, equipment, and command?
____	____	Does the plan include a location for the operations center?
____	____	Does the plan designate alternate locations for the operations center?
____	____	Does the plan allow for the use of visual aids (chalkboards, maps with overlays, and bulletin boards) to provide status reports and countermeasures?
____	____	Does the plan create and designate a location for a media center?
____	____	Have the operations and media centers been activated together within the last quarter?
____	____	Does the operations center have SOPs covering communications and reports to higher headquarters?
____	____	Does the operations center offer protection from a terrorist attack?
		9. Reaction-Force Training.
____	____	Has the force been trained and exercised under realistic conditions?
____	____	Has corrective action been applied to shortcomings and deficiencies?
____	____	Has the reaction force been formed and mission-specified trained (for example, building entry and search techniques, vehicle assault operations, countersniper techniques, and equipment)?
____	____	Has the reaction force been tested quarterly (alert procedures, response time, and overall preparedness)?
____	____	Has responsibility been fixed for the negotiation team? Has the negotiation team been trained and exercised under realistic conditions?
____	____	Does the negotiation team have the proper equipment?
		10. General Observations.
____	____	Was the plan developed as a coordinated staff effort?
____	____	Does the plan outline reporting requirements (logs, journals, and after-action reports)?

Figure D-2. Sample Crisis-Management-Plan Checklist (continued)

Yes	No	
____	____	Does the plan address the media's presence?
____	____	Does the plan include communication procedures and communication nets?
____	____	Does the plan consider the possible need for interpreters?
____	____	Does the plan consider the need for a list of personnel with various backgrounds to provide cultural profiles on foreign subjects and victims as well as to assist with any negotiation efforts?
____	____	Does the plan provide for and identify units that will augment MP assets?
____	____	Does the plan delineate specific taskings for each member of the operations center?
____	____	Does the plan provide for a response force for each phase of antiterrorism activity (initial response, negotiation, and assault)?
____	____	Does the plan designate service-support communications?
____	____	Does the plan make provisions for the notification of an accident-and-incident control officer?
____	____	Does the plan provide for EOD support?
____	____	Does the plan take into consideration the movement from various locations, including commercial airports, of civilian and military advisory personnel with military transportation assets?
____	____	Does the plan allow for the purchase or use of civilian vehicles, supplies, and food (if needed)? (This includes items used to satisfy a hostage demand.) Does the plan make provisions for paying civilian employees overtime if they are involved in a special-threat situation?
____	____	Does the plan take into consideration the messing, billeting, and transportation of civilian personnel?
____	____	Do appropriate personnel have the necessary language training?
____	____	Is WMD support available?

Figure D-2. Sample Crisis-Management-Plan Checklist (continued)

Appendix E

Office Security Measures

The office environment should afford executives the greatest degree of physical security. Executives usually work in facilities where attackers must pass by guards, security checkpoints, office workers, aides, or secretaries before reaching them. Unfortunately, the high media value of attacking executives in security strongholds where they are clearly associated with government activity increases the value of such attacks to terrorists. Hence, there may be a need to add security measures to offset the escalating capability of attack on more secure office areas by terrorist groups.

PHYSICAL-SECURITY SURVEY

E-1. A thorough physical-security survey of an office facility should be conducted. Offices of defense components attached to US embassies abroad should have these surveys performed by the DOS. Other DOD facilities should have surveys performed by the cognizant physical-security and facilities-engineering staffs. The best way to approach a physical-security site survey is to think like an intruder. Consider how approaches to the installation or facility could be made, how access to the building that houses executive offices could be gained, and how attacks on offices or other frequently used facilities could be mounted.

SECURITY-ENGINEERING ASSESSMENT

E-2. The next step in evaluating the need for supplemental physical-security measures is a thorough and detailed assessment of the weapons and tactics that terrorists might use to attack the structure in which DOD executives work. Security engineers and architects need technical threat data or assessments containing the following information—

- The mode of attack, such as—
  - Standoff weapons (man-portable AT/antiaircraft weapons, sniper rifles, rock grenades, and mortars).
  - Close combat weapons (submachine guns, pistols, knives, and garrotes).
  - Contact weapons (bombs, incendiary devices, and mines).
- Perimeter penetration aids (such as power tools, hand tools, or explosives), if used.
- The time of attack.
- The attacking force's size.
- The anticipated degree of outside support or autonomy.

E-3. Engineering design requirements are developed from the security engineering assessments. The data is used to—

- Assess the ability of building components to resist the effects of the threat.
- Identify appropriate security window-glazing materials and window treatments to determine what is required to achieve the desired penetration resistance times for anticipated threats.
- Calculate the total amount of delay time. This time is achieved by using camouflage, deception, barriers, and security devices to permit response forces to reach the scene of a terrorist attack in time to thwart the attack, capture or eliminate the terrorists, and rescue executives and their staffs or dependents.

TECHNICAL ASSESSMENT OF RESPONSES

E-4. After establishing a basic-design threat, engineers need data on the anticipated performance of response forces to be arrayed against the design threat and the expected or desired behavior of the protected executive. Some specific information needed includes—

- The response force's size, capability, supporting weapons, response time, and estimated effectiveness against the range of attacks.
- The desired options for the executive's protection—evacuate on warning, on detection, only if attacked, or only if forced to capitulate or do not evacuate.

E-5. Security planners need to know how long the structure that houses executives can withstand an attack before help arrives. Matching threat capabilities and anticipated operations by response forces establishes significant physical-security-system performance parameters. These can be quantified and used to develop detailed plans, drawings, and physical-security equipment-acquisition plans.

PHYSICAL-SECURITY ENHANCEMENT MEASURES

E-6. Several physical-security measures intended to provide additional protection for executives can be considered based on the requirements defined through the detailed analyses outlined above. The primary purpose of such measures should be to increase the time required by persons outside an installation to reach the executives housed at an installation. A secondary purpose of such measures should be to reduce or eliminate hazards to executives that might result from violence in the vicinity. Examples of physical-security measures to consider are—

- Increase the threat-detection time by installing sensors on perimeters and barriers. This includes—
  - Combining surveillance systems including seismic, acoustic, and IR sensors at or beyond the outer perimeter.
  - Supplementing surveillance systems with CCTV/imaging IR systems tied into the alert response-force staging area.

- Extending restricted areas or exclusion zones and relocating access-control points from the executive office area to a point closer to the installation's boundary.
- Enlarging and extending intrusion-detection sensors from within the installation to its perimeter, allowing the IDS to collect additional data necessary and sufficiently classify and identify an intrusion before the response force arrives.
- Enhancing both the number and the phenomenology of surveillance and detection systems within the executive office area as well as approaches leading to and from it in conjunction with measures listed below.

- Increase the threat's delay time between the perimeter and the executive office building. This includes—
  - Installing vehicle barriers and realigning roadways to eliminate straight, level stretches of road in excess of 50 meters in length.
  - Increasing concentric rings of fences, Jersey barricades, planters, bollards, and vehicle/personnel barriers.
  - Enhancing access-control areas supplemented by fire doors/security doors kept in a closed condition between the entrance to the building that houses executive offices and the executive office area.
- Confuse, camouflage, and deceive observers by hiding an executive's location. Accomplish this by—
  - Relocating executives to buildings not usually associated with office activities (barracks, motor pools, research and development [R&D] facilities, and so forth).
  - Constructing office areas in the barracks, motor pool, R&D facilities, and so forth.
  - Adding executive styles, decorative lighting, and window treatments to several different areas of office buildings to minimize the differences in external appearances between executive and nonexecutive offices.
- Increase the delay time between the entrance to the building that houses executives and the executive office area. Execute this by—
  - Adding fire doors, access-control points, dead-end corridors, and midcorridor physical barriers to complicate access to the executive office areas.
  - Adding security devices that, when activated, disrupt the intruder's ability to retain his thought processes (for example, flashing strobe lights, fog generators, noise generators, sirens, and fire-extinguishing systems).
- Increase the delay time by making access more difficult within the executive office structure. This may be accomplished by—
  - Substituting high-security doors and door frames for standard doors in areas leading to or from executive offices.
  - Installing high-security grating, wire mesh, or other materials to bar access to the executive office area through utility tunnels or conduits.

- Strengthening walls, floors, and ceilings by substituting steel-plate, concrete-filled, steel-reinforced cinder blocks or other ballistic-resistant materials for plaster/lath or wallboard room dividers, thereby protecting against explosive devices that are used as tools to breach a barrier.

- Increase the protection for building occupants against weapons and explosives effects. This includes—
  - Substituting blast- or bullet-resistant panels for glass windows or adding a fragment-retention film at least 4 millimeters thick to the interior of glass windows.
  - Adding exterior screens/plates to cover window areas and protect against gunfire and grenade/bomb fragments.
  - Installing blast curtains, metal blinds, metal shutters, or other window treatments in executive offices to protect interior space from glass shards and other small projectiles.
  - Strengthening walls to resist weapons and explosives effects by adding steel plates, reinforced concrete, or other retrofitting measures.
  - Adding steel plates or other ballistic materials in crawl spaces above dropped ceilings or extending walls separating the executive office area from other portions of an office building from floor to floor, thereby preventing unobserved and undetected access to the space between dropped ceilings.
- Increase the hold time to contain penetrators by—
  - Adding positive-action controls to a facility's doors and gates so that gates default to a closed and locked condition unless manually released.
  - Adding positive-action controls to access-control areas so that persons inside an access-control area can neither advance nor withdraw without affirmative action by a security officer posted outside the access-control area.

E-7. These measures are used to facilitate the apprehension of terrorists. There may be some instances when defeating terrorist attempts to gain access to the executive enhances the security of the executive and the response force. This is accomplished by channeling the terrorists out of the facility and installation along one route, leaving alternative routes available to evacuate executives and other key personnel.

E-8. Install emergency executive-support facilities (including a safe haven and an emergency evacuation facility) by—

- Installing helicopter landing aids on a structure's roof or on an adjacent field far removed from parking areas.
- Installing a safe haven or other reinforced security structure adjacent to a helicopter landing facility to provide a secure waiting place for executives until a rescue helicopter with additional supporting air and ground units can extract the executives.

Appendix F

Physical-Security Plan

It is essential and in the best interest of security that each installation, unit, or activity maintains and uses a detailed physical-security plan. The plan should include at least special and general guard orders, access and material control, protective barriers/lighting systems, locks, and IDSs. All physical-security plans have the potential of being classified documents and must be treated accordingly. Figure F-1 depicts a sample physical-security plan.

Map Reference

Copy No. ________
Issuing Headquarters
Place of Issue
Date of Issue

Physical-Security Plan

1. **Purpose.** State the plan's purpose.

2. **Area Security.** Define the areas, buildings, and other structures considered critical and establish priorities for their protection.

3. **Control Measures.** Define and establish restrictions on access and movement into critical areas.

   a. Categorize restrictions as to personnel, materials, and vehicles:

      (1) Personnel access:

         (a) Establishment of controls pertinent to each area or structure.

- Authority for access.
- Criteria for access.
  - Unit personnel.
  - Visitors.
  - Maintenance personnel.
  - Contractor personnel.
  - National guard.
  - Emergency response teams (police, fire, ambulance, and so forth).

Figure F-1. Sample Physical-Security Plan

(b) Identification and control.

- Description of the system to be used in each area. If a badge system is used, a complete description covering all aspects should be used in disseminating requirements for ID and control of personnel conducting business on the installation.
- Application of the system:
  - Unit personnel.
  - Visitors to restricted areas.
  - Visitors to administrative areas.
  - Vendors, tradesmen, and so forth.
  - Contractor personnel.
  - Maintenance or support personnel.
  - Fail-safe procedures during power outages.

(2) Material control.

(a) Incoming.

- Requirements for admission of material and supplies.
- Search and inspection of material for possible sabotage hazards.
- Special controls on delivery of supplies or personal shipments in restricted areas.

(b) Outgoing.

- Documentation required.
- Controls, as outlined in paragraph 3a(2a).
- Classified shipment not involving nuclear/chemical material.

(c) Nuclear/chemical material.

- Controls on movement of warheads/chemicals on the installation.
- Controls on shipments or movement of training warheads/chemicals.
- Controls on pickup or delivery of warheads/chemicals outside the installation.

(3) Vehicle control.

(a) Policy on search of military and privately owned vehicles.

(b) Parking regulations.

Figure F-1. Sample Physical-Security Plan (continued)

(c) Controls for entrance into restricted and administrative areas:

- Military vehicles.
- POVs.
- Emergency vehicles.
- Vehicle registration.

b. Indicate the manner in which the following security aids will be implemented on the installation:

(1) Protective barriers:

(a) Definition.

(b) Clear zones.

- Criteria.
- Maintenance.

(c) Signs.

- Types.
- Posting.

(d) Gates.

- Hours of operation.
- Security requirements.
- Lock security.
- Barrier plan.

(2) Protective lighting system:

(a) Use and control.

(b) Inspection.

(c) Action taken in case of commercial power failure.

(d) Action taken in case of failure of alternate power source.

(3) Emergency lighting system:

(a) Stationary.

(b) Portable.

Figure F-1. Sample Physical-Security Plan (continued)

(4) IDSs:

(a) Security classification.

(b) Inspection.

(c) Use and monitoring.

(d) Action taken in case of alarm conditions.

(e) Maintenance.

(f) Alarm logs or registers.

(g) Tamper-proof provisions.

(h) Monitor-panel locations.

(5) Communications:

(a) Locations.

(b) Use.

(c) Tests.

(d) Authentication.

(6) Security forces: General instructions that would apply to all security-force personnel (fixed and mobile). Detailed instructions such as special orders and SOP information should be attached as annexes. Security-force facets include—

(a) Composition and organization.

(b) Tour of duty.

(c) Essential posts and routes.

(d) Weapons and equipment.

(e) Training.

(f) Use of MWD teams.

(g) Method of challenging with signs and countersigns.

(h) Alert forces:

- Composition.
- Mission.
- Weapons and equipment.
- Location.
- Deployment concept.

Figure F-1. Sample Physical-Security Plan (continued)

(7) Contingency plans: Required actions in response to various emergency situations. Detailed plans for situations (counterterrorism, bomb threats, hostage negotiations, disaster, fire, and so forth) should be attached as annexes.

(a) Individual actions.

(b) Alert-force actions.

(c) Security-force actions.

(8) Use of air surveillance.

(9) Coordinating instructions. Matters that require coordination with other military and civil agencies such as—

(a) Adjacent installations or units.

(b) State and local agencies.

(c) Similar host-country agencies.

(d) Federal agencies.

The coordination/interaction allows for an exchange of intelligence information on security measures being used, contingency plans, and any other information to enhance local security.

On an installation, the host activity shall assume responsibility for coordinating physical-security efforts of all tenants, regardless of the components represented, as outlined in the support agreements and the host-activity security plan. Applicable provisions shall be included in, or be an appendix to, the support agreement. A formal agreement will contain definite assignment of physical-security responsibility for the items stored. The agreement should address—

- Maximum qualities to be stored.
- Physical safeguards to be used.
- Frequency of, and responsibility for, physical inventories or reconciliations.
- Reporting of losses for investigation.
- Lock and key control.
- The unit that has overall responsibility.

Procedures for authorization and ID of individuals to receipt for and physically take custody of Army property. The purpose of such coordination is protection in depth. Authority, jurisdiction, and responsibility must be set forth in a manner that ensures protection and avoids duplication of effort.

/s/ Commander

Figure F-1. Sample Physical-Security Plan (continued)

ANNEXES

F-1. Annexes to the plan should include, but are not limited to, the following. More information can be found in AR 190-13.

- **Annex A.** The installation threat statement (intelligence). This annex should contain the Terrorism Counteraction Plan (refer to AR 190-13).
- **Annex B.** A bomb-threat plan. As a minimum, the bomb-threat plan should provide guidance for—
  - Control of the operation.
  - Evacuation.
  - Search.
  - Finding the bomb or suspected bomb.
  - Disposal.
  - Detonation and damage control.
  - Control of publicity.
  - After-action report.
- **Annex C.** An installation closure plan.
- **Annex D.** A natural-disaster plan. This plan will be coordinated with natural-disaster plans of local jurisdictions. At a minimum, the natural-disaster plan should provide guidance for—
  - Control of the operation.
  - Evacuation.
  - Communication.
  - Control of publicity.
  - After-action report.
- **Annex E.** A civil-disturbance plan. It is the commander's responsibility to formulate a civil-disturbance plan based on local threats. (For example, commanders of chemical facilities should anticipate the need to develop crowd-control procedures to handle antichemical demonstrations.)
- **Annex F.** A resource plan to meet the minimum-essential physical-security needs for the installation or activity.
- **Annex G.** A communication plan. This plan is required to establish communications with other federal agencies and local law-enforcement agencies to share information about possible threats. The communications plan should address all communication needs for annexes B through F above.
- **Annex H.** A list of designated restricted areas.
- **Annex I.** A list of installation MEVAs.
- **Annex J.** A contingency plan. In most instances, it will be necessary to increase security for AA&E and other sensitive property, assets, and facilities during periods of natural disasters, natural emergencies, or increased threat from terrorists or criminal elements. Therefore, CONPLANs should include provisions for increasing the physical-security measures and procedures based on the local commander's assessment of the situation. Such contingencies may include hostage negotiations, protective services, and special-reaction teams. These

provisions should be designed for early detection of an attempted intrusion, theft, or interruption of normal security conditions.

- **Annex K.** Work-stoppage plan. This is a requirement for conducting a physical-security survey.

TACTICAL-ENVIRONMENT CONSIDERATIONS

F-2. In a tactical environment, the development of a physical-security plan is based on METT-TC (using the OPORD format and the higher headquarters' order). The order may be specific about the tasks the unit will perform. Time available may be limited and the scheme of maneuver may be dictated, but the leader must still evaluate the mission in terms of METT-TC to determine how MP elements can best carry out the commander's order.

F-3. Consider each of the following factors and compare courses of action to form a base for the physical-security plan. When the plan is firm, issue it as an order.

- Concepts for reconnaissance, coordination with adjacent and/or supporting units, and troop movement.
- Physical-security installation configurations and facilities. Areas to consider may include drop zones, landing zones, ranges, and training areas.

MISSION

F-4. The mission is usually the emplacement of defensive security rings to protect the populace against insurgents. The number of defensive security rings depends on the particular site and situation. The following questions must be evaluated:

- What is the mission?
- What specific and implied tasks are there to accomplish the mission?
- What is the commander's intent?

ENEMY

F-5. The commander identifies insurgent units operating in the area and tries to determine the type and size of the unit; the enemy's tactics, weapons, equipment, and probable collaborators; and the inhabitants' attitudes toward the insurgents. The following questions must be evaluated:

- What is known about the enemy?
- Where is the enemy and how strong is he?
- What weapons does the enemy have?
- What is the enemy doing?
- What can the enemy do in response to MP actions?
- How can we exploit the enemy's weaknesses?

TERRAIN AND WEATHER

F-6. The commander can use observation and fields of fire, cover and concealment, obstacles, key terrain, and avenues of approach (OCOKA) to plan for the physical-security defensive sites. The following questions must be evaluated:

- How will the terrain and weather affect the operation?
- How fast can movement be accomplished, and how much space does the terrain and unit formations take up?
- Will the weather affect the terrain or personnel?
- Has the weather already affected the terrain?

TROOPS

F-7. The commander must consider available equipment, the reaction time, reaction forces, communication assets, organization of troops, and medical support (if available). The following questions must be evaluated:

- What are the present conditions of vehicles and personnel?
- What is the status of ammunition and supplies?
- Who is best able to do a specific task?
- How much sleep have the soldiers had in the past 24 hours?
- What other assets are available to support the mission?
- How many teams/squads are available?
- What supplies and equipment are needed?
- What fire support is available and how can it be obtained?

TIME AVAILABLE

F-8. This factor is critical since the inhabitants must be ready to respond to an insurgent attack with little or no warning. The following questions must be evaluated:

- How much time is available to conduct planning?
- How long will it take to reach the objective?
- How long will it take to prepare the position?
- How much time do subordinates need?
- How long will it take the enemy to reposition forces?

CIVILIAN CONSIDERATIONS

F-9. The commander also must consider nonbelligerent third parties (such as dislocated civilians, personnel of international businesses and relief organizations, and the media). Every commander must prepare a site overlay that shows, as a minimum, the following:

- The attitude of the HN toward US forces.
- The population density near the objective.
- The condition of the local civilians.
- The possible effect of refugees and dislocated civilians on the mission.

Appendix G
Personal-Protection Measures

Terrorists frequently emulate military organizations as they develop, plan, train, and carry out terrorist attacks against DOD assets. Terrorists have a critical need for information regarding the whereabouts, habits, working environments, home environments, and other potential points of leverage against their targets. The three intelligence-collection methods used by terrorists against potential targets are human intelligence (HUMINT), photographic intelligence (PHOTINT), and signal intelligence (SIGINT).

PERSONAL PROTECTION

G-1. The measures that follow are useful in providing personal protection for US government employees and DOD civilian contractors in CONUS or OCONUS facilities.

OVERCOME ROUTINES

G-2. The reduced probability of success in kidnapping or killing a target makes the target far less desirable. Perform the following measures to prevent daily routines from being observed:

- Vary your route to and from work and your arrival and departure times.
- Vary your exercise schedule, using different routes and distances. It is best not to exercise alone.
- Do not divulge family or personal information to strangers.
- Enter and exit buildings through different doors, if possible.
- Avoid other routines.

MAINTAIN A LOW PROFILE

G-3. Americans are easy to identify in an overseas area. Perform the following measures to reduce easy ID:

- Dress and behave in public in a manner consistent with local customs. Items that are distinctively American should not be worn or displayed outside American compounds.
- Reduce visibility in the local community.
- Avoid flashing large sums of money, expensive jewelry, or luxury items.
- Avoid public disputes or confrontations, and report any trouble to the proper authorities.
- Ensure that personal information (home address, phone number, or family information) is not divulged.

PREPARE FOR UNEXPECTED EVENTS

G-4. All DOD personnel, contractors, and their family members should implement the following general measures:

- Get into the habit of checking in with friends and family.
- Know how to use the local phone system.
- Know the locations of civilian police, military police, government agencies, and the US embassy.
- Know certain key phrases in the local language.
- Set up simple signal systems that can alert family members or associates that danger is present.
- Carry ID showing your blood type and any special medical conditions.
- Keep personal affairs in good order.
- Avoid carrying sensitive or potentially embarrassing items.

WORKING ENVIRONMENT

G-5. The working environment is not immune from attempted acts by criminals or terrorists. DOD installations in CONUS and OCONUS usually provide a level of basic security comparable or superior to the basic level of security provided in the surrounding community. The following are general practices that can help reduce the likelihood of a terrorist attack:

- Establish and support an effective security program.
- Discourage the use of office facilities to store objects of significant intrinsic value unless it is mission essential.
- Train personnel to be alert for suspicious activities, persons, or objects.
- Arrange office interiors so that strange or foreign objects left in the room will be recognized immediately.
- Provide for security systems on exterior doors and windows.
- Ensure that access-control procedures are rigorously observed at all times for access to—
  - The installation.
  - Buildings within an installation.
  - Restricted or exclusion areas within buildings.
- Use an ID badge system containing a photograph.
- Identify offices by room number, color, or object name and not by rank, title, or the name of the incumbent.
- Avoid using nameplates on offices and parking places.

OFFICE PROCEDURES

G-6. In an office, the following steps can be taken to make intelligence collection and targeting more difficult for terrorists:

- Telephone and mail procedures:
  - When answering the telephone, avoid using ranks or titles.
  - When taking telephone messages, do not reveal the whereabouts or activities of the person being sought.

- When leaving telephone messages, place them in unmarked folders; do not leave them exposed for observers to identify caller names and phone numbers, persons called, and messages left.
- When opening mail, use a checklist to help identify letter bombs or packaged IEDs.

- Visitor-control procedures:
  - Place strict limitations on access to the executive office area.
  - Lock doors (from the inside) from the visitor-access area to executive offices or other restricted areas of a facility.
  - Ensure that receptionists clear all visitors before they enter inner offices.
  - Permit workmen or visitors access to restricted areas or exclusion areas under escort and only with proper ID. Confirm the work to be done before admitting workmen to restricted areas of the facility.
  - Limit publicity in public waiting areas to information that does not identify personnel by name, position, or office location.
  - Avoid posting unit rosters, manning boards, or photo boards where visitors or local contractors can view them.
  - Restrict the use of message boards, sign-in/-out boards, and other visual communications to general statements of availability.
- General working procedures:
  - Avoid carrying attaché cases, briefcases, or other courier bags unless necessary.
  - Avoid carrying items with markings that identify the owner by rank or title, even within the office environment.
  - Avoid working alone late at night and on days when the remainder of the staff is absent.
  - Ensure that office doors are locked when the office is vacant for any lengthy period, at night, and on weekends. If late-night work is necessary, work in conference rooms or internal offices where outside observation is not possible.
  - Ensure that the security office retains the office keys.
  - Ensure that papers, correspondence, communications materials, and other documents are not left unattended overnight.
  - Ensure that maintenance activity and janitorial services in key offices, production offices, or maintenance facilities are performed under the supervision of security personnel.
  - Prohibit the removal of property, material, or information stored on any media from the facility without proper written authorization.
  - Consider prohibiting the importation of property, material, or information stored on any media into the facility unless such items have been properly inspected.
  - Lock offices not in use to prohibit unauthorized access of stored material that could be used to hide IEDs or intelligence-collection devices.
  - Minimize the use of vehicles or vehicle markings that make it possible to readily identify the vehicle and its occupants as US-government or DOD-contractor personnel.

- Ensure that all personnel have access to some sort of duress alarm to annunciate and warn of a terrorist attack.
- Ensure that secretaries and guard posts are equipped with covert duress alarms that can be used to alert backup forces.
- Avoid placing office furnishings directly in front of exterior windows.

SPECIAL PROCEDURES FOR EXECUTIVE ASSISTANTS

G-7. The following suggestions are intended to be a guide for secretaries and executive assistants who may find themselves performing personnel-security duties as collateral duty. Executive assistants and security personnel should regularly train and exercise procedures used in case they must evacuate mission-critical personnel to safe havens.

- Request the installation of physical barriers (such as electromagnetically operated doors) to separate offices of senior executives from other offices.
- Request the installation of a silent trouble-alarm button with a signal terminating in the security department.
- Admit visitors into the executive area when they are positively screened in advance or are personally recognized.
- Do not inform unknown callers of an executive's whereabouts, home address, or telephone number.
- Store a fire extinguisher, a first aid kit, and an oxygen bottle in the office area.
- Remain calm and listen carefully when receiving a threatening call.
- Do not accept packages from strangers until satisfied with the individual's identity and the nature of the parcel.
- Keep travel itineraries for all personnel confidential.
- Distribute daily schedules for senior officers and civilian officials on a limited basis.

HOME ENVIRONMENT

G-8. The following discussion is intended to assist personnel in formulating plans to obtain housing outside US government compounds or DOD facilities. Personnel assigned to government housing may also find the antiterrorism and security tips presented below helpful in reducing the threat of violence and loss of property.

G-9. For general residential-security routines, discuss with family members the importance of—

- Varying routines in their daily activities.
- Blending in with the local environment.
- Avoiding unnecessary publicity and photographs that identify individual family members.
- Being alert to individuals, parked or abandoned vehicles, unusual utility work, or gatherings of people inconsistent with the residential environment.

SECURITY PRACTICES AT HOME

G-10. The following measures are specifically recommended for residential implementation. These measures are an extension of office antiterrorism-security practices.

- Do not use nameplates or uniquely American symbols on the exterior of residences occupied by DOD personnel overseas.
- Do not use nameplates on parking places, and avoid parking private or government vehicles in the same location day after day.
- Ensure that all family members answer the telephone politely but that they provide no information as to the name of the occupants until the caller's identity has been established.
- Treat all telephone conversations as though anyone who wanted to listen in was doing so.
- Examine carefully all mail delivered to the residence.

SOCIAL AND RECREATIONAL ACTIVITIES

G-11. DOD personnel are encouraged to participate in many social and recreational activities. The following precautions are recommended:

- Respond to formal social invitations in person (where possible) or by direct telephone contact.
- Be attentive to the security environment of social gatherings.
- Avoid the development of patterns with respect to time of arrival or departure at social events.
- Avoid prolonged presence at social functions where there is a high concentration of persons thought to be terrorist targets.
- Refrain from excessive use of alcohol at social functions; remain clearheaded and unimpaired.
- Vary routes to and from social events held at a central facility.
- Minimize appearances in uniform or formal attire.
- Decline invitations to appear in publicity photos.
- Participate in recreational activities within the American compound or at a DOD installation whenever possible.

NOTE: Refer to DOD 0-2000.12-H, Graphic Training Aid (GTA) 19-4-3, and Joint Services (JS) Guide 5260 for further guidance and explanation regarding protective measures.

Appendix H

Bombs

Terrorists have frequently used homemade devices or IEDs to carry out their attacks against DOD personnel, facilities, and assets. The IEDs are ideal terrorist weapons. They are relatively inexpensive to make, and the components of many IEDs are common items that can be obtained from many sources and are difficult to trace. The IEDs can be large or small and be designed so that they are transported to the attack site in components for last-minute assembly. Such design concepts make detection more difficult and provide an additional increment of personal safety to the terrorists.

GENERAL

H-1. The use of IEDs can enhance the violence that gives terrorist groups their ability to intimidate or coerce a target population. The detonation itself creates a highly visual, newsworthy scene, even hours after the detonation occurs. Bombs can detonate anywhere, without apparent reason and without warning. The use of bombs in a terror campaign emphasizes the authorities' inability to safeguard the public and maintain law and order. Bombs are ideal weapons because they can be designed to give terrorists opportunities to escape from the scene of their crimes.

CONCEALING BOMBS

H-2. Given the question, "Where have terrorists placed bombs in the past, and where should we look for them?" results in no easy answer. Table H-1, page H-2, lists a few obvious locations that should be examined. Terrorists who use bombs as their weapons of choice can be very creative in designing and placing their weapons.

H-3. Bombs can be found anywhere people can place them. Without becoming paranoid and seeing a bomb under every rock and behind every tree, the practical answer to the above questions is: "Where they can be easily placed without the bomber being caught."

DAMAGE AND CASUALTY MECHANISMS

H-4. The IEDs and other explosive devices inflict casualties in a variety of ways, including the following:

- Blast over pressure (a crushing action on vital components of the body; eardrums are the most vulnerable).
- Falling structural material.
- Flying debris (especially glass).
- Asphyxiation (lack of oxygen).

Table H-1. Potential IED Hiding Places

Outside Areas	
• Trash cans • Dumpsters • Mailboxes • Bushes	• Street drainage systems • Storage areas • Parked cars
Inside Buildings	
• Mail parcels or letters • Inside desks/storage containers • Ceilings with removable panels • Areas hidden by drapes or curtains • Recent repaired/patched segments of walls, floors, or ceilings	• Restrooms • Trash receptacles • Utility closets • Boiler rooms • Under stairwells
In Plain Sight	

- Sudden body translation against rigid barriers or objects (being picked up and thrown by a pressure wave).
- Bomb fragments.
- Burns from incendiary devices or fires resulting from blast damage.
- Inhalation of toxic fumes resulting from fires.

H-5. It is impossible to calculate a single minimum safe distance from an IED or other explosive device. The safe distance varies with each device and its placement. As a rule, the farther away from a bomb, the safer the intended or collateral targets are. Blast effects, fragmentation injuries, and injuries resulting from flying debris diminish greatly as the distance between a bomb and possible targets increase. The amount of material in the device, the type of explosive material, the manner in which the device is constructed, and the location or the container in which it is placed all have a bearing on the specific destructive potential for each IED.

H-6. The following are four general rules to follow to avoid injury from an IED:

- Move as far from a suspicious object as possible without being in further danger from other hazards such as traffic or secondary sources of explosion (such as POL storage).
- Stay out of the object's LOS, thereby reducing the hazard of injury because of direct fragmentation.
- Keep away from glass windows or other materials that could become flying debris.
- Remain alert for additional or secondary explosive devices in the immediate area, especially if the existence of a bomb-threat evacuation assembly area has been highly publicized.

H-7. Some terrorists have used two especially devious tactics in the past to intensify the magnitude of casualties inflicted by bombing attacks. In some instances, they have detonated a small device to lure media attention and curiosity seekers to the site; a larger, more deadly device has detonated some time after the first device, thereby inflicting a large number of casualties.

H-8. Other terrorists have used a real or simulated device to force the evacuation of a facility only to detonate a much more substantial device in identified bomb-threat evacuation assembly areas. These attacks are especially harmful because the evacuation assembly areas often concentrate government or commercial office workers more densely than they are when dispersed throughout their usual workplaces.

TELEPHONIC THREATS

H-9. When receiving a telephonic threat, treat the call seriously. Often, an anonymous telephone call is made regarding a bomb or an IED. See Figure H-1, page H-4, for information to record/obtain when receiving these calls

H-10. When an anonymous warning or threat is received, initiate the bomb-threat data card and notify the PMO, security police, security forces, or other law-enforcement/security offices immediately. Local SOPs will determine subsequent actions. Immediate action may include a search without evacuation, the movement of personnel within the establishment, a partial evacuation, or a total evacuation. The following criteria helps determine what immediate action to take:

- Factors favoring a search before the movement of personnel:—
  - There is a high incidence of hoax telephone threats.
  - Effective security arrangements have been established.
  - Information in the warning is imprecise or incorrect.
  - The caller sounded intoxicated, amused, or very young.
  - The prevailing threat of terrorist activity is low.
- Factors favoring movement of personnel before searching:
  - The area (post or base) is comparatively open.
  - Information in the warning is precise as to the matters of location, a description of the device, the timing, and the motive for the attack.
  - A prevailing threat of terrorist activity is high.

EVACUATION DRILLS

H-11. Evacuation and search drills should be performed periodically under the supervision of the installation's or unit's senior officer. The drills should be held in cooperation with local police if possible. Personnel in adjacent buildings should be informed of drills to avoid causing unnecessary alarm.

H-12. Evacuation procedures depend on the circumstances. Prepare, publicize, and rehearse evacuation plans in advance. Address alarm systems, assembly areas, routes to assembly areas, personnel-evacuation responses, building and area clearances, and evacuation drills.

PERSONNEL-EVACUATION RESPONSE

H-13. The bomb-threat alarm system should be easily distinguished from the fire alarm. When the alarm sounds, personnel should—

- Lock up or secure all classified materials.
- Conduct a quick visual search of their immediate working area.

Instructions: Be calm. Be courteous. Listen, do not interrupt the caller. Notify supervisor/security officer by prearranged signal while caller is on line.

Name of Operator________________________________Time__________Date________

Caller's Identity

Sex: ❑ Male ❑ Female ❑ Adult ❑ Juvenile Approximate age: Years____________

Origin of Call

❑ Local ❑ Booth ❑ Internal (From within bldg)
❑ Long Distance If internal, leave line open for tracing the call.

Voice Characteristics		Speech		Language	
❑ Loud	❑ Soft	❑ Fast	❑ Slow	❑ Excellent	❑ Good
❑ High Pitch	❑ Deep	❑ Distinct	❑ Distorted	❑ Fair	❑ Poor
❑ Raspy	❑ Pleasant	❑ Stutter	❑ Nasal	❑ Foul	❑ Other ____
❑ Intoxicated	❑ Other____	❑ Slurred	❑ Lisp		
			❑ Other ____		

Accent	Manner		Background Noises	
❑ Local	❑ Calm	❑ Angry	❑ Factory Machines	❑ Trains
❑ Not Local	❑ Rational	❑ Irrational	❑ Bedlam	❑ Animals
Region____	❑ Coherent	❑ Incoherent	❑ Music	❑ Quiet
❑ Foreign	❑ Deliberate	❑ Emotional	❑ Office Machines	❑ Voices
Race____	❑ Righteous	❑ Laughing	❑ Mixed	❑ Airplanes
			❑ Street Traffic	❑ Party Atmosphere

Bomb Facts

Pretend difficulty with your hearing. Keep caller talking.

If caller seems agreeable to further conversation, ask questions like —
When will it go off? Certain Hour - Time Remaining - What kind of bomb? - Where are you now?
How do you know so much about the bomb? - What is your name and address?

If building is occupied, inform caller that detonation could cause injury or death.
Did caller appear familiar with plant or building by his description of the bomb location?

Write out the message in its entirety and any other comments on a separate sheet of paper and attach to this checklist.

Action To Take Immediately After Call

Notify your supervisor/security officer as instructed. Talk to no one other than as instructed by your supervisor/security officer.

Figure H-1. Sample Bomb-Threat Data Card

- Open windows (wherever possible).
- Leave the building, taking only valuable personal belongings.
- Leave doors open and immediately proceed to the assembly area.

H-14. Opening the building will reduce internal damage due to blast effects. It will also somewhat mitigate the extent of debris flying out of or falling from the building should a detonation occur.

ASSEMBLY AREAS

H-15. Choose the routes to the assembly area so that personnel do not approach the IED at any time. Preselect the routes to the assembly area, but devise a system to inform personnel of the location of the suspected IED and alternate routes. Routes prevent confusion and bunching and avoid potential hazards (such as plate glass, windows, and likely locations of additional IEDs).

H-16. Assembly areas should be preselected and well known to personnel. Establish a clearly defined procedure for controlling, marshaling, and checking personnel within the assembly area. If buildings or establishments are in a public area, coordinate the assembly areas with local police. Assembly areas are selected using the following criteria:

- Locate assembly areas at least 100 meters from the likely target or building (if possible).
- Locate assembly areas in areas where there is little chance of an IED being hidden. Open spaces are best. Avoid parking areas because IEDs can be easily hidden in vehicles.
- Select alternate assembly areas to reduce the likelihood of ambush with a second device or small-arms fire. If possible, search the assembly area before personnel occupy the space.
- Avoid locating assembly areas near expanses of plate glass or windows. Blast effects can cause windows to be sucked outward rather than blown inward.
- Select multiple assembly areas (if possible) to reduce the concentration of key personnel. Drill and exercise personnel to go to different assembly areas to avoid developing an evacuation and emergency pattern that can be used by terrorists to attack identifiable key personnel.

BUILDING AND AREA CLEARANCE

H-17. Establish procedures to ensure that threatened buildings and areas are cleared. Prevent personnel from reentering the building. Establish a cordon to prevent personnel from entering the danger area. Establish an initial control point (ICP) as the focal point for the PMO and for MP control.

H-18. Cordon suspicious objects to a distance of at least 100 meters, and cordon suspicious vehicles to a distance of at least 200 meters. Ensure that nobody enters the cordoned area. Establish an ICP on the cordon to control access; relinquish ICP responsibility to the PMO or local police upon their arrival. Maintain the cordon until the PMO, security police, security forces, or local police have completed their examination or stated that the cordon may stand down.

SEARCHING FOR A SUSPECTED IED

H-19. Searches are conducted in response to a telephonic threat or a report of an unidentified object on or near premises occupied by DOD personnel. The following types of searches may be used when searching for a suspected bomb or IED:

- An occupant search is used when the threat's credibility is low. Occupants search their own areas. The search is completed quickly because occupants know their area and are most likely to notice anything unusual.
- A team search is used when the threat's credibility is high. The search is very thorough and places the minimum number of personnel at risk. Evacuate the area completely, and ensure that it remains evacuated until the search is complete. Search teams will make a slow, thorough, systematic search of the area.

H-20. The following procedures should be followed if a search for explosive devices must be conducted before qualified EOD teams arrive:

- Make an audio check, listening for unusual sounds.
- Sweep the area visually up to the waist, then sweep up to the ceiling. Do not forget the tops of cabinets and cupboards.
- Perform a thorough and systematic search in and around containers and fixtures.
- Pass search results as quickly as possible to the leader responsible for controlling the search area. Do not use a radio; it may detonate the explosive.

H-21. Circumstances might arise in the case of a very short warning period. In other instances, a threat of a bomb against some facilities (if true) might necessitate the evacuation of a very large area. In these circumstances, searching for the presence of an explosive device to identify its location, appearance, and possible operating characteristics may be warranted.

H-22. Personnel who have not been trained in IED search and ID techniques should not search for explosive devices. Two types of errors are very common—the false ID of objects as IEDs and the incorrect ID of IEDs as benign objects. Depending on the devices used to arm and trigger an IED, the search process could actually result in an explosion.

SEARCH ORGANIZATION

H-23. The person controlling the search should have a method of tracking and recording the search results (such as a diagram of the area). Delegate areas of responsibility to the search-team leader, who should report to the person controlling the search when each area has been cleared. Pay particular attention to entrances, toilets, corridors, stairs, unlocked closets, storage spaces, rooms and areas not checked by usual occupants, external building areas, window ledges, ventilators, courtyards, and spaces shielded from normal view.

DISCOVERY OF A SUSPECTED IED

H-24. When a suspicious object has been found, report its location and general description immediately to the nearest law-enforcement or supervisory person. Do not touch or move a suspicious object. Instead, perform the following steps:

- If an object appears in an area associated with a specific individual or a clearly identified area—
  - Ask the individual/occupant to describe objects they have brought to work in the past few days.
  - Ask for an accounting of objects.
  - Ask for a verbal description/ID of objects.
- If an object's presence remains inexplicable—
  - Evacuate buildings and surrounding areas, including the search team.
  - Ensure that evacuated areas are at least 100 meters from the suspicious object.
  - Establish a cordon and an ICP.
  - Inform personnel at the ICP that an object has been found.
  - Keep the person who located the object at the ICP until questioned.
  - Avoid reentering the facility to identify an object that may or may not be an IED.

REACTING TO AN EXPLODED IED

H-25. The following procedures should be taken when an explosive/IED detonates at a DOD facility:

- For explosions without casualties—
  - Maintain the cordon. Allow only authorized personnel into the explosion area.
  - Fight any fires threatening undamaged buildings without risking personnel.
  - Report the explosion to the PMO, security police, security forces, or local police if they are not on the scene.
  - Report the explosion to the installation operations center even if an EOD team is on its way. Provide as much detail as possible, such as the time of the explosion, the number of explosions, the color of smoke, and the speed and spread of fire.
  - Ensure that a clear passage for emergency vehicles (fire trucks, ambulances, and so forth) and corresponding personnel is maintained.
  - Refer media inquiries to the PAO.
  - Establish a separate information center to handle inquiries from concerned friends and relatives.
- For explosions with casualties—
  - Select a small number of personnel to help search for casualties.
  - Assign additional personnel the responsibility for maintaining the cordon to keep additional volunteers searching for casualties.

Maintain the cordon until the EOD team verifies no further presence of bombs/IEDs at the site and the fire marshal determines that risk of additional injury to searchers from falling debris is acceptable.

- Prepare a casualty list for notification of next of kin; delay publication of the list until its accuracy is determined.
- Arrange for unaffected personnel to contact their next of kin immediately.

H-26. Civilian management officials and subordinate military commanders continue to have important personal roles to fulfill during a bomb/IED attack on DOD personnel, facilities, and assets. Perform the following procedures when reporting an attack:

- Pass available information to the operations center.
- Avoid delaying reports due to lack of information; report what is known. Do not take risks to obtain information.
- Include the following information in the report:
  - Any warning received and if so, how it was received.
  - The identity of the person who discovered the device.
  - How the device was discovered (casual discovery or organized search).
  - The location of the device (give as much detail as possible).
  - The time of discovery.
  - The estimated length of time the device has been in its location.
  - A description of the device (give as much detail as possible).
  - Safety measures taken.
  - Suggested routes to the scene.
  - Any other pertinent information.

H-27. Perform the following procedures when providing emergency assistance to authorities:

- Ensure that the PMO, security police, security forces, and other emergency-response units from local police, fire and rescue, and EOD teams are not impeded from reaching the ICP. Help maintain crowd control and emergency services' access to the site.
- Evacuate through the doors and windows of buildings.
- Assist the on-scene commander by obtaining a building diagram showing detailed plans of the public-service conduits (gas, electricity, central heating, and so forth), if possible. If unavailable, a sketch can be drawn by someone with detailed knowledge of the building.
- Locate, identify, and make witnesses available to investigative agency representatives when they arrive on the scene. Witnesses include the person who discovered the device, witnessed the explosion, or possesses detailed knowledge of the building or area.

H-28. Performing the above steps will provide substantial assistance to the crisis-management team and give other personnel constructive, supportive actions to take in resolving the crisis. Care must be exercised, however, that additional explosive devices are not concealed for detonation during the midst of rescue operations. These attacks add to the physical damage and emotional devastation of bomb/IED attacks.

H-29. The use of bombs and IEDs during terrorist attacks against DOD personnel, facilities, and assets is a common occurrence. The procedures outlined in this appendix are intended to help a DOD facility respond to an attack before an explosive device detonates. The procedures are also intended to help mitigate the consequences of an attack in case efforts to find an explosive device and render it inoperable are not successful. Incurring the costs to DOD facilities and installations of detecting an explosive device and terminating a terrorist incident before the device can detonate are almost always preferable rather than exercising plans and options to respond to a detonation. Several of the security measures discussed will help reduce the likelihood of a successful bomb/IED attack against DOD assets.

Appendix I

Executive Protection

DOD Directive 2000.12 recognizes a need to provide protection to military officers and DOD civilians who are assigned to high-risk billets, who are (by the nature of their work) high-risk personnel, or who are assigned to facilities identified as high-risk targets. The directive defines these terms as follows:

- High-risk billet. Authorized personnel billet (identified and recommended by appropriate authority) that because of grade, assignment, travel itinerary, or symbolic value may make personnel filling them an especially attractive or accessible terrorist target.
- High-risk personnel. US personnel and their family members whose grade, assignment, travel itinerary, or symbolic value may make them an especially attractive or accessible terrorist target.
- High-risk target. US facilities and material resources that, because of mission sensitivity, ease of access, isolation, or symbolic value may be an especially attractive or accessible terrorist target.

NOTE: For purposes of this appendix, the term executive will be applied to all persons requiring additional security protection who are assigned to high-risk billets, designated as high-risk personnel, or identified as high-risk targets.

SUPPLEMENTAL SECURITY MEASURES

I-1. The specific supplemental security measures that may be furnished to executives are subject to a wide range of legal and policy constraints. US law establishes stringent requirements that must be met before certain security measures may be implemented. DOD regulations, instructions, and legal opinions may further constrain the implementation of some protective measures described in this chapter. The SOFAs and MOUs between the US and a foreign government will also limit the use of some supplemental security measures. Leases and other conditions imposed by contract for purchase of land or buildings by the US for DOD use may also limit the application of certain security techniques. All of these constraints should be carefully considered when conducting security surveys, developing plans, and implementing additional security measures to protect high-risk personnel.

EXECUTIVE PROTECTION GOALS

I-2. In the discussion that follows, several measures are outlined that can afford senior military officers and DOD personnel additional protection against terrorist acts. The purpose underlying these measures is to—

- Increase the interval of time between detecting a threat and the onset of hostile action against executives and their dependents.
- Increase the amount of time required by terrorists to gain physical access to executives from the onset of hostile actions, whether the executives are at home, at the office, or in transit.

I-3. The implementation of supplemental security measures should strive to achieve the following prioritized goals:

- Enhancements should hold the terrorist threat at bay until a response force arrives (delay at a distance).
- Enhancements in physical security should enable executives to flee to safety (delay to permit flight).
- Enhancements should permit the executive to retreat into a safe haven of sufficient strength and survivability. This should enable a response force to wage an effective counterattack to liberate executives and others accompanying them to a safe haven, including family members at home and colleagues and visitors at work (delay, hold, and counterattack).

I-4. The following supplemental measures should be applied with care. There is a clear trade-off between increasing the level of physical security at the office and at home and preserving the anonymity of executives, thereby avoiding telltale signs of activity that point to prominence or criticality. These measures can be expensive. Expense can be measured not just in terms of dollars, but also in terms of changes to organizational routine. Therefore, three questions must be resolved before implementing bold, disruptive, and expensive supplemental security enhancements:

- What are the most cost-effective means of enhancing the security of executives at risk?
- How many changes in organizational routines and personal behaviors will have to be made for security measures to be effective in reducing the risk of terrorist attacks and the vulnerability of executives to such attacks?
- What are the anticipated costs of additional security measures in terms of dollars, organizational functionality, and mission capability?

I-5. Security enhancements can be made to improve the security of executives and can be even more effective if executives and their families take full advantage of and reinforce those measures. If executives do not change their behavior to accommodate additional security and protective measures, then the behaviors can effectively defeat the purpose of additional protection. Additional increments of security can be obtained to defeat virtually any threat. However, there is a point at which it is no longer cost-effective to add layer upon layer of protective measures to defeat a threat.

RESIDENTIAL SECURITY MEASURES

I-6. While terrorist groups conduct intelligence operations to identify targets, mistakes have been made in the past. DOD personnel should avoid leasing residences previously used by representatives of governments or organizations known to be targets of various terrorist groups. DOD personnel leasing residences formerly used by representatives of such governments may be placing themselves unnecessarily at risk of being attacked as a result of mistaken identity.

I-7. An executive's entire lifestyle should be included in security surveys used to assess the need for supplemental physical-security measures at the office. The executive's home and transportation from home to office and back should also be examined for risk and vulnerability. The same principles used to identify supplemental security improvements in an office environment apply to an executive's home environment as well. The purposes of physical-security enhancements are to—

- Increase the amount of time terrorists need to initiate and complete an attack on executives while at home, thereby giving response forces more time to rescue executives and their dependents.
- Reduce potential harm to executives and their families because of a terrorist assault mounted against the residence.

I-8. The goals of enhanced residential physical-security measures are to—

- Increase the amount of time between detection of a threat and the onset of hostile actions.
- Delay the terrorists as long as possible. Prevent terrorist access to executives and their family members and make it difficult to leave the scene to escape prosecution. These measures should not further jeopardize the lives of executives and their family members.
- Provide a safe haven where executives and their family members may flee for security pending the arrival of a response force on the scene.

I-9. The following measures can be implemented selectively to help security personnel achieve these objectives:

- Increase the time interval between threat detection and the onset of hostile terrorist acts by—
  - Ensuring that all door locks and window clasps are working.
  - Ensuring that all doors and windows are properly secured to their frames and that the frames are properly anchored to the residential structure.
  - Locking driveway gates with a security lock to prevent entry.
  - Installing a through-door viewing device or visitor intercom.
  - Installing security lights to aid in viewing entrances.
- Increase the number of physical barriers between the outer perimeter of the residence and the interior of the residence by—
  - Adding heavy, remotely operated gates to all fences, walls, and perimeter barriers consistent with the penetration resistance of the barrier between the residence, the street, and adjacent neighbors.
  - Creating a vestibule or air lock between living quarters and the exterior of a residence, ensuring that no one can enter the residence directly from the outside.
  - Adding fire doors or security doors or gates between the residence's bedrooms and living areas.

- Increase the time required to penetrate exterior structural walls by explosives, hand-held power tools, and hand tools by—
  - Adding additional armor covered by aesthetically pleasing materials to exterior walls.
  - Adding a separate reinforced masonry wall around the residence.
- Increase the surveillance of the residence and decrease response time by—
  - Installing CCTV systems to permit remote viewing of all doors and windows accessible from the ground, nearby structures, trees, or easily acquired platforms (such as a van parked next to a wall).
  - Installing area IDSs between the residence's perimeter and the residence itself, varying the number and types of sensors, and adding backup communication channels between the IDS and a surveillance assessment/response dispatch center.
- Increase the residence's durability and survivability to a terrorist attack by—
  - Fitting windows with either venetian blinds or thick curtains to reduce the observability of activities within the residence and to reduce hazards of flying glass in case of nearby explosions or gunfire.
  - Installing backup power systems for security devices (surveillance systems, communication systems, and access-control systems).
  - Ensuring that backup communication is available with the installation or embassy's security department via a secure landline or two-way radio.
  - Fitting a panic-alarm bell to the outside of the house with switches on all floor levels. Such an alarm should also annunciate at the local police and cognizant DOD or DOS security office.
  - Installing a safe haven in the home.

TRANSPORTATION MEASURES

I-10. High-risk personnel are most accessible to terrorists while in transit in official or privately owned vehicles. Specific steps can be taken to reduce the vulnerability of executives in transit.

SPECIAL TRANSPORTATION IN TRANSIT FROM DOMICILE TO DUTY

I-11. As a rule, Congress has strongly opposed the provision of domicile-to-duty transportation by the federal government to its officers and employees. Only 16 officials are entitled by statute to such assistance. Congress did, however, grant authority to the President and the heads of executive agencies and departments to provide domicile-to-duty transportation under certain circumstances. According to the statute, "a passenger carrier may be used to transport between residence and place of employment an officer or employee with regard to whom the head of a Federal agency makes a determination, [provided] that highly unusual circumstances present a clear and present danger, that an emergency exists, or that compelling operational considerations make such transportation essential to the conduct of official business."

I-12. The phrase "highly unusual circumstances which present a clear and present danger" is understood to mean that—

- The perceived danger is real, not imaginary.
- The perceived danger is immediate or imminent, not merely potential.
- Proof is provided that the use of a government vehicle would provide protection not otherwise available.

I-13. Such a danger would exist where there is an explicit threat of terrorist attacks or riot conditions and such transportation would be the only means of providing safe passage to and from work.

I-14. The phrase "emergency exists" means that there is an immediate, unforeseeable, temporary need to provide home-to-work transportation for an agency's essential employees. The phrase "similarly compelling operational considerations" means that there is an element of gravity or importance for the need of government-furnished transportation comparable to the gravity or importance associated with a clear and present danger or an emergency. Congress suggested further, "in such instances, [it is expected] that home-to-work transportation would be provided only for those employees who are essential to the operation of the government."

I-15. The Secretary of Defense has the statutory authority to allow a CINC to use government-owned or -leased vehicles to provide transportation in an area outside of the US for members of the uniformed services and other DOD personnel under certain circumstances. These circumstances include and are limited to a determination by the CINC that public or private transportation in the area is unsafe or is not available. Under these circumstances, DOD may provide transportation (usually in government buses or passenger vans) to personnel and their family members if it will help the CINC and his subordinate commanders maintain the capability to perform or undertake assigned missions. This transportation is not intended for transporting personnel from their residences to their places of work. The Secretary of Defense and the Service Secretaries also have the statutory authority to provide transportation from home to duty stations and back on a limited basis. This authority is usually implemented by providing a nontactical armored vehicle (NTAV) to protect personnel.

I-16. It is a DOD policy to make NTAVs available where necessary to enhance the security of DOD personnel consistent with the requirements and limitations found in the statute. DOD issuances, service regulations, and CINC guidance stipulate detailed procedures by which DOD manages NTAV programs. The statute also establishes a procedure for Presidential waiver of the "buy American" requirement; DOD and service regulations provide for the delegation of Presidential authority from the President to the Secretary of Defense; to the Director, Defense Security Assistance Agency; and to the Director, DIA. DOD Instruction 5210.84 authorizes DOS acquisition and installation of light vehicle armoring to DOS specifications in local defense-component vehicles on a reimbursable basis. The level of protection provided to the Defense Component Office will comply with approved overseas security policy group armored-vehicle standards.

I-17. The DOD recognizes two classes of NTAVs—heavy and light. Heavy NTAVs are fully armored vehicles intended to protect occupants from attack by bombs; IEDs; grenades; and high-velocity, small-arms projectiles. Light NTAVs are less than fully armored vehicles and are intended to protect occupants from attack by medium-velocity, small-arms projectiles and at least some types of IEDs.

I-18. The dividing lines between heavy and light NTAVs have become less distinct over time as armoring techniques and materials have given greater capability to NTAVs that are not classified as heavy. As a practical matter, add-on vehicle-armoring kits are now in production which (when properly installed in an appropriately powered and suspended vehicle) will provide a level of protection approaching that of the heavy NTAVs.

Heavy NTAVs

I-19. Heavy NTAVs may be assigned to US personnel upon certification by a Service Secretary only under the following conditions:

- Highly unusual circumstances present a clear and present danger to the health and safety of a nominated protectee.
- Compelling operational considerations make such transportation essential to conducting official business.

I-20. If the physical-security survey concludes that a heavy NTAV is warranted, the nominated protectee's Service Secretary shall, on the advice and recommendation of a combatant commander, determine whether the use of a heavy NTAV is warranted. If so, the Service Secretary shall authorize the use of a vehicle for a renewable 90- to 360-day period. At the end of the period, the requirement will be reexamined and a recertification for the protection shall be issued by the Service Secretary.

I-21. Each of the services manages a portion of the DOD's NTAV program. Each service has issued supplementary mandatory guidance for processing requests for, as well as allocation and use of, these scarce assets.

I-22. Heavy NTAVs are complex systems requiring specialized maintenance and operation. Normally, they will be assigned to DOD personnel with a driver who has been properly trained in the operation and maintenance of the vehicle. The operator is not a chauffeur; he is an integral part of a supplemental security package provided by DOD to meet its obligations in protecting key assets.

Light NTAVs

I-23. Light NTAVs may also be provided to US employees and officers where highly unusual circumstances present a clear and present danger to the health and safety of a nominated protectee or compelling operational considerations warrant their use. This category of NTAV features add-on armoring. While they are a less-complex armoring system than those used in heavy NTAVs, light NTAVs afford substantial protection to occupants against a variety of threats. New developments in after-manufacture armoring kits for vehicles are occurring at a rapid pace, increasing the number of vehicle manufacturers and models for which other NTAV modifications are suitable. Each service and the DIA have instructions for implementing DOD policy that authorizes the use of other NTAVs to enhance personnel protection of high-risk persons.

PRIVATELY OWNED VEHICLES

I-24. High-risk personnel may wish to forego the use of POVs during periods of extreme risk. Considerations include selecting measures that—

- Deter secret entry, making undetected placement of IEDs in or under the vehicle difficult for terrorists to accomplish.
- Enhance the vehicle's ability to increase distance between it and pursuers.
- Assist response forces in case of an incident.
- Make the vehicle appear little different than its standard models.

INDIVIDUAL PROTECTIVE MEASURES

I-25. Executives can enhance their personal security in the office environment by—

- Discouraging staff members who are taking telephone messages from disclosing their whereabouts.
- Ensuring that caution is used when opening mail and being especially careful with letters or packages that might contain IEDs.
- Ensuring that access is strictly limited to their office area.
- Limiting publicity and keeping official biographies short. This includes using outdated photographs if a publicity photograph is essential.
- Ensuring that they are not working alone late at night and on days when the remainder of the staff is absent.
- Working in conference rooms or internal offices where outside observation is not possible if late-night work is necessary. Security officers should be notified of the work so that they can periodically look in.
- Ensuring that office furnishings are not placed directly in front of exterior windows.

OFFICIAL BUSINESS AWAY FROM THE OFFICE

I-26. The following suggestions reinforce efforts by executives to maintain the high level of security provided in the home or office environment while on official business outside of these locations:

- Discuss security requirements with the person planning the function.
- Travel to and from the function with escorts.
- Choose the route carefully.
- Avoid publicizing planned attendance at official functions (unless required).
- Attempt to sit away from both public areas and windows.
- Encourage the function's sponsor to close the curtains to minimize the likelihood that anyone outside will be able to see inside and determine who is attending the function. This is extremely important for an evening function, when a well-lit interior can be easily viewed from a darkened exterior.
- Request that external floodlights be used to illuminate the area around the building where an evening function will occur.

LOCAL OFFICIAL AND UNOFFICIAL TRAVEL

I-27. Executives can greatly enhance their personal security when conducting official and unofficial travel by following these general practices:

- Vary daily patterns, such as leaving and returning at different times.
- Consider escorts to and from work or travel with a neighbor.
- Establish a simple oral or visual duress procedure between executives and drivers (for example, a phrase or movement used by the executive or driver only if something is amiss).
- Vary taxi companies. Ensure that the ID photo on the license matches the driver. If uneasy for any reason, take another taxi.
- Attend social functions with others, if possible.
- Examine the car before entering to see if there has been any interference. A small mirror on a rod is a cheap and effective method to inspect underneath cars. Do not touch the vehicle until it has been thoroughly checked (inside, around, and under).
- Avoid leaving personal items exposed in the car (uniform items, service-issued maps, official briefcases, and so forth).

SECURITY PRACTICES WHILE DRIVING

I-28. Executives can take the following measures to enhance security while driving:

- Keep car doors locked. Do not open windows more than a few inches.
- Avoid overloading a vehicle, and wear seat belts.
- Park vehicles in parking areas that are either locked or monitored. Never park overnight on the street. Before entering vehicles, check for signs of tampering.
- Keep the trunk locked.
- Drive in the inner lanes to keep from being forced to the curb.
- Use defensive and evasive driving techniques. Drill with your driver by watching for suspicious cars and taking evasive action.
- Avoid driving close behind other vehicles (especially service trucks), and be aware of activities and road conditions two to three blocks ahead.
- Beware of minor accidents that could block traffic in suspect areas such as crossroads. Crossroads are preferred areas for terrorist or criminal activities because they offer escape advantages.

I-29. If a terrorist roadblock is encountered, use the shoulder or curb (hit at a 30- to 45-degree angle) of the road to go around it or ram the terrorist's blocking vehicle. Blocking vehicles should be rammed in a nonengine area, at a 45-degree angle, in low gear, and at a constant moderate speed. The goal is to knock the blocking vehicle out of the way. In all cases, do not stop and never allow the executive's vehicle to be boxed in with a loss of maneuverability. Whenever a target vehicle veers away from the terrorist vehicle, it gives adverse maneuvering room and presents a better target to gunfire.

INTERURBAN, NATIONAL, AND INTERNATIONAL TRAVEL SECURITY PRACTICES AND PROCEDURES

I-30. To enhance security in interurban, national, and international circumstances, executives should—

- Book airline seats at the last moment. Consider using an alias.
- Restrict the use of ranks or titles.
- Avoid allowing unknown visitors in the hotel room or suite.
- Keep staff and family members advised of the itinerary and subsequent changes. Clearly and emphatically restrict this information to those having a need to know.

HOME SECURITY PRACTICES AND PROCEDURES

I-31. To enhance security at home, executives should—

- Check the ID of persons entering the premises (electricians, plumbers, telephone-maintenance personnel, and so forth). When in doubt, call their office to verify their identity before allowing them in your home.
- Avoid opening the door to a caller at night until he is visually identified through a window or a door viewer.
- Close curtains in a room before turning on lights.
- Consider placing the telephone where you will not be seen from doors or windows when answering.
- Investigate the household staff (especially temporary staff members).
- Stay alert and be on the lookout for the unusual. Ensure that the home is locked and secure whenever the residence is unattended. Be cautious upon return and look for the movement of furniture or the placement of unusual wires.
- Note and report suspicious persons.
- Control house keys strictly.
- Park the car in a locked garage.
- Consider installing a panic-alarm bell to the outside of the house with switches located on all floor levels.
- Clear the area around the house of dense foliage or shrubbery.
- Test duress alarms (if available). Make certain that family members understand how they work as well as the importance of their use.
- Cooperate with law-enforcement personnel, and abide by their security recommendations.

SECURITY AT SOCIAL AND RECREATIONAL ACTIVITIES

I-32. The risk of terrorist incidents is always present for high-risk personnel or personnel assigned to high-risk billets. The following measures are intended to permit executives to live a close-to-normal life while still remaining mindful of the risks to their security.

- Ensure that the host is aware of and takes appropriate measures for your security.
- Have your personal staff assist a civilian host, if required.
- Arrange for visitors to be subject to adequate security control.

- Screen the invitation list, if possible.
- Vary the times of sporting activities (golfing, jogging, and so forth).

COMBATING-TERRORISM TRAINING FOR EXECUTIVES

I-33. Combatant commanders annually compile a list of high-risk billets in their AO. These lists are forwarded through the appropriate service personnel channels, enabling each service to identify, plan, and provide resources to meet training requirements. All personnel and adult family members en route to high-risk billets must attend the Individual Terrorism Awareness Course (INTAC) conducted at the US Army John F. Kennedy Special Warfare Center at Fort Bragg, North Carolina. During this one-week course, personnel will receive instruction in defensive-driving techniques and survival shooting as well as individual protective measures and hostage survival. These individuals should also attend the appropriate regional orientation course (Middle East, Asia/Pacific, Latin America, or Africa) offered at the US Air Force Special Operations School at Hurlburt Air Force Base (AFB), Florida. The service member whose duties will require frequent vehicle operation should attend an appropriate evasive-driving course. Information on current offerings may be obtained by contacting the service representative to the DOD Antiterrorism Coordinating Committee or the Combating Terrorism Branch in the Office of the Assistant Secretary of Defense (OASD) Special Operations/ Low-Intensity Conflict (SO/LIC).

TRAVEL TO POTENTIAL PHYSICAL-THREAT RISK AREAS

I-34. Personnel en route to potential physical-threat risk areas (as identified by the OASDSO/LIC) should attend one of the following courses:

- The Dynamics of International Terrorism Course conducted at the US Air Force Special Operations School at Hurlburt AFB, Florida. During this one-week course, personnel will receive lectures on threats by region (Europe, Middle East, Latin America, Asia/Pacific, and Africa), the history and psychology of terrorism, personnel combating-terrorism measures (vehicle, personal, airline, and physical security), and hostage survival.
- A Regional Orientation Course (Middle East, Latin America, Africa, Asia/Pacific) at the US Air Force Special Operations School at Hurlburt AFB, Florida. This one-week course offers personnel instruction in cultural, political/military, and individual security factors associated with the region.

I-35. Installation security personnel may also receive the above training if they have completed the Antiterrorism Instructor Qualification Course (AIQC) at Fort Bragg, North Carolina.

PROTECTIVE SECURITY DETAILS

I-36. Each service can provide bodyguards for key senior military officers, DOD civilians, other US officials, or foreign dignitaries requiring personal protection. Each Service Secretary is responsible for assigning protective security details (PSDs) to service members based on the recommendation of their counterintelligence and/or law-enforcement investigation staffs. The PSDs are assigned to DOD personnel who meet requirements established by service regulations. In general, PSDs may be assigned only to those personnel whose position or assignment places them at risk and whose continued availability to the National Command Authorities and the CINCs is vital to DOD's mission execution.

I-37. A PSD provides high levels of security to an executive by establishing a series of protective cordons around him. The establishment of defense in depth often means that the innermost protective layer is in close contact with the executive at all hours of the day and night.

I-38. A PSD is trained to maintain a low profile. It is concerned about the executive's visibility and its ability to blend into his surroundings. There is nothing more damaging to the security of an executive than the obvious, detectable presence of a PSD when all other measures to have him blend into the local environment have been successful. A PSD will strive to keep travel routes and means of transportation from being publicized. If this cannot be accomplished, the PSD may suggest editorial changes to the itinerary scheduled for release to limit details of planned travel from public disclosure. For example, routes to and from announced appointments usually do not need to be revealed.

I-39. During the course of a PSD's mission, its members may be asked to perform several different security functions. They may, for example, perform direct or indirect protection or escort duty. Direct protection is open and obvious; indirect is generally a surveillance measure. The security-guard unit may operate as an interior guard and may consist of one or more men stationed at fixed posts. A PSD's members should know the identity of each individual in the party of a protected official; executives can help by introducing them to each member of the official party.

I-40. The protected person's attitude is critical to the success of the PSD's mission. Executives have a right and a responsibility to make their wishes known with respect to their personal security. They also have an obligation to listen carefully to the head of the PSD who is trained and highly qualified to help make reasonable judgments about manageable risks. A PSD's members understand that their function is inherently intrusive and that executives can easily resent the loss of privacy that accompanies the protection offered. On the other hand, PSDs have jobs to do, not merely to protect executives, but to help safeguard mission-critical assets—senior military and civilian leaders.

I-41. One of the most demanding functions placed on a PSD is to limit the ability of individuals to circulate and approach the executive. This is often very frustrating to executives who wish to shake hands, engage in close conversations with visitors, and move freely and without impediment in a social situation. The PSDs are trained to strictly enforce limitations on the circulation of individuals, carefully checking each person for ID and ascertaining that he is authorized to be present at the occasion.

I-42. DOD personnel who are provided with PSDs and must conduct official business or hold social engagements in large rooms can take steps to minimize the disruptions to such functions. These steps include—

- Providing advance attendee lists to the head of the PSD.
- Having one or more members of the staff who know the attendees stand with PSD members and identify the attendees as they arrive.
- Informing attendees that they will be admitted only at specified entrances.

I-43. The PSD's members are highly trained security specialists. While in the company of executives, they will be accommodating and helpful. Executives should remember, however, that the primary function of the PSD's members is to protect them, not perform errands or carry out personal services. A PSD's members who are performing valet or other chores cannot effectively protect the senior officers or civilian officials to whom they have been assigned.

EXECUTIVE-PROTECTION SYSTEM INTEGRATION

I-44. This appendix has focused on supplemental security measures used to address terrorist threats to senior high-risk personnel within the DOD. Various methods and measures have been discussed that provide increments of security over and above the base level of security provided to all DOD personnel assigned to an installation, facility, activity, or unit. In making decisions to allocate protective resources to enhance the security of senior officers and senior DOD officials, it is essential to remember that measures must be applied systematically. Additional security measures implemented to protect high-risk personnel in the office environment must be carried over to official functions conducted outside the office. The security measures must also be extended to protected persons' private lives and, depending on the nature of the threat, the lives of their family members.

I-45. The converse is equally true. It makes no sense to provide domicile-to-duty transportation for a high-risk person and make no provision for additional protection at home, at the office, and at official business and social functions. In view of the total costs of security measured in dollars, time, and inconvenience to protected persons, their staffs, colleagues, and families, it may be more prudent to radically alter living and working arrangements than to try to augment security in a piecemeal manner. For example, it might be prudent to house high-risk personnel within a DOD installation rather than to try to secure a detached, private residence at a substantial distance from the operations base of a response force. The key to successful executive protection is to ensure that the level of protection afforded by physical-security measures, operational procedures in the office and at home, and PSDs is constant. The level of protection must be matched to the threat and must be sustainable. Executives have a special responsibility to set a personal example of combating-terrorism awareness; of attention to personal, family, office, information, and OPSEC concerns; and of combating-terrorism security measures implementation. By doing so, they make their colleagues and subordinates more aware and more conscious of their security environment and less likely to be victimized by terrorist attacks.

Appendix J

Resource Management

Programs need annual funding to operate. This funding is obtained by devising and documenting a resource program that looks seven years out. Looking ahead one or two years is generally not a problem. Commanders must be able to describe future requirements in the out years so that money will be available when the program arrives in future years.

FUNDING PROGRAMS

J-1. Physical security is dependent on integrated systems with budgetary constraints. This appendix serves to inform commanders of three basic funding programs contributing to physical security—

- The RJC6, which resources physical-security equipment purchase and maintenance.
- The QLPR, which resources law-enforcement operations (to include security guards and special-reaction teams).
- The VTER, which provides resources for projects and temporary programs that enhance any type of security due to an increase in terrorist threat.

PROJECTED REQUIREMENTS

J-2. Recurring VTER requirements are usually shifted to QLPR as standard requirements after a period of a few years. Installations send seven-year projected requirements to the MACOMs for submission into the Army's Program Objective Memorandum (POM), which is an annual significant event for resource managers at all levels. Installation requirements should be included in the POM build; otherwise, the installation's program loses visibility from the start of the funding process. Input format is determined locally, but a generic example of the format and the type of information requested is provided at Figure J-1, page J-2.

OBLIGATION PLAN

J-3. Budget execution deals with the current fiscal year (October through September), with the exception of some types of dollars (appropriations) that are multiyear (such as procurement dollars that are executable for three years). Once the resource manager notifies an installation of its available annual funding, an obligation plan by month or quarter is developed to display how and when the funds will be spent. This obligation plan is also used to forecast when the program will run out of money, which in turn will justify the submission of an unfinanced requirement (UFR) to obtain the proper level of resources. Examples of an annual obligation plan and UFR are located at Figures J-2 and J-3, page J-3.

Program Objective Memorandum (POM)

Installation: Fort McClellan, AL
Activity: Military Police School
Appropriation: Operation & Maintenance, Army (OMA)
Army Management Structure (AMS): 321731
Management Decision Package (MDEP): VTER
Element of Resource (EOR): 2500
Issue: Force Protection/Antiterrorism Courses

Resources	FY00	FY01	FY02	FY03	FY04	FY05	FY06
($000)	500	510	520	530	540	550	560

Justification/Impact:
Funding is required to continue the eight functional contract training courses conducted by the US Army Military Police School in Force Protection/Antiterrorism for the Army and selected DOD services and agencies. Courses focus on the five pillars of force protection and provide students with a variety of force-protection expertise to prevent, react to, and
contain a terrorist event. If funds are not provided, the projected annual student load of 2,500 cannot be trained.

Figure J-1. Sample POM

TYPES OF APPROPRIATIONS

J-4. The most common types of appropriations that managers will be involved with are day-to-day operations or mission (OMA) funds and procurement (OPA) dollars that must be used for major projects or equipment buys over the $100,000 threshold. Army Materiel Command (AMC) uses research, development, and engineering (RD&E) dollars for operations at US Army Test and Evaluation Command (TECOM) facilities and for testing new equipment. Construction projects over $500,000 use military construction (MILCON) funds.

J-5. Resources needed for physical security are the result of planning. Commanders include physical-security programs and improvements as a part of all annual budgets. When the situation changes based on METT-TC, physical-security programs are reviewed, updated and, when necessary, approved. Commanders allocate resources consistent with the threat. Force-protection requirements provide the fundamental reasons for resourcing physical security. Security managers, force-protection officers, and PMs help identify security requirements and prioritize expenditures. When physical-security improvements are not properly planned, integrated, and budgeted for, commanders accept risks for physical-security failures.

Annual Obligation Plan
FY 99

Installation: Fort McClellan, AL
Activity: Military Police School
Appropriation: Operation & Maintenance, Army (OMA)
Army Management Structure (AMS): 321731
Management Decision Package (MDEP): VTER
Element of Resource (EOR): 2500
Issue: Force Protection/Antiterrorism

Phased Monthly Obligation Plan (Cumulative):

Month	$000
Oct	40
Nov	80
Dec	120
Jan	160
Feb	200
Mar	240
Apr	280
May	320
Jun	360
Jul	360
Aug	360
Sep	360

Figure J-2. Sample Annual Obligation Plan

Unfinanced Requirement (UFR)
FY 99

Installation: Fort McClellan, AL
Activity: Military Police School
Appropriation: Operation & Maintenance, Army (OMA)
Army Management Structure (AMS): 321731
Management Decision Package (MDEP): VTER
Element of Resource (EOR): 2500
Issue: Force Protection/Antiterrorism

Justification/Impact:
Funds in the amount of $360K have been provided to conduct the eight Force-Protection/Antiterrorism contract courses in the US Army Military Police School. This amount will only support training through June 99. An additional $120K is required to conduct training during 4th quarter FY99. Unless additional funds are provided, 625 students will not be trained.

Figure J-3. Sample UFR

Appendix K

Vulnerability Assessment

After a commander has obtained a threat analysis, he proceeds to complete the analysis by conducting the vulnerability and criticality assessments. (This appendix will discuss only the VA.) This process considers a mission review and analysis of the installation, base, unit, or port in relation to the terrorist threat. The review should assess the cost of antiterrorism measures in terms of lost or reduced mission effectiveness. It should then assess the level of acceptable risk to facilities and personnel given the estimated erosion of mission effectiveness. Often the best operational method and routine may be the worst to counter potential terrorist activities. This review and analysis is performed routinely and particularly for deployment.

ASSESSMENT CONSIDERATIONS

K-1. The installation, base, unit, or port assessment is derived from the results of the vulnerability and criticality assessments. The assessment provides the staff with the overall vulnerability to terrorist attack. The staff then develops the crisis-management plan, which addresses all terrorist threat levels regardless of the present level. The THREATCONs are then applied according to the local threat. The considerations are—

- **Vulnerability.** The VA is a self-assessment tool used to evaluate its vulnerability to terrorist attack. The more vulnerable an installation, base, unit, or port is, the more attractive it becomes to terrorist attack.
- **Criticality.** The criticality assessment identifies key assets and infrastructures located on and adjacent to the installation, base, unit, or port. These assets are normally symbolic targets that traditionally appeal to a specific terrorist group (such as headquarters buildings and monuments). It addresses the impact of the temporary or permanent loss of key assets or infrastructures to the ability of the installation, base, unit, or port to perform its mission. The staff determines and prioritizes critical assets. The commander approves the prioritized list. The assessment—
  - Selects key assets.
  - Determines whether critical functions can be duplicated under various attack scenarios.
  - Determines the time required to duplicate key assets or infrastructure efforts if temporarily or permanently lost.
  - Determines the vulnerability of key assets or infrastructures to bombs, vehicle crashes, armed assault, and sabotage.
  - Determines the priority of response to key assets and infrastructures in case of fire, multiple bombings, or other terrorist acts.

- **Damage.** The damage assessment determines the ability of the installation, base, unit, or port to plan for and respond to a terrorist attack against key assets and infrastructures.
- **Recovery procedures.** The recovery-procedures assessment determines the capability to recover from the temporary or permanent loss of key assets and infrastructures. Based on this assessment, the staff establishes recovery procedures to ensure the continued ability to perform the mission.

THREATCON LEVELS

K-2. Specific security measures should be directly linked with THREATCON levels. These considerations are—

- **THREATCON Normal.** This THREATCON level exists when a general threat of possible terrorist activity exists but warrants only a routine security posture.
- **THREATCON Alpha.** This THREATCON applies when there is a general threat of possible terrorist activity against personnel and facilities (the nature and extent of which are unpredictable) and when circumstances do not justify full implementation of THREATCON Bravo measures. It may be necessary to implement measures from higher THREATCONs either resulting from intelligence or as a deterrent. The measures in this THREATCON must be capable of being maintained indefinitely.
- **THREATCON Bravo.** This THREATCON applies when an increased and more predictable threat of terrorist activity exists. The measures in this THREATCON must be capable of being maintained for weeks without causing undue hardship, affecting operational capability, or aggravating relations with local authorities. While in Bravo, the installation should bring manning levels and physical-protection levels to the point where the installation can instantly transition to THREATCON Charlie or Delta.
- **THREATCON Charlie.** The transition to THREATCON Charlie must be done on short notice. It is a result of an incident occurring or the receipt of intelligence indicating that some form of terrorist action against personnel and facilities is imminent. Charlie measures should primarily focus on manning adjustments and procedural changes. Security forces will usually enhance their security presence by acquiring additional manning or by adjusting the work-rest ratio (such as moving from a 3:1 to a 6:1 ratio). At Charlie, off-installation travel should be minimized.
- **THREATCON Delta.** Since the transition to THREATCON Delta is immediate, Delta measures should primarily focus on manning adjustments and procedural changes. THREATCON Delta applies in the immediate area where a terrorist attack has occurred or when intelligence has been received that terrorist action against a specific location or person is likely. The security force's manning level is usually peaked in Charlie; therefore, Delta's additional manning will usually come from an augmentation force. Once in Delta, nonessential operations will cease in order to enhance the security and response

posture. Normally, this THREATCON is declared as a localized condition.

K-3. With exception of THREATCON Normal, all THREATCON levels have certain measures associated with them. These measures are listed in JP 3-07.2, Appendix J, and AR 525-13.

ASSESSING VULNERABILITY

K-4. A VA addresses the consequences of terrorist attacks in terms of the ability of units, installations, commands, or activities to accomplish their assignments successfully, even if terrorists have inflicted casualties or destroyed or damaged DOD assets. The VA focuses on two broad areas—

- Preventing and, failing that, substantially mitigating the effects of a terrorist act.
- Maintaining emergency preparedness and crisis response.

K-5. The VA provides the commander with a tool to assess the potential vulnerability of an installation, base, unit, or port activity; but it is not a substitute for sound judgment. The VA must stand on its own and be supported by valid considerations. Typically, a small group of knowledgeable individuals develop the VA. The VA team consists of personnel with required areas of expertise. Some of these team members are the—

- Assessment-team chief.
- Physical-security specialist.
- Structural engineer.
- Infrastructure engineer.
- Operations-readiness specialist.
- Intelligence and/or counterintelligence specialist.

K-6. The functions and responsibilities of each team member are outlined in JP 3-07.2. The following paragraphs provide basic information regarding these areas:

- The assessment-team chief's key responsibilities include overseeing the management, training, and performance of the vulnerability-team members; finalizing the assessment-team out briefing; and preparing the population dynamics and risk assessment.
- The physical-security specialist is responsible for the security and safety of the installation, facility, and personnel.
- The structural engineer examines a variety of potential terrorist weapon effects and structural responses. This function serves to better protect personnel from shocks and blasts by reducing damage through the technically appropriate use of standoff measures, hardening, blast shielding, and shatter-resistant window film (such as Mylar®) as described in Chapter 3. The structural engineer's main responsibility is threat and damage assessment from terrorist weapons estimates and suggestions for threat protection or damage-mitigation measures.
- The infrastructure engineer examines protection against the effects of WMD, protection against terrorist-incident induced fires, and utility systems that can be used to minimize terrorist-incident casualties

(including elements of power, environmental control, and life-support systems). The primary responsibilities include infrastructure security and fire, safety, and damage control.

- The operations-readiness specialist examines plans, procedures, and capabilities for crisis response, consequence management, and recovery operations should a terrorist incident occur. The main responsibilities of this position include emergency-medical and individual-readiness assessments.
- The intelligence and/or counterintelligence specialist has the primary responsibility of performing logical analyses and preparing possible conclusions regarding terrorist targets and target vulnerabilities. These are based on processed intelligence information and knowledge of terrorist capabilities and methods in view of US installation, facility, and personnel safety and security practices.

CONDUCTING THE ASSESSMENT

K-7. Upon its arrival, the assessment team provides an in briefing for the commander, staff, and designated technical point of contact. Site personnel should conduct a site-familiarization briefing and tour. Administrative activities may include establishing the team support area, setting up equipment, scheduling team and technical points of contact meetings and discussions, ensuring classified-material control, establishing a personnel locator, and organizing materials for the out briefing and site folder. Each assessment-team member conducts the assessment based on the specific responsibilities for each function within his area.

POST-ASSESSMENT ACTIVITIES

K-8. Within 30 days of the visit, a summary narrative report and an annotated briefing should be delivered to the installation commander. Follow-on assistance for the commander may be applicable in areas of technical characteristics of improvement options, cost estimates, and generic sources of materials and equipment.

DRILLS AND EXERCISES

K-9. Multiechelon war gaming of possible terrorist attacks is the best test, short of an actual incident, to analyze the response of an installation, base, unit, or port. Drills and exercises test suspected vulnerabilities and antiterrorist measures. These exercises and drills also train the staff as well as reaction-force leadership and help maintain a valid threat assessment by identifying and adjusting to changing threat capabilities.

Glossary

AA&E arms, ammunition, and explosives

ABCS Army Battle Command System

AC alternating current

admin administration

ADP automated data processing

AF Air Force

AFB Air Force base

AFM Air Force manual

AFMAN Air Force manual

AFOSI Air Force Office of Special Investigations

AFR Air Force regulation

AIQC Antiterrorism Instructor Qualification Course

AIS automated information system

AL Alabama

AM amplitude-modulated

AMC Army Materiel Command

AMS Army management structure

AO area of operations

AP armor piercing

Apr April

AR Army regulation

AR-PERSCOM Army Reserve Personnel Command

ASP ammunition supply point

AT antitank

AT/FP antiterrorism/force protection

attn attention

Aug August

AWG American wire gauge

bldg building

BMS balanced magnetic switch

BTO barbed-tape obstacle

BUPERS Bureau of Naval Personnel

C^2 command and control

C^3 command, control, and communications

cav cavalry

CB citizen's band

CCB Community Counterterrorism Board

CCIR commander's critical information requirements

CCTV closed-circuit television

CD-ROM compact-disk, read-only memory

cdr commander

CG command guidance

chap chapter

CIA Central Intelligence Agency

CID Criminal Investigation Division

CINC commander in chief

CISO counterintelligence support officer

CMU concrete-masonry unit

CONEX container express

CONPLAN contingency plan

CONUS continental United States

CP command post

CPWG crime-prevention working group

CQ charge of quarters

CRIMP Crime Reduction Involving Many People

CTA common table of allowance

DA Department of the Army

DARE Drug Abuse Resistance and Education

DC direct current

DC District of Columbia

Dec December

DIA	Defense Intelligence Agency
DOD	Department of Defense
DOE	Department of Energy
DOJ	Department of Justice
DOS	Department of State
DOT	Department of Transportation
DS	direct support
DTM	data-transmission media
DTOC	division tactical operations center
EDM	emergency-destruct measures
EECS	electronic entry-control system
EOD	explosive-ordnance disposal
EOR	element of resource
EPW	enemy prisoner of war
equip	equipment
ESS	electronic security system
FAA	Federal Aviation Administration
FBI	Federal Bureau of Investigation
FCC	Federal Communications Commission
Feb	February
FIS	foreign-intelligence services

FM field manual

FM frequency-modulated

ft foot, feet

FY fiscal year

G2 Assistant Chief of Staff, G2 (Intelligence)

GHz gigahertz

GTA graphic training aid

HN host nation

HQ headquarters

HUD Housing and Urban Development Administration

HUMINT human intelligence

Hz hertz

IAW in accordance with

ICP initial control point

ID identification

IDS intrusion-detection system

IED improvised explosive device

IG inspector general

IID improvised incendiary device

in inch(es)

INSCOM US Army Intelligence and Security Command

INTAC Individual Terrorism Awareness Course

IPB intelligence preparation of the battlefield

IR infrared

ISS information systems security

J2 Intelligence Directorate (Joint Command)

Jan January

JS Joint Service

JSAT Joint Security Assistance Training

JSCP Joint Strategic Capabilities Plan

JSIIDS Joint-Service Interior Intrusion-Detection System

Jul July

Jun June

K one thousand

kHz kilohertz

LED light-emitting diode

liq liquid

LOS line of sight

LOTS logistics over the shore

LP listening post

LRA local reproduction authorized

MACOM major Army command

maint maintenance

Mar March

MCO Marine Corps order

MDEP management decision package

MDMP military decision-making process

METT-TC mission, enemy, terrain, troops, time available, and civilian considerations

MEVA mission-essential or vulnerable area

MI military intelligence

MILCON military construction

MILPO military personnel office

MILVAN military van

min minimum

mm millimeter(s)

MO modus operandi

MOU memorandum of understanding

MP military police

MPACS Military Police Automated Control System

MPMIS Military Police Management Information System

MPR military-police report

MS-DOS Microsoft®-disk operating system

MWD military working dog

N/A	not applicable
naut	nautical
NAVATAC	Navy Antiterrorism Analysis Center
NBC	nuclear, biological, and chemical
NCIC	National Crime Information Center
NCO	noncommissioned officer
NISCOM	Naval Investigative Service Command
No.	number
Nov	November
NSA	National Security Agency
NTAV	nontactical armored vehicle
NVD	night-vision device
O	official
OASD	Office of the Assistant Secretary of Defense
OCOKA	observation and fields of fire, cover and concealment, obstacles, key terrain, and avenues of approach
OCONUS	outside the continental United States
Oct	October
OMA	operations and maintenance, Army
OPA	operations and procurement, Army
OPLAN	operation plan
OPORD	operations order

OPSEC operations security

Pam pamphlet

PAO public affairs office(r)

PD probability of detection

PERSCOM Personnel Command

PHOTINT photographic intelligence

PI police intelligence

PIR passive infrared

PM provost marshal

PMO provost marshal office

POL petroleum, oil, and lubricants

POM Program Objective Memorandum

POV privately owned vehicle

PS physical security

PSD protective security detail

PSI physical-security inspector

PTO Parent-Teacher Organization

PX post exchange

R&D research and development

RD&E research, development, and engineering

ref reference(s)

RF	radio frequency
RII	relevant information and intelligence
ROI	report of investigation
RORO	roll on/roll off
RPG	rocket-propelled grenade
/s/	signed
S2	Intelligence Officer (US Army)
SAW	squad automatic weapon
SDNCO	staff duty noncommissioned officer
SDO	staff duty officer
Sep	September
SIGINT	signal intelligence
SJA	staff judge advocate
SO/LIC	Special Operations/Low-Intensity Conflict
SOFA	status of forces agreement
SOP	standing operating procedure
St.	Saint
STANO	surveillance, target acquisition, and night observation
stat	statute
STC	sound-transmission coefficient
STD	standard
STU	secure telephone unit

TAACOM Theater Army Area Command

TB technical bulletin

TDY temporary duty

TECOM US Army Test and Evaluation Command

TEMPEST Terminal Electromagnetic-Pulse Emanation Standard

THREATCON threat conditions

TM technical manual

TMDE test, measurement, and diagnostic equipment

TNT trinitrotoluene

TRADOC US Army Training and Doctrine Command

TSC triple-standard concertina

UCMJ Uniform Code of Military Justice

UFR unfinanced requirement

US United States

USACE US Army Corps of Engineers

USACIDC US Army Criminal Investigation Command

USAMPS US Army Military Police School

USCG US Coast Guard

VA vulnerability assessment

VIP very important person

vol volume

WMD weapons of mass destruction

Bibliography

AR 12-15. *Joint Security Assistance Training (JSAT) Regulation* (Navy Instructions 4950.4; AFR 50-29). 28 February 1990.

AR 50-5. *Nuclear and Chemical Weapons and Material—Nuclear Surety.* 3 October 1986.

AR 50-6. *Nuclear and Chemical Weapons and Materiel, Chemical Surety.* 1 February 1995.

AR 190-11. *Physical Security of Arms, Ammunitions, and Explosives.* 30 September 1993.

AR 190-12. *Military Police Working Dogs.* 30 September 1993.

AR 190-13. *The Army Physical Security Program.* 30 September 1993.

AR 190-14. *Carrying of Fire Arms and Use of Force for Law Enforcement and Security Duties.* 12 March 1993.

AR 190-22. *Searches, Seizures and Disposition of Property.* 1 January 1983.

AR 190-27. *Army Participation in National Crime Information Center.* 28 May 1993.

AR 190-51. *Security of Unclassified Army Property (Sensitive and Nonsensitive).* 30 September 1993.

AR 190-56. *The Army Civilian Police and Security Guard Program.* 21 June 1995.

AR 190-59. *Chemical Agent Security Program.* 27 June 1994.

AR 380-5. *Department of the Army Information Security Program.* 25 February 1988.

AR 380-10. *Technology Transfer, Disclosure of Information and Contacts with Foreign Representatives.* 30 December 1994.

AR 380-19. *Information Systems Security.* 27 February 1998.

AR 380-67. *The Department of the Army Personnel Security Program.* 9 September 1988.

AR 381-20. *The Army Counterintelligence Program.* 15 November 1993.

AR 405-20. *Federal Legislative Jurisdiction.* 1 August 1973.

AR 525-13. *Antiterrorism Force Protection (AT/FP): Security of Personnel, Information, and Critical Resources.* 10 September 1998.

AR 600-8-14. *Identification Cards for Members of the Uniformed Services, Their Family Members, and Other Eligible Personnel* (AF 36-3026[I]; BUPERS I 1750.10A; MCO P5512.1B; CG M5512.1; Manual 29.2, Instructions 1 and 2). 1 March 1998.

AR 604-5. *Personnel Security Clearance, Department of the Army Personnel Security Program Regulation* 1 February 1984.

DA Form 2028. *Recommended Changes to Publications and Blank Forms.* 1 February 1974.

DA Form 2806-R. *Physical Security Survey Report.* April 1985.

DA Form 2806-1-R. *Physical Security Inspection Report (LRA).* April 1985.

DA Form 2819. *Law Enforcement and Discipline Report.* May 1988.

DA Form 3975. *Military Police Report.* May 1988.

DA Pam 190-12. *Military Working Dog Program.* 30 September 1993.

DA Pam 190-51. *Risk Analysis for Army Property.* 30 September 1993.

DOD 0-2000.12-H. *Protection of DOD Personnel and Activities Against Acts of Terrorism and Political Turbulence.* February 1993.

DOD 4160.21-M. *Defense Reutilization and Marketing Manual.* August 1997.

DOD Directive 2000.12. *DOD Antiterrorism/Force Protection (AT/FP) Program.* 13 April 1999.

DOD Instruction 2000.16. *DOD Combating Terrorism Program Standards.* 10 May 1999.

DOD Instruction 5210.84. *Security of DOD Personnel at US Missions Abroad.* 22 January 1992.

Federal Specification FF-L-2740. *Locks, Combination.* 12 October 1989.

FM 5-34. *Engineer Field Data.* 30 August 1999.

FM 19-4. *Military Police Battlefield Circulation Control, Area Security, and Enemy Prisoner of War Operations.* 7 May 1993.

FM 55-20. *Army Rail Transport Units and Operations.* 1 June 00.

GTA 19-4-3. *Individual Protective Measures.* July 1997.

JP 3-07.2. *Joint Tactics, Techniques, and Procedures for Antiterrorism.* 17 March 1998.

JS Guide 5260. *Service Member's Personal Protection Guide: A Self-Help Handbook to Combating Terrorism.* July 1996.

TB 9-2300-422-20. *Security of Tactical Wheeled Vehicles.* 27 August 1988.

TM 5-805-8. *Builders' Hardware.* 20 January 1992.

TM 5-820-4. *Drainage for Areas Other Than Airfields* (AFM 88-5, Chap 4). 14 October 1983.

TM 5-853-1. *(O) Security Engineering Project Development* (AFMAN 32-1071, Vol 1). 12 May 1994.

TM 5-853-2. *(O) Security Engineering Concept Design* (AFMAN 32-1071, Vol 2). 12 May 1994.

TM 5-853-3. *(O) Security Engineering Final Design* (AFMAN 32-1071, Vol 3). 12 May 1994.

TM 5-853-4. *Security Engineering Electronic Security Systems.* 12 May 1994.

USACE STD 872-50-01. *Entry Points for US Army Installations.* To be published within the next six months.

USACE STD 872-90-02. *FE5 Chain-Link Security Fence Details.* May 1992.

USACE STD 872-90-03. *FE6 Chain-Link Security Fence Details.* May 1992.

USACE STD 872-90-04. *FE7 Chain-Link Security Fence Details for Non Sensored Fence.* May 1992.

USACE STD 872-90-05. *FE7 Chain-Link Security Fence Details for Sensored Fence.* May 1992.

USACE STD 872-90-06. *FE8 Chain-Link Security Fence Details for Sensored Fence.* May 1992.

USACE STD 872-90-07. *FE5 Chain-Link Fence Gate Details.* May 1992.

USACE STD 872-90-08. *FE6 Chain-Link Fence Gate Details.* May 1992.

USACE STD 872-90-09. *FE8 Chain-Link Fence Gate Details.* May 1992.

Index

Made in United States
Orlando, FL
04 November 2026

90219742R00351